METHODS OF MODERN MATHEMATICAL PHYSICS

I: FUNCTIONAL ANALYSIS

Revised and Enlarged Edition

METHODS OF MODERN MATHEMATICAL PHYSICS

I: FUNCTIONAL ANALYSIS

Revised and Enlarged Edition

MICHAEL REED

Department of Mathematics
Duke University

BARRY SIMON

Departments of Mathematics
and Physics
Princeton University

ACADEMIC PRESS, INC.
(Harcourt Brace Jovanovich, Publishers)

Orlando   San Diego   New York   London
Toronto   Montreal   Sydney   Tokyo

To
R. S. Phillips and A. S. Wightman,
Mentors, Colleagues, Friends

COPYRIGHT © 1980, BY ACADEMIC PRESS, INC.
ALL RIGHTS RESERVED.
NO PART OF THIS PUBLICATION MAY BE REPRODUCED OR
TRANSMITTED IN ANY FORM OR BY ANY MEANS, ELECTRONIC
OR MECHANICAL, INCLUDING PHOTOCOPY, RECORDING, OR ANY
INFORMATION STORAGE AND RETRIEVAL SYSTEM, WITHOUT
PERMISSION IN WRITING FROM THE PUBLISHER.

ACADEMIC PRESS, INC.
Orlando, Florida 32887

United Kingdom Edition published by
ACADEMIC PRESS, INC. (LONDON) LTD.
24/28 Oval Road, London NW1 7DX

Library of Congress Cataloging in Publication Data

Reed, Michael.
    Methods of modern mathematical physics.

    Vol. 1 Functional analysis, revised and enlarged edition.
    Includes bibliographical references.
    CONTENTS: v. 1. Functional analysis.–v. 2. Fourier analysis, self–adjointness.–v. 3. Scattering theory.–v. 4. Analysis of operators.
    1. Mathematical physics. I. Simon, Barry, joint author. II. Title.
QC20.R37   1972   530.1'5   75–182650
ISBN 0–12–585050–6 (v. 1)

AMS (MOS) 1970 Subject Classifications: 46–02, 47–02, 42–02

PRINTED IN THE UNITED STATES OF AMERICA

87 88   9 8 7 6 5 4 3

Preface

This book is the first of a multivolume series devoted to an exposition of functional analysis methods in modern mathematical physics. It describes the fundamental principles of functional analysis and is essentially self-contained, although there are occasional references to later volumes. We have included a few applications when we thought that they would provide motivation for the reader. Later volumes describe various advanced topics in functional analysis and give numerous applications in classical physics, modern physics, and partial differential equations.

This revised and enlarged edition differs from the first in two major ways. First, many colleagues have suggested to us that it would be helpful to include some material on the Fourier transform in Volume I so that this important topic can be conveniently included in a standard functional analysis course using this book. Thus, we have included in this edition Sections IX.1, IX.2, and part of IX.3 from Volume II and some additional material, together with relevant notes and problems. Secondly, we have included a variety of supplementary material at the end of the book. Some of these supplementary sections provide proofs of theorems in Chapters II–IV which were omitted in the first edition. While these proofs make Chapters II–IV more self-contained, we still recommend that students with no previous experience with this material consult more elementary texts. Other supplementary sections provide expository material to aid the instructor and the student (for example, "Applications of Compact Operators"). Still other sections introduce and develop new material (for example, "Minimization of Functionals").

It gives us pleasure to thank many individuals:

The students who took our course in 1970–1971 and especially J. E. Taylor for constructive comments about the lectures and lecture notes.

L. Gross, T. Kato, and especially D. Ruelle for reading parts of the manuscript and for making numerous suggestions and corrections.

F. Armstrong, E. Epstein, B. Farrell, and H. Wertz for excellent typing.
M. Goldberger, E. Nelson, M. Simon, E. Stein, and A. Wightman for aid and encouragement.

MIKE REED
BARRY SIMON

April 1980

Introduction

Mathematics has its roots in numerology, geometry, and physics. Since the time of Newton, the search for mathematical models for physical phenomena has been a source of mathematical problems. In fact, whole branches of mathematics have grown out of attempts to analyze particular physical situations. An example is the development of harmonic analysis from Fourier's work on the heat equation.

Although mathematics and physics have grown apart in this century, physics has continued to stimulate mathematical research. Partially because of this, the influence of physics on mathematics is well understood. However, the contributions of mathematics to physics are not as well understood. It is a common fallacy to suppose that mathematics is important for physics only because it is a useful tool for making computations. Actually, mathematics plays a more subtle role which in the long run is more important. When a successful mathematical model is created for a physical phenomenon, that is, a model which can be used for accurate computations and predictions, the mathematical structure of the model itself provides a new way of thinking about the phenomenon. Put slightly differently, when a model is successful it is natural to think of the physical quantities in terms of the mathematical objects which represent them and to interpret similar or secondary phenomena in terms of the same model. Because of this, an investigation of the internal mathematical structure of the model can alter and enlarge our understanding of the physical phenomenon. Of course, the outstanding example of this is Newtonian mechanics which provided such a clear and coherent picture of celestial motions that it was used to interpret practically all physical phenomena. The model itself became central to an understanding of the physical world and it was difficult to give it up in the late nineteenth century, even in the face of contradictory evidence. A more modern example of this influence of mathematics on physics is the use of group theory to classify elementary particles.

INTRODUCTION

The analysis of mathematical models for physical phenomena is part of the subject matter of mathematical physics. By analysis is meant both the rigorous derivation of explicit formulas and investigations of the internal mathematical structure of the models. In both cases the mathematical problems which arise lead to more general mathematical questions not associated with any particular model. Although these general questions are sometimes problems in pure mathematics, they are usually classified as mathematical physics since they arise from problems in physics.

Mathematical physics has traditionally been concerned with the mathematics of classical physics: mechanics, fluid dynamics, acoustics, potential theory, and optics. The main mathematical tool for the study of these branches of physics is the theory of ordinary and partial differential equations and related areas like integral equations and the calculus of variations. This classical mathematical physics has long been part of curricula in mathematics and physics departments. However, since 1926 the frontiers of physics have been concentrated increasingly in quantum mechanics and the subjects opened up by the quantum theory: atomic physics, nuclear physics, solid state physics, elementary particle physics. The central mathematical discipline for the study of these branches of physics is functional analysis, though the theories of group representations and several complex variables are also important. Von Neumann began the analysis of the framework of quantum mechanics in the years following 1926, but there were few attempts to study the structure of specific quantum systems (exceptions would be some of the work of Friedrichs and Rellich). This situation changed in the early 1950's when Kato proved the self-adjointness of atomic Hamiltonians and Gårding and Wightman formulated the axioms for quantum field theory. These events demonstrated the usefulness of functional analysis and pointed out the many difficult mathematical questions arising in modern physics. Since then the range and breadth of both the functional analysis techniques used and the subjects discussed in modern mathematical physics have increased enormously. The problems range from the concrete, for example how to compute or estimate the point spectrum of a particular operator, to the general, for example the representation theory of C^*-algebras. The techniques used and the general approach to the subject have become more abstract. Although in some areas the physics is so well understood that the problems are exercises in pure mathematics, there are other areas where neither the physics nor the mathematical models are well understood. These developments have had several serious effects not the least of which is the difficulty of communication between mathematicians and physicists. Physicists are often dismayed at the breadth of background and increasing mathematical sophistication which are required to understand the models. Mathematicians are often frustrated by their own inability to understand the physics and the inability of physicists to formulate the problems in a way that mathematicians can understand.

A few specific remarks are appropriate. The prerequisite for reading this volume is roughly the mathematical sophistication acquired in a typical undergraduate mathematics education in the United States. Chapter I is intended as a review of background material. We expect that the reader will have some acquaintance with parts of the material covered in Chapters II–IV and have occasionally omitted proofs in these chapters when they seem uninspiring and unimportant for the reader.

The material in this book is sufficient for a two-semester course. Although we taught most of the material in a special one-semester course at Princeton which met five days a week, we do not recommend a repetition of that, either for faculty or students. In order that the material may be easily adapted for lectures, we have written most of the chapters so that the earlier sections contain the basic topics while the later sections contain more specialized and advanced topics and applications. For example, one can give students the basic ideas about unbounded operators in nine or ten lectures from Sections 1–4 of Chapter VIII. On the other hand, by doing the details of the proofs and by adding material from the notes and problems, Chapter VIII could easily become a one-semester course by itself.

Each chapter of this book ends with a long set of problems. Some of the problems fill gaps in the text (these are marked with a dagger). Others develop alternate proofs to the theorems in the text or introduce new material. We have also included harder problems (indicated by a star) in order to challenge the reader. We strongly encourage students to do the problems. It is trite but true that mathematics is learned by doing it, not by watching other people do it.

We hope that these volumes will provide physicists with an access to modern abstract techniques and that mathematicians will benefit by learning the advanced techniques side by side with their applications.

Contents

Preface *v*
Introduction *vii*
Contents of Other Volumes *xv*

I: PRELIMINARIES

1. Sets and functions	*1*
2. Metric and normed linear spaces	*3*
Appendix Lim sup and lim inf	*11*
3. The Lebesgue integral	*12*
4. Abstract measure theory	*19*
5. Two convergence arguments	*26*
6. Equicontinuity	*28*
Notes	*31*
Problems	*32*

II: HILBERT SPACES

1. The geometry of Hilbert space	*36*
2. The Riesz lemma	*41*
3. Orthonormal bases	*44*
4. Tensor products of Hilbert spaces	*49*
5. Ergodic theory: an introduction	*54*
Notes	*60*
Problems	*63*

III: BANACH SPACES

1. Definition and examples ... 67
2. Duals and double duals ... 72
3. The Hahn–Banach theorem ... 75
4. Operations on Banach spaces ... 78
5. The Baire category theorem and its consequences ... 79
Notes ... 84
Problems ... 86

IV: TOPOLOGICAL SPACES

1. General notions ... 90
2. Nets and convergence ... 95
3. Compactness ... 97
Appendix The Stone–Weierstrass theorem ... 103
4. Measure theory on compact spaces ... 104
5. Weak topologies on Banach spaces ... 111
Appendix Weak and strong measurability ... 115
Notes ... 117
Problems ... 119

V: LOCALLY CONVEX SPACES

1. General properties ... 124
2. Fréchet spaces ... 131
3. Functions of rapid decease and the tempered distributions ... 133
Appendix The N-representation for $\mathscr{S}$ and $\mathscr{S}'$ ... 141
4. Inductive limits: generalized functions and weak solutions of partial differential equations ... 145
5. Fixed point theorems ... 150
6. Applications of fixed point theorems ... 153
7. Topologies on locally convex spaces: duality theory and the strong dual topology ... 162
Appendix Polars and the Mackey–Arens theorem ... 167
Notes ... 169
Problems ... 173

VI: BOUNDED OPERATORS

1. Topologies on bounded operators — 182
2. Adjoints — 185
3. The spectrum — 188
4. Positive operators and the polar decomposition — 195
5. Compact operators — 198
6. The trace class and Hilbert–Schmidt ideals — 206
Notes — 213
Problems — 216

VII: THE SPECTRAL THEOREM

1. The continuous functional calculus — 221
2. The spectral measures — 224
3. Spectral projections — 234
4. Ergodic theory revisited: Koopmanism — 237
Notes — 243
Problems — 245

VIII: UNBOUNDED OPERATORS

1. Domains, graphs, adjoints, and spectrum — 249
2. Symmetric and self-adjoint operators: the basic criterion for self-adjointness — 255
3. The spectral theorem — 259
4. Stone's theorem — 264
5. Formal manipulation is a touchy business: Nelson's example — 270
6. Quadratic forms — 276
7. Convergence of unbounded operators — 283
8. The Trotter product formula — 295
9. The polar decomposition for closed operators — 297
10. Tensor products — 298
11. Three mathematical problems in quantum mechanics — 302
Notes — 305
Problems — 312

THE FOURIER TRANSFORM

1.	The Fourier transform on $\mathscr{S}(\mathbb{R}^n)$ and $\mathscr{S}'(\mathbb{R}^n)$, convolutions	318
2.	The range of the Fourier transform: Classical spaces	326
3.	The range of the Fourier transform: Analyticity	332
Notes		338
Problems		339

SUPPLEMENTARY MATERIAL

II.2.	Applications of the Riesz lemma	344
III.1.	Basic properties of L^p spaces	348
IV.3.	Proof of Tychonoff's theorem	351
IV.4.	The Riesz–Markov theorem for $X = [0, 1]$	353
IV.5.	Minimization of functionals	354
V.5.	Proofs of some theorems in nonlinear functional analysis	363
VI.5.	Applications of compact operators	368
VIII.7.	Monotone convergence for forms	372
VIII.8.	More on the Trotter product formula	377
Uses of the maximum principle		382
Notes		385
Problems		387

List of Symbols	393
Index	395

Contents of Other Volumes

Volume II: Fourier Analysis, Self-Adjointness

 IX *The Fourier Transform*
  X *Self-Adjointness and the Existence of Dynamics*

Volume III: Scattering Theory

 XI *Scattering Theory*

Volume IV: Analysis of Operators

 XII *Perturbation of Point Spectra*
XIII *Spectral Analysis*

I: Preliminaries

The beginner... should not be discouraged if... he finds that he does not have the prerequisites for reading the prerequisites.

P. Halmos

I.1 Sets and functions

We assume that the reader is familiar with sets and functions but it is appropriate to standardize our terminology and to introduce here abbreviations that will occur throughout the book.

If X is a set, $x \in X$ means that x is an element of X; $x \notin X$ means that x is not in X. The clause "for all x in X" is abbreviated ($\forall x \in X$) and "there exists an $x \in X$ such that" is abbreviated ($\exists x \in X$). The symbol $\{x \mid P(x)\}$ stands for the set of x obeying the condition (or conditions) $P(x)$. If A is a subset of X (denoted $A \subset X$), the symbol $X \backslash A$ represents the complement of A in X, that is $X \backslash A = \{x \in X \mid x \notin A\}$. More generally, if A and B are subsets of X, then $A \backslash B = \{x \mid x \in A, x \notin B\}$. When we discuss sets with a topology, $\bar{A}$ will always denote the closure of the set A. Finally, the set of ordered pairs $\{\langle x, y \rangle \mid x \in X, y \in Y\}$ is called the **Cartesian product** of X and Y and is denoted $X \times Y$.

We will use the words "function" and "mapping" interchangeably. In order to emphasize that certain functions f depend on two variables, we will sometimes write $f(\cdot, \cdot)$. The symbol $f(\cdot, y)$ denotes the function of one variable obtained by picking a fixed value of y for the second variable. A linear function will also be called an **operator** or a **linear transformation**. Our functions will always be single valued; so a function from a set X to another set Y, denoted by $f\colon X \to Y$ or $X \xrightarrow{f} Y$ or $x \mapsto f(x)$ will have one and only one value in Y for each $x \in X$. If $A \subset X$, then $f[A] = \{f(x) \mid x \in A\}$ is a subset of Y and $f^{-1}[B] = \{x \mid f(x) \in B\}$ is a subset of X if $B \subset Y$. $f[X]$ will usually be called the **range** of f and will be denoted Ran f. X is called the **domain** of f. A function f will be called **injective** (or one-one) if for each $y \in \mathrm{Ran}\ f$ there is at most $x \in X$ such that $f(x) = y$; f is called surjective (or onto) if Ran $f = Y$. If f is both injective and **surjective**, we will say it is **bijective**. The restriction of f to a subset A of its domain will be denoted by $f \upharpoonright A$.

If $X \supset A$ we define the **characteristic function** $\chi_A(x)$ as

$$\chi_A(x) = \begin{cases} 1 & \text{if } x \in A \\ 0 & \text{if } x \notin A \end{cases}$$

There are two set theoretic notions which are slightly deeper than mere notation, so we will discuss them to some extent. A **relation** R on a set X is a subset R of $X \times X$; if $\langle x, y \rangle \in R$, we say that x is related (or R-related) to y and write xRy.

Definition A relation R is called an **equivalence relation** if it satisfies:

(i) $(\forall x \in X)\ xRx$ [reflexive]
(ii) $(\forall x, y \in X)\ xRy$ implies yRx [symmetric]
(iii) $(\forall x, y, z \in X)\ xRy$ and yRz implies xRz [transitive]

The set of elements in X that are related to a given $x \in X$ is called the **equivalence class** of x, denoted usually as $[x]$.

It is easy to prove:

Theorem I.1 Let R be an equivalence relation on a set X. Then each $x \in X$ belongs to a unique equivalence class.

Thus, under an equivalence relation, a set divides up in a natural way into disjoint subsets.

Example 1 (the integers mod 3) Let X be the integers and write xRy if $x - y$ is a multiple of 3. This equivalence relation divides the integers into three equivalence classes:

$$[0] = \{\ldots, -6, -3, 0, 3, 6, \ldots\}$$
$$[1] = \{\ldots, -5, -2, 1, 4, 7, \ldots\}$$
$$[2] = \{\ldots, -4, -1, 2, 5, 8, \ldots\}$$

Example 2 (the real projective line) Let $\mathbb{R}$ denote the real line and let X be the nonzero vectors in $\mathbb{R}^2$ ($= \mathbb{R} \times \mathbb{R}$). We write xRy if there is some $\alpha \in \mathbb{R}$ with $x = \alpha y$. The equivalence classes are lines through the origin (with $\langle 0, 0 \rangle$ removed).

Next, we discuss Zorn's lemma.

Definition A relation on a set X which is reflexive, transitive, and antisymmetric (that is, xRy and yRx implies $x = y$) is called a **partial ordering**. If R is a partial ordering, we often write $x \prec y$ instead of xRy.

Example 3 Let X be the collection of all subsets of a set Y. Define $A \prec B$ if $A \subset B$. Then $\prec$ is a partial ordering.

We use the word "partial" in the above definition because two elements of X need not obey $x \prec y$ or $y \prec x$. If for all x and y in X, either $x \prec y$ or $y \prec x$, X is said to be **linearly ordered**. For example, $\mathbb{R}$ with its usual order $\leq$ is linearly ordered.

Now suppose X is partially ordered by $\prec$ and $Y \subset X$. An element $p \in X$ is called **upper bound** for Y if $y \prec p$ for all $y \in Y$. If $m \in X$ and $m \prec x$ implies $x = m$, we say m is a **maximal element** of X.

Depending on one's starting point, Zorn's lemma is either a basic assumption of set theory or else derived from the basic assumptions (it is equivalent to the *axiom* of choice). We take Zorn's lemma and the rest of set theory as given.

Theorem I.2 (Zorn's lemma) Let X be a nonempty partially ordered set with the property that every linearly ordered subset has an upper bound in X. Then each linearly ordered set has some upper bound that is also a maximal element of X.

Finally, we will use Halmos' ∎ to indicate the conclusion of a proof.

I.2 Metric and normed linear spaces

Throughout this work, we will be dealing with sets of functions or operators or other objects and we will often need a way of measuring the distance

4   I: PRELIMINARIES

between the objects in the sets. It is reasonable to define a notion of distance that has the most important properties of ordinary distance in $\mathbb{R}^3$.

Definition   A **metric space** is a set M and a real-valued function $d(\cdot, \cdot)$ on $M \times M$ which satisfies:

(i) $d(x, y) \geq 0$
(ii) $d(x, y) = 0$ if and only if $x = y$
(iii) $d(x, y) = d(y, x)$
(iv) $d(x, z) \leq d(x, y) + d(y, z)$ [triangle inequality]

The function d is called a **metric** on M.

We often call the elements of a metric space points. Notice that a metric space is a set M together with a metric function d; in general, a given set X can be made into a metric space in different ways by employing different metric functions. When it is not clear from the context which metric we are talking about, we will denote the metric space by $\langle M, d \rangle$, so that the metric is explicitly displayed.

Example 1   Let $M = \mathbb{R}^n$ with the distance between two points $x = \langle x_1, \ldots, x_n \rangle$ and $y = \langle y_1, \ldots, y_n \rangle$ given by

$$d(x, y) = \sqrt{(x_1 - y_1)^2 + \cdots + (x_n - y_n)^2}$$

Example 2   Let M be the unit circle in $\mathbb{R}^2$, that is, the set of all pairs of real numbers $\langle \alpha, \beta \rangle$ with $\alpha^2 + \beta^2 = 1$, and let

$$d_1(\langle \alpha, \beta \rangle, \langle \alpha', \beta' \rangle) = \sqrt{(\alpha - \alpha')^2 + (\beta - \beta')^2}$$

Another possible metric is $d_2[p, p'] =$ arc length between the points p, p' (see Figure I.1).

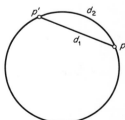

FIGURE I.1   The metrics d_1 and d_2.

I.2 Metric and normed linear spaces

Example 3 Let $M = C[0, 1]$, the continuous real-valued functions on $[0, 1]$ with either of the metrics

$$d_1(f, g) = \max_{x \in [0, 1]} |f(x) - g(x)| \qquad d_2(f, g) = \int_0^1 |f(x) - g(x)| \, dx$$

Now that we have a notion of distance, we can say what we mean by convergence.

Definition A sequence of elements $\{x_n\}_{n=1}^\infty$ of a metric space $\langle M, d \rangle$ is said to **converge** to an element $x \in M$, if $d(x, x_n) \to 0$ as $n \to \infty$. We will often denote this by $x_n \xrightarrow{d} x$ or $\lim_{n \to \infty} x_n = x$. If x_n does not converge to x, we will write $x_n \not\xrightarrow{d} x$.

In Example 2, $d_1(p, p') \leq d_2(p, p') \leq \pi d_1(p, p')$ which we will write $d_1 \leq d_2 \leq \pi d_1$. Thus $p_n \xrightarrow{d_1} p$ if and only if $p_n \xrightarrow{d_2} p$. But in Example 3, the metrics induce distinct notions of convergence. Since $d_2 \leq d_1$, $f_n \xrightarrow{d_1} f$ implies $f_n \xrightarrow{d_2} f$, but the converse is false. A counterexample is given by the functions g_n defined in Figure I.2, which converge to the zero function in the metric d_2

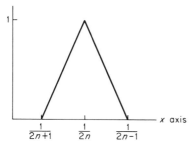

FIGURE I.2 The graph of $g_n(x)$.

but which do not converge in the metric d_1. This may be seen by introducing the important notion of Cauchy sequence.

Definition A sequence of elements $\{x_n\}$ of a metric space $\langle M, d \rangle$ is called a **Cauchy sequence** if $(\forall \varepsilon > 0)(\exists N)$ $n, m \geq N$ implies $d(x_n, x_m) < \varepsilon$.

Proposition Any convergent sequence is Cauchy.

Proof Given $x_n \to x$ and ε, find N so $n \geq N$ implies $d(x_n, x) < \varepsilon/2$. Then $n, m \geq N$ implies $d(x_n, x_m) \leq d(x_n, x) + d(x, x_m) < \tfrac{1}{2}\varepsilon + \tfrac{1}{2}\varepsilon$. ∎

We now return to the functions in Figure I.2. It is easy to see that if $n \neq m$, $d_1(g_n, g_m) = 1$. Thus g_n is not a Cauchy sequence in $\langle C[0, 1], d_1 \rangle$ and therefore not a convergent sequence. Thus, the sequence $\{g_n\}$ converges in $\langle C[0, 1], d_2 \rangle$ but not in $\langle C[0, 1], d_1 \rangle$.

Although every convergent sequence is a Cauchy sequence, the following example shows that the converse need not be true. Let $\mathbb{Q}$ be the rational numbers with the usual metric (that is, $d(x, y) = |x - y|$) and let x^* be any irrational number (that is, $x^* \in \mathbb{R} \backslash \mathbb{Q}$). Find a sequence of rationals x_n with $x_n \to x^*$ in $\mathbb{R}$. Then x_n is a Cauchy sequence of numbers in $\mathbb{Q}$, but it cannot converge in $\mathbb{Q}$ to some $y \in \mathbb{Q}$ (for, if $x_n \to y$ in $\mathbb{Q}$, then $x_n \to y$ in $\mathbb{R}$, so we would have $y = x^*$).

Definition A metric space in which all Cauchy sequences converge is called **complete**.

For example, $\mathbb{R}$ is complete, but $\mathbb{Q}$ is not. It can be shown (Sections I.3 and I.5) that $\langle C[0, 1], d_1 \rangle$ is complete but $\langle C[0, 1], d_2 \rangle$ is not. The example of $\mathbb{Q}$ and $\mathbb{R}$ suggests what we need to do to an incomplete space X to make it complete. We need to enlarge X by adding "all possible limits of Cauchy sequences." The original space X should be dense in the larger space $\tilde{X}$ where:

Definition A set B in a metric space M is called **dense** if every $m \in M$ is a limit of elements in B.

Of course, if the incomplete space is not already contained in a larger complete space (like $\mathbb{Q}$ is contained in $\mathbb{R}$) it is not clear what "all possible limits" means. That this "completion" can be done is the content of a theorem that we shall shortly state; but first some definitions:

Definition A function f from a metric space $\langle X, d \rangle$ to a metric space $\langle Y, \rho \rangle$ is called **continuous at x** if $f(x_n) \xrightarrow{\langle Y, \rho \rangle} f(x)$ whenever $x_n \xrightarrow{\langle X, d \rangle} x$.

We have already had an example of a sequence of elements in $C[0, 1]$ with $f_n \xrightarrow{d_2} 0$ but $f_n \xrightarrow{d_1} \!\!\!\!\!/\;\; 0$. Thus the identity function from $\langle C[0, 1], d_2 \rangle$ to $\langle C[0, 1], d_1 \rangle$ is *not* continuous but the identity from $\langle C[0, 1], d_1 \rangle$ to $\langle C[0, 1], d_2 \rangle$ is continuous.

Definition A bijection h from $\langle X, d \rangle$ to $\langle Y, \rho \rangle$ which preserves the metric, that is,

$$\rho(h(x), h(y)) = d(x, y)$$

is called an **isometry**. It is automatically continuous. $\langle X, d \rangle$ and $\langle Y, \rho \rangle$ are said to be **isometric** if such an isometry exists.

Isometric spaces are essentially identical as metric spaces; a theorem concerning only the metric structure of $\langle X, d \rangle$ will hold in all spaces isometric to it.

We now state precisely in which sense an incomplete space can be fattened out to be complete:

Theorem I.3 If $\langle M, d \rangle$ is an incomplete metric space, it is possible to find a complete metric space $\tilde{M}$ so that M is isometric to a dense subset of $\tilde{M}$.

Sketch of proof Consider the Cauchy sequences $\{x_n\}$ of elements of M. Call two sequences, $\{x_n\}$, $\{y_m\}$, equivalent if $\lim_{n \to \infty} d(x_n, y_n) = 0$. Let $\tilde{M}$ be the family of equivalence classes of Cauchy sequences under this equivalence relation. One can show that for any two Cauchy sequences $\lim_{n \to \infty} d(x_n, y_n)$ exists and depends only on the equivalence classes of $\{x_n\}$ and $\{y_n\}$. This limit defines a metric on $\tilde{M}$ and $\tilde{M}$ is complete. Finally, map M into $\tilde{M}$ by taking x into the constant sequence in which each x_n equals x. M is dense in $\tilde{M}$ and the map is isometric. ∎

To complete our discussion of metric spaces, we want to introduce the notions of open and closed sets. The reader should keep the example of open and closed sets on the real line in mind.

Definition Let $\langle X, d \rangle$ be a metric space:

(a) The set $\{x \mid x \in X, d(x, y) < r\}$ is called the **open ball**, $B(y; r)$, of radius r about the point y.

(b) A set $O \subset X$ is called **open** if $(\forall y \in O)(\exists r > 0)\ B(y; r) \subset O$.

(c) A set $N \subset X$ is called a **neighborhood** of $y \in N$ if $B(y; r) \subset N$ for some $r > 0$.

(d) Let $E \subset X$. A point x is called a **limit point** of E, if $(\forall r > 0)$ $B(x; r) \cap (E \setminus \{x\}) \neq \varnothing$, that is, x is a limit point of E if E contains points other than x arbitrarily near x.

(e) A set $F \subset X$ is called **closed** if F contains all its limit points.

(f) If $G \subset X$, $x \in G$ is called an **interior point** of G, if G is a neighborhood of x.

The reader can prove for himself the following collection of elementary statements:

Theorem I.4 Let $\langle X, d\rangle$ be a metric space:

(a) A set, O, is open if and only if $X\backslash O$ is closed.
(b) $x_m \xrightarrow{d} x$ if and only if for each neighborhood N of x, there exists an M so that $m \geq M$ implies $x_m \in N$.
(c) The set of interior points of a set is open.
(d) The union of a set E with its limit points is a closed set (denoted by $\bar{E}$ and called the **closure** of E).
(e) A set is open if and only if it is a neighborhood of each of its points.

One of the main uses of open sets is to check for convergence using Theorem I.4.b and in particular to check for continuity via the following criteria, the proof of which we leave as an exercise:

Theorem I.5 A function $f(\cdot)$ from a metric space X to another space Y is continuous if and only if for all open sets $O \subset Y$, $f^{-1}[O]$ is open.

Finally, we warn the reader that often in incomplete metric spaces, closed sets may not appear to be closed at first glance. For example, $[\frac{1}{2}, 1)$ is closed in $(0, 1)$ (with the usual metric).

We complete this section with a discussion of two of the central concepts of functional analysis: normed linear spaces and bounded linear transformations.

Definition A **normed linear space** is a vector space, V, over $\mathbb{R}$ (or $\mathbb{C}$) and a function, $\|\cdot\|$ from V to $\mathbb{R}$ which satisfies:

(i) $\|v\| \geq 0$ for all v in V
(ii) $\|v\| = 0$ if and only if $v = 0$
(iii) $\|\alpha v\| = |\alpha|\,\|v\|$ for all v in V and α in $\mathbb{R}$ (or $\mathbb{C}$)
(iv) $\|v + w\| \leq \|v\| + \|w\|$ for all v and w in V

Definition A **bounded linear transformation** (or bounded operator) from a normed linear space $\langle V_1, \|\ \|_1\rangle$ to a normed linear space $\langle V_2, \|\ \|_2\rangle$ is a function, T, from V_1 to V_2 which satisfies:

(i) $T(\alpha v + \beta w) = \alpha T(v) + \beta T(w)$ $(\forall v, w \in V)(\forall \alpha, \beta \in \mathbb{R}$ or $\mathbb{C})$
(ii) For some $C \geq 0$, $\|Tv\|_2 \leq C\|v\|_1$

The smallest such C is called the **norm of** T, written $\|T\|$ or $\|T\|_{1,2}$. Thus

$$\|T\| = \sup_{\|v\|_1 = 1} \|Tv\|_2$$

Since we will study these concepts in detail later, we will not give many examples now but merely note that $\mathbb{R}^n$ with the norm

$$\|\langle x_1, \ldots, x_n \rangle\| = \sqrt{|x_1|^2 + \cdots + |x_n|^2}$$

and $C[0, 1]$ with either the norm

$$\|f\|_\infty = \sup_{x \in [0,1]} |f(x)| \quad \text{or} \quad \|f\|_1 = \int_0^1 |f(x)|\, dx$$

are normed linear spaces. Observe also that any normed linear space $\langle V, \|\cdot\| \rangle$ is a metric space when given the distance function $d(v, w) = \|v - w\|$. There is thus a notion of continuity of functions, and for linear functions this is precisely captured by bounded linear transformations. The proof of this fact is left to the reader.

Theorem I.6 Let T be a linear transformation between two normed linear spaces. The following are equivalent:

(a) T is continuous at one point.
(b) T is continuous at all points.
(c) T is bounded.

Definition We say $\langle V, \|\cdot\| \rangle$ is **complete** if it is complete as a metric space in the induced metric.

If $\langle X, \|\cdot\| \rangle$ is a normed linear space, then X has a completion as a metric space by Theorem I.3. Using the fact that X is dense in $\tilde{X}$, it is easy to see that $\tilde{X}$ can be made into a normed linear space in exactly one natural way. All these concepts are well illustrated by the following important theorem and its proof:

Theorem I.7 (the B.L.T. theorem) Suppose T is a bounded linear transformation from a normed linear space $\langle V_1, \|\cdot\|_1 \rangle$ to a complete normed linear space $\langle V_2, \|\cdot\|_2 \rangle$. Then T can be uniquely extended to a bounded linear transformation (with the same bound), $\tilde{T}$, from the completion of V_1 to $\langle V_2, \|\cdot\|_2 \rangle$.

Proof Let $\tilde{V}_1$ be the completion of V_1. For each x in $\tilde{V}_1$, there is a sequence of elements $\{x_n\}$ in V_1 with $x_n \to x$ as $n \to \infty$. Since x_n converges, it is Cauchy, so given ε, we can find N so that $n, m > N$ implies $\|x_n - x_m\|_1 \leq \varepsilon/\|T\|$. Then $\|Tx_n - Tx_m\|_2 = \|T(x_n - x_m)\|_2 \leq \|T\|\,\|x_n - x_m\|_1 \leq \varepsilon$ which proves that Tx_n is a Cauchy sequence in V_2. Since V_2 is complete, $Tx_n \to y$ for some y. Set $\tilde{T}x = y$. We must first show that this definition is independent of the sequence $x_n \to x$ chosen. If $x_n \to x$ and $x'_n \to x$, then the sequence $x_1, x'_1, x_2, x'_2, \ldots \to x$ so $Tx_1, Tx'_1, \ldots \to \hat{y}$ for some $\hat{y}$ by the above argument. Thus $\lim Tx'_n = \hat{y} = \lim Tx_n$. Moreover, we can show $\tilde{T}$ so defined is bounded because

$$\|\tilde{T}x\|_2 = \lim_{n \to \infty} \|Tx_n\|_2 \quad \text{(see Problem 8)}$$
$$\leq \varlimsup_{n \to \infty} C\|x_n\|_1 \quad \text{(see Appendix to I.2)}$$
$$= C\|x\|_1$$

Thus $\tilde{T}$ is bounded. The proofs of linearity and uniqueness are left to the reader. ∎

We can use this theorem to give a very elegant definition of the Riemann integral. Let $PC[a, b]$ be the family of bounded piecewise continuous functions on $[a, b]$, which are continuous from the right, that is, $\lim_{x \downarrow y} f(x) = f(y)$ and for which $\lim_{x \uparrow y} f(x)$ exists at each y and is equal to $f(y)$ for all but finitely many y. Norm PC with the norm

$$\|f\|_\infty = \sup_{x \in [a, b]} |f(x)|$$

Let $x_0, \ldots, x_n$ be a partition of the interval $[a, b]$, $x_0 = a$, $x_n = b$. Let $\chi_i(x)$ be the characteristic function of $[x_{i-1}, x_i)$ except for $\chi_n(x)$ which is the characteristic function of $[x_{n-1}, x_n]$. A function on $[a, b]$ of the form $\sum_{i=1}^n s_i \chi_i(x)$ with s_i real is called a **step function** (to see why, draw its graph). The set of all step functions for all possible finite partitions is a normed linear space with the norm

$$\left\| \sum_{i=1}^n s_i \chi_i(x) \right\|_\infty = \sup_{x \in [a, b]} \left| \sum s_i \chi_i(x) \right| = \max_{i=1, \ldots, n} |s_i|$$

Denote this space by $S[a, b]$. It is a nice exercise (Problem 10) to prove that $S[a, b]$ is dense in $PC[a, b]$. For any step function, $\sum_{i=1}^n s_i \chi_i$, we define

$$I\left(\sum_{i=1}^n s_i \chi_i(x) \right) = \sum_{i=1}^n s_i(x_i - x_{i-1})$$

the intuitive value of the integral $\int [\sum s_i \chi_i(x)]\, dx$. I is a linear transformation from $S[a, b]$ to the real numbers, and because

$$\left| I\left(\sum_{i=1}^{n} s_i \chi_i\right)\right| = \left|\sum s_i(x_i - x_{i-1})\right|$$
$$\leq \max |s_i| \sum_{i=1}^{n} |x_i - x_{i-1}|$$
$$\leq \left\|\sum s_i \chi_i\right\|_\infty (b-a)$$

I is a *bounded* linear transformation. Since the real numbers are complete, I can be uniquely extended to $\tilde{S}$, the completion of S (by the B.L.T. theorem). The extended transformation $\tilde{I}(f)$, restricted to PC is called the Riemann integral and is denoted by

$$I(f) = \int_a^b f\,dx$$

While this method does not appear as the most intuitive definition of the Riemann integral, it will be seen upon reflection that the proof is really just the "usual" proof put into the language of completion and the B.L.T. theorem. It illustrates a main point of general philosophy in functional analysis: In order to define something on a normed linear space, it is often convenient to define it on a dense set and extend it by the B.L.T. theorem. The reader should try his hand at constructing the Riemann–Stieltjes integral (Problem 11). By using the same method, we can define the Riemann integral for continuous functions taking values in any *complete* normed linear space, in particular, for complex-valued functions.

Appendix to 1.2 Lim sup and lim inf

Lim sup and lim inf are notions which may be unfamiliar to the reader, so we summarize their definition and properties.

Definition Let $A \subset \mathbb{R}$ be a nonfinite bounded set. Let $\lim \text{pt}(A)$ = set of limit points of A. Then the **limit superior of** A is defined by

$$\limsup A \equiv \overline{\lim} A = \sup\{x \,|\, x \in \lim \text{pt}(A)\}$$

Similarly

$$\liminf A \equiv \underline{\lim}(A) = \inf\{x \,|\, x \in \lim \text{pt}(A)\}$$

Remarks 1. When A is bounded, $\lim \text{pt}(A)$ is always nonempty by the Bolzano–Weierstrass theorem.

2. If A is not bounded above, one defines $\overline{\lim} A = +\infty$. If A is bounded above and lim pt$(A) = \emptyset$ one defines $\overline{\lim} A = -\infty$.

3. $\overline{\lim} A$ is actually in lim pt(A). For let $b = \overline{\lim} A$ and let $\varepsilon > 0$ be given. We can find $a \in \lim \text{pt}(A)$ so $|b - a| < \varepsilon/2$. Since $a \in \lim \text{pt}(A)$, we can find $d \in A$ with $|a - d| < \varepsilon/2$; so given ε, we find $d \in A$ with $|b - d| < \varepsilon$, that is, $b \in \lim \text{pt}(A)$.

$\overline{\lim} A$ has a very simple alternative characterization, whose proof we leave to the reader.

Proposition Let $b = \overline{\lim} A$. Then for $\varepsilon > 0$, $A \cap \{a \mid a > b + \varepsilon\}$ is finite and $A \cap \{a \mid a > b - \varepsilon\}$ is infinite.

For a sequence $\{a_n\}$, we say $b \in \lim \text{pt}\{a_n\}$ if for all N and all ε, there is an $n > N$ with $|b - a_n| < \varepsilon$. We define $\overline{\lim}(a_n) = \sup\{b \mid b \in \lim \text{pt}\{a_n\}\}$.

Finally, let us summarize the properties of $\overline{\lim}$ (all for bounded sets; it is a useful exercise to decide which extend to unbounded sets).

Proposition
(a) $\overline{\lim}(a_n + b_n) \leq \overline{\lim} a_n + \overline{\lim} b_n$
(b) $\overline{\lim} a_n b_n \leq (\overline{\lim} a_n)(\overline{\lim} b_n)$ if $a_n, b_n \geq 0$
(c) $\overline{\lim}(ca_n) = c \overline{\lim} a_n$ if $c > 0$
(d) $\overline{\lim}(ca_n) = c \underline{\lim} a_n$ if $c < 0$

I.3 The Lebesgue integral

We have just seen that $C[a, b]$ has two quite reasonable metrics on it. In Section I.5 we will see that it is a complete metric space in the metric

$$d_1(f, g) = \sup_{x \in [a, b]} |f(x) - g(x)|$$

In the other metric we considered, $d_2(f, g) = \|f - g\|_1$ with $\|h\|_1 = \int_a^b |h(x)| \, dx$, $C[a, b]$ is *not complete*. To see this for $C[0, 1]$, let f_n be given as in Figure I.3. It is not hard to see that f_n is Cauchy in $\|\cdot\|_1$, but it does not converge to any function in $C[a, b]$; rather, in an *intuitive sense*, it "converges" to the characteristic function of $[\frac{3}{4}, \frac{1}{4}]$ (which is, of course, not in $C[0, 1]$!).

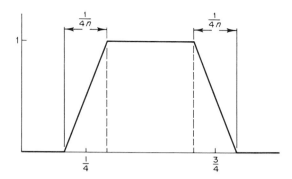

FIGURE I.3 The graph of f_n.

We can always complete $C[a, b]$ in $\|\cdot\|_1$ realizing elements of the completion as equivalence classes of Cauchy sequences of continuous functions; this realization is not noteworthy for its transparency. The example above suggests we might also be able to realize elements of the completion as functions. If we do realize them as functions, we should be able to define the integral $\int_a^b |f(x)|\,dx$ (merely as $d_2(f, 0)$!) for any f in the completion.

The simplest way to realize elements of the completion as functions is to turn the above analysis around: one introduces an extended notion of integral on a bigger space than $C[a, b]$; call it $L^1[a, b]$. We will prove L^1 is complete, so by general arguments the closure of C in L^1 is complete (and it turns out $\bar{C} = L^1$).

Now, how can one extend the notion of Riemann integral? The usual definition of the Riemann integral is based on dividing the *domain* of f into finer and finer pieces. For "nasty" functions, this method does not work and

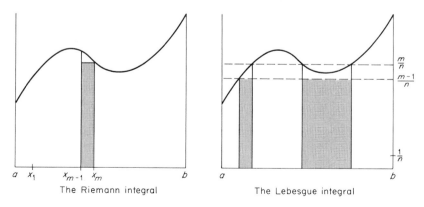

FIGURE I.4

so a different method is needed—the simplest modification is to divide the range into finer and finer pieces (Figure I.4). This method depends more on the function and so has the possibility of working for more types of functions. We are thus interested in sets $f^{-1}[a, b]$ and their size. We suppose we have a size function μ on sets which generalizes $\mu([a, b]) = b - a$. We will shortly return to this size function and see that not all sets have a "size." We will then restrict the types of f by demanding that $f^{-1}[a, b]$ have a "size." Looking at Figure I.4, we define for $f \geq 0$

$$\sum_n (f) = \sum_{m=0}^{\infty} \frac{m}{n} \mu\left(f^{-1}\left[\left[\frac{m}{n}, \frac{m+1}{n}\right)\right]\right)$$

Then $\sum_{2n} (f) \geq \sum_n (f)$ so that $\lim_{n \to \infty} \sum_{2^n} (f) = \sup_n (\sum_{2^n} (f))$ exists (it may be ∞). This limit is defined to be $\int f \, dx$. We remark that for technical purposes (that is, proving theorems!) one makes a different definition which can be shown to agree with this definition only after a lot of work. The definition as $\lim \sum_{2^n} (f)$ is however the best to keep in mind when thinking intuitively.

Thus, we have transferred the problem to one of defining an extended notion of size. We must first decide what sets are to have a size. Why not all sets? There is a classical example (see also Problem 13) which shows that not all sets in $\mathbb{R}^3$ can have a size if we want that size to be invariant under rotations and translations (and not to be trivial, such as assigning zero to all sets): it is possible to break up a unit ball into a finite number of wild pieces, move the pieces around by rotation and translation and reassemble the pieces to get two balls of radius one (Banach–Tarski paradox). Thus, all sets cannot have a size, and so some family $\mathscr{B}$ of sets will be the "measurable sets." What properties do we want $\mathscr{B}$ to have? We would like both $f^{-1}[[0, a)]$ and $f^{-1}[[a, \infty)]$ to be measurable ($f \geq 0$) so we would like $\mathscr{B}$ to have the property: $A \in \mathscr{B}$ implies $\mathbb{R} \backslash A \in \mathscr{B}$. Also, when f is continuous, we want $f^{-1}[(a, b)]$ to be in $\mathscr{B}$, so $\mathscr{B}$ should contain the open sets. Finally, we want to have

$$\mu\left(\bigcup_{n=1}^{\infty} A_n\right) = \sum_{n=1}^{\infty} \mu(A_n)$$

if the A_n are mutually disjoint (to meet our intuitive notion of size) so we would like $\bigcup_{n=1}^{\infty} A_n \in \mathscr{B}$ if each A_n is in $\mathscr{B}$.

Definition The **Borel sets** of $\mathbb{R}$ is the smallest family of subsets of $\mathbb{R}$ with the following properties:

(i) The family is closed under complements.
(ii) The family is closed under countable unions.
(iii) The family contains each open interval.

I.3 The Lebesgue integral

To see that such a *smallest* family exists we note that if $\{\mathscr{B}_\alpha\}_{\alpha \in A}$ is a collection of families obeying (i), (ii), and (iii), then so does $\bigcap_{\alpha \in A} \mathscr{B}_\alpha$. Thus the intersection of all families obeying (i)–(iii) is the smallest such family.

Now we define the Lebesgue measures of sets in $\mathscr{B}$, the Borel sets in $\mathbb{R}$.

Definition Let $\mathscr{I}$ be the family of all countable unions of disjoint open intervals (which is just the family of open sets) and let

$$\mu\left(\bigcup_{i=1}^{\infty} (a_i, b_i)\right) \equiv \sum_{i=1}^{\infty} (b_i - a_i)$$

(which may be infinite). For any $B \in \mathscr{B}$, define

$$\mu(B) = \inf_{\substack{I \in \mathscr{I} \\ B \subset I}} \mu(I)$$

This notion of size has four crucial properties:

Theorem I.8
(a) $\mu(\varnothing) = 0$
(b) If $\{A_n\}_{n=1}^{\infty} \subset \mathscr{B}$ and the A_n are mutually disjoint ($A_n \cap A_m = \varnothing$, all $m \neq n$), then $\mu(\bigcup_{n=1}^{\infty} A_n) = \sum_{n=1}^{\infty} \mu(A_n)$.
(c) $\mu(B) = \inf\{\mu(I) \mid B \subset I, I \text{ is open}\}$
(d) $\mu(B) = \sup\{\mu(C) \mid C \subset B, C \text{ is compact}\}$

The infinite sum in (b) contains only positive terms, so it either converges to a finite number or diverges to infinity, in which case we set it equal to ∞. (c) and (d) say that any Borel set can be approximated "from the outside" by open sets and from the inside by compact sets. We remind the reader that *on the real line* a set is compact if and only if it is closed and bounded.

We have thus extended the usual notion of size of intervals and we define the family of functions we will consider in the obvious way:

Definition A function f is called a **Borel function** if and only if $f^{-1}[(a, b)]$ is a Borel set for all a, b.

It is often convenient to allow our functions to take the values $\pm \infty$ on small sets in which case we require $f^{-1}[\{\pm \infty\}]$ to be Borel.

Proposition f is a Borel if and only if, for all $B \in \mathscr{B}$, $f^{-1}[B] \in \mathscr{B}$ (see Problem 14).

This last proposition implies that the composition of two Borel functions is Borel. Many books deal with a slightly larger class of functions than the Borel class. They first define a set M to be measurable if one can write $M \cup A_1 = B \cup A_2$ where B is Borel and $A_i \subset B_i$ with B_i Borel and $\mu(B_i) = 0$ (thus they add and subtract "unimportant" sets from Borel sets). A measurable function is then defined as a function, f, for which $f^{-1}[(a, b)]$ is always measurable. It is no longer true that $f \circ g$ is measurable if f and g are, and many technical problems arise. *In any event, we deal only with Borel sets and functions and use the words Borel and measurable interchangeably.*

Borel functions are closed under many operations:

Proposition (a) If f, g are Borel, then so are $f + g$, fg, $\max\{f, g\}$ and $\min\{f, g\}$. If f is Borel and $\lambda \in \mathbb{R}$, λf is Borel.

(b) If each f_n is Borel, $n = 1, 2, \ldots$, and $f_n(p) \to f(p)$ for all p, then f is Borel.

Since $|f| = \max\{f, -f\}$, $|f|$ is measurable if f is.

As we sketched above, given $f \geq 0$, one can define $\int f\, dx$ (which may be ∞). If $\int |f|\, dx < \infty$, we write $f \in \mathscr{L}^1$ and define $\int f\, dx = \int f_+\, dx - \int f_-\, dx$ where $f_+ = \max\{f, 0\}$; $f_- = \max\{-f, 0\}$. $\mathscr{L}^1(a, b)$ is the set of functions on (a, b) which are in $\mathscr{L}^1$ if we extend them to the whole real line by defining them to be zero outside of (a, b). If $f \in \mathscr{L}^1(a, b)$, we write $\int f\, dx = \int_a^b f\, dx$. We then have:

Theorem I.9 Let f and g be measurable functions. Then

(a) If $f, g \in \mathscr{L}^1(a, b)$, so are $f + g$ and λf, for all $\lambda \in \mathbb{R}$.
(b) If $|g| \leq f$ and $f \in \mathscr{L}^1$, then $g \in \mathscr{L}^1$.
(c) $\int (f + g)\, dx = \int f\, dx + \int g\, dx$ if f and g are in $\mathscr{L}^1$.
(d) $|\int f\, dx| \leq \int |f|\, dx$ if f is in $\mathscr{L}^1$.
(e) If $f \leq g$, then $\int f\, dx \leq \int g\, dx$, if f and g are in $\mathscr{L}^1$.
(f) If f is bounded and measurable on $-\infty < a < b < \infty$, then $f \in \mathscr{L}^1$ and

$$\left| \int_a^b f\, dx \right| \leq |b - a| \left(\sup_{a \leq x \leq b} |f(x)| \right)$$

This theorem shows that $\int$ has all the nice properties of the Riemann integral even though it is defined for a larger class of functions.

The properties that make the space L^1 (which we will shortly define) complete are the following absolutely essential convergence theorems:

I.3 The Lebesgue integral

Theorem I.10 (monotone convergence theorem) Let $f_n \geq 0$ be measurable. Suppose $f_n(p) \to f(p)$ for each p and that $f_{n+1}(p) \geq f_n(p)$ all p and n (in which case we write $f_n \nearrow f$). If $\int f_n(p) \, dp < C$ for all n, then $f \in \mathscr{L}^1$ and $\int |f(p) - f_n(p)| \, dp \to 0$ as $n \to \infty$.

Theorem I.11 (dominated convergence theorem) Let $f_n(p) \to f(p)$ for each p and suppose $|f_n(p)| \leq G(p)$ for all n and some $G \in \mathscr{L}^1$. Then $f \in \mathscr{L}^1$ and $\int |f(p) - f_n(p)| \, dp \to 0$ as $n \to \infty$.

In the latter case, we say G dominates the pointwise convergence. That a dominating function exists is crucial. For example, let $f_n(x) = (1/n)\chi_{[-n, n]}(x)$. Then $f_n(x) \to 0$ for each x, but $\int |f_n| \, dx = 2$ so $\int |f_n(x)| \, dx$ does not go to zero. In this case, it is not hard to see that $\sup_n |f_n(x)| = G(x)$ is not in $\mathscr{L}^1$.

We are almost ready to define $\mathscr{L}^1$ as a metric space by letting $\rho(f, g) = \int |f - g| \, dx$. We cannot quite do this because $\int |f - g| \, dx = 0$ does not imply $f \equiv g$ (for example, f and g might differ at a single point). Thus, we first define the notion of almost everywhere (a.e.):

Definition We say a condition $C(x)$ holds almost everywhere (a.e.) if $\{x \mid C(x) \text{ is false}\}$ is a subset of a set of measure zero.

Definition We say two functions $f, g \in \mathscr{L}^1$ are equivalent if $f(x) = g(x)$ a.e. (this is the same as saying $\int |f - g| \, dx = 0$).

Definition The set of equivalence classes in $\mathscr{L}^1$ is denoted by as L^1. L^1 with the norm $\|f\|_1 = \int |f| \, dx$ is a normed linear space.

Thus an element of L^1 is an equivalence class of functions equal a.e. In particular when $f \in L^1$, the symbol $f(x)$ for a particular x does not make sense. Nevertheless we continue to write "$f(x)$" but only in situations where statements are independent of a choice from the equivalence class. Thus, for example, $f_n(x) \to f(x)$ *for almost all x* is independent of the representatives chosen for f and f_n. By this replacement of pointwise convergence with pointwise convergence almost everywhere, the two convergence theorems carry over from $\mathscr{L}^1$ to L^1.

Having cautioned the reader that $f(x)$ is "technically meaningless" for $f \in L^1$, we remark that in *certain special cases* it is meaningful. Suppose $f \in L^1$

has a representative $\tilde{f}$ (that is, $\tilde{f}$ is a function; f an equivalence class of functions) which is continuous. Then no other representative of f is continuous, so it is natural to write $f(x)$ for $\tilde{f}(x)$.

The critical fact about L^1 is:

Theorem I.12 (Riesz–Fisher) L^1 is complete.

Proof Let f_n be Cauchy in L^1. It is enough to prove some subsequence converges (see Problem 3) so pass to a subsequence (also labeled f_n) with $\|f_n - f_{n+1}\|_1 \leq 2^{-n}$. Let

$$g_m(x) = \sum_{n=1}^{m} |f_n(x) - f_{n+1}(x)|$$

Let g_∞ be the infinite sum (which may be ∞). Then $g_m \nearrow g_\infty$ and $\int |g_m| \leq \sum_{n=1}^{m} \|f_n - f_{n+1}\| \leq 1$, so by the monotone convergence theorem, $g_\infty \in L^1$. Thus $|g_\infty(x)| < \infty$ a.e. As a result

$$f_m(x) = f_1(x) - \sum_{n=1}^{m-1} (f_n(x) - f_{n+1}(x))$$

converges pointwise a.e. to a function $f(x)$. Moreover, $|f_m(x)| \leq |f_1(x)| + g_\infty(x) \in L^1$ so $f_n \to f$ in L^1 by the dominated convergence theorem. ∎

This proof has a corollary (see Problem 17):

Corollary   If $f_n \to f$ in L^1, then some *subsequence* f_{n_i} converges pointwise a.e. to f.

As a final result which brings us full circle to our original motivation:

Proposition $C[a, b]$ is dense (in $\|\cdot\|_1$) in $L^1[a, b]$, i.e. L^1 is the completion of C.

Proof See Problem 18.

We defined $L^1[a, b]$ as a space of real-valued functions. It is often convenient to deal with complex-valued functions, f, whose real and imaginary parts are in $L^1[a, b]$. When no confusion arises, we will denote this space, with the norm

$$\|f\|_1 = \int_a^b |f|\, dx$$

also by $L^1[a, b]$. The integral of a complex-valued function is defined by

$$\int f\, dx = \int \mathrm{Re}(f)\, dx + i \int \mathrm{Im}(f)\, dx$$

I.4 Abstract measure theory

One of the most important tools which one combines with abstract functional analysis in the study of various concrete models is "general" measure theory, that is, the theory of the last section extended to a more abstract setting.

The simplest way to generalize the Lebesgue integral is to work with functions on the real line and with Borel sets but to generalize the underlying measure; we consider this special case of abstract measure theory first. Recall that the Lebesgue integral was constructed as follows. We started with a notion of size for intervals, $\mu([a, b]) = b - a$, and extended this in a unique way to a notion of size for arbitrary Borel sets. Armed with this notion of size for Borel sets, the integral of Borel functions was obtained by measuring sets of the form $f^{-1}([a, b])$. We found the vector space $L^1([0, 1], dx)$ constructed in the last section is just the completion of $C[0, 1]$ with the metric $d_2(f, g) = \int_0^1 |f(x) - g(x)|\, dx$, where we needed only the Riemann integral to define d_2.

Now suppose an arbitrary monotone function $\alpha(x)$ is given (that is, $x > y$ implies $\alpha(x) \geq \alpha(y)$). It is not hard to see that the limit from the right, $\lim_{\varepsilon \to 0} \alpha(x + |\varepsilon|)$ and the limit from the left, $\lim_{\varepsilon \to 0} \alpha(x - |\varepsilon|)$ exist; we write them as $\alpha(x + 0)$ and $\alpha(x - 0)$ respectively. Since (a, b) does not include the points a and b, it is natural to define $\mu_\alpha((a, b)) = \alpha(b - 0) - \alpha(a + 0)$. From this notion of size for intervals, one can construct a measure μ_α on Borel sets of $\mathbb{R}$, that is, a map $\mu_\alpha: \mathscr{B} \to [0, \infty]$ with $\mu_\alpha(\bigcup B_i) = \sum_{i=1}^\infty \mu_\alpha(B_i)$ if $B_i \cap B_j = \varnothing$ and $\mu_\alpha(\varnothing) = 0$. By construction, this measure has the regularity property

$$\mu_\alpha(B) = \sup\{\mu(C) \mid C \subset B, \ C \text{ compact}\}$$
$$= \inf\{\mu(O) \mid B \subset O, \ O \text{ open}\}$$

Also, $\mu(C) < \infty$ for any compact set C. A measure with these two regularity properties is called a **Borel measure**. In particular, $\mu_\alpha([a, b]) = \alpha(b + 0) - \alpha(a - 0)$. One can then construct an integral $f \to \int f \, d\mu_\alpha$ (we will also write $\int f \, d\alpha$) which has properties (a)–(e) of Theorem I.9; it is called a Lebesgue–Stieltjes integral. $L^1([a, b], d\alpha)$ and $L^1(\mathbb{R}, d\alpha)$ can be formed as before. These spaces of equivalence classes of functions are complete in the metric $\rho(f, g) = \int |f - g| \, d\alpha$, and analogues of the monotone and dominated convergence theorems hold. The continuous functions $C[a, b]$ form a dense subspace of $L^1([a, b], d\alpha)$; put differently, $L^1([a, b], d\alpha)$ is the completion of $C[a, b]$ with the metric $\rho_\alpha(f, g) = \int_a^b |f - g| \, d\alpha$ where we need only use the Riemann–Stieltjes integral to define ρ_α (see Problem 11).

Let us consider three examples which illustrate the variety of Lebesgue–Stieltjes measures.

Example 1 Suppose α is continuously differentiable. Then $\mu_\alpha(a, b) = \int_a^b (d\alpha/dx) \, dx$ where dx is Lebesgue measure, so it is to be expected (and is indeed true!) that

$$\int f \, d\alpha = \int f \left(\frac{d\alpha}{dx} \right) dx$$

Thus, these measures can essentially be described in terms of Lebesgue measure.

Example 2 Suppose that $\alpha(x)$ is the characteristic function of $[0, \infty)$. Then $\mu_\alpha(a, b) = 1$ if $0 \in (a, b)$ and is 0 if $0 \notin (a, b)$. The measure one gets out is very easy to describe: $\mu_\alpha(B) = 1$ if $0 \in B$, and $\mu_\alpha(B) = 0$ if $0 \notin B$. The reader is invited to construct explicitly the integral and convince himself that

$$\int f \, d\alpha = f(0)$$

This measure $d\alpha$ is known as the Dirac measure (since it is just like a δ function). Let us consider $L^1(\mathbb{R}, d\alpha)$ in this case. In $\mathscr{L}^1$ we have $\rho(f, g) = |f(0) - g(0)|$ so $\rho(f, g) = 0$ if and only if $f(0) = g(0)$. As a result, we see that the equivalence classes in L^1 are completely described by the value $f(0)$ so that $L^1(\mathbb{R}, d\alpha)$ is just a one-dimensional vector space! Notice how different this is from the case of $L^1(\mathbb{R}, dx)$ where the value of a "function" at a single point is not defined (since elements of L^1 are equivalence classes).

Example 3 Our last example makes use of a fairly pathological function, $\alpha(x)$, which we first construct. Let S be the subset of $[0, 1]$

$$S = (\tfrac{1}{3}, \tfrac{2}{3}) \cup (\tfrac{1}{9}, \tfrac{2}{9}) \cup (\tfrac{7}{9}, \tfrac{8}{9}) \cup (\tfrac{1}{27}, \tfrac{2}{27}) \cup \cdots$$

I.4 Abstract measure theory 21

FIGURE I.5 The Cantor set.

that is, remove the middle third of what is not in S at each stage and add it to S, see Figure I.5. The Lebesgue measure of S is $\frac{1}{3} + 2(\frac{1}{9}) + 4(\frac{1}{27}) + \cdots = 1$. Let $C = [0, 1]\setminus S$. It has Lebesgue measure 0. C, which is known as the Cantor set, is easy to describe if we write each $x \in [0, 1]$ in its base three decimal expansion. Then $x \in C$ if and only if this base 3 expansion has no 1's. Thus C is an uncountable set of measure 0. To see this, map C in a one-one way *onto* $[0, 1]$ by changing 2's into 1's and viewing the end result as a base 2 number. Now construct $\alpha(x)$ as follows: set $\alpha(x) = \frac{1}{2}$ on $(\frac{1}{3}, \frac{2}{3})$; $\alpha(x) = \frac{1}{4}$ on $(\frac{1}{9}, \frac{2}{9})$; $\alpha(x) = \frac{3}{4}$ on $(\frac{7}{9}, \frac{8}{9})$, etc.; see Figure I.6. Extend α to $[0, 1]$ by making it con-

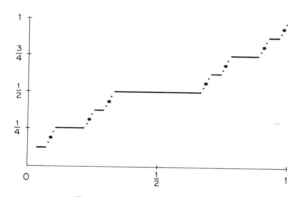

FIGURE I.6 The Cantor function.

tinuous. Then α is a nonconstant continuous function with the strange property that $\alpha'(x)$ exists a.e. (with respect to Lebesgue measure) and is zero a.e. Now, we can form the measure μ_α. Since α is continuous, $\mu_\alpha(\{p\}) = 0$ for any set $\{p\}$ with only one point. Nevertheless, μ_α is concentrated on the set C in the sense that $\mu_\alpha([0, 1]\setminus C) = \mu_\alpha(S) = 0$. On the other hand, the Lebesgue measure of C is zero. Thus μ_α and Lebesgue measure "live" on completely different sets.

In a sense we now make precise, these three examples are models of the most general Lebesgue–Stieltjes measures. Suppose μ is a Borel measure on $\mathbb{R}$.

First, let $P = \{x \mid \mu(\{x\}) \neq 0\}$, that is, P is the set of **pure points** of μ. Since μ is Borel [$\mu(C) < \infty$ for any compact set], P is a countable set. Define

$$\mu_{\text{pp}}(X) = \sum_{x \in P \cap X} \mu(\{x\}) = \mu(P \cap X)$$

Then μ_{pp} is a measure and $\mu_{\text{cont}} = \mu - \mu_{\text{pp}}$ is positive. μ_{cont} has the property $\mu_{\text{cont}}(\{p\}) = 0$ for all p, that is, it has no pure points and μ_{pp} has only pure points in the sense that $\mu_{\text{pp}}(X) = \sum_{x \in X} \mu_{\text{pp}}(\{x\})$.

Definition A Borel measure μ on $\mathbb{R}$ is called **continuous** if it has no pure points. μ is called a **pure point measure** if $\mu(X) = \sum_{x \in X} \mu(x)$ for any Borel set X.

Thus, we have seen:

Theorem I.13 Any Borel measure can be decomposed uniquely into a sum $\mu = \mu_{\text{pp}} + \mu_{\text{cont}}$ where μ_{cont} is continuous and μ_{pp} is a pure point measure.

We have thus generalized Example 2 by allowing sums of Dirac measures. Is there any generalization of Examples 1 and 3?

Definition We say that μ is **absolutely continuous with respect to (w.r.t.) Lebesgue measure** if there is a function, f, locally L^1 (that is, $\int_a^b |f(x)| \, dx < \infty$ for any finite interval (a, b)) so that

$$\int g \, d\mu = \int gf \, dx$$

for any Borel function g in $L^1(\mathbb{R}, d\mu)$. We then write $d\mu = f \, dx$.

This definition generalizes Example 1; we will eventually make a different (but equivalent!) definition of absolute continuity.

Definition We say μ is **singular relative to Lebesgue measure** if and only if $\mu(S) = 0$ for some set S where $\mathbb{R} \backslash S$ has Lebesgue measure 0.

The fundamental result is:

Theorem I.14 (Lebesgue decomposition theorem) Let μ be a Borel measure. Then $\mu = \mu_{\text{ac}} + \mu_{\text{sing}}$ in a unique way with μ_{ac} absolutely continuous w.r.t. Lebesgue measure and with μ_{sing} singular relative to Lebesgue measure.

I.4 Abstract measure theory

Thus Theorems I.13 and I.14 tell us that any measure μ on $\mathbb{R}$ has a canonical decomposition $\mu = \mu_{pp} + \mu_{ac} + \mu_{sing}$ where μ_{pp} is pure point, μ_{ac} absolutely continuous with respect to Lebesgue measure, and μ_{sing} is *continuous and singular* relative to Lebesgue measure. This decomposition will recur in a quantum-mechanical context where any state will be a sum of bound states, scattering states, and states with no physical interpretation (one of our hardest jobs will be to show that this last type of state does not occur; that is, that certain measures have $\mu_{sing} = 0$; (see Chapter XIII).)

This completes our study of measures on $\mathbb{R}$. The next level of generalization involves measures on sets with some underlying topological structure; we will return to study this case of intermediate generality in Section IV.4. The most general setting lets us deal with an arbitrary set. We first need an abstraction of Borel sets:

Definition A nonempty family $\mathscr{R}$ of subsets of a set M is called a σ-**ring** if and only if

(a) $A_i \in \mathscr{R}$, $i = 1, 2, \ldots$ implies $\bigcup_{i=1}^{\infty} A_i \in \mathscr{R}$.
(b) If $A, B \in \mathscr{R}$, then $A \backslash B \in \mathscr{R}$.

If $M \in \mathscr{R}$, we say that $\mathscr{R}$ is a σ-**field**.

The definition of measure is obvious(!):

Definition A **measure** on a set M with σ-ring $\mathscr{R}$ is a map $\mu: \mathscr{R} \to [0, \infty]$ with the properties:

(a) $\mu(\varnothing) = 0$
(b) $\mu\left(\bigcup_{i=1}^{\infty} A_i\right) = \sum_{i=1}^{\infty} \mu(A_i)$, if $A_i \cap A_j = \varnothing$ for all $i \neq j$.

We shall often speak of the measure space $\langle M, \mu \rangle$ without explicitly mentioning $\mathscr{R}$, but the σ-ring is a crucial element of the definition. Occasionally, we will write $\langle M, \mathscr{R}, \mu \rangle$. For certain pathologically "big" spaces, one wants to use the notion of σ-ring rather than σ-field, but to keep things simple, we will consider measures on σ-fields and will suppose the whole space isn't too big in the sense:

Definition A measure μ on a σ-field $\mathscr{F}$ is called σ-**finite** if and only if $M = \bigcup_{i=1}^{\infty} A_i$ with each $\mu(A_i) < \infty$.

We will suppose all our underlying measures are σ-finite.

Definition Let M, N be sets with σ-fields $\mathcal{R}$ and $\mathcal{F}$. A map $T\colon M \to N$ is called **measurable** (w.r.t. $\mathcal{R}$ and $\mathcal{F}$) if and only if $\forall A \in \mathcal{F}$, $T^{-1}[A] \in \mathcal{R}$. A map $f\colon M \to \mathbb{R}$ is called measurable if it is measurable w.r.t. $\mathcal{R}$ and the Borel sets of $\mathbb{R}$.

Given a measure μ on a measure space M, we can define $\int f\,d\mu$ for any positive real-valued measurable function on M and we can form $\mathscr{L}^1(M, d\mu)$, the set of integrable functions and $L^1(M, d\mu)$, the equivalence classes of functions in $\mathscr{L}^1$ equal a.e.$[\mu]$. As in the case $\langle M, d\mu \rangle = \langle \mathbb{R}, dx \rangle$, the following crucial theorems hold:

Theorem I.15 (monotone convergence theorem) If $f_n \in \mathscr{L}^1(M, d\mu)$, $0 \leq f_1(x) \leq f_2(x) \leq \cdots$ and $f(x) = \lim_{n \to \infty} f_n(x)$, then $f \in \mathscr{L}^1$ if and only if $\lim_{n \to \infty} \|f_n\|_1 < \infty$ and in that case $\lim_{n \to \infty} \|f - f_n\|_1 = 0$ and $\lim_{n \to \infty} \|f_n\|_1 = \|f\|_1$.

Theorem I.16 (dominated convergence theorem) If $f_n \in L^1(M, d\mu)$, $\lim_{n \to \infty} f_n(x) = f(x)$ a.e.$[\mu]$, and if there is a $G \in L^1$ with $|f_n(x)| \leq G(x)$ a.e.$[\mu]$, for all n, then $f \in L^1$ and $\lim_{n \to \infty} \|f - f_n\|_1 = 0$.

Theorem I.17 (Fatou's lemma) If $f_n \in \mathscr{L}^1$, each $f_n(x) \geq 0$ and if $\varliminf \|f_n\|_1 < \infty$, then $f(x) = \varliminf f_n(x)$ is in $\mathscr{L}^1$ and $\|f\|_1 \leq \varliminf \|f_n\|_1$.

Note In Fatou's lemma nothing is said about $\lim_{n \to \infty} \|f - f_n\|_1$.

Theorem I.18 (Riesz–Fisher theorem) $L^1(M, d\mu)$ is complete.

One also has the idea of mutually singular:

Definition Let μ, ν be two measures on a space M with σ-field $\mathcal{R}$. We say that μ and ν are mutually singular if there is a set $A \in \mathcal{R}$ with $\mu(A) = 0$, $\nu(M \backslash A) = 0$.

It is useful to take a weaker looking definition of absolute continuity which is essentially the opposite of singular:

Definition We say ν is absolutely continuous w.r.t. μ if and only if $\mu(A) = 0$ implies $\nu(A) = 0$.

That this definition is the same as the previous one is a consequence of:

Theorem I.19 (Radon–Nikodym theorem) v is absolutely continuous w.r.t. μ if and only if there is a measurable function f so that

$$v(A) = \int f(x)\chi_A(x)\,d\mu(x)$$

for any measurable set A. f is uniquely determined a.e. (w.r.t. μ).

Finally the Lebesgue decomposition theorem has an abstract form:

Theorem I.20 (Lebesgue decomposition theorem)   Let μ, v be two measures on a measure space $\langle M, \mathscr{R} \rangle$. Then v can be written uniquely as $v = v_{ac} + v_{sing}$ where μ and v_{sing} are mutually singular and v_{ac} is absolutely continuous w.r.t. μ.

There is one final subject in measure theory which we must consider and that involves changing the order of integration in a multiple integral. We first must consider what functions can be multiply integrated:

Definition   Let $\langle M, \mathscr{R} \rangle$, $\langle N, \mathscr{F} \rangle$ be two sets with associated σ-fields. Then the σ-field, $\mathscr{R} \otimes \mathscr{F}$ of subsets of $M \times N$ is defined to be the smallest σ-field containing $\{R \times F \mid R \in \mathscr{R}, F \in \mathscr{F}\}$.

Notice that if $f: M \times N \to \mathbb{R}$ is measurable (w.r.t. $\mathscr{R} \otimes \mathscr{F}$), then for any $m \in M$, the function $n \mapsto f(m, n)$ is measurable (w.r.t. $\mathscr{F}$). If v is a measure on N such that $\int f(m, n)\,dv(n)$ exists for all m, then one can show that $m \mapsto \int f(m, n)\,dv(n)$ is measurable (w.r.t. $\mathscr{R}$). There is a direct analogue of the fact that absolute convergent sums can be rearranged at will:

Theorem I.21 (Fubini's theorem)   Let f be a measurable function on $M \times N$. Let μ be a measure on M, v a measure on N. Then

$$\int_M \left(\int_N |f(m, n)|\,dv(n) \right) d\mu(m) < \infty$$

if and only if

$$\int_N \left(\int_M |f(m, n)|\,d\mu(m) \right) dv(n) < \infty$$

and if one (and thus both) of these integrals is finite, then

$$\int_N \left(\int_M f(m, n)\,d\mu(m) \right) dv(n) = \int_M \left(\int_N f(m, n)\,dv(n) \right) d\mu(m)$$

In Problem 25, the reader will see that the finiteness of the integral of the absolute value is critical.

Fubini's theorem can be put into perspective by the notion of product measure:

Theorem I.22 Let μ be a σ-finite measure on $\langle M, \mathscr{R} \rangle$ and ν a σ-finite measure on $\langle N, \mathscr{F} \rangle$. Then, there is a unique measure $\mu \otimes \nu$ on $\langle M \times N, \mathscr{R} \otimes \mathscr{F} \rangle$ obeying

$$(\mu \otimes \nu)(R \times F) = \mu(R)\nu(F)$$

(where $0 \cdot \infty = 0$). If f is a measurable function on $M \times N$, then

$$\int_M \left(\int_N |f(m, n)| \, d\nu(n) \right) d\mu(m) < \infty$$

if and only if

$$\int_{M \times N} |f| \, d(\mu \otimes \nu) < \infty$$

and in that case

$$\int_{M \times N} f \, d(\mu \otimes \nu) = \int_M \left(\int_N f \, d\nu \right) d\mu$$

One can describe the measure $\mu \otimes \nu$ quite explicitly. If $M \in \mathscr{R} \times \mathscr{F}$ and $M \subset \bigcup_{i=1}^{\infty} R_i \times F_i$ we have $(\mu \otimes \nu)(M) \leq \sum_{i=1}^{\infty} \mu(R_i)\nu(F_i)$. In fact, for any $M \in \mathscr{R} \times \mathscr{F}$,

$$(\mu \otimes \nu)(M) = \inf \left\{ \sum_{i=1}^{\infty} \mu(R_i)\nu(F_i) \, \middle| \, M \subset \bigcup_{i=1}^{\infty} R_i \times F_i \right\}$$

In particular, we can approximate M with a countable union of rectangles making an arbitrarily small error.

I.5 Two convergence arguments

In this section we single out two "tricks" which we will have occasion to use over and over. While they are elementary and the reader may well have seen them, it seems reasonable to discuss them explicitly.

The first argument, which we will call the $\varepsilon/3$ argument, is best seen in the proof of:

1.5 Two convergence arguments

Theorem I.23 Let $C[a, b]$ be the continuous functions on $[a, b]$ with the metric
$$d_1(f, g) = \sup_{a \leq x \leq b} |f(x) - g(x)|$$
induced by the norm $\|f\|_\infty = d_1(f, 0)$. Then $C[a, b]$ with the norm $\|\cdot\|_\infty$ is complete.

Proof Let f_n be a $\|\cdot\|_\infty$-Cauchy sequence. Then, for any fixed $x \in [a, b]$, $|f_n(x) - f_m(x)| \leq \|f_n - f_m\|_\infty \to 0$ as $n, m \to \infty$ so $f_n(x)$ is a Cauchy sequence of real numbers. Since the reals are complete, for each x there is a number, $f(x)$, with $f_n(x) \to f(x)$. Given ε, find N so $n, m \geq N$ implies $\|f_n - f_m\|_\infty \leq \varepsilon$. Then
$$\sup_{a \leq x \leq b} |f(x) - f_N(x)| = \sup_{a \leq x \leq b} \lim_{n \to \infty} |f_n(x) - f_N(x)|$$
$$\leq \sup_{a \leq x \leq b} \sup_{n \geq N} |f_n(x) - f_N(x)|$$
$$= \sup_{n \geq N} \|f_n - f_N\|_\infty \leq \varepsilon$$

Thus, if we can show that $f \in C[a, b]$, we can conclude that $\|f - f_n\|_\infty \to 0$ so $f_n \to f$ in $C[a, b]$.

We are thus left with proving that f is continuous, or put differently that "a uniform limit of continuous functions is continuous." Fix $x \in [a, b]$ and $\varepsilon > 0$. We want to find δ so $|x - y| < \delta$ implies $|f(x) - f(y)| < \varepsilon$. Pick n so that $\|f_n - f\|_\infty < \varepsilon/3$. Now, since f_n is continuous, pick δ so that $|x - y| < \delta$ implies $|f_n(x) - f_n(y)| < \varepsilon/3$. Then $|x - y| < \delta$ implies
$$|f(x) - f(y)| \leq |f(x) - f_n(x)| + |f_n(x) - f_n(y)| + |f_n(y) - f(y)|$$
$$< \tfrac{1}{3}\varepsilon + \tfrac{1}{3}\varepsilon + \tfrac{1}{3}\varepsilon$$

Thus f is continuous. ∎

What is the essence of the $\varepsilon/3$ argument? We had a family of convergent sequences $f_n(x) \to f(x)$ for each x and had *uniform control on the rate of convergence*, that is, control independent of the object x that parametrized the family. We also had some information on the behavior of $f_n(x)$ for fixed n as the parameter x varied *but this information was not necessarily uniform in n*. What we did is pictorially indicated in Figure I.7; one could also call the $\varepsilon/3$ argument the "up, over, and around" proof. In the next section we will consider what happens when one has no uniform information on the rate of convergence but instead has uniform control on how $f_n(x)$ behaves as x varies (uniform in n). There we will see an $\varepsilon/3$ argument also works. For further examples of the $\varepsilon/3$ trick, see Problems 27, 29.

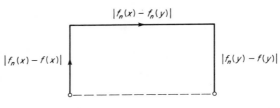

FIGURE I.7   The $\varepsilon/3$ argument.

The second argument which we refer to as the "diagonal sequence trick" is illustrated in:

Theorem I.24   Let $f_n(m)$ be a sequence of functions on the positive integers which is uniformly bounded, i.e. $|f_n(m)| \leq C$ for all n, m. Then there is a subsequence $\{f_{\hat{n}(i)}(m)\}_{i=1}^{\infty}$ so that for each fixed m, $f_{\hat{n}(i)}(m)$ converges as $i \to \infty$.

Proof   Consider the sequence $f_n(1)$. It is a bounded set of numbers, so we can find a subsequence $f_{n_1(i)}$ so $f_{n_1(i)}(1) \to f_\infty(1)$, for some number $f_\infty(1)$. Now consider the sequence $f_{n_1(i)}(2)$. We can find a subsequence $f_{n_2(i)}(2) \to f_\infty(2)$ as $i \to \infty$. Proceeding inductively, we find successive subsequences, $f_{n_k(i)}$ so that (a) $f_{n_{k+1}(i)}$ is a subsequence of $f_{n_k(i)}$ and (b) $f_{n_k(i)}(k) \to f_\infty(k)$ as $i \to \infty$. Thus, in particular, $f_{n_k(i)}(j) \to f_\infty(j)$ as $i \to \infty$ for $j = 1, 2, \ldots, k$. To get a subsequence $f_{\hat{n}(i)}$ converging for each j, one is tempted to try to take the limit of the horizontal sequence (see Figure I.8a) but that won't work! (for it may happen $n_k(1) \to \infty$). The simple way out is to take the diagonal sequence $\hat{n}(k) = n_k(k)$. Then $f_{\hat{n}(k)}, f_{\hat{n}(k+1)}, \ldots$ is a subsequence of $f_{n_k(i)}$ so $f_{\hat{n}(i)}(k) \to f_\infty(k)$ as $i \to \infty$ for any k. ∎

I.6 Equicontinuity

We have just seen that one can control the x dependence of $\lim_{n \to \infty} f_n(x)$ if one is given information on the approach to the limit which is uniform in x. In this section we study what happens when the given information is instead uniform in n; what we will see is that one can obtain not only information about the x behavior of the limit but that one can also turn weak information about the approach to the limit into stronger information. We first isolate the notion of "control on the x behavior uniform in n."

I.6 Equicontinuity

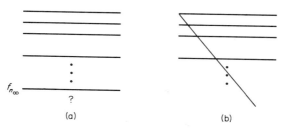

FIGURE I.8 The diagonal trick.

Definition Let $\mathcal{F}$ be a family of functions from a metric space $\langle X, \rho \rangle$ to another metric space $\langle Y, d \rangle$. We say $\mathcal{F}$ is an **equicontinuous family** if and only if

$$(\forall \varepsilon)(\forall x \in X)(\exists \delta)(\forall f \in \mathcal{F})\ \rho(x, x') < \delta \quad \text{implies} \quad d(f(x), f(x')) < \varepsilon.$$

We say $\mathcal{F}$ is a **uniformly equicontinuous family** if and only if

$$(\forall \varepsilon)(\exists \delta)(\forall x \in X)(\forall f \in \mathcal{F})\ \rho(x, x') < \delta \quad \text{implies} \quad d(f(x), f(x')) < \varepsilon.$$

For comparison sake, note that to say all $f \in \mathcal{F}$ are continuous means $(\forall \varepsilon)(\forall x \in X)(\forall f \in \mathcal{F})(\exists \delta)\rho(x, x') < \delta$ implies $d(f(x), f(x')) < \varepsilon$. Thus, for mere continuity, δ can depend on f and x (as well as ε), while equicontinuity says δ is independent of f; finally uniform equicontinuity says δ is dependent only on ε.

As promised, it is easy to turn information about $f_n(x)$, uniform in n into information about the limit:

Theorem I.25 Let f_n be a sequence of functions from one metric space to another with the property that the family $\{f_n\}$ is equicontinuous. Suppose that $f_n(x) \to f(x)$ pointwise for each x. Then f is continuous.

Proof Given ε and x, choose δ so $\rho(x, x') < \delta$ implies $d(f_n(x), f_n(x')) < \tfrac{1}{2}\varepsilon$ for all n. Since d is continuous, we have $d(f(x), f(x')) = \lim_{n \to \infty} d(f_n(x), f_n(x'))$ so $\rho(x, x') < \delta$ implies $d(f(x), f(x')) \leq \varepsilon/2 < \varepsilon$. ∎

The proof makes it clear that if $\{f_{n,m}\}$ is an equicontinuous family and for each m, $\lim_{n \to \infty} f_{n,m} \equiv f_m$ exists, then $\{f_m\}$ is an equicontinuous family. Equicontinuity has a crucial consequence which we will see combines nicely with the diagonal sequence trick:

Theorem I.26 Let $\{f_n\}$ be an equicontinuous family of functions from one metric space $\langle X, \rho \rangle$ to another $\langle Y, d \rangle$ with Y complete. Suppose that for a dense set $D \subset X$, we know $f_n(x)$ converges for all $x \in D$. Then $f_n(x)$ converges for all $x \in X$. (Note that by Theorem I.25, the limit function is continuous.)

Proof See Problem 29.

Theorem I.26 tells us that, in general, pointwise convergence on a dense set combined with equicontinuity implies pointwise convergence everywhere. More spectacularly, for a sequence of functions on [0, 1] (see Problem 30), uniform equicontinuity and pointwise convergence imply *uniform convergence*:

Theorem I.27 Let $\{f_n\}$ be a uniformly equicontinuous family of functions on [0, 1]. Suppose that $f_n(x) \to f(x)$ for each x in [0, 1]. Then $f_n(x) \to f(x)$ uniformly in x.

Proof Let ε be given. Choose δ so that $|x - y| < \delta$ implies $|f_n(x) - f_n(y)| < \varepsilon/3$ for all n. Now choose $y_1, \ldots, y_m$ so that every point of [0, 1] is within δ of some y_i. Since $y_1, \ldots, y_m$ is a finite set, we can find n so $n > N$ implies $|f_n(y_i) - f(y_i)| < \varepsilon/3$; $i = 1, \ldots, m$. By an $\varepsilon/3$ argument, $\|f_n - f\|_\infty < \varepsilon$ for all $n > N$. ∎

For functions on [0, 1], every equicontinuous family is uniformly equicontinuous (Problem 31).

We can combine the convergence theorems (Theorems I.26 and I.27), the diagonalization trick of Section I.5 and the above remark (Problem 31) to prove the beautiful:

Theorem I.28 (Ascoli's theorem) Let f_n be a family of uniformly bounded equicontinuous functions on [0, 1]. Then some subsequence $f_{n(i)}$ converges uniformly on [0, 1].

Proof Let $q_1, q_2, \ldots$ be a numbering of the rationals. Since the f_n's are uniformly bounded, $|f_n(q_m)| \leq C$ for all m and n. Thus, by the diagonalization trick, we can find a subsequence with $f_{n(i)}(q_m)$ converging as $i \to \infty$ for each m. By Theorem I.26, the $f_{n(i)}$ converge pointwise everywhere and then by Theorem I.27, they are uniformly convergent. ∎

At this point, we discuss no applications in detail. However, we mention two examples to which we shall return which show the variety of applications: In Section V.1, we define a metric on all functions, $\mathcal{O}_D$, analytic in a region D. In Theorem V.25 we use equicontinuity arguments to prove that certain subsets of $\mathcal{O}_D$ are compact. In Chapter XX, we will discuss the limit of the free energy per unit volume of a "lattice gas" in a box as the volume goes to infinity. Our proof that this limit exists for a large class of interactions will proceed in three steps: (1) The interactions will be given a metric and it will be shown that strictly finite range interactions are dense in all "allowable" interactions in this metric. (2) If for fixed volume Λ the free energy per unit volume F_Λ is treated as a function on the metric space of allowable interactions, the $\{F_\Lambda\}$ are equicontinuous. (3) The $\lim_{\Lambda \to \infty} F_\Lambda(\Phi)$ will be shown to exist if Φ is a finite range interaction. Then equicontinuity arguments will be used to tell us that $\lim_{\Lambda \to \infty} F_\Lambda(\Phi)$ exists for any allowable interaction Φ (and the limit will be continuous in Φ).

NOTES

Section I.1 For a discussion of the subtleties of Zorn's lemma, the axiom of choice, etc. intended for the novice, we recommend: P. R. Halmos, *Naive Set Theory*, Van Nostrand–Reinhold, Princeton, New Jersey, 1960.

Section I.2 For additional discussion of metric space notions, see A. Gleason, *Introduction to Abstract Analysis*, Addison-Wesley, Reading, Massachusetts, 1966, or A. Kolmogorov and S. Fomin, *Elements of the Theory of Functional Analysis*, Vol. I, Graylock Press, 1957. For normed linear spaces (and in particular for our discussion of the Riemann integral) see J. Dieudonné, *Foundations of Modern Analysis*, Academic Press, New York and London, 1960 or L. Loomis and S. Sternberg, *Advanced Calculus*, Addison–Wesley, Reading, Massachusetts, 1968.

Sections I.3, I.4 For a discussion of Lebesgue Integration, see J. Williamson, *Introduction to the Lebesgue Integral*, Holt, New York, 1962 or W. Rudin, *Principles of Modern Analysis*, McGraw-Hill, New York, 1963. For abstract measure theory we particularly recommend S. K. Berberian, *Measure and Integration*, Macmillan, New York, 1965 and H. Royden, *Real Analysis*, Macmillan, New York, 1968. See also P. Halmos: *Measure Theory*, Van Nostrand, Reinhold, Princeton, New Jersey, 1950; N. Dunford and J. Schwartz, *Linear Operators*, Chapter 3, Wiley (Interscience), New York, 1958.

For a discussion of the Banach–Tarski paradox, see R. Rosenblum, *Elements of Mathematical Logic*, p. 150, Dover, New York, 1950, or R. Robinson, *Fund. Math.*, 34 (1947), 246.

We note that one can construct all Borel sets as follows: Start with the open sets and their complements, the closed sets. Add countable unions of closed sets, called F_σ sets and their complements (countable intersections of open sets) called G_δ sets. Then add countable unions of G_δ's called $G_{\delta\sigma}$'s and their complements $F_{\sigma\delta}$'s. Next add $G_{\delta\sigma\delta}$, etc. *After countably many steps one is still not done*, for a union of one G_δ, one $F_{\sigma\delta}$, one $G_{\delta\sigma\delta}$, ... may not have been included. In the end transfinite induction up to the first uncountable ordinal is needed.

As the problems show, the Borel functions are the smallest family closed under pointwise limits and containing all continuous functions. As with the Borel sets, their construction requires transfinite induction. We note however that, given any Borel measure μ, any Borel function f is equal almost everywhere (w.r.t. μ) to a pointwise limit of continuous functions (Problems 18 and 19).

By removing the middle 2^n-ths at the nth step in $[0, 1]$, one can construct a closed set of positive measure with an empty interior.

The approach to measures on topological spaces (rather than abstract spaces) which we discuss in Section IV.4, is fashionable with the French School. See N. Bourbaki, *Intégration*, Chapters 1–8, Hermann, Paris, (1952, 1956, 1959, 1963) or (for a beautiful and brief discussion) L. Nachbin, *The Haar Integral*, Chapter I, Van Nostrand–Reinhold, Princeton, New Jersey, 1965.

Section I.6 The natural setting for Ascoli's theorem is functions on an arbitrary compact metric space, or more generally, a second countable compact uniform space (which is actually always metrizable), for example, a compact topological group.

The idea of using equicontinuity in establishing the existence of the thermodynamic limit goes back at least as far as R. B. Griffiths: "A Proof that the Free Energy of a Spin System Is Extensive," *J. Math. Phys.* **5** (1964), 1215–1222. The proof we outlined for lattice gases, which we discuss in Chapter XX, is due to G. Gallavotti and S. Miracle: "Statistical Mechanics of Lattice Systems," *Commun. Math. Phys.* **5** (1967), 317–324.

In analytic function theory, equicontinuity is really behind one of the proofs of the Riemann mapping theorem, see e.g. L. Ahlfors, *Complex Analysis*, pp. 172–174, McGraw-Hill, New York, 1953, where sets of equicontinuous functions are called "normal families."

PROBLEMS

1. Find a counterexample to the statement: Every symmetric, transitive relation is reflexive. What is wrong with the proof "xRy and yRx implies xRx (by transitivity)"?

†*2.* Verify that the proposed metrics of Examples 1–3 in Section 1.2 are in fact metrics.

3. Let x_n be a Cauchy sequence in a metric space $\langle X, \rho \rangle$. Suppose that for some subsequence $x_{n(i)}$, $x_{n(i)} \xrightarrow{i \to \infty} x_\infty$. Prove that $x_n \to x_\infty$.

4. Let x_n be a sequence in a metric space and let x_∞ be given. Suppose that every subsequence of x_n has a sub-subsequence converging to x_∞. Prove that $x_n \to x_\infty$.

†*5.* Fill in the details of the proof of Theorem I.3.

†*6.* Prove Theorems I.4 and I.5.

†*7.* Prove Theorem I.6.

8. Prove: If $x_n \to x_\infty$ in a metric space $\langle X, d \rangle$, then for any x, $\lim_{n \to \infty} d(x, x_n) = d(x, x_\infty)$.

†*9.* Complete the proof of Theorem I.7.

†*10.* Prove $S[a, b]$ is dense in $PC[a, b]$ in the $\| \cdot \|_\infty$ norm.

11. (a) Let α be a function on $[0, 1]$. α is said to be of **bounded variation** if there is a C so that
$$\sum_{i=1}^{n-1} |\alpha(x_{i+1}) - \alpha(x_i)| \leq C$$
for any sequence $0 \leq x_1 \leq x_2 \leq \cdots \leq x_n \leq 1$. Prove that any monotone function is of bounded variation.

(b) Define I_α on $S[0, 1]$ by
$$I_\alpha\left(\sum_{i=1}^{n} s_i \chi_i\right) = \sum_{i=1}^{n} s_i[\alpha(x_i) - \alpha(x_{i-1})]$$
Prove I_α is a B.L.T. if and only if α is of bounded variation.

(c) Let α be of bounded variation on $[0, 1]$. Construct a Riemann–Stieltjes integral $\int f \, d\alpha$.

†*12.* Prove the various properties of $\overline{\lim}$ and $\underline{\lim}$ given in the appendix to Section I.2.

13. Construct a set V (the Vitali set) as follows: Call two numbers $x, y \in [0, 1)$ equivalent if $x - y$ is rational. Let V consist of exactly one number from each equivalence class. Let μ be Lebesgue measure. Prove V is not Lebesgue measurable. (Hint: Prove $[0, 1)$ is a disjoint union of "translates" of V.)

†*14.* (a) Let f be a Borel function Prove $f^{-1}[B] \in \mathscr{B}$ for any $B \in \mathscr{B}$.
(b) Let f and g be Borel functions. Prove that $f \circ g$ is Borel.

15. (a) Let $\lim_{n \to \infty} r_n = r$ for r_n, r real numbers. Prove that
$$r = \sup_{m}(\inf_{n > m} r_n)$$
(b) Prove that if f_n is a sequence of functions and $f(x) = \inf f_n(x)$, then
$$f^{-1}[[a, \infty)] = \bigcap_{n=1}^{\infty} f_n^{-1}[[a, \infty)] \qquad f^{-1}[(a, \infty)] = \bigcup_{m=1}^{\infty} f^{-1}[[a + 1/m, \infty)]$$
Conclude that the infimum of any sequence of Borel functions is Borel.
(c) Using (a) and (b), prove that any pointwise limit of a sequence of Borel functions is a Borel function.
(d) Using (a), prove the dominated convergence theorem from the monotone convergence theorem.

*16. Prove that the bounded Borel functions on $[0, 1]$ are the smallest family $\mathscr{F}$ which includes $C[0, 1]$ and has the property: If f_n is a *sequence* of uniformly bounded functions in $\mathscr{F}$ and $f_n \to f$ pointwise, then $f \in \mathscr{F}$.

†*17.* Prove the corollary to Theorem I.12.

†*18.* (a) Prove that for any open set A in $[0, 1]$, χ_A is an L^1 limit of continuous functions.
(b) Let B be a Borel set in $[0, 1]$. Prove χ_B is an L^1 limit of functions χ_A with A open (use the regularity of Lebesgue measure).
(c) Prove $C[a, b]$ is L^1 dense in $L^1[a, b]$.

*19. (a) Let $f_n \to f$ pointwise with f_n continuous. Prove $f^{-1}[(a, \infty)]$ is a F_σ set (that is, a union of closed sets). (Hint: Use Problem 15a.)

(b) Prove that every Borel set on the real line is equal a.e. (w.r.t. Lebesgue measure) to an F_σ set and also equal a.e. to a G_δ set.

*20. A function f is called a jump function if it is monotone and continuous except at a countable number of points (jumps) and if f is piecewise constant on the complement of these jumps. f is called singular if f' exists a.e. and is zero a.e. f is called absolutely continuous if, given ε, we can find δ so that for any n

$$\sum_{i=0}^{n} |x_{2i+1} - x_{2i}| < \delta$$

for points $x_0 < x_1 < \ldots < x_{2n+1}$ implies

$$\sum_{i=0}^{n} |f(x_{2i+1}) - f(x_{2i})| < \varepsilon$$

Prove that any monotone function α on $[0, 1]$ can be written (uniquely) $\alpha = \alpha_{pp} + \alpha_{sing} + \alpha_{ac}$ where α_{pp} is a jump function, α_{sing} is singular and continuous, and α_{ac} is absolutely continuous.

21. Let α be a monotone function. Suppose $\alpha'(x)$ exists a.e. Prove $\alpha(b) - \alpha(a) \geq \int_a^b \alpha'(x)\, dx$. Does equality always hold?

†22. Prove that σ-rings are closed under countable intersections.

23. (a) Let $\mathscr{S}$ be a family of subsets of M. Prove there is a smallest σ-field $\mathscr{F}$ with $\mathscr{S} \subset \mathscr{F}$. We say $\mathscr{S}$ generates $\mathscr{F}$.

(b) Let $T: M \to N$ where M, N have associated σ-fields $\mathscr{R}, \mathscr{F}$. Let $\mathscr{S}$ generate $\mathscr{F}$. Prove T is measurable if and only if $T^{-1}[S] \subset \mathscr{R}$ for all $S \in \mathscr{S}$.

*24. Let μ, ν be two finite measures. Prove ν is a.c.w.r.t. μ if and only if $(\forall \varepsilon)(\exists \delta)\, \mu(A) < \delta$ implies $\nu(A) < \varepsilon$.

25. Consider the function $f(x\ y)$ on $\mathbb{R}^2$ given by

$$f(x, y) = \begin{cases} 1, & x > 0,\ y > 0,\ 0 \leq x - y \leq 1 \\ -1, & x > 0,\ y > 0,\ 0 < y - x \leq 1 \\ 0, & \text{otherwise} \end{cases}$$

Compute $\int_{-\infty}^{\infty} (\int_{-\infty}^{\infty} f(x, y)\, dy)\, dx$ and $\int_{-\infty}^{\infty} (\int_{-\infty}^{\infty} f(x, y)\, dx)\, dy$ and comment on Fubini's theorem and Theorem I.22.

26. Construct a sequence of functions, f_n, which are continuous and pointwise convergent to a function f which is *not* continuous. Prove directly: (a) the convergence $f_n(x) \to f(x)$ is not uniform in x, (b) the f_n are not equicontinuous.

27. Use an $\varepsilon/3$ argument to prove the following: Let B be a complete normed linear space and suppose T_n is a sequence of linear maps $T_n : B \to B$ with two properties: (i) The T_n are bounded uniformly in n, that is, $\|T_n\| \leq C$ for some C independent of n. (ii) For a dense set $D \subset B$, $T_n x$ converges if $x \in D$. Then $T_n x$ converges for each x and the limiting function, Tx, so defined is a bounded linear map.

28. Construct a sequence $f_n(x)$ of bounded functions on [0, 1] converging to zero in L^1 so that f_n converges at no point in [0, 1].

†29. Use an $\varepsilon/3$ argument to prove Theorem I.26.

30. A metric space X is called *totally bounded* if for every ε, X can be covered by finitely many ε-balls. Prove that a pointwise convergent uniformly equicontinuous sequence of functions on a totally bounded metric space is uniformly convergent.

31. (a) Using the Heine–Borel property, prove that a continuous function on [0, 1] is uniformly continuous.
    (b) Prove that an equicontinuous family of functions on [0, 1] is uniformly equicontinuous.

32. Let $F(x, y)$ be a continuous function on $[0, 1] \times [0, 1]$ and consider the map $\mathscr{F}: C[0, 1] \to C[0, 1]$ given by

$$(\mathscr{F}f)(x) = \int_0^1 F(x, y)f(y)\, dy$$

Prove that $\{\mathscr{F}f \mid \|f\|_\infty \leq 1\}$ is an equicontinuous family so that any given sequence f_n with $\|f_n\| \leq 1$, all n, has a subsequence $f_{n(i)}$ with $\mathscr{F}f_{n(i)}$ uniformly convergent.

Remark. It is this last subsequence property that makes $\mathscr{F}$ what is known as a compact operator (a class we discuss in Section VI.5). The classical Fredholm theory of integral equations works because $\mathscr{F}$ is a compact operator.

*33. (a) Let D be a domain in the complex plane. Let $\mathscr{F}$ be a family of analytic function on D so that, for any compact set $C \subset D$, $\{|f(z)| \mid f \in \mathscr{F}, z \in C\}$ is bounded. Prove that $\mathscr{F}$ is an equicontinuous family by using the Cauchy integral formula.
   (b) Prove the Vitali convergence theorem: If D is a connected domain of the complex plane and if f_n is a sequence of analytic functions on D uniformly bounded on compact subsets of D and if $f_n(z)$ converges pointwise for all z lying in some subset of D with a limit point in D, then f_n converges uniformly on compact subsets to an analytic function. (Hint: Use Problem 4.)
   (c) Prove Vitali's theorem by a Taylor series and analytic continuation argument.

Remark. For a discussion of Vitali's theorem from the point of view of (c), see E. C. Titchmarsh, *Theory of Functions*, Oxford Univ. Press, London and New York, 1939, pp. 168–170.

II: Hilbert Spaces

Gentlemen: there's lots of room left in Hilbert space   S. MacLane

II.1 The geometry of Hilbert space

Finite-dimensional vector spaces have three kinds of properties whose generalizations we will study in the next four chapters: linear properties, metric properties, and geometric properties. In this chapter we study vector spaces that have an inner product, a generalization of the usual dot product on finite dimensional vector spaces. The geometric properties of these spaces follow from the notion of angle which is implicit in the definition of inner product.

Definition   A complex vector space V is called an **inner product space** if there is a complex-valued function $(\cdot, \cdot)$ on $V \times V$ that satisfies the following four conditions for all $x, y, z \in V$ and $\alpha \in \mathbb{C}$:

(i) $(x, x) \geq 0$ and $(x, x) = 0$ if and only if $x = 0$
(ii) $(x, y + z) = (x, y) + (x, z)$
(iii) $(x, \alpha y) = \alpha(x, y)$
(iv) $(x, y) = \overline{(y, x)}$

The function $(\cdot, \cdot)$ is called an inner product.

II.1 The geometry of Hilbert space

We note that (ii), (iii), and (iv) imply that $(x, \alpha y + \beta z) = \alpha(x, y) + \beta(x, z)$ and that $(\alpha x, y) = \bar{\alpha}(x, y)$. The reader should be aware that some texts use a convention different from the one introduced in (iii); they take the inner product to be linear in the *first* vector and conjugate-linear in the *second*.

Example 1 ($\mathbb{C}^n$) Let $\mathbb{C}^n$ denote the set of all n-tuples of complex numbers. For $x = \langle x_1, \ldots, x_n \rangle$ and $y = \langle y_1, \ldots, y_n \rangle$ in $\mathbb{C}^n$ define

$$(x, y) = \sum_{j=1}^{n} \bar{x}_j y_j$$

Example 2 Let $C[a, b]$ denote the complex-valued continuous functions on the interval $[a, b]$. For $f(x), g(x) \in C[a, b]$ define

$$(f, g) = \int_a^b \overline{f(x)} g(x) \, dx$$

We now develop those geometrical notions that extend to arbitrary inner product spaces.

Definition Two vectors, x and y, in an inner product space V are said to be **orthogonal** if $(x, y) = 0$. A collection $\{x_i\}$ of vectors in V is called an **orthonormal set** if $(x_i, x_i) = 1$ for all i, and $(x_i, x_j) = 0$ if $i \neq j$.

We introduce the shorthand $\|x\| = \sqrt{(x, x)}$. We will shortly see that $\|\cdot\|$ is in fact a norm.

Theorem II.1 (Pythagorean theorem) Let $\{x_n\}_{n=1}^{N}$ be an orthonormal set in an inner product space V. Then for all $x \in V$,

$$\|x\|^2 = \sum_{n=1}^{N} |(x, x_n)|^2 + \left\| x - \sum_{n=1}^{N} (x_n, x) x_n \right\|^2$$

Proof We write x as

$$x = \sum_{n=1}^{N} (x_n, x) x_n + \left(x - \sum_{n=1}^{N} (x_n, x) x_n \right)$$

A short computation using the properties of inner products shows that

$$\sum_{n=1}^{N} (x_n, x) x_n \quad \text{and} \quad x - \sum_{n=1}^{N} (x_n, x) x_n$$

are orthogonal. Thus,

$$(x, x) = \left\| \sum_{n=1}^{N} (x_n, x)x_n \right\|^2 + \left\| x - \sum_{n=1}^{N} (x_n, x)x_n \right\|^2$$

$$= \sum_{n=1}^{N} |(x_n, x)|^2 + \left\| x - \sum_{n=1}^{N} (x_n, x)x_n \right\|^2 \blacksquare$$

Corollary (Bessel's inequality) Let $\{x_n\}_{n=1}^{N}$ be an orthonormal set in an inner product space, V. Then for all $x \in V$,

$$\|x\|^2 \geq \sum_{n=1}^{N} |(x, x_n)|^2$$

Corollary (the Schwarz inequality) If x and y are vectors in an inner product space V, then

$$|(x, y)| \leq \|x\| \|y\|$$

Proof The case $y = 0$ is trivial, so suppose $y \neq 0$. The vector $y/\|y\|$ by itself forms an orthonormal set, so applying Bessel's inequality to any $x \in V$ we get

$$\|x\|^2 \geq |(x, y/\|y\|)|^2 = \frac{|(x, y)|^2}{\|y\|^2}$$

from which $|(x, y)| \leq \|x\| \|y\|$ follows. ∎

Another useful geometric equality is the parallelogram law (Problem 4):

$$\|x + y\|^2 + \|x - y\|^2 = 2\|x\|^2 + 2\|y\|^2$$

In Section I.2 we defined normed linear spaces and observed that every normed linear space is a metric space. The following theorem shows that every inner product space is a normed linear space.

Theorem II.2 Every inner product space V is a normed linear space with the norm $\|x\| = (x, x)^{1/2}$

Proof Since V is a vector space, we need only verify that $\|\cdot\|$ has all the properties of a norm. All of these properties, except the triangle inequality, follow immediately from the properties (i)–(iv) of inner products. Suppose $x, y \in V$. Then

$$\|x + y\|^2 = (x, x) + (x, y) + (y, x) + (y, y)$$
$$= (x, x) + 2 \operatorname{Re}(x, y) + (y, y)$$
$$\leq (x, x) + 2|(x, y)| + (y, y)$$
$$\leq (x, x) + 2(x, x)^{1/2}(y, y)^{1/2} + (y, y)$$

by the Schwarz inequality. Thus
$$\|x + y\|^2 \leq (\|x\| + \|y\|)^2$$
which proves the triangle inequality. ∎

This theorem shows that we have a natural metric,
$$d(x, y) = \sqrt{(x - y, x - y)}$$
in V. We thus have the notions of convergence, completeness, and density defined for metric spaces in Section I.2. In particular, we can always complete V to a normed linear space $\tilde{V}$ in which V is isometrically embedded as a dense subset. In fact, $\tilde{V}$ is also an inner product space since the inner product can be extended from V to $\tilde{V}$ by continuity (Problem 1).

Definition A complete inner product space is called a **Hilbert space**. Inner product spaces are sometimes called pre-Hilbert spaces.

Definition Two Hilbert spaces $\mathcal{H}_1$ and $\mathcal{H}_2$ are said to be **isomorphic** if there is a linear operator U from $\mathcal{H}_1$ onto $\mathcal{H}_2$ such that $(Ux, Uy)_{\mathcal{H}_2} = (x, y)_{\mathcal{H}_1}$ for all $x, y \in \mathcal{H}_1$. Such an operator is called **unitary**.

We elaborate these ideas and show the reader what types of Hilbert spaces he is likely to meet by a series of examples.

Example 2 (revisited) Define $L^2[a, b]$ to be the set of complex-valued measurable functions on $[a, b]$, a finite interval, that satisfy $\int_a^b |f(x)|^2 \, dx < \infty$. We define an inner product by
$$(f, g) = \int_a^b \overline{f(x)} g(x) \, dx$$
Observe that the inner makes sense since
$$|\overline{f(x)}g(x)| \leq \tfrac{1}{2}|f(x)|^2 + \tfrac{1}{2}|g(x)|^2$$
so that $\overline{f(x)}g(x)$ is in $L^1[a, b]$. A proof similar to the Riesz–Fisher theorem (Theorem I.12) shows that $L^2[a, b]$ is complete and is therefore a Hilbert space. It is not too difficult to show (Problem 2) that $L^2[a, b]$ is the completion of $C[a, b]$ in the norm
$$\|f\| = \left(\int_a^b |f(x)|^2 \, dx \right)^{1/2}$$

Example 3 (ℓ_2) Define ℓ_2 to be the set of sequences $\{x_n\}_{n=1}^{\infty}$ of complex numbers which satisfy $\sum_{n=1}^{\infty} |x_n|^2 < \infty$ with the inner product

$$(\{x_n\}_{n=1}^{\infty}, \{y_n\}_{n=1}^{\infty}) = \sum_{n=1}^{\infty} \bar{x}_n y_n$$

In Section II.3 we will see that any Hilbert space that has a countable dense set and is not finite dimensional is isomorphic to ℓ_2. In this sense, ℓ_2 is the canonical example of a Hilbert space.

Example 4 ($L^2(\mathbb{R}^n, d\mu)$) Let μ be a Borel measure on $\mathbb{R}^n$. $L^2(\mathbb{R}^n, d\mu)$ is the set of complex-valued measurable functions on $\mathbb{R}^n$ which satisfy $\int_{\mathbb{R}^n} |f(x)|^2 \, d\mu < \infty$. $L^2(\mathbb{R}^n, d\mu)$ is a Hilbert space under the inner product

$$(f, g) = \int_{\mathbb{R}^n} \overline{f(x)} g(x) \, d\mu$$

Example 5 (direct sum) Suppose that $\mathcal{H}_1$ and $\mathcal{H}_2$ are Hilbert spaces. Then the set of pairs $\langle x, y \rangle$ with $x \in \mathcal{H}_1$, $y \in \mathcal{H}_2$ is a Hilbert space with inner product

$$(\langle x_1, y_1 \rangle, \langle x_2, y_2 \rangle) = (x_1, x_2)_{\mathcal{H}_1} + (y_1, y_2)_{\mathcal{H}_2}$$

This space is called the **direct sum** of the spaces $\mathcal{H}_1$ and $\mathcal{H}_2$ and is denoted by $\mathcal{H}_1 \oplus \mathcal{H}_2$. If μ_1 and μ_2 are mutually singular Borel measures on $\mathbb{R}$ and $\mu = \mu_1 + \mu_2$, then $L^2(\mathbb{R}, d\mu)$ is isomorphic in a natural way to $L^2(\mathbb{R}, d\mu_1) \oplus L^2(\mathbb{R}, d\mu_2)$ (Problem 3). We can also construct countable direct sums as follows. Suppose $\{\mathcal{H}_n\}_{n=1}^{\infty}$ is a sequence of Hilbert spaces. Let $\mathcal{H}$ denote the set of sequences $\{x_n\}_{n=1}^{\infty}$, with $x_n \in \mathcal{H}_n$, which satisfy

$$\sum_{n=1}^{\infty} \|x_n\|_{\mathcal{H}_n}^2 < \infty$$

$\mathcal{H}$ is a Hilbert space under the natural inner product and is denoted by

$$\mathcal{H} = \bigoplus_{n=1}^{\infty} \mathcal{H}_n$$

Example 6 (vector-valued functions) Suppose $\langle X, \mu \rangle$ is a measure space and $\mathcal{H}'$ is a Hilbert space. Let $L^2(X, d\mu; \mathcal{H}')$ be the set of measurable functions on X with values in $\mathcal{H}'$ which satisfy

$$\int_X \|f(x)\|_{\mathcal{H}'}^2 \, d\mu(x) < \infty$$

This set is a Hilbert space with the inner product

$$(f, g) = \int_X (f(x), g(x))_{\mathscr{H}'} \, d\mu(x)$$

Of course, we have not said what it means for a vector-valued function to be measurable. For this definition and related matters see Problem 12 and the appendix to Section IV.5.

II.2 The Riesz lemma †

In the examples in Section II.1 we showed several ways of constructing new Hilbert spaces from old ones. Another way to do this is to restrict attention to a closed subspace $\mathscr{M}$ of the given Hilbert space $\mathscr{H}$. Under the natural inner product that it inherits as a subspace of $\mathscr{H}$, $\mathscr{M}$ is a Hilbert space. We denote by $\mathscr{M}^\perp$ the set of vectors in $\mathscr{H}$ which are orthogonal to $\mathscr{M}$; $\mathscr{M}^\perp$ is called the **orthogonal complement** of $\mathscr{M}$. It follows from the linearity of the inner product that $\mathscr{M}^\perp$ is a linear subspace of $\mathscr{H}$ and an elementary argument (Problem 6) shows that $\mathscr{M}^\perp$ is closed. Thus $\mathscr{M}^\perp$ is also a Hilbert space. $\mathscr{M}$ and $\mathscr{M}^\perp$ have only the zero element in common. The following theorem shows that there are vectors perpendicular to any closed proper subspace, indeed there are enough of them so that

$$\mathscr{H} = \mathscr{M} + \mathscr{M}^\perp = \{x + y \,|\, x \in \mathscr{M}, y \in \mathscr{M}^\perp\}$$

This important geometric property is one of the main reasons that Hilbert spaces are easier to handle than Banach spaces (Chapter III). In the following lemma and theorem, the reader should keep the finite-dimensional case in mind (see Figure II.1).

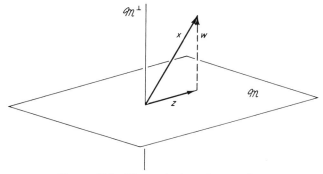

FIGURE II.1 The projection of x on $\mathscr{M}$.

† A supplement to this section begins on p. 344.

Lemma Let $\mathcal{H}$ be a Hilbert space, $\mathcal{M}$ a closed subspace of $\mathcal{H}$, and suppose $x \in \mathcal{H}$. Then there exists in $\mathcal{M}$ a unique element z closest to x.

Proof Let $d = \inf_{y \in \mathcal{M}} \|x - y\|$. Choose a sequence $\{y_n\}$, $y_n \in \mathcal{M}$, so that

$$\|x - y_n\| \to d$$

Then

$$\begin{aligned}
\|y_n - y_m\|^2 &= \|(y_n - x) - (y_m - x)\|^2 \\
&= 2\|y_n - x\|^2 + 2\|y_m - x\|^2 - \|-2x + y_n + y_m\|^2 \\
&= 2\|y_n - x\|^2 + 2\|y_m - x\|^2 - 4\|x - \tfrac{1}{2}(y_n + y_m)\|^2 \\
&\leq 2\|y_n - x\|^2 + 2\|y_m - x\|^2 - 4d^2 \\
&\underset{\substack{m \to \infty \\ n \to \infty}}{\longrightarrow} 2d^2 + 2d^2 - 4d^2 = 0
\end{aligned}$$

The second equality follows from the parallelogram law; the inequality follows from the fact that $\tfrac{1}{2}(y_n + y_m) \in \mathcal{M}$. Thus $\{y_n\}$ is Cauchy and since $\mathcal{M}$ is closed, $\{y_n\}$ converges to an element z of $\mathcal{M}$. It follows easily that $\|x - z\| = d$. Uniqueness is left as an exercise. ∎

Theorem II.3 (the projection theorem) Let $\mathcal{H}$ be a Hilbert space, $\mathcal{M}$ a closed subspace. Then every $x \in \mathcal{H}$ can be uniquely written $x = z + w$ where $z \in \mathcal{M}$ and $w \in \mathcal{M}^\perp$.

Proof Let x be in $\mathcal{H}$. Then by the lemma there is a unique element $z \in \mathcal{M}$ closest to x. Define $w = x - z$, then we clearly have $x = z + w$. Let $y \in \mathcal{M}$ and $t \in \mathbb{R}$. If $d = \|x - z\|$, then

$$\begin{aligned}
d^2 &\leq \|x - (z + ty)\|^2 = \|w - ty\|^2 \\
&= d^2 - 2t \operatorname{Re}(w, y) + t^2 \|y\|^2
\end{aligned}$$

Thus, $-2t \operatorname{Re}(w, y) + t^2 \|y\|^2 \geq 0$ for all t, which implies $\operatorname{Re}(w, y) = 0$. A similar argument using ti instead of t shows that $\operatorname{Im}(w, y) = 0$. Thus, $w \in \mathcal{M}^\perp$. Uniqueness is left as an exercise. ∎

The projection theorem sets up a natural isomorphism between $\mathcal{M} \oplus \mathcal{M}^\perp$ and $\mathcal{H}$ given by

$$\langle z, w \rangle \mapsto z + w$$

We will often suppress the isomorphism and simply write $\mathcal{H} = \mathcal{M} \oplus \mathcal{M}^\perp$.

We have already defined in Section I.2 what we mean by a bounded linear transformation from one Hilbert space $\mathcal{H}$ to another $\mathcal{H}'$. We will denote by

$\mathscr{L}(\mathscr{H}, \mathscr{H}')$ the set of such transformations. $\mathscr{L}(\mathscr{H}, \mathscr{H}')$ is clearly a vector space and it becomes a Banach space under the norm

$$\|T\| = \sup_{\|x\|_{\mathscr{H}} = 1} \|Tx\|_{\mathscr{H}'}$$

The proof of this fact, though not difficult, is postponed until Chapter III where it is proven in greater generality. For the time being we are interested in the special case where $\mathscr{H}' = \mathbb{C}$:

Definition The space $\mathscr{L}(\mathscr{H}, \mathbb{C})$ is called the **dual space** of $\mathscr{H}$ and is denoted by $\mathscr{H}^*$. The elements of $\mathscr{H}^*$ are called **continuous linear functionals**.

The following important theorem which characterizes $\mathscr{H}^*$ is due to F. Riesz and M. Fréchet.

Theorem II.4 (the Riesz lemma) For each $T \in \mathscr{H}^*$, there is a unique $y_T \in \mathscr{H}$ such that $T(x) = (y_T, x)$ for all $x \in \mathscr{H}$. In addition $\|y_T\|_{\mathscr{H}} = \|T\|_{\mathscr{H}^*}$.

Proof Let $\mathscr{N}$ be the set of $x \in \mathscr{H}$ such that $T(x) = 0$. By the continuity of T, $\mathscr{N}$ is a closed subspace. If $\mathscr{N} = \mathscr{H}$, then $T(x) = 0 = (0, x)$ for all x and we are finished; so assume $\mathscr{N}$ is not all of $\mathscr{H}$. Then by the projection theorem there is a nonzero vector x_0 in $\mathscr{N}^\perp$. Define $y_T = \overline{T(x_0)}\|x_0\|^{-2} x_0$. We will verify that the vector y_T has the right properties. First, if $x \in \mathscr{N}$, then $T(x) = 0 = (y_T, x)$. Further, if $x = \alpha x_0$, then

$$T(x) = T(\alpha x_0) = \alpha T(x_0) = (\overline{T(x_0)}\|x_0\|^{-2} x_0, \alpha x_0) = (y_T, \alpha x_0)$$

Since the functions $T(\cdot)$ and $(y_T, \cdot)$ are linear and agree on $\mathscr{N}$ and x_0, they must agree on the space spanned by $\mathscr{N}$ and x_0. But $\mathscr{N}$ and x_0 span $\mathscr{H}$ since every element $y \in \mathscr{H}$ can be written

$$y = \left(y - \frac{T(y)}{T(x_0)} x_0\right) + \frac{T(y)}{T(x_0)} x_0$$

Thus $T(x) = (y_T, x)$ for all $x \in \mathscr{H}$. If $T(x) = (y', x)$ also, then $\|y' - y_T\|^2 = T(y' - y_T) - T(y' - y_T) = 0$ so $y' = y_T$, proving uniqueness.

To prove that $\|T\|_{\mathscr{H}^*} = \|y_T\|_{\mathscr{H}}$ we observe that

$$\|T\| = \sup_{\|x\| \leq 1} |T(x)| = \sup_{\|x\| \leq 1} |(y_T, x)| \leq \sup_{\|x\| \leq 1} \|y_T\| \|x\| = \|y_T\|$$

and

$$\|T\| = \sup_{\|x\| \leq 1} |T(x)| \geq \left|T\left(\frac{y_T}{\|y_T\|}\right)\right| = \left(y_T, \frac{y_T}{\|y_T\|}\right) = \|y_T\| \quad \blacksquare$$

We note that the Schwarz inequality shows that the converse of the Riesz lemma is true. Namely, each $y \in \mathcal{H}$ defines a continuous linear functional T_y on $\mathcal{H}$ by $T_y(x) = (y, x)$. The Riesz lemma has the following corollary which is very important in applications.

Corollary Let $B(\cdot, \cdot)$ be a function from $\mathcal{H} \times \mathcal{H}$ to $\mathbb{C}$ which satisfies:
(i) $B(x, \alpha y + \beta z) = \alpha B(x, y) + \beta B(x, z)$
(ii) $B(\alpha x + \beta y, z) = \bar{\alpha} B(x, z) + \bar{\beta} B(y, z)$
(iii) $|B(x, y)| \leq C \|x\| \|y\|$

for all $x, y, z \in \mathcal{H}$, $\alpha, \beta \in \mathbb{C}$. Then there is a unique bounded linear transformation A, from $\mathcal{H}$ to $\mathcal{H}$ so that

$$B(x, y) = (Ax, y) \quad \text{for all} \quad x, y \in \mathcal{H}$$

The norm of A is the smallest constant C such that (iii) holds.

Proof Fix x, then (i) and (iii) show that $B(x, \cdot)$ is a continuous linear functional on $\mathcal{H}$. Thus by the Riesz lemma there is an $x' \in \mathcal{H}$ so that

$$B(x, y) = (x', y) \quad \text{for all} \quad y \in \mathcal{H}$$

Define $Ax = x'$. It is not difficult to show that A is a continuous linear operator with the right properties (Problem 8). ∎

A bilinear function on $\mathcal{H}$ obeying (i) and (ii) is called a **sesquilinear form**.

II.3 Orthonormal bases

We have already defined what it means for a set of vectors to be orthonormal. In this section we develop this idea further; in particular we want to extend the idea of a "basis," so useful for finite-dimensional vector spaces, to complete inner product spaces. If S is an orthonormal set in a Hilbert space $\mathcal{H}$ and no other orthonormal set contains S as a proper subset, then S is called an **orthonormal basis** (or a **complete orthonormal system**) for $\mathcal{H}$.

Theorem II.5 Every Hilbert space $\mathcal{H}$ has an orthonormal basis.

Proof Consider the collection $\mathscr{C}$ of orthonormal sets in V. We order $\mathscr{C}$ by inclusion; that is, we say $S_1 \prec S_2$ if $S_1 \subset S_2$. With this definition of $\prec$, $\mathscr{C}$ is partially ordered; it is also nonempty since if v is any element of V, the set consisting only of $v/\|v\|$ is an orthonormal set. Now let $\{S_\alpha\}_{\alpha \in A}$ be any linearly ordered subset of $\mathscr{C}$. Then $\bigcup_{\alpha \in A} S_\alpha$ is an orthonormal set which contains each S_α and is thus an upper bound for $\{S_\alpha\}_{\alpha \in A}$. Since every linearly ordered subset of $\mathscr{C}$ has an upper bound, we can apply Zorn's lemma (Theorem I.2) and conclude that $\mathscr{C}$ has a maximal element; that is, an orthonormal system not properly contained in any other orthonormal system. ∎

The following theorem shows that as in the finite-dimensional case every element of a Hilbert space can be expressed as a linear combination (possibly infinite) of basis elements.

Theorem II.6 Let $\mathscr{H}$ be a Hilbert space and $S = \{x_\alpha\}_{\alpha \in A}$ an orthonormal basis. Then for each $y \in \mathscr{H}$,

$$y = \sum_{\alpha \in A} (x_\alpha, y) x_\alpha \tag{II.1}$$

and

$$\|y\|^2 = \sum_{\alpha \in A} |(x_\alpha, y)|^2 \tag{II.2}$$

The equality in (II.1) means that the sum on the right-hand side converges (independent of order) to y in $\mathscr{H}$. Conversely, if $\sum_{\alpha \in A} |c_\alpha|^2 < \infty$, $c_\alpha \in \mathbb{C}$, then $\sum_{\alpha \in A} c_\alpha x_\alpha$ converges to an element of $\mathscr{H}$.

Proof We have already shown in Section II.1 (Bessel's inequality) that for any finite subset $A' \subset A$, $\sum_{\alpha \in A'} |(x_\alpha, y)|^2 \leq \|y\|^2$. Thus $(x_\alpha, y) \neq 0$ for at most a countable number of α's in A which we order in some way $\alpha_1, \alpha_2, \alpha_3, \ldots$. Furthermore, since $\sum_{j=1}^{N} |(x_{\alpha_j}, y)|^2$ is monotone increasing and bounded, it converges to a finite limit as $N \to \infty$. Let $y_n = \sum_{j=1}^{n} (x_{\alpha_j}, y) x_{\alpha_j}$. Then for $n > m$,

$$\|y_n - y_m\|^2 = \left\| \sum_{j=m+1}^{n} (x_{\alpha_j}, y) x_{\alpha_j} \right\|^2 = \sum_{j=m+1}^{n} |(x_{\alpha_j}, y)|^2$$

Therefore $\{y_n\}$ is a Cauchy sequence and converges to an element y' of $\mathscr{H}$. Observe that

$$(y - y', x_{\alpha_\ell}) = \lim_{n \to \infty} \left(y - \sum_{j=1}^{n} (x_{\alpha_j}, y) x_{\alpha_j}, x_{\alpha_\ell} \right)$$
$$= (y, x_{\alpha_\ell}) - (y, x_{\alpha_\ell}) = 0$$

And if $\alpha \neq \alpha_\ell$ for some ℓ we have

$$(y - y', x_\alpha) = \lim_{n \to \infty} \left(y - \sum_{j=1}^{n} (x_{\alpha_j}, y) x_{\alpha_j}, x_\alpha \right) = 0$$

Therefore $y - y'$ is orthogonal to all the x_α in S. Since S is a complete orthonormal system we must have $y - y' = 0$. Thus

$$y = \lim_{n \to \infty} \sum_{j=1}^{n} (x_{\alpha_j}, y) x_{\alpha_j}$$

and (II.1) holds. Furthermore,

$$\begin{aligned} 0 &= \lim_{n \to \infty} \left\| y - \sum_{j=1}^{n} (x_{\alpha_j}, y) x_{\alpha_j} \right\|^2 \\ &= \lim_{n \to \infty} \left(\|y\|^2 - \sum_{j=1}^{n} |(x_{\alpha_j}, y)|^2 \right) \\ &= \|y\|^2 - \sum_{\alpha \in A} |(x_\alpha, y)|^2 \end{aligned}$$

so that (II.2) holds also. We omit the easy proof of the converse statement. ∎

We note that (II.2) is called Parseval's relation. The coefficients (x_α, y) are often called the **Fourier coefficients** of y with respect to the basis $\{x_\alpha\}$. The reason for this terminology will become apparent shortly.

We now describe a useful procedure, called Gram–Schmidt orthogonalization, for constructing an orthonormal set from an arbitrary sequence of independent vectors. Suppose the independent vectors $u_1, u_2, \ldots$ are given and define

$$\begin{aligned} w_1 &= u_1, & v_1 &= w_1/\|w_1\| \\ w_2 &= u_2 - (v_1, u_1)v_k, & v_2 &= w_2/\|w_2\| \\ &\vdots & &\vdots \\ w_n &= u_n - \sum_{k=1}^{n-1} (v_k, u_n)v_k, & v_n &= w_n/\|w_n\| \\ &\vdots & &\vdots \end{aligned}$$

The family $\{v_j\}$ is an orthonormal set and has the property that for each m, $\{u_j\}_{j=1}^{m}$ and $\{v_j\}_{j=1}^{m}$ span the same vector space. In particular, the set of finite linear combinations of all the v's is the same as the finite linear combinations of the u's (see Figure II.2).

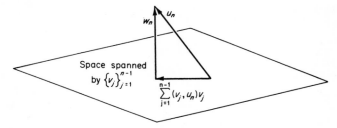

FIGURE II.2 Gram–Schmidt orthogonalization.

We remark that the Legendre polynomials (up to constant multiples) are obtained by applying the Gram–Schmidt process to the functions $1, x, x^2, x^3, \ldots$, on the interval $[-1, 1]$ with the usual L^2 inner product.

Definition A metric space which has a countable dense subset is said to be **separable**.

Most Hilbert spaces that arise in practice are separable. The following theorem characterizes them up to isomorphism.

Theorem II.7 A Hilbert space $\mathcal{H}$ is separable if and only if it has a countable orthonormal basis S. If there are $N < \infty$ elements in S, then $\mathcal{H}$ is isomorphic to $\mathbb{C}^N$. If there are countably many elements in S, then $\mathcal{H}$ is isomorphic to ℓ_2 (Example 3, Section II.1).

Proof Suppose $\mathcal{H}$ is separable and let $\{x_n\}$ be a countable dense set. By throwing out some of the x_n's we can get a subcollection of independent vectors whose span (finite linear combinations) is the same as the $\{x_n\}$ and is thus dense. Applying the Gram–Schmidt procedure to this subcollection we obtain a countable complete orthonormal system. Conversely, if $\{y_n\}$ is a complete orthonormal system for a Hilbert space $\mathcal{H}$ then it follows from Theorem II.6 that the set of finite linear combinations of the y_n with rational coefficients is dense in $\mathcal{H}$. Since this set is countable, $\mathcal{H}$ is separable.

Suppose $\mathcal{H}$ is separable and $\{y_n\}_{n=1}^\infty$ is a complete orthonormal system. We define a map $\mathcal{U} : \mathcal{H} \to \ell_2$ by

$$\mathcal{U} : x \to \{(y_n, x)\}_{n=1}^\infty$$

Theorem II.6 shows that this map is well defined and onto. It is easy to show it is unitary. The proof that $\mathcal{H}$ is isomorphic to $\mathbb{C}^N$ if S has N elements is similar. ∎

Notice that in the separable case, the Gram–Schmidt process allows us to construct an orthonormal basis without using Zorn's lemma.

We conclude this section with an example that shows how Hilbert spaces arose naturally from problems in classical analysis. If $f(x)$ is an integrable function on $[0, 2\pi]$ we can define the numbers

$$c_n = \frac{1}{(2\pi)^{1/2}} \int_0^{2\pi} e^{-inx} f(x) \, dx$$

The formal series $\sum_{n=-\infty}^{\infty} c_n (2\pi)^{-1/2} e^{inx}$ is called the Fourier series of f. The classical problem is: for which f and in what sense does the Fourier series of

f converge to f? This problem which originated with Fourier in 1811 has had a rich and eventful history. It has given rise to an entire branch of modern mathematics (abstract harmonic analysis). Furthermore, some of the nicest results on the classical case have just been proven recently (see the Notes). As an example of a classical result we state (Problems 14 and 15):

Theorem II.8 Suppose that $f(x)$ is periodic of period 2π and is continuously differentiable. Then the functions $\sum_{-M}^{M} c_n e^{inx}$ converge uniformly to $f(x)$ as $M \to \infty$.

This theorem gives sufficient conditions for the Fourier series of a function to converge uniformly. But, finding the exact class of functions whose Fourier series converge uniformly or converge pointwise has proved to be a hard problem. We can, however, get a nice answer to this question if we change our notion of "convergence" and this is just where Hilbert spaces come in. The collection of functions, $\{(2\pi)^{-1/2} e^{inx}\}_{-\infty}^{\infty}$, is clearly an orthonormal set in $L^2[0, 2\pi]$. If we knew that it was a *complete* orthonormal set, then Theorem II.6 would allow us to conclude that for all functions in $L^2[0, 2\pi]$,

$$f(x) = \lim_{M \to \infty} \sum_{-M}^{M} (2\pi)^{-1/2} c_n e^{inx}$$

where convergence means convergence in the L^2 norm. In fact, $\{(2\pi)^{-1/2} e^{inx}\}_{-\infty}^{\infty}$ is complete. We will give a proof that relies on the classical theorem stated above.

Theorem II.9 If $f \in L^2[0, 2\pi]$, then $\sum_{-M}^{M} c_n (2\pi)^{-1/2} e^{inx}$ converges to f in the L^2 norm as $M \to \infty$.

Proof We need to know that the periodic, continuously differentiable functions $C_p^1[0, 2\pi]$ are dense in $L^2[0, 2\pi]$. In Problem 2 the reader is asked to show that the step functions are dense. But a step function can be approximated (in L^2) by a $C_p^1[0, 2\pi]$ function by rounding off the corners in a smooth way and by changing it at one end to make it periodic. The reader should convince himself that this can be done so that the resulting function is arbitrarily close to the step function in L^2 norm.

To show that $\{(2\pi)^{-1/2} e^{inx}\}_{-\infty}^{\infty}$ is a complete set we need only show that $(e^{inx}, g) = 0$ for all n implies $g = 0$. Suppose $f \in C_p^1[0, 2\pi]$, then by Theorem II.8

$$\sum_{-M}^{M} c_n (2\pi)^{-1/2} e^{inx} \to f$$

uniformly and thus in the L^2 sense also. Therefore,

$$(f, g) = \lim_{M \to \infty} \left(\sum_{-M}^{M} c_n (2\pi)^{-1/2} e^{inx}, g \right) = 0$$

if $(e^{inx}, g) = 0$ for all n. But, then g is orthogonal to all f in the dense set $C_p^1[0, 2\pi]$ which implies $g = 0$. Thus $\{(2\pi)^{-1/2} e^{inx}\}_{-\infty}^{\infty}$ is a complete orthonormal set, and it follows from Theorem II.6 that the Fourier series of every $L^2[0, 2\pi]$ function converges in the L^2-norm to the function. ∎

This theorem shows that the "natural" notion of convergence for Fourier series is L^2 convergence and illustrates one of the basic principles of functional analysis: namely, to choose an abstract space and a notion of convergence that is appropriate to the problem at hand, a space in which one can prove nice theorems. By doing this one avoids some hard problems; this has both advantages and disadvantages.

II.4 Tensor products of Hilbert spaces

We described in Sections II.1 and II.2 several ways of making new Hilbert spaces from old ones. In this section we describe the tensor product $\mathscr{H}_1 \otimes \mathscr{H}_2$ of two Hilbert spaces $\mathscr{H}_1$ and $\mathscr{H}_2$. The construction of the tensor product which we use is not the most elegant, but is very direct. The reader can easily extend our proofs to construct the tensor product $\mathscr{H}_1 \otimes \mathscr{H}_2 \otimes \cdots \otimes \mathscr{H}_n$ of finitely many Hilbert spaces.

Let $\mathscr{H}_1$ and $\mathscr{H}_2$ be Hilbert spaces. For each $\varphi_1 \in \mathscr{H}_1$, $\varphi_2 \in \mathscr{H}_2$, let $\varphi_1 \otimes \varphi_2$ denote the conjugate bilinear form which acts on $\mathscr{H}_1 \times \mathscr{H}_2$ by

$$(\varphi_1 \otimes \varphi_2)\langle \psi_1, \psi_2 \rangle = (\psi_1, \varphi_1)(\psi_2, \varphi_2)$$

Let $\mathscr{E}$ be the set of finite linear combinations of such conjugate linear forms; we define an inner product $(\cdot, \cdot)$ on $\mathscr{E}$ by defining

$$(\varphi \otimes \psi, \eta \otimes \mu) = (\varphi, \eta)(\psi, \mu)$$

and extending by linearity to $\mathscr{E}$.

Proposition 1 $(\cdot, \cdot)$ is well defined and positive definite.

Proof To show that $(\cdot, \cdot)$ is well defined, we must show that (λ, λ') does not depend on which finite linear combinations are used to express λ and λ'. To do this it is sufficient to show that if μ is a finite sum which is the zero form, then $(\eta, \mu) = 0$ for all $\eta \in \mathscr{E}$. To see that this is true, let $\eta = \sum_{i=1}^{N} c_i(\varphi_i \otimes \psi_i)$, then

$$(\eta, \mu) = \left(\sum_{i=1}^{N} c_i(\varphi_i \otimes \psi_i), \mu \right)$$

$$= \sum_{i=1}^{N} c_i \mu \langle \varphi_i, \psi_i \rangle$$

$$= 0$$

since μ is the zero form. Thus, $(\cdot, \cdot)$ is well defined.

Now, suppose $\lambda = \sum_{k=1}^{M} d_k(\eta_k \otimes \mu_k)$. Then $\{\eta_k\}_{k=1}^{M}$ and $\{\mu_k\}_{k=1}^{M}$ span subspaces $M_1 \subset \mathscr{H}_1$ and $M_2 \subset \mathscr{H}_2$ respectively. If we let $\{\varphi_j\}_{j=1}^{N_1}$ and $\{\psi_\ell\}_{\ell=1}^{N_2}$ be orthonormal bases for M_1 and M_2, we can express each η_k in terms of the φ_j's and each μ_k in terms of the ψ_ℓ's obtaining

$$\lambda = \sum_{\substack{j=1 \\ \ell=1}}^{M_1, M_2} c_{j\ell}(\varphi_j \otimes \psi_\ell)$$

But,

$$(\lambda, \lambda) = \left(\sum c_{j\ell}(\varphi_j \otimes \psi_\ell), \sum c_{im}(\varphi_i \otimes \psi_m) \right)$$

$$= \sum \overline{c_{j\ell}} c_{im} (\varphi_j, \varphi_i)(\psi_\ell, \psi_m)$$

$$= \sum_{j\ell} |c_{j\ell}|^2$$

so if $(\lambda, \lambda) = 0$, then all the $c_{j\ell} = 0$ and λ is the zero form. Thus $(\cdot, \cdot)$ is positive definite. ∎

Definition We define $\mathscr{H}_1 \otimes \mathscr{H}_2$ to be the completion of $\mathscr{E}$ under the inner product $(\cdot, \cdot)$ defined above. $\mathscr{H}_1 \otimes \mathscr{H}_2$ is called the **tensor product** of $\mathscr{H}_1$ and $\mathscr{H}_2$.

Proposition 2 If $\{\varphi_k\}$ and $\{\psi_\ell\}$ are orthonormal bases for $\mathscr{H}_1$ and $\mathscr{H}_2$ respectively, then $\{\varphi_k \otimes \psi_\ell\}$ is an orthonormal basis for $\mathscr{H}_1 \otimes \mathscr{H}_2$.

Proof To simplify notation, we consider the case in which both $\mathscr{H}_1$ and $\mathscr{H}_2$ are infinite dimensional and separable. The other cases are similar. The set $\{\varphi_k \otimes \psi_\ell\}$ is clearly orthonormal and therefore we need only show that $\mathscr{E}$ is contained in the closed space S spanned by $\{\varphi_k \otimes \psi_\ell\}$. Let $\varphi \otimes \psi \in \mathscr{E}$. Since $\{\varphi_k\}$ and $\{\psi_\ell\}$ are bases, $\varphi = \sum c_k \varphi_k$ and $\psi = \sum d_\ell \psi_\ell$ where $\sum |c_k|^2 < \infty$ and $\sum |d_\ell|^2 < \infty$. Thus $\sum_{k,\ell} |c_k d_\ell|^2 < \infty$. Therefore by Theorem II.6, there

II.4 Tensor products of Hilbert spaces

is a vector $\mu = \sum_{k,\ell} c_k d_\ell \varphi_k \otimes \psi_\ell$ in S. By direct computation

$$\left\| \varphi \otimes \psi - \sum_{\substack{k < M \\ \ell < N}} c_k d_\ell \varphi_k \otimes \psi_\ell \right\| \to 0$$

as $M, N \to \infty$. ∎

To show how the tensor product arises naturally, we will show how it is related to Hilbert spaces with which the reader is already familiar. First, let $\langle M_1, \mu_1 \rangle$ and $\langle M_2, \mu_2 \rangle$ be measure spaces. We suppose that $L^2(M_1, d\mu_1)$ and $L^2(M_2, d\mu_2)$ are separable (see Problems 24 and 25 of this chapter and Problem 43 of Chapter IV). Let $\{\varphi_k(x)\}$ and $\{\psi_\ell(y)\}$ be bases for $L^2(M_1, d\mu_1)$ and $L^2(M_2, d\mu_2)$ respectively. Then $\{\varphi_k(x)\psi_\ell(y)\}$ is certainly an orthonormal set in $L^2(M_1 \times M_2, d\mu_1 \otimes d\mu_2)$. The fact that $\{\varphi_k(x)\psi_\ell(y)\}$ is actually a basis can be seen as follows. Suppose that $f(x, y) \in L^2(M_1 \times M_2, d\mu_1 \otimes d\mu_2)$, and

$$\iint_{M_1 \times M_2} \overline{f(x, y)} \varphi_k(x) \psi_\ell(y) \, d\mu_1(x) \, d\mu_2(y) = 0$$

for all k and ℓ. By Fubini's theorem this can be rewritten

$$\int_{M_2} \left(\int_{M_1} \overline{f(x, y)} \varphi_k(x) \, d\mu_1(x) \right) \psi_\ell(y) \, d\mu_2(y) = 0$$

Since $\{\psi_\ell\}$ is a basis for $L^2(M_2, \mu_2)$, this implies that

$$\int_{M_1} \overline{f(x, y)} \varphi_k(x) \, d\mu_1(x) = 0$$

except on a set $S_k \subset M_2$ with $\mu_2(S_k) = 0$. Thus, for $y \notin \bigcup S_k$, $\int_{M_1} f(x, y) \times \varphi_k(x) \, d\mu_1(x) = 0$ for all k, which implies that $f(x, y) = 0$, a.e. $[\mu_1]$. Thus, $f(x, y) = 0$ a.e. $[\mu_1 \otimes \mu_2]$. So, $\{\varphi_k(x)\psi_\ell(y)\}$ is a basis for

$$L^2(M_1 \times M_2, d\mu_1 \otimes d\mu_2)$$

Now, let

$$U: \varphi_k \otimes \psi_\ell \to \varphi_k(x) \psi_\ell(y)$$

Then U takes an orthonormal basis for $L^2(M_1, d\mu_1) \otimes L^2(M_2, d\mu_2)$ onto an orthonormal basis for $L^2(M_1 \times M_2, d\mu_1 \otimes d\mu_2)$ and extends uniquely to a unitary mapping of

$$L^2(M_1, d\mu_1) \otimes L^2(M_2, d\mu) \quad \text{onto} \quad L^2(M_1 \times M_2, d\mu_1 \otimes d\mu_2).$$

Notice that if $f \in L^2(M_1, d\mu_1)$, $g \in L^2(M_2, d\mu_2)$, then

$$U(f \otimes g) = U(\sum c_k \varphi_k \otimes \sum d_\ell \psi_\ell)$$
$$= U\left(\sum_{k,\ell} c_k d_\ell \varphi_k \otimes \psi_\ell\right)$$
$$= \sum_{k,\ell} c_k d_\ell \varphi_k(x) \psi_\ell(y)$$
$$= f(x)g(y)$$

Because of this property, we often say that $L^2(M_1 \times M_2, d\mu_1 \otimes d\mu_2)$ and $L^2(M_1, d\mu_1) \otimes L^2(M_2, d\mu_2)$ are "naturally" isomorphic. Let $M_i = \mathbb{R}$ and $\mu_i =$ Lebesgue measure, then we have shown that $L^2(\mathbb{R}^2)$ is naturally isomorphic to $L^2(\mathbb{R}) \otimes L^2(\mathbb{R})$.

Let us return for a moment to Example 6 of Section II.1: $\langle M, \mu \rangle$ is a measure space and $\mathscr{H}'$ a separable Hilbert space with basis $\{\varphi_k\}$. In Problem 12, the reader is asked to show that each $g \in L^2(M, d\mu; \mathscr{H}')$ is a limit

$$g(x) = \lim_{N \to \infty} \sum_{k=1}^{N} (\varphi_k, g(x))_{\mathscr{H}'} \varphi_k$$

of finite linear combinations of vectors of the form $f_k(x)\varphi_k$, $f_k(x) \in L^2(M, d\mu)$. We now define

$$U: \sum_{k=1}^{N} f_k(x) \otimes \varphi_k \to \sum_{k=1}^{N} f_k(x)\varphi_k$$

Then U is a well-defined map from a dense set in $L^2(M, d\mu) \otimes \mathscr{H}'$ onto a dense set in $L^2(M, d\mu; \mathscr{H}')$ which preserves norms, so U extends uniquely to a unitary operator from $L^2(M, d\mu) \otimes \mathscr{H}'$ to $L^2(M, d\mu; \mathscr{H}')$. Notice that under this map, $U(f(x) \otimes \varphi) = f(x)\varphi$ for all $\varphi \in \mathscr{H}'$. In this sense, U is called the natural isomorphism between $L^2(M, d\mu) \otimes \mathscr{H}'$ and $L^2(M, d\mu; \mathscr{H}')$. We summarize this discussion in a theorem:

Theorem II.10 Let $\langle M_1, \mu_1 \rangle$ and $\langle M_2, \mu_2 \rangle$ be measure spaces so that $L^2(M_1, d\mu_1)$ and $L^2(M_2, d\mu_2)$ are separable. Then

(a) There is a unique isomorphism from $L^2(M_1, d\mu_1) \otimes L^2(M_2, d\mu_2)$ to $L^2(M_1 \times M_2, d\mu_1 \otimes d\mu_2)$ so that $f \otimes g \mapsto fg$.

(b) If $\mathscr{H}'$ is a separable Hilbert space, then there is a unique isomorphism from $L^2(M_1, d\mu_1) \otimes \mathscr{H}'$ to $L^2(M_1, d\mu_1; \mathscr{H}')$ so that $f(x) \otimes \varphi \mapsto f(x)\varphi$.

(c) There is a unique isomorphism from $L^2(M_1 \times M_2, d\mu_1 \otimes d\mu_2)$ to $L^2(M_1, d\mu_1; L^2(M_2, d\mu_2))$ such that $f(x, y)$ is taken into the function $x \mapsto f(x, \cdot)$.

Example 1 The Hilbert space in the quantum-mechanical description of a single Schrödinger particle of spin one-half is $L^2(\mathbb{R}^3, dx; \mathbb{C}^2)$, that is, the set of pairs $\{\psi_1(x), \psi_2(x)\}$ of square-integrable functions (dx is Lebesgue measure). By what we have shown above, $L^2(\mathbb{R}^3, dx; \mathbb{C}^2)$ is naturally isomorphic to $L^2(\mathbb{R}^3) \otimes \mathbb{C}^2$.

Example 2 (Fock spaces) Let $\mathcal{H}$ be a Hilbert space and denote by $\mathcal{H}^n$ the n-fold tensor product $\mathcal{H}^n = \mathcal{H} \otimes \mathcal{H} \otimes \cdots \otimes \mathcal{H}$. Set $\mathcal{H}^0 = \mathbb{C}$ and define

$$\mathcal{F}(\mathcal{H}) = \bigoplus_{n=0}^{\infty} \mathcal{H}^n$$

$\mathcal{F}(\mathcal{H})$ is called the Fock space over $\mathcal{H}$; it will be separable if $\mathcal{H}$ is. For example, if $\mathcal{H} = L^2(\mathbb{R})$, then an element $\psi \in \mathcal{F}(\mathcal{H})$ is a sequence of functions

$$\psi = \{\psi_0, \psi_1(x_1), \psi_2(x_1, x_2), \psi_2(x_1, x_2, x_3), \ldots\}$$

so that

$$|\psi_0|^2 + \sum_{n=1}^{\infty} \int_{\mathbb{R}^n} |\psi_n(x_1, \ldots, x_n)|^2 \, dx_1 \cdots dx_n < \infty$$

Actually, it is not $\mathcal{F}(\mathcal{H})$ itself, but two of its subspaces which are used most frequently in quantum field theory. These two subspaces are constructed as follows: Let $\mathcal{P}_n$ be the permutation group on n elements and let $\{\varphi_k\}$ be a basis for $\mathcal{H}$. For each $\sigma \in \mathcal{P}_n$, we define an operator (which we also denote by σ) on basis elements of $\mathcal{H}^{(n)}$ by

$$\sigma(\varphi_{k_1} \otimes \varphi_{k_2} \otimes \cdots \otimes \varphi_{k_n}) = \varphi_{k_{\sigma(1)}} \otimes \varphi_{k_{\sigma(2)}} \cdots \otimes \varphi_{k_{n(p)}}$$

σ extends by linearity to a bounded operator (of norm one) on $\mathcal{H}^n$ so we can define $S_n = (1/n!) \sum_{\sigma \in \mathcal{P}_n} \sigma$. It is an easy exercise (Problem 23) to show that $S_n^2 = S_n$ and $S_n^* = S_n$, so S_n is an orthogonal projection (the reader unfamiliar with adjoints and orthogonal projections should look up their definitions and elementary properties in Chapter VI). The range of S_n is called the n-fold symmetric tensor product of $\mathcal{H}$. In the case where $\mathcal{H} = L^2(\mathbb{R})$ and $\mathcal{H}^n = L^2(\mathbb{R}) \otimes \cdots \otimes L^2(\mathbb{R}) = L^2(\mathbb{R}^n)$, $S_n \mathcal{H}^n$ is just the subspace of $L^2(\mathbb{R}^n)$ of all functions left invariant under any permutation of the variables. We now define

$$\mathcal{F}_s(\mathcal{H}) = \bigoplus_{n=0}^{\infty} S_n \mathcal{H}^n$$

$\mathcal{F}_s(\mathcal{H})$ is called the **symmetric Fock space over** $\mathcal{H}$ or the **Boson Fock space over** $\mathcal{H}$.

Let $\varepsilon(\cdot)$ be the function from P_n to $\{1, -1\}$ which is one on even permutations and minus one on odd permutations. Define $A_n = (1/n!) \sum_{\sigma \in \mathscr{P}_n} \varepsilon(\sigma)\sigma$; then A is an orthogonal projection on $\mathscr{H}^n$. $A_n \mathscr{H}^n$ is called the *n*-fold antisymmetric tensor product of $\mathscr{H}$. In the case where $\mathscr{H} = L^2(\mathbb{R})$, $A_n \mathscr{H}^n$ is just the subspace of $L^2(\mathbb{R}^n)$ consisting of those functions odd under interchange of two coordinates. The subspace

$$\mathscr{F}_a(\mathscr{H}) = \bigoplus_{n=0}^{\infty} A_n \mathscr{H}^n$$

is called the **antisymmetric Fock space over** $\mathscr{H}$ or the **Fermion Fock space over** $\mathscr{H}$.

II.5 Ergodic theory: an introduction

In this section we give a brief introduction to ergodic theory. For our discussion we need several concepts not formally defined until Chapter VI: adjoint operators, projection operators, and the kernel and range of an operator. Any reader not already familiar with these concepts should consult Chapter VI. We give this discussion here because ergodic theory illustrates nicely the power and limitations of Hilbert-space methods and serves as a nice example of the main theme of these books, namely the interplay between functional analysis and mathematical physics. We will see that it is useful to reformulate the question of why macroscopic systems approach equilibrium in terms of abstract spaces, but that one must pay a price: The natural question in the abstract setting is slightly different from the original question and one may be tempted to accept weaker results.

The statement "any system approaches an equilibrium state" is sometimes known as the zeroth law of thermodynamics. From a microscopic point of view it is perhaps surprising that any system should approach equilibrium since microscopically there is no steady state and therefore no equilibrium. Nevertheless, any attempt at a microscopic justification of thermodynamics must explain why the zeroth laws holds macroscopically. There is far from universal agreement among physicists as to what constitutes a justification of the zeroth law, but we would like to avoid a discussion of the pros and cons of the many different approaches which have been suggested (however, see the Notes). The approach that we use is generally accepted by most physicists.

The first basic idea is that thermodynamical systems undergo fluctuations (see Problem 17); put differently, by their very nature the laws of thermodynamics are not absolute statements about a system at a fixed time but are statements about measurements made over time periods long with respect to some characteristic times such as relaxation or collision times. Thus, thermodynamics deals with average measurements of observables over a time period T. Since the collision times, etc., are dependent on the dynamics, one can only hope to prove thermodynamic statements about the limit as $T \to \infty$. How large T has to be for the average over the interval T to be approximately equal to the limit is a detailed dynamical question, but in specific cases one would hope to be able to prove something.

Let us suppose that we describe the state of a classical mechanical system by a point in some phase space Γ. For each time t; there is a map $T_t: \Gamma \to \Gamma$, where $T_t x$ is the state which results by taking a state x at t_0 and waiting until $t_0 + t$ (we are assuming time-translation invariance so t_0 never enters). Obviously, $T_{t+s} = T_t T_s$. In classical mechanics, the observables of the system like energy or angular momentum are functions on phase space. So, our discussion above suggests that we study

$$\lim_{T \to \infty} (1/T) \int_0^T f(T_t x)\, dt$$

We would like to show that the limit exists, at least for continuous functions. Typically Γ is a metric space, so "continuous" has a meaning. Not only would we like the limit to exist, but it should be independent of the initial point x or at least only dependent on a few "macroscopic" observables we can associate with an equilibrium state. For systems which are time-translation independent, the energy is a conserved quantity, so the average energy is the initial energy—thus we cannot hope for measurements to be independent of the initial energy. Therefore, for each energy E we look at the constant energy surface, Ω_E, in phase space, and for each $w \in \Omega_E$ and each continuous function f on Ω_E we hope that

$$\lim_{T \to \infty} (1/T) \int_0^T f(T_t w)\, dt$$

exists and is a number, $\mu(f)$, independent of w. The map $f \mapsto \mu(f)$ clearly has three properties:

(a) $\mu(1) = 1$
(b) μ is linear.
(c) $\mu(f) \geq 0$ if $f \geq 0$.

We will eventually see (Section IV.4) that such a μ is always associated with a measure $\hat{\mu}$ on Ω_E with $\hat{\mu}(\Omega_E) = 1$, so that

$$\mu(f) = \int_{\Omega_E} f(w)\, d\hat{\mu}(w)$$

From now on we denote the linear functional μ and the measure $\hat{\mu}$, by the same letter μ.

To summarize: We have shown that if

$$\lim_{T \to \infty} (1/T) \int_0^T f(T_t w) \, dt$$

exists for each fixed w and is independent of $w \in \Omega_E$, then there is a measure μ on Ω_E so that

$$\lim_{T \to \infty} (1/T) \int_0^T f(T_t w) \, dt = \int_{\Omega_E} f(w) \, d\mu(w) \qquad (\text{II.3})$$

The measure μ has a very important property. Let s be fixed and suppose χ_F is the characteristic function of a measurable set $F \subset \Omega_E$. Then

$$(1/T) \int_0^T \chi_{T_s^{-1}F}(T_t w) \, dt = (1/T) \int_0^T \chi_F(T_s T_t w) \, dt$$

so if the $\lim_{T \to \infty}$ exists, then $\mu(T_s^{-1} F) = \mu(F)$, that is, the measure is **invariant**. We also say that T is **measure preserving**. Classical mechanical systems come equipped with a natural invariant measure: if $\Gamma = \mathbb{R}^{6N}$ (N is the number of particles), the measure $d^{3N}q \, d^{3N}p$ is known to be invariant under the Hamiltonian flow (Liouville's theorem). This measure has a restriction to Ω_E given formally by

$$\mu_E(F) = \int_F \delta(H(p,q) - E) \, d^{3N}p \, d^{3N}q$$

where $H(p, q)$ is the Hamiltonian. Explicitly, if we pick a set of local coordinates at $x \in \Omega_E$, say $Q_1, \ldots, Q_{6N-1}$, which are orthogonal and normalized, then

$$d\mu_E = C \, d^{6N-1}Q / |\text{grad } H|$$

C is picked so that $\mu_E(\Omega_E) = 1$. Thus, the goal in justifying the zeroth law is to consider

$$M_T(f)(w) = (1/T) \int_0^T f(T_t w) \, dt$$

and to prove that in a suitable sense the function $(M_T f)(w)$ converges as $T \to \infty$ to the constant function with value

$$\int_{\Omega_E} f(w) \, d\mu_E(w)$$

Notice that if we can prove this, we will have proven much more; not only will we have shown that measurements over long periods of time are independent

II.5 Ergodic theory: an introduction

of the initial conditions (except for the energy), but we will have shown that the equilibrium state is described by a measure in phase space and this measure is

$$\int_F \delta(H(p,q) - E) \, d^{3N}p \, d^{3N}q$$

the "microcanonical ensemble."

Hilbert space methods are so powerful that as soon as one has a measure, it is tempting to try to reformulate the problem in terms of $L^2(\Omega_E, d\mu_E)$. Therefore, if $f \in L^2(\Omega_E, \mu_E)$, we define a map $f \xrightarrow{U_t} f \circ T_t$, that is,

$$(U_t f)(w) = f(T_t w)$$

Lemma (Koopman's lemma) U_t is a unitary map of $L^2(\Omega_E, d\mu_E)$ onto $L^2(\Omega_E, d\mu_E)$.

Proof
$$(U_t f, U_t g) = \int_{\Omega_E} \overline{f(T_t w)} g(T_t w) \, d\mu_E(w)$$
$$= \int_{\Omega_E} \overline{f(y)} g(y) \, d\mu_E(T_t^{-1} y) = \int_{\Omega_E} \overline{f(y)} g(y) \, d\mu_E(y)$$
$$= (f, g)$$

where we have used the invariance of the measure μ_E. Since $U_t U_{-t} = U_0 = I$, U is invertible and thus unitary. ∎

We want to study

$$\frac{1}{T} \int_0^T (U_t f)(w) \, d\mu(w)$$

but it is simpler to consider the discrete analogue

$$\frac{1}{N} \sum_{m=0}^{N-1} U^m f$$

The following elegant result settles the convergence question in the discrete case. Problem 18 extends the discrete result to the continuous case.

Theorem II.11 (mean ergodic theorem, or von Neumann's ergodic theorem) Let U be a unitary operator on a Hilbert space $\mathcal{H}$. Let P be the orthogonal projection onto $\{\psi \mid \psi \in \mathcal{H}, U\psi = \psi\}$. Then, for any $f \in \mathcal{H}$,

$$\lim_{N \to \infty} \frac{1}{N} \sum_{n=0}^{N-1} U^n f = Pf$$

We first prove an elementary technical lemma:

Lemma   (a) If U is unitary, $Uf = f$ if and only if $U^*f = f$.
(b)   For any operator on a Hilbert space $\mathcal{H}$, $(\operatorname{Ran} A)^\perp = \operatorname{Ker} A^*$.

Proof   To prove (a), notice that both conditions are equivalent to $f = U^{-1}f$.

To prove (b), observe that $\psi \in \operatorname{Ker} A^*$ means that $(\varphi, A^*\psi) = 0$ for all φ in $\mathcal{H}$. But, $\psi \in (\operatorname{Ran} A)^\perp$ means that $(A\varphi, \psi) = 0$ for all $\varphi \in \mathcal{H}$. (b) now follows from the definition of adjoint. ∎

Proof of the mean ergodic theorem   First let $f = g - Ug$, that is, $f \in \operatorname{Ran}(I - U)$. Then,

$$\left\| \frac{1}{N} \sum_{n=0}^{N-1} U^n f \right\| = \left\| \frac{1}{N} (g - U^N g) \right\| \leq \frac{2\|g\|}{N} \to 0$$

as $N \to \infty$. By an $\varepsilon/3$ argument

$$\frac{1}{N} \sum_{n=0}^{N-1} U^n f \to 0$$

for any $f \in \overline{\operatorname{Ran}(I - U)}$. By the lemma, $(\operatorname{Ran}(I - U))^\perp = \operatorname{Ker}(I - U^*) = \{\psi \mid U^*\psi = \psi\} = \{\psi \mid U\psi = \psi\}$. Therefore, $Pf = 0$ if and only if $f \in \overline{\operatorname{Ran}(I - U)}$. Now, suppose $Pf = f$. Trivially,

$$\frac{1}{N} \sum_{n=0}^{N-1} U^n f = f$$

converges to $f = Pf$. Thus the limit statement holds on $\overline{\operatorname{Ran}(I - U)}$ and on $\operatorname{Ker}(I - U^*)$ and therefore on $\overline{\operatorname{Ran}(I - U)} \oplus \operatorname{Ker}(I - U^*)$, which is all of $\mathcal{H}$ by the projection theorem and (b) above. ∎

In the continuous case $U_t f = f \circ T_t$, what are the functions in $L^2(\Omega_E, d\mu_E)$ which satisfy $U_t f = f$? Clearly, the constant functions are invariant.

Definition T_t is called **ergodic** if the constant functions are the only functions in $L^2(\Omega_E, d\mu_E)$ for which $f \circ T_t = f$ (as L^2 functions) for all t.

Given the continuous analogue of the mean ergodic theorem (Problem 18) we have:

Corollary   Let T_t be ergodic. Then for any $f \in L^2(\Omega_E, d\mu_E)$,

$$L^2 - \lim_{T \to \infty} \frac{1}{T} \int_0^T f(T_t w) \, dt = \int_{\Omega_E} f(y) \, d\mu_E(y) \tag{II.4}$$

II.5 Ergodic theory: an introduction

Proof In this case $\{\psi \,|\, U\psi = \psi\}$ is one dimensional. Thus $P\psi$ is a constant C and

$$C = (1, \psi) = \int_{\Omega_E} \psi(w) \, d\mu_E(w) \quad \blacksquare$$

Notice that if (II.4) holds then $P\psi$ is constant so that T_t must be ergodic; thus ergodicity is necessary and sufficient for (II.4) to hold.

It is sometimes useful to express ergodicity in terms of the measure.

Proposition T_t is ergodic if and only if for all measurable sets $F \subset \Omega_E$, $T_t^{-1}F = F$ for all t implies $\mu_E(F) = 0$ or $\mu_E(F) = 1$.

Proof Suppose T_t is ergodic and $T_t^{-1}F = F$ for all t. Then $f = \chi_F$ is an invariant function so χ_F is constant a.e., which implies $\mu(F) = 0$ or $\mu(F) = 1$.

Conversely, suppose that the second condition holds. Then $\{w \,|\, f(w) < a\}$ is invariant under T_t so $f(w) < a$ a.e. or $f(w) \geq a$ a.e. Since this is true for all a, $f(w)$ is constant a.e. $\blacksquare$

The condition that $T_t^{-1}F = F$ implies $\mu_E(F) = 0$ or $\mu_E(F) = 1$ is sometimes called **metric transitivity**.

Let us take stock of what we have proven. We have derived a necessary and sufficient condition on the flow T_t so that

$$\lim_{T \to \infty} \frac{1}{T} \int_0^T f(T_t w) \, dt$$

is precisely what we want it to be, but not in the sense of convergence for each w; instead, we have L^2 convergence of $(1/T) \int_0^T f(T_t w) \, dt$ to the constant function

$$\int_{\Omega_E} f(w) \, d\mu(w)$$

This is not surprising since pointwise convergence is not an L^2 notion. By using Hilbert space methods we have given up the chance of proving that

$$\frac{1}{T} \int_0^T f(T_t w) \, dt$$

converges pointwise for each w as $T \to \infty$. Actually, the pointwise limit does exist but this must be proven by entirely different methods. We state the result:

Theorem II.12 (individual or Birkhoff ergodic theorem)   Let T be a measure preserving transformation on a measure space $\langle \Omega, \mu \rangle$. Then for any $f \in L^1(\Omega, \mu)$,

$$\lim_{N \to 0} \frac{1}{N} \sum_{n=0}^{N-1} f(T^n x)$$

exists pointwise a.e. and is some function $f^\# \in L^1(\Omega, d\mu)$ satisfying $f^\#(Tx) = f^\#(x)$. If $\mu(\Omega) < \infty$, then

$$\int_\Omega f^\#(x)\, d\mu(x) = \int_\Omega f(x)\, d\mu(x)$$

Furthermore, if μ is ergodic and $\mu(\Omega) = 1$, then

$$\frac{1}{N} \sum_{n=0}^{N-1} f(T^n x) \xrightarrow[N \to \infty]{} \int_\Omega f(y)\, d\mu(y)$$

for almost all x.

This theorem is closer to what one wants to justify statistical mechanics than the von Neumann theorem, and it is fashionable to say that the von Neumann theorem is unsuitable for statistical mechanics. We feel that this is an exaggeration. If we had only the von Neumann theorem we could probably live with it quite well. Typically, initial conditions are not precisely measurable anyway, so that one could well associate initial states with measures $f\, d\mu$ where $\int f\, d\mu = 1$, in which case the von Neumann theorem suffices. However, the Birkhoff theorem does hold and is clearly a result that we are happier to use in justifying the statement that phase-space averages and time averages are equal.

Finally, one should ask whether classical mechanical flows on constant energy surfaces are in fact ergodic. Little is known about this interesting but difficult question. However, Sinai has shown recently that a gas of hard spheres in a box is an ergodic system.

NOTES

Section II.1   A good reference for material on Hilbert spaces is the first chapter of the book, *Introduction to Hilbert Space* by Paul Halmos, Chelsea, Bronx, New York, 1957. His book, *A Hilbert Space Problem Book*, Van Nostrand–Reinhold, Princeton, New Jersey, 1967, which consists of problems, hints, and solutions, is very advanced but is a useful learning device as the reader becomes more sophisticated. The standard reference, *Functional*

Analysis, Ungar, New York, 1955 by F. Riesz and B. Sz.-Nagy has applications to integral equations.

Section II.2 The Riesz Lemma was proved independently by F. Riesz, "Sur une espèce de géométrie analytiques des systèms de fonctions summable," *C. R. Acad. Sci. Paris*, **144** (1907), 1409–1411, and by M. Fréchet in "Sur les ensembles de fonctions et les opérations linéaires," *C. R. Acad. Sci. Paris*, **144** (1907), 1414–1416. The Riesz lemma can be used to give a short proof of the existence of adjoint operators in the case of Hilbert spaces. The general definition of adjoint for Banach spaces is given in Chapter VI.

Section II.3 It may seem at first a little strange that $L^2[0, 1]$ is separable since the functions take values at uncountably many points. However, these values cannot be assigned arbitrarily since the function must be measurable, a strong restriction, and furthermore we have identified functions which differ only on a set of measure zero.

The following question often puzzles students of functional analysis. If all infinite-dimensional separable Hilbert spaces are the same (that is, isomorphic to ℓ_2) why do we talk about them? That is, why worry separately about $L^2(\mathbb{R}^n, d\mu)$ if, as a Hilbert space, it is isomorphic to ℓ_2. The answer is that we are often interested not just in the space but in some other structures, for example some bounded operators on the space. It is true that under the isomorphism these operators go over into bounded operators on ℓ_2, but their structure may be easy to analyze on $L^2(\mathbb{R}, d\mu)$ while it is difficult analyze on ℓ_2. This is one of the general features of functional analysis: One tries to choose a representation of the structures with which one is dealing so that the structures are easy to analyze. As a very simple example the reader should think of the principal axis theorem (spectral theorem) for $\mathbb{C}^n$ which says that given a self-adjoint transformation, one can choose an orthonormal basis in $\mathbb{C}^n$ so that the matrix of the transformation in that basis is diagonal. That is, if one chooses the right isomorphic copy of $\mathbb{C}^n$ (change of basis) then the operator becomes especially simple. As the reader will see, this example is the first note of a rather long symphony.

The first proof of the convergence of Fourier series for a large class of functions was given by Dirichlet in 1829. A good reference for both the classical theory and the modern approach is *An Introduction to Harmonic Analysis*, Wiley, New York, 1968, by Y. Katznelson. Recently Carleson has proven the spectacular result that the Fourier series of a function in $L^2[0, 2\pi]$ converges pointwise a.e. in "On the Convergence and Growth of Partial Sums of Fourier Series," *Acta Math.* **116** (1966), 135–157, and R. Hunt has extended this result to various L^p spaces in "On the Convergence of Fourier Series" appearing in *Orthogonal Expansions and their Continuous Analogues* (D. Haimo, ed.), pp. 235–237, Southern Illinois Univ. Press, 1968.

Section II.4 A description of finite tensor products of Hilbert spaces was first given by J. von Neumann and F. Murray in "On Rings of Operators," *Ann. Math.* (2) **37**, (1936), 116–229, though tensor products of finite-dimensional spaces were known long before that. For a modern treatment of tensor products, see F. Trèves, *Topological Vector Spaces, Distributions and Kernels*, Academic Press, New York, 1967, or R. Schatten, *A Theory of Cross Spaces*, Princeton University Press, Princeton, N.J., 1950.

The definition and use of the spaces we have called Fock spaces goes back to the original paper by V. Fock: "Konfigurationsraum und Zweite Quantelung," *Z. Phys.* **75** (1932), 622–647. In Chapter X a Fock space is used in the construction of the free field, a field theory satisfying the Wightman axioms.

Section II.5 For a discussion of thermodynamics from a nonstatistical point of view, that is, as a basically empirical subject, see A. B. Pippard, *The Elements of Classical Thermodynamics*, Cambridge Univ. Press, London and New York, 1957.

For a discussion of points of view regarding the zeroth law of thermodynamics which do not embrace the ergodic theorem, see L. P. Landau and E. M. Lifshitz, *Statistical Physics*, Chapter 1, Pergamon, Oxford, 1958, or F. Strocchi: " Microscopic and Macroscopic Quantities in Statistical Mechanics," *Il Nuovo Cimento*, **65B** (1970), 239–265.

For a proof of Liouville's theorem, see M. Goldstein, *Classical Mechanics*, pp. 266–268, Addison-Wesley, Reading, Massachusetts, 1950, or R. Abraham, *Foundations of Mechanics*, p. 108, Benjamin, New York, 1967.

The idea of using Hilbert space methods to study classical mechanical systems first appeared in B. O. Koopman, "Hamiltonian Systems and Transformations in Hilbert Spaces," *Proc. Nat. Acad. Sci. (U.S.A.)* **17** (1931), 315–318.

The von Neumann ergodic theorem was first proven in J. von Neumann, "Proof of the Quasiergodic Hypothesis," *Proc. Nat. Acad. Sci. (U.S.A.)* **18** (1932) 70–82. Our proof is due to F. Riesz, "Sur la théorie ergodique," *Comm. Math. Helv.* **17** (1945), 221–239.

The Birkhoff ergodic theorem was proven by G. D. Birkhoff, "Proof of the Ergodic Theorem," *Proc. Nat. Acad. Sci. (U.S.A.)* **17** (1931), 656–660. F. Riesz (op. cit.) provided an alternate and simple proof based on the "maximal ergodic theorem" of N. Wiener, "The Ergodic Theorem," *Duke Math. J.* **5** (1939), 1–18, and of K. Yoshida and S. Kakutani, "Birkhoff's Ergodic Theorem and the Maximal Ergodic Theorem," *Proc. Imp. Acad. Tokyo* **15** (1939), 165–168. A further simplification in the proof of the maximal ergodic theorem may be found in A. M. Garsia, "A Simple Proof of E. Hopf's Maximal Ergodic Theorem," *J. Math. Mech.* **14** (1965), 381–382.

For a delightful discussion of the mathematics of ergodic theory, see P. R. Halmos, *Lectures in Ergodic Theory*, Chelsea, Bronx, New York, 1956, and for a historical summary of the subject see P. R. Halmos, "Measurable Transformations," *Bull. Amer. Math. Soc.* **55**, (1948), 1015–1034.

For a discussion of the mean ergodic theorem in a Banach space setting (which includes L^p-mean ergodic theorems for $1 < p < \infty$), see E. Lorch, *Spectral Theory*, pp. 54–56, Oxford Univ. Press, London and New York, 1962.

There are deep connections between notions from information theory and ergodic theory: for a pleasant, readable treatment, see P. Billingsley, *Ergodic Theory and Information*, Wiley, New York, 1965.

Sinai's result on the ergodicity of a hard sphere gas was announced in Ya. Sinai, "On the Foundations of the Ergodic Hypothesis for a Dynamical System of Statistical Mechanics," *Dokl. Akad. Nauk.* **153** (1963) [*Sov. Math. Dokl.* **4**, (1963), 1818–1822]. A sketch of the proof appears in Ya. Sinai, "Ergodicity of Boltzmann's Gas Model" in *Statistical Mechanics, Foundations and Applications* (T. Bak, ed.), Benjamin, New York, 1967. His proof uses important ideas of Krylov, Kolmogorov, and Anosov.

An alternative property to ergodicity which has some of its consequences is proposed in R. Prosser, "Spectral Analysis of Classical Central Force Motion," *J. Math. Phys.* **10** (1969), 2233–2239. Ideal gases with no collisions are shown to have this property.

For many purposes, one wants thermodynamical systems to possess a stronger property than ergodicity known as mixing; this stronger notion expresses the "irreversibility" of thermodynamic systems and it is this stronger notion that Sinai proves. We return to mixing briefly in Chapter VII. For a discussion of the hierarchy of notions related to ergodicity, see V. I. Arnold and A. Avez, *Ergodic Problems of Classical Mechanics*, Benjamin, New York, 1968 and A. S. Wightman, "Statistical Mechanics and Ergodic Theory: An Expository

Lecture," in *Statistical Mechanics at the Turn of the Decade*, (E. Cohen, ed.), Ungar, New York, 1970.

We have been slightly cavalier in our statement that the question of how big T must be for

$$T^{-1} \int_0^T f(w_t)\, dt$$

to be close to its limit is a detailed dynamical question. In a general ergodic system the time necessary for the limit to be reached should be a typical "recurrence time," that is, the typical time needed for the system to return close to its initial state. Usually, in macroscopic systems this time is astronomically long. Thus, an important question to ask is what properties mechanical systems have that make the "relaxation time"; that is, the time to approach equilibrium, so much smaller than the recurrence time. While this is certainly a detailed dynamical question, it suggests there is an additional mechanism at work which one would like to understand.

PROBLEMS

†*1.* (a) Let V be an inner product space. Prove that the inner product can be extended to $\tilde{V}$ as follows: First, show that if $x, y \in \tilde{V}$, $x_n, y_n \in V$ and $x_n \to x$, $y_n \to y$, then (x_n, y_n) converges. Define $(x, y) = \lim_{n \to \infty}(x_n, y_n)$ and show that it is independent of which convergent sequences are chosen. Finally, show that $(\cdot, \cdot)$ has the right properties.

(b) Prove the statement in (a) by applying the B.L.T. theorem twice.

*2. (a) A simple function is a finite linear combination of the characteristic functions of disjoint measurable sets. Show that the simple functions are dense in $L^2[a, b]$.

(b) Show that any simple function on $[a, b]$ can be approximated arbitrarily closely (in the L^2 sense) by a step function.

(c) Show that any step function can be approximated arbitrarily closely (in the L^2 sense) by a continuous function and thus conclude that $C[a, b]$ is dense in $L^2[a, b]$ in the L^2 norm.

3. Prove that if μ_1 and μ_2 are mutually singular Borel measures on $\mathbb{R}$ and $\mu = \mu_1 + \mu_2$, then $L^2(\mathbb{R}, d\mu)$ is naturally isomorphic to $L^2(\mathbb{R}, d\mu_1) \oplus L^2(\mathbb{R}, d\mu_2)$. (Hint: let A be a set with $\mu_1(A) = 0$ and $\mu_2(\mathbb{R}\setminus A) = 0$ and map f to $\langle (1 - \chi_A)f, \chi_A f \rangle$.)

4. (a) Prove that the inner product can be recovered from the norm by the **polarization identity**

$$(x, y) = \tfrac{1}{4}\{(\|x + y\|^2 - \|x - y\|^2) - i(\|x + iy\|^2 - \|x - iy\|^2)\}$$

*(b) Prove that a normed linear space is an inner product space if and only if the norm satisfies the parallelogram law.

5. Let V be an inner product space and let $\{x_n\}_{n=1}^N$ be an orthonormal set. Prove that

$$\left\| x - \sum_{n=1}^N c_n x_n \right\|$$

is minimized by choosing $c_n = (x_n, x)$.

†6. Let $\mathscr{M}$ be any linear subset of a Hilbert space $\mathscr{H}$. Prove that $\mathscr{M}^\perp$ is a closed linear subspace and that $\bar{\mathscr{M}} = (\mathscr{M}^\perp)^\perp$.

†7. Prove the uniqueness statements in Theorem II.3 and the preceding lemma.

†8. Complete the proof of the corollary to the Riesz Lemma.

9. Let $\mathscr{M}$ be a subspace of a Hilbert space $\mathscr{H}$. Let $f: \mathscr{M} \to \mathbb{C}$ be a linear functional on $\mathscr{M}$ with bound C. Prove that there is a unique extension of f to a continuous linear functional on $\mathscr{H}$ with the same bound. (We note that the existence part of this statement is just the Hahn–Banach theorem for Hilbert spaces, see Section III.3.)

10. Apply the Gram–Schmidt process to the functions $1, x, x^2, x^3$ on the interval $[-1, 1]$ with the L^2 inner product and obtain the first four Legendre polynomials (up to constant multiples).

11. Prove that $L^2(\mathbb{R})$ is separable. (Hint: see Problem 2.)

†12. (Example 6, Section II.1) We say that a vector-valued function f from a measure space $\langle X, \mu \rangle$ to a separable Hilbert space, $\mathscr{H}'$, is measurable if $(y, f(x))_{\mathscr{H}'}$ is measurable for each $y \in \mathscr{H}'$.
  (a) Show that if $f(x)$ and $g(x)$ are measurable vector-valued functions, then $\|f(x)\|^2_{\mathscr{H}'}$ and $(f(x), g(x))_{\mathscr{H}'}$ are measurable.
  (b) Let $\{\varphi_k\}_{k=1}^\infty$ be a basis for $\mathscr{H}'$. Prove that if $g \in L^2(X, d\mu; \mathscr{H}')$, then
  $$\sum_{k=1}^N (\varphi_k, g(x))_{\mathscr{H}'} \cdot \varphi_k \to g$$
  and if $f \in L^2(X, d\mu; \mathscr{H}')$
  $$(f, g) = \sum_{k=1}^\infty \int_X (f(x), \varphi_k)_{\mathscr{H}'} \cdot (\varphi_k, g(x))_{\mathscr{H}'} \, d\mu(x)$$
  (c) Assume that $L^2(X, d\mu)$ is separable and prove that $L^2(X, d\mu; \mathscr{H}')$ is separable.

13. Using direct sums, construct an inseparable Hilbert space and an uncountable orthonormal basis.

*14. The goal of this exercise is to prove that the Fourier series of a continuous function is pointwise Cesàro summable to f. View $[0, 2\pi]$ as a group with addition mod 2π and write $\int_0^{2\pi}$ as $\oint$. Let $f(\theta) \in L^2[0, 2\pi]$ and $c_n = (e^{in\theta}/\sqrt{2\pi}, f)$.
  (a) Let $S_N(f) = \sum_{-N}^N c_n e^{in\theta}/\sqrt{2\pi}$. Prove that
  $$(S_N f)(\theta) = \frac{1}{2\pi} \oint f(\theta + x) \frac{\sin(N + 1/2)x}{\sin(x/2)} \, dx$$
  (b) Let $(\sum_n(f))(\theta) = [1/(N+1)] \sum_0^N S_n(f)(\theta)$ (Cesàro sum). Prove that
  $$\left(\sum_N(f)\right)(\theta) = \frac{1}{2\pi(N+1)} \oint f(x + \theta) \frac{\sin^2[(N+1)/2]x}{\sin^2(x/2)} \, dx$$
  (c) Let
  $$K_N(x) = \frac{\sin^2[(N+1)/2]x}{2\pi(N+1)\sin^2(x/2)}$$

Prove that for any $\delta > 0$, $K_N(x) \to 0$ uniformly in $[\delta, 2\pi - \delta]$.
(d) Prove that $(\sum_N(f))(\theta_0) \to f(\theta_0)$ if f is bounded and continuous at θ_0.
(e) Prove that if f is continuous and periodic, then $(\sum_N(f))(\theta) \to f(\theta)$ uniformly in θ. (Hint: Recall that f continuous on $[0, 2\pi]$ implies that f is uniformly continuous.)
(f) Show that $\|f - S_N(f)\|_2 \leq \|f - \sum_n(f)\|_2$ and conclude that $S_N(f) \xrightarrow{L^2} f$ if f is continuous.

†*15.* Suppose $f \in C_p^1[0, 2\pi]$ and let $c_n = (e^{inx}/\sqrt{2\pi}, f)$, $b_n = (e^{inx}/\sqrt{2\pi}, f'(x))$.
(a) Prove $\sum |b_n|^2 < \infty$ and conclude that $\sum n^2 |c_n|^2 < \infty$.
(b) Prove that $\sum |c_n| < \infty$.
(c) Prove that $\sum_{-M}^{M} c_n e^{inx}/\sqrt{2\pi}$ is uniformly convergent as $M \to \infty$.
(d) Use 14(f) to conclude that $\sum_{-M}^{M} c_n e^{inx}/\sqrt{2\pi}$ is uniformly convergent to f.

16. Show that the unit ball in an infinite dimensional Hilbert space contains infinitely many *disjoint* translates of a ball of radius $\sqrt{2}/4$. Conclude that one cannot have a nontrivial translation invariant measure on an infinite dimensional Hilbert space.

17. Prove the Poincaré recurrence theorem: Given a measure preserving map, T, on a set Ω with $\mu(\Omega) < \infty$, then for any measurable set $E \subset \Omega$, $T^n x \in E$ infinitely often for almost all $x \in E$. This result says that almost every state returns arbitrarily close to its initial position infinitely often (thus assuring that fluctuations continue to take place). Hint: Let $F = \{x \in E \mid T^n x \notin E \text{ for any } n > 0\}$. Show $\{T^{-m}F\}$ are disjoint and prove thereby that F has measure zero.

18. Let T_t be a one-parameter group of measure-preserving transformations of a measure space $\langle \Omega, \mu \rangle$. Apply the discrete mean ergodic theorem to $\int_0^1 f(T_t w)\, dt$ to prove the mean ergodic theorem for $\lim_{T \to \infty} (1/T) \int_0^T f(T_t w)\, dt$.

19. Let V be an operator in a Hilbert space satisfying $\|V^n\| \leq C$ for all n. Prove that $(1/N)\sum_{n=0}^{N-1} V^n f \to Pf$ for all $f \in \mathcal{H}$ where P is a (not necessarily orthogonal) projection onto $\{f \mid Vf = f\}$.

20. Consider the unit circle, $\{z \in \mathbb{C} \mid |z| = 1\}$ with Lebesgue measure. Let $T(z) = e^{2\pi i \theta} z$. Show that T is ergodic if and only if θ is irrational.

21. Suppose Ω is a compact metric space with metric ρ and some measure μ. Let T be a measure-preserving ergodic transformation with the additional property that $\rho(Tx, Ty) = \rho(x, y)$. (For example, the map T in Problem 20 when θ is irrational.) Show that if f is a continuous function on Ω, then $(1/N)\sum_{n=0}^{N-1} f(T^n w)$ converges *uniformly* to $\int_\Omega f\, d\mu$. (Hint: Prove that the family

$$(M_N f)(w) = \frac{1}{N} \sum_{n=0}^{N-1} f(T^n w)$$

is uniformly equicontinuous and then use the mean ergodic and Ascoli theorems.)

*22. Let $\{\eta_n\}_{n=-\infty}^{\infty}$ be a set of vectors in a Hilbert space, $\mathcal{H}$, so that $a_{nm} = |(\eta_n, \eta_m)|$ is the matrix (in the natural basis) of an operator A on $\ell_2(-\infty, \infty)$. Prove that

$$\sum_{n=-\infty}^{\infty} |(f, \eta_n)|^2 \leq \|A\| \|f\|^2$$

for any $f \in \mathcal{H}$.

66   **II: HILBERT SPACES**

23. Let S_n be the operator defined in Example 2 of Section 4.
    (a) Prove that S_n is independent of the basis $\{\varphi_k\}$.
    (b) Prove that $S_n^2 = S_n$ and $S_n = S_n^*$ (Hint: show that $\sigma^* = \sigma^{-1}$).
    (c) Do parts (a) and (b) for A_n.

*24. Let $\langle M, \mathscr{R}, \mu \rangle$ be a σ-finite measure space. Let $\mathscr{R}_F = \{X \in \mathscr{R} \mid \mu(X) < \infty\}$. Call $X, Y \in \mathscr{R}_F$ equivalent if and only if $\mu(X \triangle Y) = 0$ where $X \triangle Y = (X \backslash Y) \cup (Y \backslash X)$. Let $\hat{\mathscr{R}}_F$ be the family of equivalence classes of $\mathscr{R}_F$ under this relation.
    (a) Prove that $\mu(X \triangle Y)$ only depends on the equivalence classes of X and Y in $\hat{\mathscr{R}}_F$.
    (b) Prove that $\hat{\mathscr{R}}_F$ with the function $\rho(X, Y) = \mu(X \triangle Y)$ is a metric space.
    (c) Prove that $L^2(M, d\mu)$ is separable if and only if $\mathscr{R}_F$ with the metric ρ is a separable metric space.

*25. Find a finite measure space (that is, $\langle M, \mathscr{R}, \mu \rangle$ with $\mu(M) < \infty$) with $L^2(M, d\mu)$ nonseparable. (Hint: Take an uncountable Cartesian product of sets of the form $[0, 1]$.)

26. Prove that part (c) of Theorem II.10 follows from parts (a) and (b).

27. Prove the projection theorem using the existence of orthonormal bases.

III: Banach Spaces

Reductio ad absurdum is one of a mathematician's finest weapons. It is a far finer gambit than any chess gambit: a chess player may offer the sacrifice of a pawn or even a piece, but a mathematician offers the game.

G. H. Hardy

III.1 Definition and examples †

We defined normed linear spaces in Section I.2. Since normed linear spaces are metric spaces, they may have the property of being complete.

Definition A complete normed linear space is called a **Banach space**.

Banach spaces have many of the properties of $\mathbb{R}^n$: they are vector spaces, they have a notion of distance provided by the norm, and every Cauchy sequence has a limit. In general the norm does not arise from an inner product (see Problem 4 of Chapter II), so Banach spaces are not necessarily Hilbert spaces and will not have all of the same nice geometrical properties. In order to acquaint the reader with the types of Banach spaces he is likely to encounter, we discuss several examples in detail.

Example 1 ($L^\infty(\mathbb{R})$ and its subspaces) Let $L^\infty(\mathbb{R})$ be the set of (equivalence classes of) complex-valued measurable functions on $\mathbb{R}$ such that $|f(x)| \leq M$ a.e. with respect to Lebesgue measure for some $M < \infty$ ($f \sim g$ means $f(x) = g(x)$ a.e.). Let $\|f\|_\infty$ be the smallest such M. It is an easy exercise (Problem 1) to

† A supplement to this section begins on p. 348.

show that $L^\infty(\mathbb{R})$ is a Banach space with norm $\|\cdot\|_\infty$. The bounded continuous functions $C(\mathbb{R})$ is a subspace of $L^\infty(\mathbb{R})$ and restricted to $C(\mathbb{R})$ the $\|\cdot\|_\infty$-norm is just the usual supremum norm under which $C(\mathbb{R})$ is complete (since the uniform limit of continuous functions is continuous). Thus, $C(\mathbb{R})$ is a closed subspace of $L^\infty(\mathbb{R})$.

Consider the set $\kappa(\mathbb{R})$ of continuous functions with compact support, that is, the continuous functions that vanish outside of some closed interval. $\kappa(\mathbb{R})$ is a normed linear space under $\|\cdot\|_\infty$ but is not complete. The completion of $\kappa(\mathbb{R})$ is not all of $C(\mathbb{R})$; for example, if f is the function which is identically equal to one, then f cannot be approximated by a function in $\kappa(\mathbb{R})$ since $\|f - g\|_\infty \geq 1$ for all $g \in \kappa(\mathbb{R})$. The completion of $\kappa(\mathbb{R})$ is just $C_\infty(\mathbb{R})$, the continuous functions which approach zero at $\pm\infty$ (Problem 5). Some of the most powerful theorems in functional analysis (Riesz–Markov, Stone–Weierstrass) are generalizations of properties of $C(\mathbb{R})$ (see Sections IV.3 and IV.4).

Example 2 (L^p spaces)   Let $\langle X, \mu \rangle$ be a measure space and $p \geq 1$. We denote by $L^p(X, d\mu)$ the set of equivalence classes of measurable functions which satisfy:

$$\|f\|_p \equiv \left(\int_X |f(x)|^p \, d\mu(x) \right)^{1/p} < \infty$$

Two functions are equivalent if they differ only on a set of measure zero. The following theorem collects many of the standard facts about L^p spaces.

Theorem III.1   Let $1 \leq p < \infty$, then

(a) (the Minkowski inequality)   If $f, g \in L^p(X, d\mu)$, then

$$\|f + g\|_p \leq \|f\|_p + \|g\|_p$$

(b) (Riesz–Fisher) $L^p(X, d\mu)$ is complete.

(c) (the Hölder inequality) Let p, q, and r be positive numbers satisfying $p, q, r \geq 1$ and $p^{-1} + q^{-1} = r^{-1}$. Suppose $f \in L^p(X, d\mu)$, $g \in L^q(X, d\mu)$. Then $fg \in L^r(X, d\mu)$ and

$$\|fg\|_r \leq \|f\|_p \|g\|_q$$

Proofs of many of the basic facts about L^p spaces, including these inequalities, can be found in the second supplemental section. The Minkowski inequality shows that $L^p(X, d\mu)$ is a vector space and that $\|\cdot\|_p$ satisfies the

III.1 Definition and examples

triangle inequality. Combined with (b) this shows that $L^p(X, d\mu)$ is a Banach space. We have given the proof of (b) for the case where $p = 1$, $X = \mathbb{R}$ and μ = Lebesgue measure; the proof for the general case is similar.

Example 3 (sequence spaces)   There is a nice class of spaces which is easy to describe and which we will often use to illustrate various concepts. In the following definitions,
$$a = \{a_n\}_{n=1}^{\infty}$$
always denotes a sequence of complex numbers.

$$\ell_\infty = \left\{ a \,\Big|\, \|a\|_\infty \equiv \sup_n |a_n| < \infty \right\}$$

$$c_0 = \left\{ a \,\Big|\, \lim_{n \to \infty} a_n = 0 \right\}$$

$$\ell_p = \left\{ a \,\Big|\, |a|_p \equiv \left(\sum_{n=1}^{\infty} |a_n|^p \right)^{1/p} < \infty \right\}$$

$$s = \left\{ a \,\Big|\, \lim_{n \to \infty} n^p a_n = 0 \text{ for all positive integers } p \right\}$$

$$f = \left\{ a \,\Big|\, a_n = 0 \text{ for all but a finite number of } n \right\}$$

It is clear that as sets $f \subset s \subset \ell_p \subset c_0 \subset \ell_\infty$.

The spaces ℓ_∞ and c_0 are Banach spaces with the $\|\cdot\|_\infty$ norm; ℓ_p is a Banach space with the $\|\cdot\|_p$ norm (note that this follows from Example 2 since $\ell_p = L^p(\mathbb{R}, d\mu)$ where μ is the measure with mass one at each positive integer and zero everywhere else). It will turn out that s is a Fréchet space (Section V.2). One of the reasons that these spaces are easy to handle is that f is dense in ℓ_p (in $\|\cdot\|_p$; $p < \infty$) and is dense in c_0 (in the $\|\cdot\|_\infty$ norm). Actually, the set of elements of f with only rational entries is also dense in ℓ_p and c_0. Since this set is countable, ℓ_p and c_0 are separable. ℓ_∞ is not separable (Problem 2).

Example 4 (the bounded operators)   In Section I.3 we defined the concept of a bounded linear transformation or bounded operator from one normed linear space, X, to another Y; we will denote the set of all bounded linear operators from X to Y by $\mathscr{L}(X, Y)$. We can introduce a norm on $\mathscr{L}(X, Y)$ by defining
$$\|A\| = \sup_{x \in X, \, x \neq 0} \frac{\|Ax\|_Y}{\|x\|_X}$$
This norm is often called the **operator norm**.

Theorem III.2   If Y is complete, $\mathscr{L}(X, Y)$ is a Banach space.

Proof Since any finite linear combination of bounded operators is again a bounded operator, $\mathscr{L}(X, Y)$ is a vector space. It is easy to see that $\|\cdot\|$ is a norm; for example, the triangle inequality is proven by the computation

$$\|A + B\| = \sup_{x \neq 0} \frac{\|(A + B)x\|}{\|x\|} \leq \sup_{x \neq 0} \frac{\|Ax\| + \|Bx\|}{\|x\|}$$

$$\leq \sup_{x \neq 0} \frac{\|Ax\|}{\|x\|} + \sup_{x \neq 0} \frac{\|Bx\|}{\|x\|}$$

$$= \|A\| + \|B\|$$

To show that $\mathscr{L}(X, Y)$ is complete, we must prove that if $\{A_n\}_{n=1}^{\infty}$ is a Cauchy sequence in the operator norm, then there is a bounded linear operator A so that $\|A_n - A\| \to 0$. Let $\{A_n\}_{n=1}^{\infty}$ be Cauchy in the operator norm; we construct A as follows. For each $x \in X$, $\{A_n x\}_{n=1}^{\infty}$ is a Cauchy sequence in Y. Since Y is complete, $A_n x$ converges to an element $y \in Y$. Define $Ax = y$. It is easy to check that A is a linear operator. From the triangle inequality it follows that

$$| \|A_n\| - \|A_m\| | \leq \|A_n - A_m\|$$

so $\{\|A_n\|\}_{n=1}^{\infty}$ is a Cauchy sequence of real numbers converging to some real number C. Thus,

$$\|Ax\|_Y = \lim_{n \to \infty} \|A_n x\|_Y \leq \lim_{n \to \infty} \|A_n\| \, \|x\|_X$$

$$= C\|x\|_X$$

so A is a bounded linear operator. We must still show that $A_n \to A$ in the operator norm. Since $\|(A - A_n)x\| = \lim_{m \to \infty} \|(A_m - A_n)x\|$, we have

$$\frac{\|(A - A_n)x\|}{\|x\|} \leq \lim_{n \to \infty} \|A_m - A_n\|$$

which implies

$$\|A - A_n\| = \sup_{x \neq 0} \frac{\|(A - A_n)x\|}{\|x\|} \leq \lim_{m \to \infty} \|A_m - A_n\|$$

which is arbitrarily small for n large enough. The triangle inequality shows that the norm of A is actually equal to C. ∎

It is important to have criteria to determine whether normed linear spaces are complete. Such a criterion is given by the following theorem (whose proof is left to Problem 3). A sequence of elements $\{x_n\}_{n=1}^{\infty}$ in a normed linear space X is called **absolutely summable** if $\sum_{n=1}^{\infty} \|x_n\| < \infty$. It is called **summable** if $\sum_{n=1}^{N} x_n$ converges as $N \to \infty$ to an $x \in X$.

Theorem III.3 A normed linear space is complete if and only if every absolutely summable sequence is summable.

For a typical application of this theorem, see the construction of quotient spaces in Section III.4. We conclude this introductory section with some definitions.

Definition A bounded linear operator from a normed linear space X to a normed linear space Y is called an **isomorphism** if it is a bijection which is continuous and which has a continuous inverse. If it is norm preserving, it is called an **isometric isomorphism** (any norm preserving map is called an **isometry**).

For example, we proved in Section II.3 that all separable, infinite-dimensional Hilbert spaces are isometric to ℓ_2. Two Banach spaces which are isometric can be regarded as the same as far as their Banach space properties are concerned.

We will often encounter a situation in which we have two different norms on a normed linear space.

Definition Two norms, $\|\cdot\|_1$ and $\|\cdot\|_2$, on a normed linear space X are called **equivalent** if there are positive constants C and C' such that, for all $x \in X$,

$$C\|x\|_1 \leq \|x\|_2 \leq C'\|x\|_1$$

For example, the following three norms on $\mathbb{R}^2$ are all equivalent:

$$\|\langle x, y\rangle\|_2 = \sqrt{|x|^2 + |y|^2}$$
$$\|\langle x, y\rangle\|_1 = |x| + |y|$$
$$\|\langle x, y\rangle\|_\infty = \max\{|x|, |y|\}$$

In fact, all norms on $\mathbb{R}^2$ are equivalent; see Problem 4. The usual situation we will encounter is an incomplete normed linear space with two norms. The completions of the space in the two norms will be isomorphic if and only if the norms are equivalent. An example is provided by the sequence spaces of Example 3. The completion of f in the $\|\cdot\|_\infty$ norm is c_0 while the completion in the $\|\cdot\|_p$ norm is ℓ_p. Two norms, $\|\cdot\|_1$ and $\|\cdot\|_2$, on a normed linear space X are equivalent if and only if the identity map is an isomorphism from $\langle X, \|\cdot\|_1\rangle$ to $\langle X, \|\cdot\|_2\rangle$.

III.2 Duals and double duals

In the last section we proved that the set of bounded linear transformations from one Banach space X to another Y was itself a Banach space. In the case where Y is the complex numbers, this space $\mathscr{L}(X, \mathbb{C})$ is denoted by X^* and called the **dual space** of X. The elements of X^* are called bounded linear functionals on X. In this chapter when we talk about convergence in X^* we always mean convergence in the norm given in Theorem III.2. If $\lambda \in X^*$, then

$$\|\lambda\| = \sup_{x \in X,\, \|x\| \le 1} |\lambda(x)|$$

In Section IV.5, we discuss another notion of convergence for X^*.

Dual spaces play an important role in mathematical physics. In many models of physical systems, whether in quantum mechanics, statistical mechanics, or quantum field theory, the possible states of the system in question can be associated with linear functionals on appropriate Banach spaces. Furthermore, linear functionals are important in the modern theory of partial differential equations. For these reasons, and because they are interesting in their own right, dual spaces have been studied extensively. There are two directions in which such study can proceed: either determining the dual spaces of particular Banach spaces or proving general theorems relating properties of Banach spaces to properties of their duals. In this section we study several examples of special interest and prove one general theorem. For an example of another general theorem see Theorem III.7.

Example 1 (L^p spaces)   Suppose that $1 < p < \infty$ and $p^{-1} + q^{-1} = 1$. If $f \in L^p(\mathbb{R})$ and $g \in L^q(\mathbb{R})$ then, according to the Hölder inequality (Theorem III.1), fg is in $L^1(\mathbb{R})$. Thus,

$$\int_{-\infty}^{\infty} \overline{g(x)} f(x)\, dx$$

makes sense. Let $g \in L^q(\mathbb{R})$ be fixed and define

$$G(f) = \int_{-\infty}^{\infty} \bar{g} f\, dx$$

for each $f \in L^p(\mathbb{R})$. The Hölder inequality shows that $G(\cdot)$ is a bounded linear functional on $L^p(\mathbb{R})$ with norm less than or equal to $\|g\|_q$; actually the norm is equal to $\|g\|_q$. The converse of this statement is also true. That is, every bounded linear functional on L^p is of the form $G(\cdot)$ for some $g \in L^q$. Furthermore, different functions in L^q give rise to different functionals on L^p. Thus, the mapping that assigns to each $g \in L^q$ the corresponding linear functional, $G(\cdot)$, on $L^p(\mathbb{R})$ is a (conjugate linear) isometric isomorphism of L^q onto $(L^p)^*$. In this sense, L^q is the dual of L^p. Since the roles of p and q in the expression $p^{-1} + q^{-1} = 1$ are symmetric, it is clear that $L^p = (L^q)^* = ((L^p)^*)^*$. That is, the dual of the dual of L^p is again L^p.

The case where $p = 1$ is different. The dual of $L^1(\mathbb{R})$ is $L^\infty(\mathbb{R})$ with the elements of $L^\infty(\mathbb{R})$ acting on functions in $L^1(\mathbb{R})$ in the natural way given by the above integral. However, the dual of $L^\infty(\mathbb{R})$ is not $L^1(\mathbb{R})$ but a much larger space (see Problems 7 and 8). As a matter of fact, we will prove later (Chapter XVI) that $L^1(\mathbb{R})$ is not the dual of any Banach space. The duality statements in this example hold for $L^p(X, d\mu)$ where $\langle X, \mu \rangle$ is a general measure space except that $L^1(X)$ may be the dual of $L^\infty(X)$ if $\langle X, \mu \rangle$ is trivially small.

Example 2 (Hilbert spaces) If we let $p = 2$ in Example 1, then $q = 2$ and we obtain the result that $L^2(\mathbb{R}) = L^2(\mathbb{R})^*$, that is, $L^2(\mathbb{R})$ is its own dual space. In fact, we have already shown (the Riesz lemma) in Section II.2 that this is true for all Hilbert spaces. The reader is cautioned again that the map which identifies $\mathscr{H}$ with its dual $\mathscr{H}^*$ is conjugate linear. If $g \in \mathscr{H}$, then the linear functional G corresponding to g is $G(f) = (g, f)$.

Example 3 ($\ell_\infty = \ell_1^*, \ell_1 = c_0^*$) Suppose that $\{\lambda_k\}_{k=1}^\infty \in \ell_1$. Then for each $\{a_k\}_{k=1}^\infty \in c_0$

$$\Lambda(\{a_k\}_{k=1}^\infty) = \sum_{k=1}^\infty \lambda_k a_k$$

converges and $\Lambda(\cdot)$ is a continuous linear functional on c_0 with norm equal to $\sum_{k=1}^\infty |\lambda_k|$. To see that all continuous linear functionals on c_0 arise in this way, we proceed as follows. Suppose $\lambda \in c_0^*$ and let e^k be the sequence in c_0 which has all its terms equal to zero except for a one in the kth place. Define $\lambda_k = \lambda(e^k)$ and let $f^\ell = \sum_{k=1}^\ell (|\lambda_k|/\lambda_k) e^k$. If some λ_k is zero, we simply omit that term from the sum. Then for each ℓ, $f^\ell \in c_0$ and $\|f^\ell\|_{c_0} = 1$. Since,

$$\lambda(f^\ell) = \sum_{k=1}^\ell |\lambda_k| \quad \text{and} \quad |\lambda(f^\ell)| \leq \|f^\ell\|_{c_0} \|\lambda\|_{c_0^*}$$

we have

$$\sum_{k=1}^\ell |\lambda_k| \leq \|\lambda\|_{c_0^*}$$

Since this is true for all ℓ, $\sum_{k=1}^\infty |\lambda_k| < \infty$ and

$$\Lambda(\{a_k\}_{k=1}^\infty) = \sum_{k=1}^\infty \lambda_k a_k$$

is a well-defined linear functional on c_0. However, $\lambda(\cdot)$ and $\Lambda(\cdot)$ agree on finite linear combinations of the e_k. Because such finite linear combinations are dense in c_0, we conclude that $\lambda = \Lambda$. Thus every functional in c_0^* arises from a sequence in ℓ_1, and the reader can check for himself that the norms in ℓ_1 and c_0^* coincide. Thus $\ell_1 = c_0^*$. A similar proof shows that $\ell_\infty = \ell_1^*$.

Since the dual X^* of a Banach space is itself a Banach space (Theorem III.2), it also has a dual space, denoted by X^{**}. X^{**} is called the **second dual**, the **bidual**, or the **double dual** of the space X. In Example 3, ℓ_1 is the first dual of c_0 and ℓ_∞ is the second dual. It is not a priori evident that X^* is always nonzero and if $X^* = \{0\}$ then $X^{**} = \{0\}$ too. However, this situation does not occur; dual spaces always have plenty of linear functionals in them. We prove this fact in the next section. Using a corollary also proven there we will prove that X can be regarded in a natural way as a subset of X^{**}.

Theorem III.4 Let X be a Banach space. For each $x \in X$, let $\tilde{x}(\cdot)$ be the linear functional on X^* which assigns to each $\lambda \in X^*$ the number $\lambda(x)$. Then the map $J: x \to \tilde{x}$ is an isometric isomorphism of X onto a (possibly proper) subspace of X^{**}.

Proof Since
$$|\tilde{x}(\lambda)| = |\lambda(x)| \leq \|\lambda\|_{X^*} \|x\|_X$$
$\tilde{x}$ is a bounded linear functional on X^* with norm $\|\tilde{x}\|_{X^{**}} \leq \|x\|_X$. It follows from Theorems III.5 and III.6 that, given x, we can find a $\lambda \in X^*$ so that
$$\|\lambda\|_{X^*} = 1 \quad \text{and} \quad \lambda(x) = \|x\|_X$$
This shows that
$$\|\tilde{x}\|_{X^{**}} = \sup_{\lambda \in X^*, \, \|\lambda\| \leq 1} |\tilde{x}(\lambda)| \geq \|x\|_X$$
which implies that
$$\|\tilde{x}\|_{X^{**}} = \|x\|_X$$
Thus, J is an isometry of X into X^{**}. ∎

Definition If the map J, defined in Theorem III.4, is surjective, then X is said to be **reflexive**.

The $L^p(\mathbb{R})$ spaces are reflexive for $1 < p < \infty$ since $(L^p)^{**} = (L^q)^* = L^p$, but $L^1(\mathbb{R})$ is not reflexive. All Hilbert spaces are reflexive. c_0 is not reflexive, since its double dual is ℓ_∞. The theory of reflexive spaces is developed further in Problems 22 and 26 of this chapter and Problem 15 of Chapter V.

III.3 The Hahn–Banach theorem

In dealing with Banach spaces, one often needs to construct linear functionals with certain properties. This is usually done in two steps: first one defines the linear functional on a subspace of the Banach space where it is easy to verify the desired properties; second, one appeals to (or proves) a general theorem which says that any such functional can be extended to the whole space while retaining the desired properties. One of the basic tools of the second step is the following theorem, whose variants will reappear in Section V.1 and Chapter XIV.

Theorem III.5 (Hahn–Banach theorem)   Let X be a real vector space, p a real-valued function defined on X satisfying $p(\alpha x + (1 - \alpha)y) \leq \alpha p(x) + (1 - \alpha)p(y)$ for all x and y in X and all $\alpha \in [0, 1]$. Suppose that λ is a linear functional defined on a subspace Y of X which satisfies $\lambda(x) \leq p(x)$ for all $x \in Y$. Then, there is a linear functional Λ, defined on X, satisfying $\Lambda(x) \leq p(x)$ for all $x \in X$, such that $\Lambda(x) = \lambda(x)$ for all $x \in Y$.

Proof   The idea of the proof is the following. First we will show that if $z \in X$ but $z \notin Y$, then we can extend λ to a functional having the right properties on the space spanned by z and Y. We then use a Zorn's lemma argument to show that this process can be continued to extend λ to the whole space X.

Let $\tilde{Y}$ denote the subspace spanned by Y and z. The extension of λ to $\tilde{Y}$, call it $\tilde{\lambda}$, is specified as soon as we define $\tilde{\lambda}(z)$ since

$$\tilde{\lambda}(az + y) = a\tilde{\lambda}(z) + \lambda(y)$$

Suppose that $y_1, y_2 \in Y$, $\alpha, \beta > 0$. Then

$$\beta\lambda(y_1) + \alpha\lambda(y_2) = \lambda(\beta y_1 + \alpha y_2) = (\alpha + \beta)\lambda\left(\frac{\beta}{\alpha + \beta} y_1 + \frac{\alpha}{\alpha + \beta} y_2\right)$$

$$\leq (\alpha + \beta)p\left(\frac{\beta}{\alpha + \beta}(y_1 - \alpha z) + \frac{\alpha}{\alpha + \beta}(y_2 + \beta z)\right)$$

$$\leq \beta p(y_1 - \alpha z) + \alpha p(y_2 + \beta z)$$

Thus, for all $\alpha, \beta > 0$ and $y_1, y_2 \in Y$,

$$\frac{1}{\alpha}[-p(y_1 - \alpha z) + \lambda(y_1)] \leq \frac{1}{\beta}[p(y_2 + \beta z) - \lambda(y_2)]$$

We can therefore find a real number a such that

$$\sup_{\substack{y \in Y \\ \alpha > 0}} \left[\frac{1}{\alpha}(-p(y - \alpha z) + \lambda(y))\right] \leq a \leq \inf_{\substack{y \in Y \\ \alpha > 0}} \left[\frac{1}{\alpha}(p(y + \alpha z) - \lambda(y))\right]$$

We now define $\tilde{\lambda}(z) = a$. It may be easily verified that the resulting extension satisfies $\tilde{\lambda}(x) \leq p(x)$ for all $x \in \tilde{Y}$. This shows that λ can be extended one dimension at a time.

We now proceed with the Zorn's lemma argument. Let $\mathscr{E}$ be the collection of extensions e of λ which satisfy $e(x) \leq p(x)$ on the subspace where they are defined. We partially order $\mathscr{E}$ by setting $e_1 \prec e_2$ if e_2 is defined on a larger set than e_1 and $e_2(x) = e_1(x)$ where they are both defined. Let $\{e_\alpha\}_{\alpha \in A}$ be a linearly ordered subset of $\mathscr{E}$; let X_α be the subspace on which e_α is defined. Define e on $\bigcup_{\alpha \in A} X_\alpha$ by setting $e(x) = e_\alpha(x)$ if $x \in X_\alpha$. Clearly $e_\alpha \prec e$ so each linearly ordered subset of $\mathscr{E}$ has an upper bound. By Zorn's lemma, $\mathscr{E}$ has a maximal element Λ, defined on some set X', satisfying $\Lambda(x) \leq p(x)$ for $x \in X'$. But, X' must be all of X, since otherwise we could extend Λ to a $\tilde{\Lambda}$ on a larger space by adding one dimension as above. Since this contradicts the maximality of Λ, we must have $X = X'$. Thus, the extension Λ is everywhere defined. ∎

In the theorem we have just proven, X is a real vector space. We now extend the theorem to the case where X is complex.

Theorem III.6 (complex Hahn–Banach theorem)   Let X be a complex vector space, p a real-valued function defined on X satisfying $p(\alpha x + \beta y) \leq |\alpha| p(x) + |\beta| p(y)$ for all $x, y \in X$, and $\alpha, \beta \in \mathbb{C}$ with $|\alpha| + |\beta| = 1$. Let λ be a complex linear functional defined on a subspace Y of X satisfying $|\lambda(x)| \leq p(x)$ for all $x \in Y$. Then, there exists a complex linear functional Λ, defined on X, satisfying $|\Lambda(x)| \leq p(x)$ for all $x \in X$ and $\Lambda(x) = \lambda(x)$ for all $x \in Y$.

Proof   Let $\ell(x) = \operatorname{Re}\{\lambda(x)\}$. ℓ is a real linear functional on Y and since

$$\ell(ix) = \operatorname{Re}\{\lambda(ix)\} = \operatorname{Re}\{i\lambda(x)\} = -\operatorname{Im}\{\lambda(x)\}$$

we see that $\lambda(x) = \ell(x) - i\ell(ix)$. Since ℓ is real linear and $p(\alpha x + (1 - \alpha)y) \leq \alpha p(x) + (1 - \alpha)p(y)$ for $\alpha \in [0, 1]$, ℓ has a real linear extension L to all of X obeying $L(x) \leq p(x)$ (by Theorem III.5). Define $\Lambda(x) = L(x) - iL(ix)$. Λ clearly extends λ and is real linear. Moreover, $\Lambda(ix) = L(ix) - iL(-x) = i\Lambda(x)$, so Λ is also complex linear. To complete the proof, we need only show that $|\Lambda(x)| \leq p(x)$. First, note that $p(\alpha x) = p(x)$ if $|\alpha| = 1$. If we let $\theta = \text{Arg}\{\Lambda(x)\}$ and use the fact that $\text{Re } \Lambda = L$, we see that

$$|\Lambda(x)| = e^{-i\theta}\Lambda(x) = \Lambda(e^{-i\theta}x) = L(e^{-i\theta}x)$$
$$\leq p(e^{-i\theta}x) = p(x) \quad \blacksquare$$

Corollary 1 Let X be a normed linear space, Y a subspace of X, and λ an element of Y^*. Then there exists a $\Lambda \in X^*$ extending λ and satisfying $\|\Lambda\|_{X^*} = \|\lambda\|_{Y^*}$.

Proof Choose $p(x) = \|\lambda\|_{Y^*}\|x\|$ and apply the above theorems. $\blacksquare$

Corollary 2 Let y be an element of a normed linear space X. Then there is a nonzero $\Lambda \in X^*$ such that $\Lambda(y) = \|\Lambda\|_{X^*}\|y\|$.

Proof Let Y be the subspace consisting of all scalar multiples of y and define $\lambda(ay) = a\|y\|$. By using Corollary 1, we can construct Λ with $\|\Lambda\| = \|\lambda\|$ extending λ to all of X. But, since $\Lambda(y) = \|y\|$, $\|\Lambda\| = 1$ and therefore

$$\Lambda(y) = \|\Lambda\|_{X^*}\|y\| \quad \blacksquare$$

Corollary 3 Let Z be a subspace of a normed linear space X and suppose that y is an element of X whose distance from Z is d. Then there exists a $\Lambda \in X^*$ so that $\|\Lambda\| \leq 1$, $\Lambda(y) = d$, and $\Lambda(z) = 0$ for all z in Z.

The proof of the third corollary is left to the reader (Problem 10). To show how useful these corollaries are we prove the following general theorem.

Theorem III.7 Let X be a Banach space. If X^* is separable, then X is separable.

Proof Let $\{\lambda_n\}$ be a dense set in X^*. Choose $x_n \in X$, $\|x_n\| = 1$, so that

$$|\lambda_n(x_n)| \geq \|\lambda_n\|/2$$

Let $\mathscr{D}$ be the set of all finite linear combinations of the $\{x_n\}$ with rational coefficients. Since $\mathscr{D}$ is countable, it is sufficient to show that $\mathscr{D}$ is dense in X. If $\mathscr{D}$ is not dense in X, then there is a $y \in X \backslash \mathscr{D}$ and a linear functional $\lambda \in X^*$ so that $\lambda(y) \neq 0$, but $\lambda(x) = 0$ for all $x \in \mathscr{D}$ (Corollary 3). Let $\{\lambda_{n_k}\}$ be a subsequence of $\{\lambda_n\}$ which converges to λ. Then

$$\|\lambda - \lambda_{n_k}\|_{X^*} \geq |(\lambda - \lambda_{n_k})(x_{n_k})|$$
$$= |\lambda_{n_k}(x_{n_k})| \geq \|\lambda_{n_k}\|/2$$

78 III: BANACH SPACES

which implies $\|\lambda_{n_k}\| \to 0$ as $k \to \infty$. Thus $\lambda = 0$ which is a contradiction. Therefore $\mathscr{D}$ is dense and X is separable. ∎

The example of ℓ_1 and ℓ_∞ shows that the converse of this theorem does not hold. Incidently, Theorem III.7 provides a proof that ℓ_1 is not the dual of ℓ_∞, since ℓ_1 is separable and ℓ_∞ is not.

III.4 Operations on Banach spaces

We have already seen several ways in which new Banach spaces can arise from old ones. The successive duals of a Banach space are Banach spaces and the bounded operators from one Banach space to another form a Banach space. Also, any closed linear subspace of a Banach space is a Banach space. There are two other ways of constructing new Banach spaces which we will need: direct sums and quotient spaces.

Let A be an index set (not necessarily countable), and suppose that for each $\alpha \in A$, X_α is a Banach space. Let

$$X = \{\{x_\alpha\}_{\alpha \in A} \mid x_\alpha \in X_\alpha, \sum_{\alpha \in A} \|x_\alpha\|_{X_\alpha} < \infty\}$$

Then X with the norm

$$\|\{x_\alpha\}\| = \sum_{\alpha \in A} \|x_\alpha\|_{X_\alpha}$$

is a Banach space. It is called the **direct sum** of the spaces X_α and is often written $X = \bigoplus_{\alpha \in A} X_\alpha$. We remark that the Hilbert space direct sum and the Banach space direct sum are not necessarily the same. For example, if we take a countable number of copies of $\mathbb{C}$, the Banach space direct sum is ℓ_1, while the Hilbert space direct sum is ℓ_2. However, if one has a *finite* number of Hilbert spaces, their Hilbert space direct sum and their Banach space direct sum are isomorphic in the sense of Section III.1.

Let M be a closed linear subspace of a Banach space X. If X were a Hilbert space, we could write $X = M \oplus M^\perp$. The Banach space that we now define can sometimes take the place of $M^\perp$ in the Banach space case where there is no orthogonality. If x and y are elements of X, we will write $x \sim y$ if $x - y \in M$. The relation $\sim$ is an equivalence relation; we denote the set of equivalence classes by X/M. As usual we denote the equivalence class containing x by $[x]$. We define addition and scalar multiplication of equivalence classes by

$$\alpha[x] + \beta[y] = [\alpha x + \beta y]$$

which makes sense since the equivalence class on the right only depends on the equivalence classes from which x and y are chosen, not on the elements themselves. With these operations X/M, becomes a complex vector space (the class M is the zero element). Now define

$$\|[x]\|_1 = \inf_{m \in M} \|x - m\|_X$$

It is not hard to show that $\|\cdot\|_1$ is a norm on X/M. $\|[x]\| = 0$ implies $[x] = 0$ because M is closed. We will show that X/M with this norm is complete by using Theorem III.3. Let $\{[x_n]\}_{n=1}^{\infty}$, be an absolutely summable sequence in X/M. That is,

$$\sum_{n=1}^{\infty} \inf_{m \in M} \|x_n - m\| < \infty$$

For each n, choose $m_n \in M$ so that

$$\|x_n - m_n\| \leq 2 \inf_{m \in M} \|x_n - m\|$$

Then $\{x_n - m_n\}$ is absolutely summable in X. Since X is complete, $\{x_n - m_n\}$ is summable. Let

$$y = \lim_{N \to \infty} \sum_{n=1}^{N} (x_n - m_n)$$

Then

$$\left\| \sum_{n=1}^{N} [x_n] - [y] \right\|_1 \leq \left\| \sum_{n=1}^{N} x_n - y - \sum_{n=1}^{N} m_n \right\| \to 0 \quad \text{as } N \to \infty$$

This proves that $\{[x_n]\}$ is summable. Using Theorem III.3 again we conclude that X/M is complete. X/M is called the **quotient space** of X by M. The reader should work out the easy details of the following example.

Example   Let $X = C[0, 1]$ and let $M = \{f \mid f(0) = 0\}$. Then $X/M = \mathbb{C}$.

III.5 The Baire category theorem and its consequences

Many questions in Banach space theory involve proving that sets have nonempty interiors. For example:

Proposition   Let X and Y be normed linear spaces. Then a linear map $T: X \to Y$ is bounded if and only if
$$T^{-1}[\{y \mid \|y\|_Y \leq 1\}]$$
has a nonempty interior.

Proof   Suppose that T is given and the set in question contains the ball
$$\{x \mid \|x - x_0\|_X < \varepsilon\}$$
Then $\|x\| < \varepsilon$ implies
$$\|Tx\| \leq \|T(x + x_0)\| + \|Tx_0\| \leq 1 + \|T(x_0)\|$$
since $x + x_0$ is in the ball of radius ε about x_0. Thus for all $x \in X$,
$$\|Tx\| \leq \varepsilon^{-1}(\|Tx_0\| + 1)\|x\|$$
so T is bounded. The converse is easy. ∎

It is thus of great interest to know when sets must have nonempty interiors. There is an extraordinary theorem about complete metric spaces. Before stating it, we make the following definition.

Definition   A set S in a metric space M is called **nowhere dense** if $\bar{S}$ has an empty interior.

Theorem III.8 (Baire category theorem)   A complete metric space is never the union of a countable number of nowhere dense sets.

Proof   The idea of the proof is simple. Suppose that M is the complete metric space and $M = \bigcup_{n=1}^{\infty} A_n$ with each A_n nowhere dense. We will construct a Cauchy sequence $\{x_m\}$ which stays away from each A_n so that its limit point x (which is in M by completeness) is in no A_n, thereby contradicting the statement $M = \bigcup_{n=1}^{\infty} A_n$.

Since A_1 is nowhere dense, we can find $x_1 \notin \bar{A}_1$. Pick an open ball B_1 about x_1 so that $B_1 \cap A_1 = \varnothing$ and so that the radius of B_1 is smaller than one. Since A_2 is nowhere dense, we can find $x_2 \in B_1 \backslash \bar{A}_2$. Let B_2 be an open ball about x_2 so that $\bar{B}_2 \subset B_1$, $B_2 \cap A_2 = \varnothing$, and with radius smaller than $\tfrac{1}{2}$. Proceeding inductively, we pick $x_n \in B_{n-1} \backslash \bar{A}_n$ and choose an open ball B_n about x_n satisfying $\bar{B}_n \subset B_{n-1}$, $B_n \cap A_n = \varnothing$, and having a radius smaller than 2^{1-n}. Now $\{x_n\}_{n=1}^{\infty}$ is a Cauchy sequence since $n, m \geq N$ implies that $x_n, x_m \in B_N$ so
$$\rho(x_n, x_m) \leq 2^{1-N} + 2^{1-N} = 2^{2-N} \to 0$$

III.5 The Baire category theorem and its consequences

as $N \to \infty$. Let $x = \lim_{n \to \infty} x_n$. Since $x_n \in B_N$ for $n \geq N$, we have
$$x \in \bar{B}_N \subset B_{N-1}$$
Thus $x \notin A_{N-1}$ for any N which contradicts $M = \bigcup_{n=1}^{\infty} A_n$. ∎

The Baire category theorem tells us that if $M = \bigcup_{n=1}^{\infty} A_n$, then some of the sets $\bar{A}_n$ must have nonempty interior. In practice, one rarely uses the Baire category theorem directly but rather one of the following consequences. The first is known as the Banach–Steinhaus theorem or the principle of uniform boundedness.

Theorem III.9 (principle of uniform boundedness) Let X be a Banach space. Let $\mathscr{F}$ be a family of bounded linear transformations from X to some normed linear space Y. Suppose that for each $x \in X$, $\{\|Tx\|_Y \mid T \in \mathscr{F}\}$ is bounded. Then $\{\|T\| \mid T \in \mathscr{F}\}$ is bounded.

Proof Let $B_n = \{x \mid \|Tx\| \leq n \text{ for all } T \in \mathscr{F}\}$. By the hypothesis each x is in some B_n, that is, $X = \bigcup_{n=1}^{\infty} B_n$. Moreover each B_n is closed (since each T is continuous). By the Baire category theorem, some B_n has a nonempty interior. By mimicking the argument in the proposition at the beginning of this section, we conclude that the $\|T\|$'s are uniformly bounded. ∎

As a typical application of this theorem we have (see also Problem 13):

Corollary Let X and Y be Banach spaces and let $B(\cdot, \cdot)$ be a separately continuous bilinear mapping from $X \times Y$ to $\mathbb{C}$, that is, for each fixed x, $B(x, \cdot)$ is a bounded linear transformation, and for each fixed y, $B(\cdot, y)$ is a bounded linear transformation. Then $B(\cdot, \cdot)$ is jointly continuous, that is, if $x_n \to 0$ and $y_n \to 0$ then $B(x_n, y_n) \to 0$.

Proof Let $T_n(y) = B(x_n, y)$. Since $B(x_n, \cdot)$ is continuous, each T_n is bounded. Since $x_n \to 0$ and $B(\cdot, y)$ is bounded, $\{\|T_n(y)\|\}$ is bounded for each fixed y. Therefore, there exists C so that
$$\|T_n(y)\| \leq C\|y\|$$
for all n. Thus
$$\|B(x_n, y_n)\| = \|T_n(y_n)\| \leq C\|y_n\| \to 0$$
as $n \to \infty$. ∎

We remark that even on $\mathbb{R}^2$, for nonlinear functions separate continuity does not imply joint continuity. The standard example is

$$f(x, y) = \frac{xy}{x^2 + y^2} \quad \text{if} \quad \langle x, y \rangle \neq \langle 0, 0 \rangle$$

$$f(0, 0) = 0$$

The second application of the Baire category theorem is to the following series of results.

Theorem III.10 (open mapping theorem) Let $T: X \to Y$ be a bounded linear transformation of one Banach space *onto* another Banach space Y. Then if M is an open set in X, $T[M]$ is open in Y.

Proof We make a series of remarks which will simplify the proof. We need only show that, for every neighborhood N of x, $T[N]$ is a neighborhood of $T(x)$. Since $T[x + N] = T(x) + T[N]$ we need only show this for $x = 0$. Since neighborhoods contain balls it is sufficient to show that $T[B_r^X] \supset B_{r'}^Y$ for some r' where

$$B_r^X = \{x \in X \mid \|x\| < r\}$$

However, since $T[B_r^X] = rT[B_1^X]$, we need only show that $T[B_r^X]$ is a neighborhood of zero for some r. Finally, by the "translation argument" of the proposition, it is sufficient to show that $\overline{T[B_r^X]}$ has a nonempty interior for some r.

Since T is onto,

$$Y = \bigcup_{n=1}^{\infty} T[B_n]$$

so some $\overline{T(B_n)}$ has a nonempty interior. Now the hard work begins, since we want $T(B_n)$ to have a nonempty interior. By scaling and translating we can suppose that B_ε is contained in $\overline{T[B_1]}$; we will show that $\overline{T[B_1]} \subset T[B_2]$ which will complete the proof.

Let $y \in \overline{T[B_1]}$. Pick $x_1 \in B_1$ so $y - Tx_1 \in B_{\varepsilon/2} \subset \overline{T[B_{1/2}]}$. Now pick $x_2 \in B_{1/2}$ so that

$$y - Tx_1 - Tx_2 \in B_{\varepsilon/4}$$

By induction, we choose $x_n \in B_{2^{1-n}}$ so that

$$y - \sum_{j=1}^{n} Tx_j \in B_{\varepsilon 2^{1-n}}$$

Then $x = \sum_{i=1}^{\infty} x_i$ exists and is in B_2 and

$$y = \sum_{j=1}^{\infty} Tx_i = Tx$$

Thus $y \in T[B_2]$. ∎

Theorem III.11 (inverse mapping theorem) A continuous bijection of one Banach space onto another has a continuous inverse.

Proof T is open so T^{-1} is continuous. ∎

For an application of this result see Problem 19.

Definition Let T be a mapping of a normed linear space X into a normed linear space Y. The **graph** of T, denoted by $\Gamma(T)$, is defined as

$$\Gamma(T) = \{\langle x, y \rangle \mid \langle x, y \rangle \in X \times Y, \ y = Tx\}$$

Theorem III.12 (closed graph theorem) Let X and Y be Banach spaces and T a linear map of X into Y. Then T is bounded if and only if the graph of T is closed.

Proof Suppose that $\Gamma(T)$ is closed. Then, since T is linear, $\Gamma(T)$ is a subspace of the Banach space $X \oplus Y$. By assumption $\Gamma(T)$ is closed and thus is a Banach space in the norm

$$\|\langle x, Tx \rangle\| = \|x\| + \|Tx\|$$

Consider the continuous maps Π_1, Π_2,

$$\Pi_1 : \langle x, Tx \rangle \to x, \qquad \Pi_2 : \langle x, Tx \rangle \to Tx$$

Π_1 is a bijection so by the inverse mapping theorem Π_1^{-1} is continuous. But $T = \Pi_2 \circ \Pi_1^{-1}$, so T is continuous. The converse is trivial. ∎

To avoid future confusion, we emphasize that the T in this theorem is implicitly assumed to be defined on all of X. We will later deal with transformations defined on *algebraic* subspaces of X (not all of X) with closed graphs which are not continuous. To appreciate what the closed graph theorem really does, consider the three statements:

(a) x_n converges to some element x.
(b) Tx_n converges to some element y.
(c) $Tx = y$.

A priori, to prove that T is continuous one must show that (a) implies (b) and (c). What the closed graph theorem says is that it is sufficient to prove that (a) and (b) imply (c).

The following corollary of the closed graph theorem has important consequences in mathematical physics.

Corollary (the Hellinger–Toeplitz theorem)   Let A be an everywhere-defined linear operator on a Hilbert space $\mathcal{H}$ with $(x, Ay) = (Ax, y)$ for all x and y in $\mathcal{H}$. Then A is bounded.

Proof   We will prove that $\Gamma(A)$ is closed. Suppose that $\langle x_n, Ax_n \rangle \to \langle x, y \rangle$. We need only prove that $\langle x, y \rangle \in \Gamma(A)$, that is, that $y = Ax$. But, for any $z \in \mathcal{H}$,

$$(z, y) = \lim_{n \to \infty} (z, Ax_n) = \lim_{n \to \infty} (Az, x_n)$$
$$= (Az, x) = (z, Ax)$$

Thus $y = Ax$ and $\Gamma(A)$ is closed. ∎

As we shall see, this theorem is the cause of much technical pain because in quantum mechanics there are operators (like the energy) which are unbounded but which we want to obey

$$(x, Ay) = (Ax, y)$$

in some sense. The Hellinger–Toeplitz theorem tells us that such operators cannot be everywhere defined. Thus such operators are defined on subspaces $D(A)$ of $\mathcal{H}$ and defining what one means by $A + B$ or AB may be difficult. For example, $A + B$ is a priori only defined on $D(A) \cap D(B)$ which may equal $\{0\}$ even in the case where both $D(A)$ and $D(B)$ are dense. We return to these questions in Chapters VIII and X.

NOTES

Section III.1   The name Banach space honors the important work of S. Banach on normed linear spaces during the 1920's culminating in his book, *Théorie des Opérations Linéaires*, Monografie Math., I, Warsaw, 1932. A good elementary reference for the material in this chapter is the book *Foundations of Modern Analysis* by A. Friedman, Holt, New York, 1970. In the second supplement we prove Hölder's inequality only in the case $r = 1$. To prove the general case where $p^{-1} + q^{-1} = r^{-1}$ observe that

$$|fg|^r = |f|^r |g|^r$$

and use the Hölder inequality for the special case where

$$\frac{1}{(p/r)} + \frac{1}{(q/r)} = 1$$

obtaining

$$\int |fg|^r \leq \left(\int |f|^{rp/r}\right)^{r/p} \left(\int |g|^{rq/r}\right)^{r/q}$$

or

$$\left(\int |fg|^r\right)^{1/r} \leq \left(\int |f|^p\right)^{1/p} \left(\int |g|^q\right)^{1/q}$$

Suppose X is a Banach space. One of the ways of studying the Banach space of operators from X to itself, $\mathscr{L}(X, X)$, is to use the fact that it is also an algebra. Thus, one can use algebraic notions like ideals and commutators to investigate the structure of $\mathscr{L}(X, X)$. In Section VI.6 certain important ideals of $\mathscr{L}(\mathscr{H}, \mathscr{H})$, where $\mathscr{H}$ is a separable Hilbert space, are studied. The general theory of operator algebras is studied in later volumes.

Section III.2 The proof that $(L^p)^* = L^q$ may be found in Royden's book (see the Notes for Chapter I) or may be proven using the notion of uniformly convex space (see Problems 25 and 26 of Chapter III and Problem 15 of Chapter V). In Section VI.6 we discuss the duals of several subalgebras of $\mathscr{L}(\mathscr{H}, \mathscr{H})$.

Section III.3 The Hahn–Banach theorem dates back to the work of Helly in " Über Lineare Funktional Operationen," *Sitzgsber, Akad. Wiss. Wien Math-Nat. Kl.* **121 IIa**, (1912), 265–297, and " Über Systeme linearer Gleichungen mit Unendlich Vielen Unbekannten," *Monatsh. Math. Phys.* **31**, (1921), 60–91. The modern version is due to H. Hahn, " Über lineare Gleichungssystem in linearen Räumen," *J. Reine Angew. Math.* **157**, (1926), 214–229, and S. Banach, "Sur les fonctionelles linéaires, I, II," *Studia Math.* **1** (1929), 211–216, 223–239. A nice example of the concrete applications of the Hahn–Banach theorem may be found in the book by Friedman mentioned above. There it is shown how to use the Hahn–Banach theorem to prove the existence of a Green's function for the Dirichlet problem in two dimensions.

Section III.5 The Baire theorem was proven in R. Baire, "Sur les fonctions de variables réelles," *Annali di Mat. Ser. 3* **3** (1899), 1–123. The general case is in C. Kuratowski, "La propriété de Baire dans les espaces métriques," *Fund. Math.* **16** (1930), 390–394, and S. Banach, "Théorèmes sur les ensembles de premières catégorie," *Fund. Math.* **16**, (1930), 395–398. The Banach–Steinhaus theorem was proven by S. Banach and H. Steinhaus in "Sur le principe de la condensation de singularités." *Fund. Math.* **9**, (1927), 50–61. There is a discussion of the Baire theorem and its consequences in Lorch's book (see the Notes for Section II.5). The term category comes from the following: A countable union of nowhere dense sets is called a **first category** set. All other sets are called **second category**. The Baire theorem says that any complete metric space is second category.

Complements of first category sets are often called residuals. A residual set is thus a set containing a countable intersection of dense open sets. The Baire theorem implies that any residual in a complete metric space is dense (Problem 21).

In metric spaces, one sometimes says something is "true almost everywhere" if it is true on a residual; thereby, first category sets play the role of "sets of measure zero". There are some amusing results on this notion of a.e. in G. Choquet's book, *Lectures in Analysis*, Vol. I, pp. 120–126, Benjamin, New York, 1969. Warning: There exist sets $X \subset [0, 1]$ which are first category with measure 1! Thus the two notions of a.e., Lebesgue and Baire, are quite different.

Other topological spaces besides complete metric spaces have the property that residuals are dense; such spaces are called Baire spaces. For example, every locally compact space is Baire. For additional discussion, see Choquet's book, pp. 105–120.

PROBLEMS

†*1.* Prove that $L^\infty(\mathbb{R})$ is a Banach space.

†*2.* (a) Prove that ℓ_p and c_0 are separable but ℓ_∞ is not.
  (b) Prove that $s \subset \ell_p$ for all p.

†*3.* Prove that a normed linear space is complete if and only if every absolutely summable sequence is summable. (Hint for the "if" part: To show that a Cauchy sequence converges it is only necessary to show that a subsequence converges.)

**4.* Prove that all norms on $\mathbb{R}^n$ are equivalent. (Hint: Use the fact that the unit sphere is compact in the Euclidean topology.)

†*5.* Prove that $C_\infty(\mathbb{R})$ is the completion of $\kappa(\mathbb{R})$.

6. Prove that if $\{\lambda_k\}_{k=1}^\infty \in \ell_1$ then the linear functional on c_0 given by
$$\Lambda(\{a_k\}_{k=1}^\infty) = \sum \lambda_k a_k$$
has norm $\sum_{k=1}^\infty |\lambda_k|$.

7. Prove that $\ell_\infty = \ell_1^*$ but that $\ell_\infty^* \neq \ell_1$ by using the Hahn–Banach theorem.

8. (a) Prove that there is a nonzero bounded linear functional on $L^\infty(\mathbb{R})$ which vanishes on $C(\mathbb{R})$.
  (b) Prove that there is a bounded linear functional λ on $L^\infty(\mathbb{R})$ such that $\lambda(f) = f(0)$ for each $f \in C(\mathbb{R})$.

9. Suppose that $\mathcal{H}$ is a Hilbert space and that λ is a bounded linear functional on $\mathcal{M}$, a not necessarily closed subspace. Describe the continuous extensions of λ.

†*10.* Prove the third corollary to the Hahn–Banach theorem.

11. Prove that there is a linear functional λ on $\ell_\infty(\mathbb{R})$ so that
$$\varliminf_{n \to \infty} a_n \leq \lambda(\{a_n\}_{n=1}^\infty) \leq \varlimsup_{n \to \infty} a_n$$
$\ell_\infty(\mathbb{R}) = \{a \mid a \in \ell_\infty, \quad a_n \in \mathbb{R} \quad \text{for all } n\}$.

†*12.* Prove the statement in the example at the end of Section 4.

13. Use the uniform boundedness principle to provide an alternative proof of the Hellinger–Toeplitz theorem.

**14.* Let X be a Banach space. Give an example of an everywhere-defined but discontinuous linear functional λ. Show directly that λ is not closed.

15. Let $\mathscr{H}$ be a separable Hilbert space with an orthonormal basis $\{x_n\}_{n=1}^{\infty}$. Let $\{y_n\}$ be a sequence of elements of $\mathscr{H}$ and prove that the following two statements are equivalent.
  (a) $(x, y_n) \underset{n \to \infty}{\to} 0, \forall x \in \mathscr{H}$.
  (b) $(x_m, y_n) \underset{n \to \infty}{\to} 0$, for each $m = 1, 2, \ldots$, and $\{\|y_n\|\}_{n=1}^{\infty}$ is bounded.

16. A subset S of a Banach space is called **weakly bounded** if and only if for all $\lambda \in X^*$, $\sup_{x \in S} |\lambda(x)| < \infty$. S is called **strongly bounded** if and only if $\sup_{x \in S} \|x\| < \infty$. Prove that a set is strongly bounded if and only if it is weakly bounded (see Section V.7).

17. Prove that a separately continuous multilinear functional on a Banach space is jointly continuous.

18. Extend the Hellinger–Toeplitz theorem to include pairs of operators A, B satisfying: $(Ax, y) = (x, By)$.

19. Let X be a Banach space in either of the norms $\|\cdot\|_1$ or $\|\cdot\|_2$. Suppose that $\|\cdot\|_1 \leq C \|\cdot\|_2$ for some C. Prove that there is a D with $\|\cdot\|_2 \leq D \|\cdot\|_1$.

20. Why doesn't a one-point space violate the Baire theorem?

**21.* Prove that any countable intersection of dense open sets in a complete metric space is dense.

22. (a) Prove that a Banach space X is reflexive if and only if X^* is reflexive. (Hint: If $X \neq X^{**}$ find a bounded linear functional on X^{**} which vanishes on X).
  (b) Prove that whenever X is a nonreflexive Banach space, $(\cdots(X^*)^*\cdots)^*$ is not reflexive.

23. Let X be a Hilbert space and let $\mathscr{M}$ be a closed subspace. Show that the restriction of the natural map $\pi : X \to X/\mathscr{M}$ is an isomorphism of $\mathscr{M}^{\perp}$ and $X/\mathscr{M}$.

24. Let ℓ be a linear functional on a real Banach space X. Prove that $X/\ker \ell$ is isomorphic to $\mathbb{R}$ with the usual norm and that the natural projection $\pi : X \to X/\ker \ell = \mathbb{R}$ is related to ℓ by $\ell = \pm \|\ell\| \pi$.

25. A Banach space is called **uniformly convex** if for each $\varepsilon > 0$, there is a $\delta > 0$, so that $\|x\| = \|y\| = 1$ and $\|\tfrac{1}{2}(x+y)\| > 1 - \delta$ imply $\|x - y\| < \varepsilon$; thus the unit ball is uniformly round. We will see in Problem 15 of Chapter V, that every uniformly convex space is reflexive.

(a) Prove directly that $L^1(\mathbb{R})$ and $L^\infty(\mathbb{R})$ are not uniformly convex.
(b) Prove that any Hilbert space is uniformly convex.
*(c) Prove that $L^p(X, d\mu)$ is uniformly convex for $p \geq 2$. Hint: Prove that for $\alpha, \beta \in \mathbb{C}$, one has $|\alpha + \beta|^p + |\alpha - \beta|^p \leq 2^{p-1}(|\alpha|^p + |\beta|^p)$ by first proving

$$(|\alpha + \beta|^p + |\alpha - \beta|^p)^{1/p} \leq \sqrt{2}(|\alpha|^2 + |\beta|^2)^{1/2}$$

Notes: 1. L^p is actually uniformly convex for all $1 < p < \infty$, but the proof for $1 < p < 2$ is harder; c.f. G. Köthe: *Topological Vector Spaces, I*, Springer (1969), 358–359.
2. Uniformly convex spaces were introduced by J. Clarkson, "Uniformly convex spaces," *Trans. A.M.S.* **40** (1936), 396–414.
3. M. Day has given examples of reflexive Banach spaces which are not uniformly convex in "Reflexive Banach spaces not isomorphic to uniformly convex spaces," *Bull. A.M.S.* **47** (1941), 313–317; see also Köthe, pp. 360–363.

26. (a) A pair of Banach spaces, X and Y, are said to be in **strict duality** if there is a map $f: X \to Y^*$ which is isometric, so that the induced map $f^*: Y \to X^*$ is also isometric. Prove that if X and Y are in strict duality and X is reflexive, then $Y = X^*$ and $X = Y^*$. (Hint: Use the Hahn–Banach theorem.)
(b) Prove that $L^p(X, d\mu)$ and $L^q(X, d\mu)$ are in strict duality if $p^{-1} + q^{-1} = 1$.
(c) Prove that $L^p(X, d\mu)^* = L^q(X, d\mu)$ if $1 < p < \infty$ and $p^{-1} + q^{-1} = 1$. (Hint: Use Problem 25 and Problem 15 of Chapter V).

*27. Prove the Banach–Schauder theorem: Let T be a continuous linear map, $T: E \to F$, where E and F are Banach spaces. Then either $T[A]$ is open in $T[E]$ for each open $A \subset E$, or $T[E]$ is of first category in $\overline{T[E]}$ (see the notes to Section 5 for the definition of first category).

28. (a) Prove that every quotient of ℓ_2 by a closed subspace is isometrically isomorphic to either ℓ_2 or $\mathbb{C}^N$ for some N.
(b) Prove that ℓ_1 is not topologically isomorphic to any quotient space of ℓ_2.

29. Let X be a separable Banach space. Let $\{x_1, \ldots, x_n, \ldots\}$ be a dense subset of the unit ball in X. Map $\ell_1 \to X$ by

$$A: \langle \alpha_1, \ldots, \alpha_n, \ldots \rangle \to \sum_{n=1}^\infty \alpha_n x_n$$

(a) Prove that A is well defined and continuous.
(b) Prove that Ker A is closed and that A "lifts" to a continuous map $\hat{A}: \ell_1/\text{Ker } A \to X$.
(c) Prove Ran $\hat{A} = $ Ran A is all of X. Hint: Given x with $\|x\| = 1$, choose $x_{n(i)}$ recursively by requiring

$$\left\| x - \sum_{i=1}^k 2^{-i+1} x_{n(i)} \right\| \leq 2^{-k}$$

(d) Conclude that any separable Banach space is topologically isomorphic to some quotient space of ℓ_1.
(e) By using (c) with 2 replaced by 3, 4, ..., show that $\hat{A}$ is actually an isometry.

30. Let X be a Banach space and let Y be a closed subspace of X. Let Y^o in X^* be defined by
$$Y^o = \{\ell \in X^* \,|\, \ell \upharpoonright Y = 0\}$$
Given a bounded linear functional f on X/Y, define $\pi^*(f) \in X^*$ by $[\pi^*(f)](x) = f([x])$. Prove that π^* is an isometric isomorphism of $(X/Y)^*$ onto Y^o.

31. (a) Let E be a Banach space with separable dual and $\langle M, \mu \rangle$ a measure space with $L^p(M, d\mu)$ separable for all $1 < p < \infty$. Develop the theory of $L^p(M, d\mu; E)$ analogous to the theory of $L^2(M, d\mu; \mathcal{H})$ discussed in Sections II.1 and II.4.
   (b) Prove $L^p(M \times N, d\mu \otimes d\nu)$ and $L^p(M, d\mu; L^p(N, d\nu))$ are naturally isomorphic.
*(c) Let E^{**} be a separable Banach space and let $1 \leq p < \infty$. Prove that $L^p(M, d\mu, E)^*$ is naturally isometrically isomorphic to $L^q(M, d\mu, E^*)$ (Hint: First show that it is enough to prove that every bounded linear transformation T of E into $L^q(M, d\mu)$ is of the form $[T(x)](m) = [f(m)](x)$ for some $f \in L^q(M, d\mu; E^*)$. Prove this in the special case where $E = \ell_1$. Finally use Problems 29 and 30 to treat the general separable Banach space, E.)

*32. Let S be a closed linear subspace of $L^1[0, 1]$. Suppose that $f \in S$ implies that $f \in L^p[0, 1]$ for some $p > 1$. Prove that $S \subset L^p[0, 1]$ for some $p > 1$.

IV: Topological Spaces

Everyone knows what a curve is, until he has studied enough mathematics to become confused through the countless number of possible exceptions.

F. Klein

IV.1 General notions

The abstract notions of limit and convergence are the bread and butter of functional analysis. The purely metric space formulations that we have used thus far are sadly lacking in some cases, so it is necessary to introduce more general concepts. It is possible to describe what is known as a topological space purely in terms of convergence, but it is very awkward. Instead, one usually defines a topological space by abstracting the notion of open sets in metric spaces. Convergence then becomes a derived concept. We discuss convergence in Section IV.2.

This section consists primarily of definitions as we introduce an extensive language needed to describe topological notions. We urge the reader to learn the language by returning to this section when necessary rather than by brute memorization.

Definition A **topological space** is a set S with a distinguished family of subsets $\mathcal{T}$ called **open sets** with the properties:

(i) $\mathcal{T}$ is closed under finite intersections, that is, if $A, B \in \mathcal{T}$, then $A \cap B \in \mathcal{T}$.

(ii) $\mathcal{T}$ is closed under *arbitrary* unions, that is, if $A_\alpha \in \mathcal{T}$ for all α in some index set I, then $\bigcup_{\alpha \in I} A_\alpha \in \mathcal{T}$.
(iii) $\emptyset \in \mathcal{T}$ and $S \in \mathcal{T}$.

$\mathcal{T}$ is called a **topology** for S. We will occasionally write $\langle S, \mathcal{T} \rangle$ for a topological space.

In contradistinction to Borel structure, topological structures are *not* symmetric between intersection and union and involve not merely countable operations but arbitrary operations.

The prime example of a topological space is a metric space. The open sets, $\mathcal{T}$, are those sets, $M \subset S$, with the property $(\forall x \in M)(\exists r > 0)$ $\{y \mid \rho(x, y) < r\} \subset M$. After discussing continuous functions, we will describe another family of examples. We first mention, however, two trivial examples: Given a set S, the family of all subsets of S is a topology; it is called the discrete topology. $\mathcal{T} = \{\emptyset, S\}$ is also a topology; it is called the indiscrete topology.

The family of all topologies on a set S is ordered in a natural way $\mathcal{T}_1 \prec \mathcal{T}_2$ if $\mathcal{T}_1 \subset \mathcal{T}_2$ in the sense of set-theoretic inclusion. If $\mathcal{T}_1 \prec \mathcal{T}_2$, we say $\mathcal{T}_1$ is a **weaker** topology than $\mathcal{T}_2$. (The term weaker comes from the fact that more sequences converge in $\mathcal{T}_1$ than in $\mathcal{T}_2$; so $\mathcal{T}_1$ convergence is a weaker notion than $\mathcal{T}_2$ convergence.)

Definition A family $\mathcal{B} \subset \mathcal{T}$ is called a **base** if and only if any $T \in \mathcal{T}$ is of the form $T = \bigcup_\alpha B_\alpha$ for some family $\{B_\alpha\} \subset \mathcal{B}$.

For example, the balls in a metric space are always a base. We now take a whole family of definitions directly from metric spaces:

Definition A set N is called a **neighborhood** of a point $x \in S$, a topological space, if there exists an open set U with $x \in U \subset N$.

A family $\mathcal{N}$ of subsets of S, a topological space, is called a **neighborhood base at x** if each $N \in \mathcal{N}$ is a neighborhood of x and if given any neighborhood M of x, there is an $N \in \mathcal{N}$ with $N \subset M$. Equivalently, $\mathcal{N}$ is a neighborhood base at x if and only if $\{M \mid N \subset M \text{ for some } N \in \mathcal{N}\}$ is the family of all neighborhoods of x. For example, if $\mathcal{B}$ is a base for $\mathcal{T}$, $\{N \in \mathcal{B} \mid x \in N\}$ is a neighborhood base at x. We emphasize that neighborhoods need not be open. In a metric space, the *closed* balls of radius greater than zero are a neighborhood base.

IV: TOPOLOGICAL SPACES

Definition   A set $C \subset S$, a topological space, is called **closed** if and only if it is the complement of an open set.

The properties of the family of all closed sets can be read off from the properties of $\mathcal{T}$.

Definition   Let $A \subset S$, a topological space. The **closure of** A, $\bar{A}$, is the smallest closed set containing A. The **interior of** A, $A°$, is the largest open set contained in A. The **boundary** of A is the set $\bar{A}\backslash A° \equiv \bar{A} \cap \overline{[S\backslash A]}$.

That a smallest closed set containing A exists follows from the fact that $\mathcal{T}$ is closed under arbitrary unions.

As examples, we consider several topologies on $\mathbb{R}^2$:

Example 1   The ordinary metric topology.

Example 2   Consider the family of sets of the form $\{\langle x, y\rangle \mid x \in O\}$ where y is *fixed* and O is an open set of $\mathbb{R}$ in the usual topology. This family of sets is the base for a topology whose open sets are the sets C such that for each $y \in \mathbb{R}$, $\{x \mid \langle x, y\rangle \in C\}$ is open in $\mathbb{R}$ in the usual topology. In an intuitive sense, which we shall shortly make precise, this topology is the "product" of the usual topology in one factor and the discrete topology in the other factor.

Example 3   Let $\mathcal{T}$ consist of the empty set and all sets containing $\langle 0, 0\rangle$. A neighborhood base for $\langle x, y\rangle$ in this topology is the single (!) set $\{\langle 0, 0\rangle, \langle x, y\rangle\}$.

Our experience with metric spaces suggests that continuous functions will play a major role.

Definition   Let $\langle S, \mathcal{T}\rangle$ and $\langle T, \mathcal{U}\rangle$ be two topological spaces. A function $f: S \to T$ is called **continuous** if $f^{-1}[A] \in \mathcal{T}$ for every $A \in \mathcal{U}$; that is, if the inverse image of any open set is open. f is called **open** if $f[B]$ is open for each $B \in \mathcal{T}$. If f is open and continuous, it is called **bicontinuous**. A bicontinuous bijection is called a **homeomorphism**.

Homeomorphisms are the "isomorphisms" of topological spaces. A topological notion is some notion (or object) invariant under homeomorphism. As an example, the intervals $(-\infty, \infty)$ and $(-1, 1)$ are homeomorphic under the homeomorphism $x \mapsto x/(1 + x^2)$. They are not isometric in the usual metric; in fact, only one of them is complete. This demonstrates that completeness is not a topological notion. However, most metric space notions that are useful in analysis are topological notions.

Continuity is often used to define topologies:

Definition Let $\mathcal{K}$ be a family of functions from a set S to a topological space $\langle T, \mathcal{U} \rangle$. The $\mathcal{K}$-**weak** (or simply **weak**) topology on S is the weakest topology for which all the functions $f \in \mathcal{K}$ are continuous.

To construct the $\mathcal{K}$-weak topology, take the family of all finite intersections of sets of the form $f^{-1}[U]$ where $f \in \mathcal{K}$ and $U \in \mathcal{U}$. These sets form a base for the $\mathcal{K}$-weak topology. If $\mathcal{K}$ is a family of functions on a set S but with values in different topological spaces, we define the $\mathcal{K}$-weak topology in the obvious way.

Example 4 Consider $C[a, b]$, the continuous functions on $[a, b]$. The **topology of pointwise convergence** on $C[a, b]$ is the weak topology given by the family of functions $f \mapsto f(x)$. That is, for each $x \in [a, b]$, let $E_x(f) = f(x)$ so the $E_x(\cdot)$ are maps of $C[a, b]$ to $\mathbb{R}$. As we will see, the topology of pointwise convergence is the topology on $C[a, b]$ for which $f_n \to f$ if and only if $f_n(x) \to f(x)$ for each x.

Example 5 Let $\mathcal{H}$ be a Hilbert space. The "weak topology" is the weakest topology making $\varphi \mapsto (\psi, \varphi)_{\mathcal{H}}$ continuous for each ψ in $\mathcal{H}$. A neighborhood base for 0 is given explicitly by the sets

$$N(\psi_1, \ldots, \psi_n; \varepsilon_1, \ldots, \varepsilon_n) = \{\varphi \,|\, |(\psi_i, \varphi)| < \varepsilon_i, \quad i = 1, \ldots, n\}$$

where $\varepsilon_i > 0$, $\psi_1, \ldots, \psi_n$ are arbitrary, and $n = 1, 2, \ldots$. Thus, the neighborhoods in the weak topology are cylinders in all but finitely many dimensions. That is, there is a subspace M (the orthogonal complement of $\psi_1, \ldots, \psi_n$) whose complement, $M^\perp$, is finite dimensional and so that $\varphi \in N, \eta \in M$ implies $\varphi + \eta \in N$.

Example 6 On $\mathbb{R}^2$ consider the maps π_1, π_2 given by $\pi_1(x, y) = x$; $\pi_2(x, y) = y$. The weak topology defined by π_1 and π_2 and the usual topology on $\mathbb{R}$ has rectangles $(a, b) \times (c, d)$ as a base for its open sets and thus the weak topology is the "usual" topology on $\mathbb{R}^2$.

Example 7 The weak topology can be used to topologize Cartesian products. Recall if $\{S_\alpha\}_{\alpha \in I}$ is a family of sets, $S = \bigtimes_{\alpha \in I} S_\alpha$ is the family of all $\{x_\alpha\}_{\alpha \in I}$ with $x_\alpha \in S_\alpha$. For each α, we have a map $\pi_\alpha: S \to S_\alpha$ given by $\pi_\alpha(\{x_\beta\}_{\beta \in I}) = x_\alpha$. If each S_α has a topology $\mathcal{T}_\alpha$, we define the **product topology**, $\bigtimes_{\alpha \in I} \mathcal{T}_\alpha$ as the weak topology generated by the projections π_α.

We now return to our listing of definitions by classifying spaces by how well open sets separate points and closed sets:

Definition
(a) A topological space is called a T_1 **space** if and only if for all x and y, $x \neq y$, there is an open set O with $y \in O$, $x \notin O$. Equivalently, a space is T_1 if and only if $\{x\}$ is closed for each x.
(b) A topological space is called **Hausdorff** (or T_2) if and only if for all x and y, $x \neq y$, there are open sets O_1, O_2 such that $x \in O_1$, $y \in O_2$, and $O_1 \cap O_2 = \emptyset$.
(c) A topological space is called **regular** (or T_3) if and only if it is T_1 and for all x and C, closed, with $x \notin C$, there are open sets O_1, O_2 such that $x_1 \in O_1$, $C \subset O_2$, and $O_1 \cap O_2 = \emptyset$. Equivalently, a space is T_3 if the closed neighborhoods of any point are a neighborhood base.
(d) A topological space is called **normal** (or T_4) if and only if it is T_1 and for all C_1, C_2, closed, with $C_1 \cap C_2 = \emptyset$, there are open sets O_1, O_2 with $C_1 \subset O_1$, $C_2 \subset O_2$, and $O_1 \cap O_2 = \emptyset$.

Obviously:

Proposition $\quad T_4 \Rightarrow T_3 \Rightarrow T_2 \Rightarrow T_1$

We remark that the two most important notions are Hausdorff and normal. At this time, we avoid discussing another way of separating sets, namely with continuous functions. Urysohn's lemma (Theorem IV.7) deals with this question.

We next consider various countability criteria:

Definition
(i) A topological space S is called **separable** if and only if it has a countable dense set.
(ii) A topological space S is called **first countable** if and only if each point $x \in S$ has a countable neighborhood base.
(iii) A topological space S is called **second countable** if and only if S has a countable base.

The relation between these topological notions and metric spaces is set forward in the elementary:

Proposition (a) Every metric space is first countable.
(b) A metric space is second countable if and only if it is separable.
(c) Any second countable topological space is separable.

Warning There are separable spaces that are not second countable (see Problem 7). To add to the confusion, some authors use "separable" to mean second countable. *By separable we always mean that there exists a countable dense set.*

The geometric idea of being connected has a topological formulation:

Definition A topological space S is called **disconnected** if and only if it contains a nonempty proper subset, C, which is both open and closed; equivalently, S is disconnected if and only if it can be written as the union of two disjoint nonempty closed sets. If S is not disconnected, it is called **connected**.

We examine connectivity in Problems 3 and 6. As a final topological notion, we consider restricting topologies to subsets.

Definition Let $\langle S, \mathcal{T} \rangle$ be a topological space and let $A \subset S$. The **relative topology** on A is the family of sets $\mathcal{T}_A = \{O \cap A \mid O \in \mathcal{T}\}$. A subset $B \subset A$ is called **relatively open** if $B \in \mathcal{T}_A$ and **relatively closed** if $A \backslash B \in \mathcal{T}_A$.

IV.2 Nets and convergence

In this section we introduce new objects, called nets, in order to handle limit operations in general topological spaces. Although nets seem on first acquaintance to be bizarre, the propositions in this section show how natural they are.

Definition A **directed system** is an index set I together with an ordering $\prec$ which satisfies:

(i) If $\alpha, \beta \in I$, then there exists $\gamma \in I$ so that $\gamma \succ \alpha$ and $\gamma \succ \beta$.
(ii) $\prec$ is a partial ordering.

96   IV: TOPOLOGICAL SPACES

Definition   A **net** in a topological space S is a mapping from a directed system I to S; we denote it by $\{x_\alpha\}_{\alpha \in I}$.

If we choose the positive integers with the usual order as a directed system, the nets on that directed system are just sequences in S, so nets are a generalization of the notion of sequence. If $P(\alpha)$ is a proposition depending on an index α in a directed set I we say $P(\alpha)$ is **eventually true** if there is a β in I with $P(\alpha)$ true if $\alpha \succ \beta$. We say $P(\alpha)$ is **frequently true** if it is not eventually false, that is, if for any β there is an $\alpha \succ \beta$ with $P(\alpha)$ true.

Definition   A net $\{x_\alpha\}_{\alpha \in I}$ in a topological space S is said to **converge** to a point $x \in S$ (written $x_\alpha \to x$) if for any neighborhood N of x, there is a $\beta \in I$ so that $x_\alpha \in N$ if $\alpha \succ \beta$.

Thus $x_\alpha \to x$ if and only if x_α is eventually in any neighborhood of x. If x_α is frequently in any neighborhood of x, we say that x is a **cluster point** of $\{x_\alpha\}$. Notice that the notions of limit and cluster point generalize the same notions for sequences in a metric space.

Theorem IV.1   Let A be a set in a topological space S. Then, a point x is in the closure of A if and only if there is a net $\{x_\alpha\}_{\alpha \in I}$ with $x_\alpha \in A$, so that $x_\alpha \to x$.

Proof   We first observe that $\bar{A}$ is just the set of points x such that any neighborhood of x contains a point of A. This set certainly contains A and its complement is the largest open set not containing any points of A. Now suppose $x_\alpha \to x$ where each $x_\alpha \in A$. Then any neighborhood of x contains some x_α and hence some points of A, that is, x is a limit point of A, so $x \in \bar{A}$.

Conversely, suppose $x \in \bar{A}$. Let I be the collection of neighborhoods of x with the ordering $N_1 \prec N_2$ if $N_2 \subset N_1$. For each $N \in I$, let x_N be a point in $A \cap N$. Then $\{x_N\}_{N \in I}$ is a net and $x_N \to x$. ∎

In spaces that are first countable, we can construct the closures of sets by using only sequences. Such is the case in metric spaces. The following example is a case where sequences are not enough:

Example   Let $S = [0, 1]$; the nonempty open sets will be the subsets of $[0, 1]$ whose complements contain at most a countable infinity of points. Let $A = [0, 1)$. Then $\bar{A} = S$ since $\{1\}$ is not open. But, let $\{x_n\}_{n=1}^\infty$ be any sequence of points of $[0, 1)$. $\{x_n\}_{n=1}^\infty$ cannot converge to 1 since the complement of the points $\{x_n\}_{n=1}^\infty$ is an open set containing 1.

Although the above example seems artificial, spaces that are not first countable play a large role in functional analysis. Usually, they arise when dual spaces of Banach spaces are considered with topologies weaker than the norm topology (Section IV.5).

We state two facts about nets whose proofs are not difficult and are left as problems:

Theorem IV.2 (a) A function f from a topological space S to a topological space T is continuous if and only if for every convergent net $\{x_\alpha\}_{\alpha \in I}$ in S, with $x_\alpha \to x$, the net $\{f(x_\alpha)\}_{\alpha \in I}$ converges in T to $f(x)$.

(b) Let S be a Hausdorff space. Then a net $\{x_\alpha\}_{\alpha \in I}$ in S can have at most one limit; that is, if $x_\alpha \to x$ and $x_\alpha \to y$, then $x = y$.

Analogous to the concept of a subsequence we have the following definition:

Definition A net $\{x_\alpha\}_{\alpha \in I}$ is a **subnet** of a net $\{y_\beta\}_{\beta \in J}$ if and only if there is a function $F: I \to J$ such that

(i) $x_\alpha = y_{F(\alpha)}$ for each $\alpha \in I$.
(ii) For all $\beta' \in J$, there is an $\alpha' \in I$ such that $\alpha \succ \alpha'$ implies $F(\alpha) \succ \beta'$ (that is, $F(\alpha)$ is eventually larger than any fixed $\beta \in J$).

We then have the following proposition which shows that the above definition is the right one.

Proposition A point x in a topological space S is a cluster point of a net $\{x_\alpha\}$ if and only if some subnet of $\{x_\alpha\}$ converges to x.

Of course, subsequences are subnets of sequences. But it is also possible for a sequence in a topological space to have no convergent subsequences but to have convergent subnets (see Problem 12).

IV.3 Compactness †

The reader no doubt remembers the special role that closed bounded subsets of $\mathbb{R}^n$ played in elementary analysis. In this section we will study the topological abstraction of this concept:

† A supplement to this section begins on p. 351.

Definition We say a topological space $\langle S, \mathcal{T}\rangle$ is **compact** if any open cover of S has a finite subcover, that is, if for any family $\mathcal{U} \subset \mathcal{T}$ with $S = \bigcup_{U \in \mathcal{U}} U$, there is a finite subset $\{U_1, \ldots, U_n\} \subset \mathcal{U}$ with $S = \bigcup_{i=1}^n U_i$. A subset of a topological space is called a **compact set** if it is a compact space in the relative topology.

Henceforth in our discussion we will always suppose that all compact spaces are Hausdorff, although occasionally we will repeat this condition for emphasis.

Since we have a considerable amount of material to discuss, it is perhaps useful to describe briefly the contents of the next two sections. After studying some equivalent formulations of compactness and some elementary properties of compact spaces, we turn to some of the pillars of functional analysis. We first state and discuss Tychonoff's theorem. We then turn to the study of continuous functions on compact sets. After showing that a compact Hausdorff space X has lots of continuous functions (Urysohn's lemma), we discuss the Banach space $C(X)$ of continuous functions. We state the Stone–Weierstrass theorem but defer its instructive proof to an appendix. In the next section, we determine the dual of $C(X)$. Using the Riesz–Markov theorem, we will prove that $C(X)^*$ is identical with $\mathcal{M}(X)$, the family of signed measures on X.

We first reformulate the notion of compactness by taking complements of open sets:

Definition A topological space S is said to have the **finite intersection property** (f.i.p.) if and only if any family of closed sets $\mathcal{F}$ with $\bigcap_{i=1}^n F_i \neq \emptyset$ for any finite subfamily $\{F_i\}_{i=1}^n \subset \mathcal{F}$ satisfies $\bigcap_{F \in \mathcal{F}} F \neq \emptyset$.

Proposition (f.i.p. criterion) S is compact if and only if S has the f.i.p.

Proof Let $\mathcal{F}$ be given and let $\mathcal{U} = \{S\backslash F \mid F \in \mathcal{F}\}$. Then $\mathcal{F}$ has the property that $\bigcap_{i=1}^n F_i \neq \emptyset$ if and only if $\mathcal{U}$ has no finite subcover and the property that $\bigcap_{F \in \mathcal{F}} F \neq \emptyset$ if and only if $\mathcal{U}$ is not a cover. The reader is invited to wend his way through the double negatives to complete the proof. ∎

A somewhat deeper reformulation is:

Theorem IV.3 (The Bolzano–Weierstrass theorem) A space S is compact if and only if every net in S has a convergent subnet.

Proof Suppose that every net has a convergent subnet and let $\mathcal{U}$ be an open cover. Let us suppose that $\mathcal{U}$ has no finite subcover and derive a contradiction.

Order the finite subfamilies $\mathfrak{S}$ of $\mathcal{U}$ by inclusion; $\mathfrak{S}$ is thereby a directed set. For each $\mathscr{F} \equiv \{F_1, \ldots, F_m\} \in \mathfrak{S}$, pick $x_{\mathscr{F}} \notin \bigcup_{i=1}^m F_i$. By assumption, the net $x_{\mathscr{F}}$ has a cluster point x. Since $\mathcal{U}$ is a cover, we can find $U \in \mathcal{U}$ with $x \in U$. Since $x_{\mathscr{F}}$ is frequently in U we can find a finite subfamily $\mathscr{G} \in \mathfrak{S}$ so that $\{U\} \prec \mathscr{G}$ and $x_{\mathscr{G}} \in U$. Since $\{U\} \prec \mathscr{G}$, $U \subset \bigcup_{G \in \mathscr{G}} G$, and so $x_{\mathscr{G}} \in \bigcup_{G \in \mathscr{G}} G$, which is a contradiction.

Suppose that S is compact and let $\{y_\alpha\}_{\alpha \in I}$ be a net. If $\{y_\alpha\}$ has no cluster points, then for any $x \in S$, there is an open set U_x containing x and an $\alpha_x \in I$ with $y_\alpha \notin U_x$ if $\alpha \succ \alpha_x$. The family $\{U_x \mid x \in S\}$ is an open cover of S, so we can find $x_1, \ldots, x_n$ so that $\bigcup_{i=1}^n U_{x_i} = S$. Since I is directed, we can find $\alpha_0 \succ \alpha_{x_i}$ for $i = 1, \ldots, n$. But $y_{\alpha_0} \notin U_{x_i}$, $i = 1, \ldots, n$, which is impossible since $\bigcup_{i=1}^n U_{x_i} = S$. This contradiction establishes that $\{y_\alpha\}_{\alpha \in I}$ has a cluster point and thus a convergent subnet. ∎

Second countable spaces are compact if and only if every sequence has a convergent subsequence (this can be shown by mimicking the above proof).

Example 1 The unit ball in ℓ_2 is not compact in the *metric topology*. No subset of a sequence of orthonormal elements can converge.

Example 2 Let $S = \{\{a_n\} \in \ell_2 \mid |a_n| \leq 1/n\}$. It is easy to see that a sequence of elements of S converges if and only if each component converges. Using the diagonalization trick, we conclude that every sequence has a convergent subsequence. Therefore, by the Bolzano–Weierstrass theorem, S is compact.

Warning Compact is not the same as closed and bounded in a general Banach space. In fact the unit ball in a Banach space is compact (in the norm topology) if and only if the space is finite dimensional (see Problem 4 of Chapter V).

We now mention two simple "hereditary" properties of compact spaces (see Problem 38):

Proposition   (a) A closed subset of a compact space is compact in the relative topology.
  (b) A continuous image of a compact space is compact.

Corollary Any continuous function on a compact space takes on its maximum and minimum values. That is, there are $x_\pm$ so that

$$f(x_+) = \sup_{x \in C} f(x) \quad \text{and} \quad f(x_-) = \inf_{x \in C} f(x)$$

The following theorem is often useful:

Theorem IV.4 Let S and T be compact Hausdorff spaces; let $f: S \to T$ be a continuous bijection. Then f is a homeomorphism.

We need the following lemma:

Lemma If T is Hausdorff and $S \subset T$ is compact, then S is closed.

Proof. Let $x \in \bar{S}$. We can find a net $\{x_\alpha\}_{\alpha \in I}$ in S with $x_\alpha \to x$. Since limits are unique in Hausdorff spaces, x is the only cluster point of the net. But since S is compact, the net has a cluster point in S, that is, $x \in S$. Thus $S = \bar{S}$. ∎

Proof of Theorem IV.4 We need only prove f is open or equivalently, since f is a bijection, that $f[C]$ is closed if C is closed. But if $C \subset S$ is closed, then C is compact. By the last proposition, $f[C]$ is compact. The result now follows from the lemma. ∎

Proposition If $\{A_i\}_{i=1}^n$ is a family of compact sets, then $\bigtimes_{i=1}^n A_i$ with the product topology is compact.

Proof Let $\{x_\alpha\}_{\alpha \in I}$ be a net in $A = \bigtimes_{i=1}^n A_i$, $x_\alpha = \langle x_\alpha^1, x_\alpha^2, \ldots, x_\alpha^n \rangle$. Since A_1 is compact, we can find a subnet $\{x_{\alpha(i)}\}_{i \in D_1}$ so that $\{x_{\alpha(i)}^1\}$ converges to an $x_1 \in A_1$. By a finite induction, we can find a subnet $\{x_{\alpha(i)}\}_{i \in D_n}$ so that $x_{\alpha(i)}^j$ converges to an $x_j \in A_j$ for each j. Then $\{x_{\alpha(i)}\}$ converges in A to $x = \langle x_1, \ldots, x_n \rangle$, so A is compact by the Bolzano–Weierstrass criterion. ∎

This last proposition is not deep; what is deep is that it remains true for an arbitrary product of compact spaces:

Theorem IV.5 (Tychonoff's theorem) Let $\{A_\alpha\}_{\alpha \in I}$ be a collection of compact spaces. Then $\bigtimes_{\alpha \in I} A_\alpha$ is compact in the product (that is weak) topology.

Since this theorem has a mildly complicated proof well-treated in the textbook literature, we refer the reader to the references given in the Notes. Let us, however, make several comments. We first remark that it is this theorem that supports the feeling that the weak topology is the "natural" topology for $\bigtimes_\alpha A_\alpha$. Another a priori candidate, the "box topology," which is generated by sets of the form $\bigtimes_\alpha U_\alpha$, where each U_α is open in A_α is not a topology for which Tychonoff's theorem holds. Secondly, we note that this theorem depends crucially on the axiom of choice (Zorn's lemma). In fact it is known that, set theoretically speaking, Tychonoff's theorem implies Zorn's lemma. Finally, we note that in the special case of countably many *metric* spaces, Theorem IV.5 can be proven by the method of the proposition and the diagonal trick of Section I.5.

Next, we would like to discuss functions on compact Hausdorff spaces. We first show that compact Hausdorff spaces have strong separation properties in the sense of separating closed sets with open sets. We then use these separation properties to construct continuous functions:

Theorem IV.6 Any compact Hausdorff space X is normal (T_4).

Proof We first prove X is regular (T_3). Let $p \in X$ and let $C \subset X$ be closed with $p \notin C$. Since X is Hausdorff, we can find, for any $y \in C$, open and disjoint sets, U_y and V_y, so that $y \in U_y$, and $p \in V_y$. The $\{U_y\}_{y \in C}$ cover C, which is compact. Thus $U_{y_1}, \ldots, U_{y_n}$ cover C. Let $U = \bigcup_{i=1}^n U_{y_i}$; $V = \bigcap_{i=1}^n V_{y_i}$. Then U and V are open and disjoint with $C \subset U$ and $p \in V$. This shows that X is regular. Now let C, D be closed and disjoint. By repeating the above argument with D replacing p and "since X is regular" replacing "since X is Hausdorff," we prove that X is normal. ∎

Normal spaces always have lots of continuous functions for:

Theorem IV.7 (Urysohn's lemma) Let C and D be closed disjoint sets in a normal space, X. Then, there is a continuous function from X to $\mathbb{R}$ with $0 \leq f(x) \leq 1$ for all x such that $f(x) = 0$ if $x \in C$ and $f(x) = 1$ if $x \in D$.

Sketch of proof Using the normality of X, one constructs by induction for each dyadic rational (that is, $r = k/2^n$, k, n integers, $0 \leq k \leq 2^n$) open sets, U_r, with $C \subset U_r \subset \bar{U}_r \subset U_s \subset \bar{U}_s \subset X \backslash D$ if $r < s$. One uses the U_r to define a function with $f(x) < r$ if and only if $x \in U_r$. f can be shown to be continuous. For details, see the references discussed in the notes. ∎

We will see below that one can prove even stronger function theoretic results (Theorem IV.11).

As a final result about the general properties of functions on X we will prove that certain families are dense in $C_\mathbb{R}(X)$, the family of all real-valued continuous functions on X. We first note that our proof in Section I.5 for $C[a, b]$ holds on any compact set:

Theorem IV.8 Let $C(X)$ be the family of all continuous complex-valued functions on a compact Hausdorff space, X, endowed with the norm $\|f\|_\infty = \sup_{x \in X} |f(x)|$. Let $C_\mathbb{R}(X) = \{f \in C(X) | f \text{ is real-valued}\}$. Then $C(X)$ is a complex Banach space and $C_\mathbb{R}(X)$ is a real Banach space.

The density theorem we state generalizes a classical theorem of Weierstrass which says that any real-valued continuous function on $[0, 1]$ is a uniform limit (on $[0, 1]$) of polynomials (see Problems 19 and 20). Note that $C_\mathbb{R}(X)$ has a natural multiplication given by $(fg)(x) = f(x)g(x)$. A **subalgebra** of $C_\mathbb{R}(X)$ is a subspace closed under multiplication:

Theorem IV.9 (Stone–Weierstrass theorem) Let B be a subalgebra of $C_\mathbb{R}(X)$ which is closed in $\|\cdot\|_\infty$. We say that B separates points if, given any $x, y \in X$, we can find $f \in B$ with $f(x) \neq f(y)$. If B separates points, then either $B = C_\mathbb{R}(X)$ or for some $x_0 \in X$, $B = \{f \in C_\mathbb{R}(X) | f(x_0) = 0\}$. If $1 \in B$, and B separates points, $B = C_\mathbb{R}(X)$.

We defer the instructive lattice-theoretic proof to an appendix.

The fact that we deal with $C_\mathbb{R}(X)$ and not $C(X)$ is crucial (see Problem 15), but, by adding an extra hypothesis we can easily extend Theorem IV.9 to the complex case.

Theorem IV.10 (complex Stone–Weierstrass theorem) Let B be a subalgebra of $C(X)$ with the property that if f is in B, then the complex conjugate, $\bar{f}$, is in B also. If B is closed and separates points, then $B = C(X)$ or $B = \{f | f(x) = 0\}$ for some fixed x.

The complex conjugate condition is crucial. For example, let D be the unit disc in the complex plane. The functions analytic in the interior of D, continuous on all of D, are a closed subalgebra of D containing 1 and separating points which is not $C(D)$. It is, however, not closed under complex conjugation.

As an example of how to use the Stone–Weierstrass theorem as well as an example of how several functional analytic theorems can combine in a very powerful way, we prove an extension theorem for functions in $C(Y)$ for $Y \subset X$ when X is compact and Y is closed. Actually, this theorem is true if X is merely normal (Problem 18):

Theorem IV.11 (Tietze extension theorem) Let X be a compact space and let $Y \subset X$ be closed. Let f be any continuous real-valued function on Y. Then there is an $\tilde{f} \in C_\mathbb{R}(X)$ so that $f(y) = \tilde{f}(y)$ for all $y \in Y$.

Proof Consider the map $\rho: C_\mathbb{R}(X) \to C_\mathbb{R}(Y)$ given by $\rho(f) = f \upharpoonright Y$. The theorem is equivalent to the statement that ρ is onto. Clearly, Ran ρ is a subalgebra of $C_\mathbb{R}(Y)$ and $1 \in \text{Ran } \rho$. Moreover, by Urysohn's lemma, Ran ρ separates points. If we can show that Ran ρ is closed in $\|\cdot\|_{C(Y)}$ we can complete the proof by using the Stone–Weierstrass theorem.

Let $I = \text{Ker } \rho$. Then I is clearly closed in $C_\mathbb{R}(X)$, so we can form the quotient Banach space $C_\mathbb{R}(X)/I$. By elementary algebra, ρ "lifts" to a bijection, $\tilde{\rho}: C_\mathbb{R}(X)/I \to \text{Ran } \rho$. If we can prove $\|\tilde{\rho}([f])\|_{C_\mathbb{R}(Y)} = \|[f]\|_{C_\mathbb{R}(X)/I}$, Ran ρ will be a Banach space and thus closed.

Clearly, $\|\rho(f)\|_{C_\mathbb{R}(Y)} \leq \|f\|_{C_\mathbb{R}(X)}$, so $\|\tilde{\rho}([f])\|_{C_\mathbb{R}(Y)} \leq \|[f]\|_{C_\mathbb{R}(X)/I}$. Thus, it is enough to show that given $g \in \text{Ran } \rho$, we can find $f \in C_\mathbb{R}(X)$ with $g = \rho(f)$ and $\|g\|_{C_\mathbb{R}(Y)} = \|f\|_{C_\mathbb{R}(X)}$ (remember the definition of quotient norm!). Since $g \in \text{Ran } \rho$, we know that $g = \rho(h_1)$ for some $h_1 \in C_\mathbb{R}(X)$. Let

$$h_2 = \min\{\|g\|_{C_\mathbb{R}(Y)}, h_1\}$$

so that $\rho(h_2) = g$ and $h_2(x) \leq \|g\|_{C_\mathbb{R}(Y)}$ for all x. Let $h_3 = \max\{-\|g\|_{C_\mathbb{R}(X)}, h_2\}$. Then, $\|h_3\|_{C_\mathbb{R}(X)} = \|g\|_{C_\mathbb{R}(Y)}$ and $\rho(h_3) = g$. This completes the proof. ∎

Appendix to IV.3 The Stone–Weierstrass theorem

In this appendix we prove Theorem IV.9 in the case $1 \in B$. The general proof is left as an exercise. Interestingly enough, the first step in the proof is the proof of the classical Weierstrass theorem (which is a special case of the general theorem!)

Lemma 1 The polynomials are dense in $C_\mathbb{R}[a, b]$ for any finite real numbers a, b.

Proof See Problems 19 and 20.

This can now be used to prove that B is a lattice, where:

Definition A subset $S \subset C_\mathbb{R}(X)$ is called a lattice if for all $f, g \in S$, $f \wedge g = \min\{f, g\}$ and $f \vee g = \max\{f, g\}$ are in S.

Lemma 2 Any closed subalgebra B of $C_\mathbb{R}(X)$ with $1 \in B$ is a lattice.

Proof We show that if $f \in B$, then $|f| \in B$. The result then follows from the formulas: $f \vee g = \frac{1}{2}|f-g| + \frac{1}{2}(f+g)$, $f \wedge g = -[(-f) \vee (-g)]$. Without loss suppose that $\|f\|_\infty \leq 1$. By the classical Weierstrass theorem, we can find a sequence of polynomials $P_n(x)$ converging uniformly to $|x|$ on $[-1, 1]$, for example $|P_n(x) - |x|| < 1/n$ for all x in $[0, 1]$. Since $\|f\|_\infty \leq 1$, it follows that $\|P_n(f) - |f|\|_\infty < 1/n$, i.e. $|f| = \lim_{n \to \infty} P_n(f)$. Since B is an algebra with $1 \in B$, $P_n(f) \in B$. Since B is closed, $|f| \in B$. ∎

Finally, the full Stone–Weierstrass theorem is a consequence of Lemma 2 and the following theorem which is of some interest in itself:

Theorem IV.12 (Kakutani–Krein theorem) Let X be a compact Hausdorff space. Any lattice $\mathscr{L} \subset C_\mathbb{R}(X)$ which is a closed subspace containing 1 and which separates points is all of $C_\mathbb{R}(X)$.

Proof Let $h \in C_\mathbb{R}(X)$ and let ε be given. We seek $f \in \mathscr{L}$ with $\|h - f\| < \varepsilon$. Suppose we can show for any $x \in X$, there is $f_x \in \mathscr{L}$ with $f_x(x) = h(x)$ and $h \leq f_x + \varepsilon$. Then for each x, find U_x, an open neighborhood of x with $h(y) \geq f_x(y) - \varepsilon$ for all $y \in U_x$ (by the continuity of $h - f_x$). The U_x cover X so let $U_{x_1}, \ldots, U_{x_n}$ be a subcover. Then $f = f_{x_1} \wedge \cdots \wedge f_{x_m}$ obeys $f(y) + \varepsilon = \min_i \{f_{x_i}(y) + \varepsilon\} \geq h(y)$. Moreover, since any $y \in U_{x_i}$ for some i, $f(y) - \varepsilon \leq f_{x_i}(y) - \varepsilon \leq h(y)$. Thus $\|f - h\|_\infty < \varepsilon$.

It remains to find some f_x with the desired properties. Since $\mathscr{L}$ separates points and $1 \in \mathscr{L}$, for any x and y in X, we can find $f_{xy} \in \mathscr{L}$ with $f_{xy}(x) = h(x)$ and $f_{xy}(y) = h(y)$. For each y, we can find V_y, an open set about y with $f_{xy}(z) + \varepsilon \geq h(z)$ for $z \in V_y$. $V_{y_1}, \ldots, V_{y_n}$ will cover X for suitable $y_1, \ldots, y_n$. If we take $f_x = f_{xy_1} \vee \cdots \vee f_{xy_n}$, then $f_x(x) = h(x)$, and for any $z \in X$

$$f_x(z) + \varepsilon = \max_{i=1,\ldots,n} \{f_{xy_i}(z) + \varepsilon\} \geq h(z)$$

This completes the proof. ∎

IV.4 Measure theory on compact spaces †

In this section, we wish to discuss several aspects of measure theory which are special for compact spaces. In particular, we will see that the dual of $C(X)$ can be interpreted as a space of measures (the Riesz–Markov theorem). Since many of the measure-theoretic proofs are not enlightening, we will not prove all of the theorems.

† A supplement to this section begins on p. 353.

IV.4 Measure theory on compact spaces

The first question that arises is what to pick as the σ-field of measurable sets. Let us begin with a minimal family. We clearly want to integrate continuous functions $f \in C(X)$. This might lead one to suspect that we want to allow all closed (and open) sets to be measurable but this is not necessary:

Definition A G_δ set is a set which is a countable intersection of open sets.

Proposition Let X be a compact Hausdorff space and let $f \in C_\mathbb{R}(X)$. Then $f^{-1}([a, \infty))$ is a compact G_δ set.

Proof $f^{-1}([a, \infty))$ is closed and thus compact. Since

$$f^{-1}([a, \infty)) = \bigcap_{n=1}^{\infty} f^{-1}((a - 1/n, \infty))$$

it is a G_δ. ∎

Thus, to integrate continuous functions, we need only have compact G_δ's in our σ-field.

Definition The σ-field generated by the compact G_δ's in a compact space X is called the family of **Baire sets**. The functions $f: X \to \mathbb{R}$ (or $\mathbb{C}$) measurable relative to this σ-field are called **Baire functions**. A measure on the Baire sets is called a **Baire measure** if in addition it is finite, that is $\mu(X) < \infty$.

As in the case of the finite intervals of the real line and Lebesgue measure:

Theorem IV.13 If μ is a Baire measure, then $C(X) \subset L^p(X, d\mu)$ for all p and $C(X)$ is dense in $L^1(X, d\mu)$ or any L^p space for $p < \infty$ (but *not* L^∞ except in pathological cases where $C(X)$ is already all of L^∞!).

Despite the fact that Baire sets are all that are needed, the reader no doubt wants to repress G_δ's and consider all **Borel** sets, i.e. the σ-field generated by all open sets. The question of extending Baire measures to Borel measures, that is, measures on all Borel sets, is answered by the following remarks:

(1) Every Baire measure is automatically **regular**, that is,

$$\mu(Y) = \inf\{\mu(O) \mid Y \subset O, O \text{ open and Baire}\}$$
$$= \sup\{\mu(C) \mid C \subset Y, C \text{ compact and Baire}\}$$

(2) In general, a Baire measure has many extensions to all Borel sets but there is exactly one regular extension to a Borel measure. A Borel measure is called regular if

$$\mu(Y) = \inf\{\mu(O) \mid Y \subset O, O \text{ open}\}$$
$$= \sup\{\mu(C) \mid C \subset O, C \text{ compact and } Borel\}$$

Thus there is a one-one correspondence between Baire measures and *regular* Borel measures.

(3) If μ is a Borel measure, then $C(X)$ is dense in $L^1(X, d\mu)$ if and only if μ is regular. If μ is regular, every Borel set is almost everywhere a Baire set in the sense that given a Borel set Y, there is a a Baire set $\tilde{Y}$ with

$$\int |\chi_Y - \chi_{\tilde{Y}}| \, d\mu \equiv \mu(Y \backslash \tilde{Y}) + \mu(\tilde{Y} \backslash Y) = 0$$

In addition, every Borel function is equal, after a change on a Borel set of measure zero, to a Baire function.

(4) In certain cases, every compact set is a G_δ, so the Baire and Borel sets are identical. This is the case if X is a compact metric space (see Problem 30).

Henceforth, we will use the word measure in the context of a compact set, X, to mean Baire (or equivalently regular Borel) measure unless we specifically indicate otherwise.

Now, let X be compact and let μ be a measure on X. Consider the map $C(X) \to \mathbb{C}$ given by $f \mapsto \ell_\mu(f) \equiv \int f \, d\mu$. ℓ_μ is clearly linear and

$$|\ell_\mu(f)| \leq \int |f| \, d\mu \leq \|f\|_\infty \mu(X)$$

so ℓ_μ is a continuous linear functional on $C(X)$. In fact, $\|\ell_\mu\|_{C(X)^*} \equiv \mu(X)$, for take $f = 1$. Moreover, ℓ_μ is positive in the sense:

Definition A **positive linear functional** on $C(X)$ is a (not necessarily a priori continuous) linear functional ℓ with $\ell(f) \geq 0$ for all f with $f \geq 0$ pointwise.

In the more general context of C^*-algebras, positive linear functionals will again arise; see Chapter XVII. They have the following nice property (for other properties of positivity, see Problem 37):

Proposition Let ℓ be a positive linear functional. Then ℓ is continuous and $\|\ell\|_{C(X)^*} = \ell(1)$.

Proof Suppose first that f is real. Since $-\|f\|_\infty \leq f \leq \|f\|_\infty$, we have $-\ell(1)\|f\|_\infty \leq \ell(f) \leq \ell(1)\|f\|_\infty$; that is, $|\ell(f)| \leq \|f\|_\infty \ell(1)$. If f is arbitrary, $\ell(f) = e^{i\phi}r$ with r real and positive, so

$$|\ell(f)| = \ell(\text{Re}[e^{-i\phi}f]) \leq \|\text{Re}(e^{-i\phi}f)\|_\infty \ell(1) \leq \ell(1)\|f\|_\infty \quad \blacksquare$$

We have seen that any Baire measure provides an example of a positive linear functional on $C(X)$; that these are the only examples is the content of:

Theorem IV.14 (the Riesz–Markov theorem) Let X be a compact Hausdorff space. For any positive linear functional ℓ on $C(X)$ there is a unique Baire measure μ on X with

$$\ell(f) = \int f\,d\mu$$

While we will not give a detailed proof, let us show how μ may be recovered from ℓ_μ. A similar process allows one to construct a measure from any positive linear functional, even if one does not know a priori that it is of the form ℓ_μ. Since μ is inner regular (that is, $\mu(Y) = \sup\{\mu(C)\mid C \subset Y, C \text{ compact}\}$), we need only find $\mu(C)$ for C compact to "recover" μ. We claim $\mu(C) = \inf\{\ell_\mu(f)\mid f \in C(X), f \geq \chi_C\}$. Since μ is positive, it is clear that $\mu(C) \leq \ell_\mu(f)$ if $f \geq \chi_C$; thus, we need only show that, given ε, we can find $f \in C(X)$ with $\chi_C \leq f$ and $\ell_\mu(f) \leq \mu(C) + \varepsilon$. Since μ is outer regular, given ε we can find O open with $\mu(O\setminus C) < \varepsilon$ and $C \subset O$. By Urysohn's lemma, we can find $f \in C(X)$ with $0 \leq f \leq 1$, $f(x) = 1$ if $x \in C$ and $f(x) = 0$ if $x \in X\setminus O$. Thus $\ell(f) \leq \mu(O) < \mu(C) + \varepsilon$. This shows μ can be recovered from ℓ_μ, and so it is not too surprising that a measure can be constructed from an arbitrary ℓ.

The Riesz-Markov theorem is the usual way that measures arise in functional analysis. For example, we have already intimated that measures on $\mathbb{R}$ are associated with quantum mechanical Hamiltonians and they, in turn, arise from certain positive linear functions and the use of the Riesz–Markov theorem (or rather its extension to locally compact spaces which we will discuss shortly).

In general, a pointwise limit of a net of Baire functions is not a Baire or even a Borel function (Problem 13). However, if $\{f_\alpha\}_{\alpha \in I}$ is a net of functions with each f_α continuous and $\{f_\alpha\}$ is increasing in the sense that $f_\alpha \geq f_\beta$ if $\alpha \succ \beta$, then $f = \lim_\alpha f_\alpha = \sup_\alpha f_\alpha$ is a Borel function because

$$f^{-1}[(a, \infty)] = \bigcup_\alpha f_\alpha^{-1}[(a, \infty)]$$

is open. The monotone convergence theorem has the following net generalization:

Theorem IV.15 (monotone convergence theorem for nets) Let μ be a regular Borel measure on a compact Hausdorff space X. Let $\{f_\alpha\}_{\alpha \in I}$ be an increasing net of continuous functions. Then $f = \lim_\alpha f_\alpha \in L^1(X, d\mu)$ if and only if $\sup_\alpha \|f_\alpha\|_1 < \infty$ and in that case $\lim_\alpha \|f - f_\alpha\|_1 = 0$.

Before leaving measure theory on compact spaces, we should identify the dual space of $C(X)$. Of course, not every continuous linear functional on $C(X)$ is a positive linear functional, but the major result we are heading toward is that any $\ell \in C_\mathbb{R}(X)^*$ is the difference of two positive linear functionals. This depends on a simple "lattice-theoretic" result about $C_\mathbb{R}(X)$:

Lemma Let $f, g \in C_\mathbb{R}(X)$ with $f, g \geq 0$. Suppose $h \in C_\mathbb{R}(X)$ and $0 \leq h \leq f + g$. Then, we can write $h = h_1 + h_2$ with $0 \leq h_1 \leq f$, $0 \leq h_2 \leq g$, $h_1, h_2 \in C_\mathbb{R}(X)$.

Proof Let $h_1 = \min\{f, h\}$. Then $0 \leq h_1 \leq f$ and if $h_2 \equiv h - h_1$, then $h_2 \geq 0$. Moreover, if $h_1(x) = h(x)$, then $h_2(x) = 0 \leq g(x)$ and if $h_1(x) = f(x)$, then $h_2(x) = h(x) - f(x) \leq f(x) + g(x) - f(x) = g(x)$, so $h_2 \leq g$. ∎

Theorem IV.16 Let $\ell \in C_\mathbb{R}(X)^*$. Then ℓ can be written $\ell = \ell_+ - \ell_-$ with ℓ_+ and ℓ_- positive linear functionals. Moreover, $\ell_+(1) + \ell_-(1) = \|\ell\|$ and this uniquely determines ℓ_+ and ℓ_-.

Proof For $f \in C(X)_+ \equiv \{f \in C(X) \mid f \geq 0\}$, define $\ell_+(f) = \sup\{\ell(h) \mid h \in C(X); 0 \leq h \leq f\}$. Since $|\ell(h)| \leq \|\ell\| \|h\|_\infty \leq \|\ell\| \|f\|_\infty$, this supremum is finite. Clearly $\ell_+(tf) = t\ell_+(f)$ for any scalar $t > 0$ and $\ell_+(f) \geq \ell(0) = 0$ for all $f \in C(X)_+$. Let $f, g \in C(X)_+$. Then, by the lemma:

$$\ell_+(f + g) = \sup\{\ell(h) \mid 0 \leq h \leq f + g\}$$
$$= \sup\{\ell(h_1) + \ell(h_2) \mid 0 \leq h_1 \leq f, \; 0 \leq h_2 \leq g\}$$
$$= \ell_+(f) + \ell_+(g)$$

For any $f \in C(X)$, define $f_+ = \max\{f, 0\}$ and $f_- = -\min\{f, 0\}$, so $f = f_+ - f_-$. Define $\ell_+(f) = \ell_+(f_+) - \ell_+(f_-)$. It is then easy to show ℓ_+ is linear on $C(X)$. By definition $\ell_+(f) \geq \ell(f)$ if $f \geq 0$ so $\ell_-(f) \equiv \ell_+(f) - \ell(f)$ is a positive linear functional. We have thus written $\ell = \ell_+ - \ell_-$ as the difference of positive linear functionals.

To prove $\ell_+(1) + \ell_-(1) = \|\ell\|$, we note first $\|\ell\| \leq \|\ell_+\| + \|\ell_-\| = \ell_+(1) + \ell_-(1)$. For the inequality in the other direction, we first rewrite ℓ_- in a way symmetric to ℓ_+. For $f \geq 0$

$$\ell_-(f) = \sup\{\ell(h) - \ell(f) \mid 0 \leq h \leq f\}$$
$$= \sup\{\ell(k) \mid -f \leq k \leq 0\}$$

where $k = h - f$. Thus:

$$\ell_+(1) + \ell_-(1) = \sup\{\ell(h) \mid 0 \leq h \leq 1\} + \sup\{\ell(k) \mid -1 \leq k \leq 0\}$$
$$= \sup\{\ell(g) \mid -1 \leq g \leq 1\}$$
$$\leq \|\ell\| \sup\{\|g\|_\infty \mid -1 \leq g \leq 1\}$$
$$= \|\ell\|$$

The proof of uniqueness is left to the reader (Problem 31). ∎

Definition A **complex Baire measure** is a finite linear complex combination of Baire measures.

An easy consequence of Theorem IV.14 and Theorem IV.16 is:

Theorem IV.17 Let X be a compact space. Then the dual $C(X)^*$ of $C(X)$ is the space of all complex Baire measures.

Definition We write $\mathcal{M}(X) = C(X)^*$; $\mathcal{M}_+(X) = \{\ell \in \mathcal{M}(X) \mid \ell$ is a positive linear functional$\}$ and $\mathcal{M}_{+,1}(X) = \{\ell \in \mathcal{M}_+ \mid \|\ell\| = 1\}$.

In some cases, it is important to think of measures not merely as individual objects but instead as elements of $\mathcal{M}(X)$, so that we can employ geometric ideas. To give the reader a feel for this sort of reasoning we conclude our discussion of $\mathcal{M}(X)$ by a simple convexity theorem.

Definition A set A in a vector space Y is called **convex** if x and $y \in A$ and $0 \leq t \leq 1$ implies $tx + (1-t)y \in A$. Thus A is convex if the line segment between x and y is in A whenever x and y are in A (Figure IV.1). A is called a **cone** if $x \in A$ implies $tx \in A$ for all $t > 0$. If A is convex and a cone, it is called a **convex cone**.

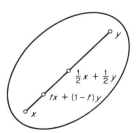

FIGURE IV.1 A convex set.

Proposition Let X be a compact Hausdorff space. Then $\mathscr{M}_{+,1}(X)$ is convex and $\mathscr{M}_+(X)$ is a convex cone.

Proof A convex combination of positive linear functionals is clearly a positive linear functional. Moreover, $\|\ell\| = \ell(1)$ if ℓ is a positive linear functional so $\|t\ell_1 + (1-t)\ell_2\| = 1$, if $\ell_1, \ell_2 \in \mathscr{M}_{+,1}$. ∎

At first sight, this geometric fact may appear a little strange since the reader is used to thinking of the unit sphere, $\{x \mid \|x\| = 1\}$ as "round" and here we are saying a piece of it is absolutely flat! The moral is that every norm is not the Euclidean norm (the parallelogram law implies that in a Hilbert space, if $\|x\| = \|y\| = 1$, and $x \neq y$, then $\|tx + (1-t)y\| < 1$). In fact, $\mathbb{R}^n$ with the norm $\|\langle x_1, \ldots, x_n \rangle\| = \sum_{i=1}^n |x_i|$ has a unit sphere with flat faces, see Figure IV.2. This is not a coincidence; $\{1, \ldots, n\}$ is a compact set when given the discrete topology, and $\mathbb{R}^n$ with the norm considered is precisely $\mathscr{M}(\{1, \ldots, n\})$.

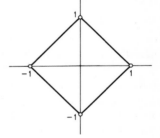

FIGURE IV.2 The unit sphere in R^2 when $\|\langle x, y \rangle\| = |x| + |y|$.

Now, we want to extend "topological measure theory" to a larger class of spaces:

Definition A topological space, X is called **locally compact** if and only if every point $p \in X$, has a compact neighborhood.

By thinking of Lebesgue measure on $\mathbb{R}$, we realize that we want to relax the condition $\mu(X) < \infty$ which we required when X was compact. We first define the Baire sets in X, a locally compact space, to be the σ-ring $\mathscr{B}$ generated by compact G_δ sets. Note that, in general, X may not be a set of $\mathscr{B}$. However, if X is σ-compact, that is, a countable union of compact sets, X is in $\mathscr{B}$.

Definition A **Baire measure** on X, a locally compact space, is a measure on the Baire sets for which $\mu(C) < \infty$ for any compact Baire set C.

Given any Baire measure μ on X, and given a compact G_δ set $C \subset X$, there is induced by restriction a Baire measure μ_C on C. Conversely, it is easy to see that a family of measures $\{\mu_C\}$, one for each compact G_δ set, with the property that $\mu_C(Y) = \mu_D(Y)$ if $Y \subset C \cap D$, defines a Baire measure. This association allows us to prove theorems in the locally compact case from their compact case analogues.

Definition   Let X be a locally compact space. $\kappa(X)$, the algebra of **continuous functions of compact support**, is the set of functions that vanish outside some compact set. $C_\infty(X)$, the algebra of continuous functions vanishing at ∞, is the set of $f \in C(X)$ with the property that for any $\varepsilon > 0$, there is a compact set $D_\varepsilon \subset X$ such that $|f(x)| < \varepsilon$ if $x \notin D_\varepsilon$. Thus

$$\kappa(X) \subset C_\infty(X) \subset C(X)$$

With this definition, Theorem IV.14 implies

Theorem IV. 18 (Riesz–Markov)   Let X be a locally compact space. A positive linear functional on $\kappa(X)$ is of the form $\ell(f) = \int f\, d\mu$ for some Baire measure, μ. A positive linear functional on $C_\infty(X)$ comes from a measure μ with total finite mass, that is, $\sup_{A \in \mathscr{B}} \mu(A) < \infty$.

In the next chapter, we will find a topology on $\kappa(X)$ for which the dual is just the complex Baire measures. Notice that this topology is *not* given by $\|\cdot\|_\infty$. $\kappa(X)$ is not complete in the norm $\|\cdot\|_\infty$; its completion is $C_\infty(X)$ and its dual in $\|\cdot\|_\infty$ is the *finite* measures.

IV.5 Weak topologies on Banach spaces †

Definition   Let X be a Banach space with dual space X^*. The **weak topology on** X is the weakest topology on X in which each functional ℓ in X^* is continuous.

Thus a neighborhood base at zero for the weak topology is given by the sets of the form

$$N(\ell_1, \ldots, \ell_n; \varepsilon) = \{x \mid |\ell_i(x)| < \varepsilon;\ i = 1, \ldots, n\}$$

† A supplement to this section begins on p. 354.

that is, neighborhoods of zero contain cylinders with finite-dimensional open bases. A net $\{x_\alpha\}$ converges weakly to x, written $x_\alpha \xrightarrow{w} x$, if and only if $\ell(x_\alpha) \to \ell(x)$ for all $\ell \in X^*$.

For infinite dimensional Banach spaces, the weak topology does not arise from a metric. This is one of the main reasons we have introduced topological spaces. Before considering examples, let us note three elementary properties of the weak topology:

Proposition (a) The weak topology is weaker than the norm topology, that is, every weakly open set is norm open.
(b) Every weakly convergent *sequence* is norm bounded.
(c) The weak topology is a Hausdorff topology.

Proof (a) follows from $|\ell(x)| \leq \|\ell\| \, \|x\|$; (b) is a consequence of the uniform boundedness principle; and (c) follows from the Hahn–Banach theorem. We leave the details to the reader. ∎

We emphasize that (b) is only true for sequences. In Problem 39, the reader is asked to construct a counterexample to the analogous net statement.

Let us consider two examples; in both of them, we will describe what it means for sequences to converge. This does *not* completely describe the topology, but it will give the reader an impressionistic view of the underlying topology.

Example 1 Let $\mathscr{H}$ be a Hilbert space. Let $\{\varphi_\alpha\}_{\alpha \in I}$ be an orthonormal basis for $\mathscr{H}$. Given a *sequence* $\psi_n \in \mathscr{H}$, let $\psi_n^{(\alpha)} = \langle \varphi_\alpha, \psi_n \rangle$ be the coordinates of ψ_n. We claim $\psi_n \to \psi$ in the weak topology if and only if (a) $\psi_n^{(\alpha)} \to \psi^{(\alpha)}$ for each α and (b) $\|\psi_n\|$ is bounded. For suppose $\psi_n \xrightarrow{w} \psi$; then (a) follows by definition and (b) comes from (ii) of the proposition. On the other hand, let (a) and (b) hold and let $F \subset \mathscr{H}$ be the subspace of finite linear combinations of the φ_α. By (a), $\langle \varphi, \psi_n \rangle \to \langle \varphi, \psi \rangle$ if $\varphi \in F$. Using the fact that X is dense, (b), and an $\varepsilon/3$ argument, the weak convergence follows.

Example 2 Let X be a compact Hausdorff space and consider the weak topology on $C(X)$. Let $\{f_n\}$ be a sequence in $C(X)$. We claim $f_n \to f$ in the weak topology if and only if (a) $f_n(x) \to f(x)$ for each $x \in X$, and (b) $\|f_n\|$ is bounded. For if $f_n \xrightarrow{w} f$, then (a) holds since $f \mapsto f(x)$ is an element of $C(X)^*$ and (b) comes from (ii) of the proposition. On the other hand, if (a) and (b) hold, then $|f_n(x)| \leq \sup_n \|f_n\|_\infty$ which is L^1 with respect to any Baire measure μ.

IV.5 Weak topologies on Banach spaces

Thus, by the dominated convergence theorem, for any $\mu \in \mathcal{M}_+(X)$, $\int f_n \, d\mu \to \int f \, d\mu$. Since any $\ell \in \mathcal{M}(X)$ is a finite linear combination of measures in $\mathcal{M}_+$, we conclude that $f_n \to f$ weakly.

We have seen that the weak topology is weaker than the norm topology; actually, it is very weak indeed! To see this, we note that having few open sets is the same as having few closed sets and this is the same as big closures. In Problem 40, the reader will prove that the weak closure of the unit sphere, $\{x \in X \mid \|x\| = 1\}$, in X is the unit ball, $\{x \mid \|x\| \leq 1\}$, in any infinite dimensional Banach space.

We will shortly study general "dual" topologies. As a special case of Theorem IV.20, we state;

Theorem IV.19 A linear functional ℓ on a Banach space is weakly continuous if and only if it is norm continuous.

While this theorem follows from Theorem IV.20, it has a simple direct proof (Problem 42).

Finally, we should like to discuss the weak-∗ topology and prove a compactness theorem which will often be of use to us. Suppose $Y = X^*$ is the dual of some Banach space X. $Y^* = X^{**}$, of course, induces the weak topology on Y, but we may instead consider the topology induced by X acting on X^*; explicitly:

Definition Let X^* be the dual of a Banach space. The **weak-∗ topology** is the weakest topology on X^* in which all the functions $\ell \mapsto \ell(x)$, $x \in X$, are continuous.

Notice that the weak-∗ topology is even weaker than the weak topology. As one might expect, X is reflexive if and only if the weak and weak-∗ topologies coincide, and many characterizations of reflexivity depend on relations involving the weak and weak-∗ topologies.

To avoid confusion and to be able to state our next theorem in its natural setting, let us introduce a new notion:

Definition Let X be a vector space and let Y be a family of linear functionals on X which separates points of X. Then the **Y-weak topology on X**, written $\sigma(X, Y)$, is the weakest topology on X for which all the functionals in Y are continuous.

Because Y is assumed to separate points, $\sigma(X, Y)$ is a Hausdorff topology on X. For example, the weak topology on X is the $\sigma(X, X^*)$ topology while the $\sigma(X^*, X)$ topology is the weak-$*$ topology on X^*. The $\sigma(X, Y)$ topology depends only on the vector space generated by Y, so we henceforth suppose that Y is a vector space.

Example The weak-$*$ topology on $\mathcal{M}(X)$, X a compact Hausdorff space, is often called the **vague** topology. To get an idea of how weak it is, let us show the linear combinations of point masses are weak-$*$ dense in $\mathcal{M}(X)$. In Problem 41, the reader is asked to show they are actually norm closed. Suppose that μ is a given measure. We must show that every weak neighborhood of μ contains a sum of point measures, or equivalently, given $f_1, \ldots, f_n$ and ε, that we can find $\alpha_1, \ldots, \alpha_m$ and $x_1, \ldots, x_m$ so that

$$|\mu(f_i) - \sum_{j=1}^{m} \alpha_j f_i(x_j)| < \varepsilon \quad \text{for } i = 1, \ldots, n$$

For then $\sum \alpha_j \delta_{x_j}$ will be in the vague neighborhood $N(f_1, \ldots, f_n, \varepsilon) + \mu$. Without loss, suppose that $f_1, \ldots, f_n$ are linearly independent. For each x, consider the vector $\mathbf{f}_x = \langle f_1(x), \ldots, f_n(x)\rangle \in \mathbb{R}^n$. If the $\{\mathbf{f}_x\}$ do not span $\mathbb{R}^n$ there is an $\mathbf{a} = \langle a_1, \ldots, a_n\rangle \neq 0 \in \mathbb{R}^n$ with $\mathbf{a} \cdot \mathbf{f}_x = 0$ for all x, that is, $\sum_{i=1}^{n} a_i f_i = 0$ contradicting linear independence. Thus, the $\mathbf{f}_x$ span $\mathbb{R}^n$. So, we can find $x_1, \ldots, x_n$ and $\alpha_1, \ldots, \alpha_n$ with

$$\langle \mu(f_1), \ldots, \mu(f_n)\rangle = \sum_{i=1}^{n} \alpha_i \mathbf{f}_{x_i}$$

So, $\mu(f_i) = \sum_{j=1}^{n} \alpha_j f_i(x_j)$, which proves our claim.

The $\sigma(X^*, X)$ topology is of course weaker than the norm topology on X^* so all the $\sigma(X^*, X)$-continuous linear functionals are in X^{**}. In general, however, not all of X^{**} is weak-$*$ continuous on X^*; in fact:

Theorem IV.20 The $\sigma(X, Y)$ continuous linear functionals on X are precisely Y; in particular the only weak-$*$ continuous functionals on X^* are the elements of X.

Proof Suppose that ℓ is a $\sigma(X, Y)$ continuous functional on X. Then $\{x \mid |\ell(x)| < 1\} \supset \{x \mid |y_i(x)| < \varepsilon; i = 1, \ldots, n\}$ for some ε and some $y_1, \ldots, y_n \in Y$. Now suppose that $y_i(x) = 0$ for $i = 1, \ldots, n$. Then $|\ell(\varepsilon^{-1}x)| < 1$ for all $\varepsilon > 0$, which implies that $\ell(x) = 0$. As a result, ℓ lifts to a functional $\tilde{\ell}$ on X/K where $K = \{x \mid y_i(x) = 0, i = 1, \ldots, n\}$. Elementary abstract algebra shows $\tilde{y}_1, \ldots, \tilde{y}_n$ span the dual space of X/K. Thus $\tilde{\ell} = \sum_{i=1}^{n} \alpha_i \tilde{y}_i$, so that $\ell = \sum_{i=1}^{n} \alpha_i y_i \in Y$. ∎

Finally, we conclude this section with its most important result, a result which is perhaps the most important consequence of Tychonoff's theorem:

Theorem IV. 21 (the Banach–Alaoglu theorem)   Let X^* be the dual of some Banach space, X. Then the unit ball in X^* is compact in the weak-$*$ topology.

Proof   For each $x \in X$, let $B_x = \{\lambda \in \mathbb{C} \mid |\lambda| \leq \|x\|\}$. Each B_x is compact, so, by Tychonoff's theorem, $B = \bigtimes_{x \in X} B_x$ is compact in the product topology. Now what is B? An element of B is just an assignment of a number $b(x) \in B_x$ for each x in X, that is, b is a function from X to $\mathbb{C}$ with $|b(x)| \leq \|x\|$. In particular, the unit ball $(X^*)_1$ is a subset of B, namely those $b \in B$ which are linear. What is the relative topology induced on $(X^*)_1$ by the product topology on B? It is precisely the weakest topology making $\ell \mapsto \ell(x)$ continuous for each x, that is, the weak-$*$ topology.

Thus, we must only show that $(X^*)_1$ is closed in the product topology. Suppose that ℓ_α is a net in $(X^*)_1$ with $\ell_\alpha \to \ell$. Since $|\ell(x)| \leq \|x\|$, we need only show ℓ is linear. But this is easy; if $x, y \in X$ and $\lambda, \mu \in \mathbb{C}$, then

$$\ell(\lambda x + \mu y) = \lim_\alpha \ell_\alpha(\lambda x + \mu y) = \lim_\alpha \lambda \ell_\alpha(x) + \mu \ell_\alpha(y)$$

$$= \lambda \ell(x) + \mu \ell(y) \quad \blacksquare$$

Appendix to IV.5   Weak and strong measurability

In Section II.1, we briefly discussed vector-valued measurable functions with values in an infinite dimensional Hilbert space $\mathcal{H}$. f was called measurable (in Problem 12 of Chapter II) if $(y, f(\cdot))$ was a complex-valued measurable function for each $y \in \mathcal{H}$. This notion might be called **weak measurability**. Another natural candidate for measurability is the a priori stronger notion of measurability which requires that $f^{-1}[C]$ be measurable for each open set $C \subset \mathcal{H}$. Throughout this book, *by a vector-valued measurable function, we will mean a function measurable in the weak sense*. However, to satisfy the reader's natural curiosity, a brief comparison of the various notions of measurability of vector-valued functions seems in order.

Definition   Let f be a function on a measure space $\langle M, \mu, \mathcal{R} \rangle$ taking values in a Banach space E.

(i) f is called **strongly measurable** if and only if there is a sequence of functions f_n so that $f_n(x) \to f(x)$ in norm for a.e. $x \in M$ and each f_n takes only finitely many values, each value being taken on a set in $\mathscr{R}$.

(ii) f is called **Borel measurable** if $f^{-1}[C] \in \mathscr{R}$ for each open set C in E (in the metric space topology on E).

(iii) f is called **weakly measurable** if and only if $\ell(f(x))$ is a complex-valued measurable function for each $\ell \in E^*$.

Proposition (a) A pointwise limit of a sequence of Borel measurable functions is a Borel measurable function.

(b) Let f be a function from M to E. If f is strongly measurable, then f is Borel measurable.

(c) Let f be a function from M to E. If f is Borel measurable, it is weakly measurable.

Proof (a) Let $f_n \to f$ pointwise in norm. Let C be an open set in E. Let $C_k = \{x \mid B_{k^{-1}}^x \subset C\}$ where B_ε^x is the ball of radius ε about x. Then,

$$f^{-1}[C] = \bigcup_{k=1}^{\infty} \bigcup_{n=1}^{\infty} \bigcap_{m>n} f^{-1}[C_k]$$

so f is Borel measurable.

(b) This is a direct consequence of (a) and the definitions.

(c) The composition of Borel functions is Borel. ∎

Theorem IV.22 Let $\mathscr{H}$ be a separable Hilbert space. Let f be a function from a measure space $\langle M, \mu, \mathscr{R} \rangle$ to $\mathscr{H}$. Then the following three statements are equivalent:

(a) f is strongly measurable.
(b) f is Borel measurable.
(c) f is weakly measurable.

Proof By the last proposition, we need only show that (c) implies (a). Let $\{\psi_n\}_{n=1}^{\infty}$ be an orthonormal basis for $\mathscr{H}$. Let $a_n = (\psi_n, f(x))$. Each a_n is a complex-valued measurable function. It is easy to construct $a_{n,m}(x)$ finite valued, $|a_{n,m}(x)| < |a_n(x)|$ for all x and $\lim_{m \to \infty} a_{n,m}(x) = a_n(x)$ for all $x \in X$. Define $f_N = \sum_{n=1}^{N} a_{n,N}(x)\psi_n$. f_N is finite valued and $f_N \to f$ in norm so f is strongly measurable. ∎

Example Let $\mathbb{C}_t$ be a copy of the complex numbers $\mathbb{C}$ and let $\mathscr{H} = \bigoplus_{t \in \mathbb{R}} \mathbb{C}_t$, that is, $\mathscr{H}$ consists of functions φ on $\mathbb{R}$, nonzero at only countably many t with $\sum_{t \in \mathbb{R}} |\varphi(t)|^2 < \infty$. Let φ_s be given by

$$\varphi_s(t) = \begin{cases} 1 & \text{if } t = s \\ 0 & \text{otherwise} \end{cases}$$

Then $\{\varphi_s\}_{s \in \mathbb{R}}$ is an orthonormal basis for $\mathcal{H}$. Let $f: \mathbb{R} \to \mathcal{H}$ be defined by $f(s) = \varphi_s$. For any $\psi \in \mathcal{H}$, $(\psi, f(s)) = 0$ except for a countable set so $(\psi, f(s))$ is measurable. Thus $f(s)$ is weakly measurable. But f is not strongly measurable; for if $f = \lim f_n$ pointwise in norm, then $\operatorname{Ran} f \in \overline{\bigcup \operatorname{Ran} f_n}$. If each f_n were finite valued, $\operatorname{Ran} f$ would be separable, which it is not.

NOTES

Section IV.1 For the reader who wishes to delve further into the realm of general point set topology, we recommend J. Kelley's *General Topology*, Van Nostrand–Reinhold, Princeton, New Jersey, 1955, most enthusiastically. The best way to read the book is to do all the problems; it is time consuming but well worth the effort if the reader can afford the time. Other good references on elementary (and sophisticated) topological notions include: K. Kuratowski, *Topology*, Vol. I, Academic Press, New York, 1966, W. Pervin, *Foundations of General Topology*, Academic Press, New York, 1964, and W. Thron, *Topological Structures*, Holt, New York, 1966.

The notion of topological spaces grew out of work of Fréchet and Hausdorff. The T_1–T_4 classification is due to P. Alexandroff and H. Hopf, in *Topologie I*, Berlin, 1935.

The concept of "Cauchy sequence" does not extend to an arbitrary topological space. However, one can add a "uniform structure" to the topological structure and thereby have spaces in which Cauchy sequence and completeness make sense. One thinks of neighborhoods of x as describing closeness to x. To have a notion of "closeness to x" uniform in x, we need a family $\mathcal{U}$ of subsets of $X \times X$ each containing the set $\Delta = \{\langle x, x \rangle \mid x \in X\}$. We need enough conditions on $\mathcal{U}$ so that $\mathcal{U}_x = \{U_x \mid U \in \mathcal{U}\}$ with $U_x = \{y \mid \langle x, y \rangle \in U\}$ is a neighborhood system for a topology. The canonical example is to let $\mathcal{U}$ be the family of all sets in $X \times X$ containing a set of the form $\{\langle x, y \rangle \mid \rho(x, y) < \varepsilon\}$ with ρ a metric. If G is a topological group (in particular, if G is a topological vector space), there is also a natural uniform structure given by $\mathcal{U} = \{U_N \mid N \in \eta\}$ where η the family of neighborhoods of the identity and $U_N = \{\langle x, y \rangle \mid xy^{-1} \in N\}$. Given a uniform structure $\mathcal{U}$, a net $\{x_\alpha \mid \alpha \in D\}$ is called a Cauchy net if and only if for each $U \in \mathcal{U}$, there is an $\alpha_0 \in D$ so that $\alpha, \beta > \alpha_0$ implies $\langle x_\alpha, x_\beta \rangle \in U$.

The notion of uniform space was first formalized in A. Weil, "Sur les espaces à structure uniforme et sur la topologie générale," *Actualités Sci. Ind.* **551**, Paris (1937). For a modern treatment of uniform spaces, see Kelley, Chapter 6, or G. Choquet, *Lectures on Analysis*, Benjamin, New York, §5.

Section IV.2 Nets were first introduced in E. H. Moore and H. L. Smith, "A General Theory of Limits," *Amer. J. Math.* **44** (1922), 102, and the theory is sometimes called Moore–Smith convergence in the older literature. See Kelley, Chapter 2, for additional discussion.

There is an alternate approach to convergence in topological spaces popularized by Bourbaki. For a discussion of this *theory of filters* see Choquet, §4, or Bourbaki, *Topologie générale*, Chapter 1. We find the filter theory of convergence very unintuitive and prefer the use of nets in all cases.

Section IV.3   It was Tychonoff who realized the utility of the product topology (and proved the Tychonoff theorem) in two fundamental papers: "Über die topologische Erweiterung von Räumen," *Math. Ann.* **102** (1929), 544–556, and "Über einen Funktionenraum," *Math. Ann.* **111** (1935), 762–766. The usual proof of Tychonoff's theorem (c.f. Kelley), depends on the f.i.p. criterion and is a little complicated. The machinery of filters, especially ultrafilters is ideal for a simpler looking proof of the theorem (cf. Choquet). This filter theoretic proof has a net theory translation which we should like to sketch. (1) A net $\{x_\alpha\}$ in a space X is called *universal* if for any $A \subset X$, $x_\alpha \in A$ eventually or $x_\alpha \in X \backslash A$ eventually. Note: A is arbitrary and the definition of universal net makes no mention of topology. (2) If x is a cluster point of a universal net, one has $x_\alpha \to x$, for it cannot happen that $x_\alpha \in A$ frequently without $x_\alpha \in A$ eventually. (3) Any net has a universal subnet. This is the technical heart of the proof and requires the axiom of choice. (4) X is compact if and only if every universal net converges. Given (3), this is just the Bolzano–Weierstrass theorem. (5) To prove Tychonoff's theorem, let $\{x_\alpha\}_{\alpha \in D}$ be a universal net in $\mathbf{X}_{i \in I} A_i$ with each A_i compact. Write $x_\alpha = \{x_\alpha^{(i)}\}_{i \in I}$ with $x_\alpha^{(i)} \in A_i$. Since $\{x_\alpha\}$ is universal, $\{x_\alpha^{(i)}\}$ is universal for each i. Since A_i is compact, $x_\alpha^{(i)} \to x^{(i)}$ for some $x^{(i)} \in A_i$. Let x be the element $\{x^{(i)}\}_{i \in I}$ in $\mathbf{X}_{i \in I} A_i$. Then $x_\alpha \to x$, so every universal net converges. We first learned this proof from O. E. Lanford, III, Les Houches lectures, 1970.

When does a topological space have a topology given by a metric? In general, there is not a simple answer, but for compact Hausdorff spaces, X is metrizable (has a topology given by a metric) if and only if it is second countable. In Section V.2, we see that a similar result holds for topological vector spaces. Both the compact and the vector space results are best understood in the context of uniform spaces; see Kelley, Chapter 6.

K. Weierstrass' original proof of the polynomial approximation theorem can be found on page 5 of Vol. 3 of his *Mathematische Werke*, Mayer and Müller, Berlin, 1903. Stone's generalization first appeared in M. H. Stone, "Applications of the Theory of Boolean Rings to General Topology," *Trans. Amer. Math. Soc.* **41** (1937), 325–481, and a simplified proof was given in his classic article "The Generalized Weierstrass Approximation Theorem," *Math. Mag.* **21** (1947/48), 167–184, 237–254.

Section IV.4   For a brief readable discussion of measure theory on compact spaces we especially recommend the first chapter of L. Nachbin, *The Haar Integral*, Van Nostrand–Reinhold, Princeton, New Jersey, 1965. For a more comprehensive discussion see N. Bourbaki, *Integration*, Chapters 1–4.

Much of our discussion on positive linear functionals goes through for vector spaces with an order allowing finite inf's and sup's, that is for vector lattices. For the deep relations between order notions and topology, see L. Nachbin, *Topology and Order*, Van Nostrand–Reinhold, Princeton, New Jersey, 1965.

For additional discussion of measure theory on locally compact spaces, see the quoted references of Nachbin and Bourbaki, or, for a discussion more similar to our approach, Choquet's book (see notes to Section IV.1).

Section IV.5   We will eventually prove a stronger result than our claim that the linear combinations of Dirac measures are vaguely dense in $\mathscr{M}(X)$. We will actually show that the linear combinations in $\mathcal{M}_{+,1}(X)$ are vaguely dense in $\mathcal{M}_{+,1}(X)$. Thus any positive measure μ with $\mu(X) = 1$ can be vaguely approximated by measures $\sum_{n=1}^{N} t_n \delta_{x_n}$ with $0 \leq t_n \leq 1$, $\sum t_n = 1$. This will follow from the Krein–Milman theorem which we discuss in Section XIV.1. The Banach–Alaoglu theorem was proven in L. Alaoglu: "Weak Topologies of Normed Linear Spaces," *Ann. Math.* **41** (1940), 252–267.

Theorem IV.22 can be extended to an arbitrary separable Banach space. More generally, one has Pettis' theorem: A vector-valued function is strongly measurable if and only if it is weakly measurable and almost separably valued (in the sense that after changing f on a set of measure zero, Ran f is separable). This theorem was first proven in B. J. Pettis, "On Integration in Vector Spaces," *Trans. Amer. Math. Soc.* **44** (1938), 277–304.

One can define the integral of a strongly measurable function by methods analogous to the methods used for real-valued functions. This *Bochner integral* is discussed in K. Yosida, *Functional Analysis*, Springer, New York, 1965 and in many other texts. It was invented by S. Bochner in "Integration von Funktionen, deren Werte die Elemente eines Vektoraumes sind," *Fund. Math.* **20** (1933), 262–276. The Bochner integral obeys a norm dominated convergence theorem. Throughout this book, we use the weak integral defined by $\ell(\int f(x)\, d\mu) = \int \ell(f(x))\, d\mu$. The Bochner integral has nicer properties than this weak integral but we will not need these extra properties so we settle for the simpler weak integral.

PROBLEMS

1. Prove that the family of all topologies on a space is a complete lattice, that is, that any family of topologies has a least upper bound and a greatest lower bound.

2. (Kuratowski closure axioms) Show that the operation $A \mapsto \bar{A}$ in a topological space has the properties:

    (i) $\overline{(\bar{A})} = \bar{A}$
    (ii) $\overline{A \cup B} = \bar{A} \cup \bar{B}$
    (iii) $A \subset \bar{A}$
    (iv) $\bar{\varnothing} = \varnothing$

    Conversely, suppose that $^- : 2^X \to 2^X$ is given ($2^X \equiv$ all subsets of X) obeying (i)–(iv). Show the family of sets B with $\overline{X \backslash B} = X \backslash B$ forms a topology for which the closure operation is $^-$.
    Reference: Kelley, pp. 42–43.

3. (a) Let 2 be the topological space $\{0, 1\}$ with the discrete topology. Prove that a topological space X is connected if and only if any continuous function $f: X \to 2$ is constant.

   (b) Prove that any product of connected spaces is connected.

   (c) Let S be a topological space. Suppose that $A, B \subset S$ are connected in the relative topology and $A \cap B \neq \varnothing$; $A \cup B = S$. Show that S is connected.

   (d) Let S be a topological space. Suppose that $S = \bar{D}$ and D is connected. Prove that S is connected.

   (e) Prove that a continuous image of a connected space is connected.

(f) Prove the intermediate value theorem of freshman calculus, that is, if f is a continuous function on $[a, b]$, then for any $f(a) < x < f(b)$, there is a $c \in [a, b]$ with $f(c) = x$.
Hint: Use (a) to prove (b)–(e).

4. (a) A topological space X is called **Lindelöf** if every open cover has a countable subcover. Prove that any second countable space is Lindelöf.
   (b) Prove that a second countable, regular (that is T_3) space is normal (that is T_4).
   Reference: Kelley, pp. 49, 113.

5. (a) Prove that $\mathbb{R}$ and $\mathbb{R}^n$ are not homeomorphic for any $n > 1$.
   (b) Prove that $\mathbb{R} \neq X \times X$ for any topological space X.
   Hint: What happens to $\mathbb{R}$ if a single point is removed?

6. A topological space X is called **arcwise connected** if given $x, y \in X$, there is a continuous function (an arc!) $f\colon [0, 1] \to X$ with $f(0) = x, f(1) = y$.
   (a) Show that if X is arcwise connected, it is connected.
   (b) Let X_0 be the graph of the function $y = \sin 1/x$ on $\mathbb{R} - \{0\}$, given the relative topology as a subset of the plane. Let $X = X_0 \cup \{\langle x, y \rangle \mid x = 0\}$. Show that Y is connected but not arcwise connected.

7. Let $X = \mathbb{R}$ with the topology $\mathcal{T}$ generated by all sets of the form $\{[a, b) \mid a, b \in \mathbb{R}\}$ which is actually a base for $\mathcal{T}$. Prove that
   (a) $\langle X, \mathcal{T} \rangle$ is separable.
   (b) $\langle X, \mathcal{T} \rangle$ is first countable.
   (c) $\langle X, \mathcal{T} \rangle$ is *not* second countable.

8. Prove that a subspace of a separable metric space is separable.

9. Let Y be $\mathbb{R}^2$ with the product topology given by taking the topology $\mathcal{T}$ of Problem 7 on each factor. Prove that:
   (a) Y is separable.
   (b) The line $x + y = 1$ is not separable in the relative topology.

10. Let X be any uncountable set and let $\mathcal{T}$ be the topology consisting of $\varnothing$ and complements of finite sets. Prove that
    (a) X is separable.
    (b) X is compact.
    (c) X is T_1 but not T_2.
    (d) X is neither first nor second countable.

†*11*. Prove Theorem IV.2.

12. Let X be the Banach space l_∞ and consider the sequence $\delta_1, \delta_2, \ldots$ in X^* given by
$$\delta_n(\{c_k\}_{k=1}^\infty) = c_n$$
Prove that $\{\delta_n\}, \ldots$ has no weak-∗ convergent subsequence but that it has a weak-∗ convergent subnet.

13. Give an example to show that a pointwise limit of a *net* of Borel functions on $\mathbb{R}$ may not be Borel.

14. Show that the space of the example in Section IV.2 is not compact but is Lindelöf (see Problem 4).

15. Let $\mathscr{A}$ be the family of continuous functions on $[0, 2\pi]$ with the property $\int_0^{2\pi} e^{ikx} f(x) = 0$ if k is a negative integer. Prove $\mathscr{A}$ is an algebra which is closed and separates points with $1 \in \mathscr{A}$ but for which $\mathscr{A} \neq C[0, 2\pi]$.

†16. Prove the conclusion of the Stone–Weierstrass theorem in the case where we do not suppose $1 \in \mathscr{B}$.

*17. Let $\mathscr{B}$ be an ideal of $C_{\mathbb{R}}(X)$ which is closed. Let $Y = \{x \in X | f(x) = 0 \text{ for all } f \in \mathscr{B}\}$. Prove that Y is closed and that $\mathscr{B} = \{f \in C_{\mathbb{R}}(X) | f = 0 \text{ on } Y\}$.

18. Prove the Tietze theorem in the case when X is merely assumed normal. (See the hints given in Kelley, Chapter 7, Problem O.)

19. Let f be a continuous function on $[-\frac{1}{2}, \frac{1}{2}]$ with $f(\frac{1}{2}) = f(-\frac{1}{2}) = 0$. Let $s_k(x)$ be a sequence of functions with $\int_{-1}^{1} s_k(x)\, dx = 1$, each $s_k \geq 0$ so that for any $\delta > 0$,

$$\lim_{k \to \infty} \int_{1 \geq |x| \geq \delta} s_k(x) = 0$$

Prove that

$$\lim_{k \to \infty} \int_{-1/2}^{1/2} s_k(x - y) f(y)\, dy = f(x)$$

for any $x \in [-\frac{1}{2}, \frac{1}{2}]$ and that the convergence is uniform.

20. Let $s_k(x) = (I_k)^{-1}(1 - x^2)^k$ where $I_k = \int_{-1}^{1} (1 - x^2)^k\, dx$. Using Problem 19, prove that any continuous function on $[-\frac{1}{4}, \frac{1}{4}]$ is a limit of polynomials uniformly on $[-\frac{1}{4}, \frac{1}{4}]$.

21. Use the Stone–Weierstrass theorem to prove that:
    (a) $\{e^{ikx}\}_{k=-\infty}^{\infty}$ are a complete *orthogonal* set for $L^2[0, 2\pi]$.
    (b) The Legendre polynomials are a complete *orthogonal* set for $L^2[-1, 1]$.
    *(c) The spherical harmonics are a complete orthonormal set for L^2 of the sphere. (Hint: Use your knowledge of Clebsch–Gordon coefficients!)

22. Prove *Dini's theorem*: Let X be a compact Hausdorff space. Suppose f_n is a monotone decreasing family of functions; let $f_n(x) \to f(x)$ pointwise. Then f_n converges uniformly if and only if f is continuous.

23. Let X be a locally compact Hausdorff space. Consider $\hat{X} = X \cup \{\infty\}$ where ∞ is a "point" not in X. Call $O \subset \hat{X}$ open if either $\infty \notin O$ and O is open in X or $\infty \in O$ and $\hat{X} \backslash O$ is compact. Prove that $\hat{X}$ is a compact Hausdorff space; it is called the **one-point compactification** of X.

24. Prove the Stone–Weierstrass theorem for a locally compact space X: If $\mathscr{A}$ is a closed subalgebra of $C_\infty(X)$, the continuous real-valued functions vanishing at ∞, and if $\mathscr{A}$ separates points and for each $x \in X$, there is $f \in \mathscr{A}$ with $f(x) \neq 0$, then $\mathscr{A} = C_\infty(X)$.

25. Let X be a locally compact Hausdorff space. Prove that for $C, D \subset X$, D closed, C compact, there is a continuous function f, $0 \leq f \leq 1$, on X with $f[C] = 0, f[D] = 1$.
Remark. Use the space $\tilde{X}$ of Problem 23 to solve 24 and 25.

26. (a) Prove that any locally compact Hausdorff space is T_3.
 (b) Prove that any second countable, locally compact Hausdorff space is normal.
 (c) Prove that any σ-compact, locally compact Hausdorff space is normal.
 Remark. There exist locally compact spaces which are Hausdorff but not normal, see Kelley, Chapter 4, Problem E.

27. A group G with a topology is called a topological group if the map $\langle x, y\rangle \mapsto xy^{-1}$ of $G \times G \to G$ is jointly continuous. A function f on a topological group G is called uniformly continuous if, for any ε, we can find a neighborhood N_ε of $e \in G$ (the identity) with $|f(x) - f(y)| < \varepsilon$ if $xy^{-1} \in N_\varepsilon$. Prove that any continuous function on a compact topological group is uniformly continuous.

*28. (a) Let $\mathscr{A}$ be an algebra of real-valued bounded continuous functions on $\mathbb{R}$ which separates points and is closed in $\|\cdot\|_\infty$. Form $X_\mathscr{A} = \times_{f \in \mathscr{A}} \{x \in \mathbb{R} \mid |x| \leq \|f\|_\infty\}$ with the product topology. Map $\mathbb{R} \to X_\mathscr{A}$ by letting x go into the point whose coordinates are $\{f(x)\}_{f \in \mathscr{A}}$. Prove that the image of $\mathbb{R}$ in $X_\mathscr{A}$ is homeomorphic to $\mathbb{R}$ if and only if $\mathscr{A}$ contains the functions of compact support.
 (b) A topological space X with a map $f : \mathbb{R} \to X$ is called a **compactification** of $\mathbb{R}$ if f is a homeomorphism of $\mathbb{R}$ and its image, if the image is dense in X and if X is a compact Hausdorff space. Two compactifications $f : \mathbb{R} \to X$ and $g : \mathbb{R} \to Y$ are considered identical if there is a homeomorphism $h : X \to Y$ with $h \circ f = g$. Prove that there is a one–one correspondence between compactifications of $\mathbb{R}$ and algebras $\mathscr{A} \subset C_\mathbb{R}$ obeying the conditions of (a).
 (c) If we take $\mathscr{A} = C(\mathbb{R})$, the compactification we obtain via the construction in (a) is called the **Stone–Čech compactification**, $\tilde{\mathbb{R}}$. Prove that $\tilde{\mathbb{R}}$ is a universal compactification of $\mathbb{R}$ in the following sense: Given any compactification $f : \mathbb{R} \to X$ and given the Stone–Čech compactification $g : \mathbb{R} \to \tilde{\mathbb{R}}$ we can find $h : \tilde{\mathbb{R}} \to X$ continuous and surjective with $h \circ g = f$.

29. Let $\langle X, d\rangle$ be a metric space with no isolated points. Suppose that every continuous function on X is uniformly continuous. Show that X is compact.

30. (a) Prove that every metric space is normal.
 (b) Prove that every closed set in a metric space is a G_δ.

†31. Prove the uniqueness statement of Theorem IV.16.

32. Let $\{a_n\}$ be a sequence of numbers with the following property: If $\sum_{n=0}^N \alpha_n x^n \geq 0$ for all $x \in [0, 1]$ then $\sum_{n=0}^N \alpha_n a_n \geq 0$. Prove that there is a unique, (positive) measure μ on $[0, 1]$ with $a_n = \int_0^1 x^n \, d\mu$.

33. Let X be a vector space with Y a family of functionals separating points. Prove that if the $\sigma(X, Y)$ topology comes from a metric, then Y has a countable *algebraic* dimension. An **algebraic basis** for Y is a subset whose *finite* linear combinations span Y. The **algebraic dimension** is the number of elements in a minimal algebraic basis.

34. Let X be a real Banach space and let C be the unit ball of X^* with the weak-$*$ topology. Prove that a continuous function on C can be uniformly approximated by polynomials in the elements of X acting as linear functionals on X^*.

35. Let X be a Banach space, X^* its dual. Let L_n, $n \geq 1$ be elements of X^* with $L_n \to L \in X^*$ in the weak-$*$ sense. Let $x_n \to x$ in norm. Is it necessarily true that $L_n(x_n) \to L(x)$?

36. Prove that X is dense in X^{**} in the $\sigma(X^{**}, X^*)$ topology.

37. Let $T: C(X) \to C(Y)$ be linear. We say T is positivity preserving (or positive) if $Tf \geq 0$ whenever $f \geq 0$. If T is positive, we write $T \geq 0$. If $S - T \geq 0$ we write $T \leq S$.
   (a) Prove that any $T \geq 0$ is automatically continuous and that $\|T\| = \|T1\|_\infty$.
   (b) Let S_n be an increasing family of maps. Prove that S_n converges in operator norm if and only if $S_n 1$ converges in function norm.

†38. Prove the first proposition in Section IV.2.

†39. Find a Banach space and a weakly convergent net which is not norm bounded.

†40. Let X be an infinite-dimensional Banach space with the weak topology. Prove that the closure of the unit sphere is the unit ball.

†41. Let X be a compact Hausdorff space. Prove that the set of convergent infinite linear combinations of point measures is norm closed in $\mathcal{M}(X)$.

†42. Prove Theorem IV.19 directly.

*43. (a) Let X be a compact set with a countable basis. Let μ be a Baire measure on X. Prove that $L^p(X, d\mu)$ is separable for all $p < \infty$. (Hint: Let A_n be a countable basis of sets. For all n, m with $\bar{A}_n \cap \bar{A}_m = \varnothing$, find $f_{n,m} \in C(X)$ with $f = 0$ on A_n, $f = 1$ on A_m. Use the $f_{n,m}$ to construct a countable dense set in $C(X)$. Then use the fact that $C(X)$ is dense in $L^p(X, d\mu)$.)
   (b) Extend the result of (a) to the case where X is only locally compact (Hint: Prove that X is σ-compact).

*44. Do any fifty problems in Kelley's book.

V: Locally Convex Spaces

Mathematicians are like Frenchmen: whatever you say to them they translate into their own language and forthwith it is something entirely different.

J. W. Goethe

V.1 General properties

We have already discussed several nonnormed topologies on vector spaces in the investigation of weak topologies in Section IV.5. We have also alluded to the fact that if X is a locally compact topological space, the Baire measures on X are the dual space of the continuous functions of compact support, $\kappa(X)$, when it is given a suitable nonnormed topology. Our goal in this chapter is to discuss a general class of topologized vector spaces which includes these examples and also the spaces of distributions which arise in a wide variety of functional situations and physical problems.

The idea behind the topologies we discuss is quite simple. Suppose that, instead of one norm, we have a family of norms $\{\rho_\alpha\}_{\alpha \in A}$ where A is some index set. We should like a topology in which a net $\{x_\beta\}$ converges to x if and only if $\rho_\alpha(x_\beta - x) \to 0$ for each fixed $\alpha \in A$. However, it is useful to weaken one condition on the norm. Recall that $\|x\| = 0$ implies $x = 0$ and that this condition is needed for limits to be unique, that is, for the induced topology to be Hausdorff. Suppose that $\{\rho_\alpha\}_{\alpha \in A}$ is a family of objects obeying all the norm conditions except $x = 0$ when $\rho_\alpha(x) = 0$ for some α. But suppose instead that $x = 0$ whenever $\rho_\alpha(x) = 0$ for all α; then it is easy to see limits are unique in a topology where convergence means $\rho_\alpha(x_\beta - x) \to 0$ for each fixed α. We thus define:

Definition A **seminorm** on a vector space V is a map $\rho: V \to [0, \infty)$ obeying:

(i) $\rho(x + y) \leq \rho(x) + \rho(y)$
(ii) $\rho(\alpha x) = |\alpha| \rho(x)$ for $\alpha \in \mathbb{C}$ (or $\mathbb{R}$).

A family of seminorms $\{\rho_\alpha\}_{\alpha \in A}$ is said to **separate points** if

(iii) $\rho_\alpha(x) = 0$ for all $\alpha \in A$ implies $x = 0$.

Definition A **locally convex space** is a vector space X (over $\mathbb{R}$ or $\mathbb{C}$) with a family $\{\rho_\alpha\}_{\alpha \in A}$ of seminorms separating points. The **natural topology** on a locally convex space is the weakest topology in which all the ρ_α are continuous and in which the operation of addition is continuous.

We temporarily defer giving examples or the explanation of the term "locally convex." We also note that many authors do not require the seminorms to separate points but add it as an extra condition. The significance of the separation condition is that it implies (Problem 6a):

Proposition The natural topology of a locally convex space (with our definition!) is Hausdorff.

A neighborhood base at 0 for the natural topology is given by the sets $\{N_{\alpha_1, \ldots, \alpha_n; \varepsilon} | \alpha_1, \ldots, \alpha_n \in A; \varepsilon > 0\}$ where

$$N_{\alpha_1, \ldots, \alpha_n; \varepsilon} = \{x | \rho_{\alpha_i}(x) < \varepsilon, i = 1, \ldots, n\}$$

Thus, a net $x_\beta \to x$ if and only if $\rho_\alpha(x_\beta - x) \to 0$ for all $\alpha \in A$. The notion of completeness extends naturally:

Definition A net $\{x_\beta\}$ in a locally convex space X is called **Cauchy** if and only if, for all $\varepsilon > 0$, and for each seminorm ρ_α there is a β_0 so that $\rho_\alpha(x_\beta - x_\gamma) < \varepsilon$ if $\beta, \gamma > \beta_0$. X is called **complete** if every Cauchy net converges.

The important structure on a locally convex space is the natural topology rather than the particular seminorms used to generate the topology. We call two families of seminorms $\{\rho_\alpha\}_{\alpha \in A}$ and $\{d_\beta\}_{\beta \in B}$ on a vector space X **equivalent** if they generate the same natural topology. It is often useful to know (Problem 6b):

Proposition Let $\{\rho_\alpha\}_{\alpha \in A}$ and $\{d_\beta\}_{\beta \in B}$ be two families of seminorms. The following statements are equivalent:

(a) The families are equivalent families of seminorms.

(b) Each ρ_α is continuous in the d-natural topology and each d_β is continuous in the ρ-natural topology.

(c) For each $\alpha \in A$, there are $\beta_1, \ldots, \beta_n \in B$ and $C > 0$ so that for all $x \in X$

$$\rho_\alpha(x) \leq C(d_{\beta_1}(x) + \cdots + d_{\beta_n}(x))$$

and for each $\beta \in B$, there are $\alpha_1, \ldots, \alpha_m \in A$ and $D > 0$ so that for all $x \in X$

$$d_\beta(x) \leq D(\rho_{\alpha_1}(x) + \cdots + \rho_{\alpha_m}(x))$$

The appearance of expressions like $C(d_{\beta_1}(x) + \cdots + d_{\beta_n}(x))$ is quite common in the theory of locally convex spaces. It is thus useful to consider families of seminorms with a special property:

Definition A family $\{\rho_\alpha\}_{\alpha \in A}$ of seminorms on a vector space V is called **directed** if and only if for all $\alpha, \beta \in A$ there is a $\gamma \in A$ and a C so that

$$\rho_\alpha(x) + \rho_\beta(x) \leq C\rho_\gamma(x)$$

for all $x \in V$. Equivalently, by induction, for all $\alpha_1, \ldots, \alpha_n \in A$ there is a γ and D so that

$$\rho_{\alpha_1}(x) + \cdots + \rho_{\alpha_n}(x) \leq D\rho_\gamma(x)$$

for all $x \in V$.

For example, if $\{\rho_\alpha\}_{\alpha \in A}$ is a directed family, then $\{\{x \mid \rho_\alpha(x) < \varepsilon\} \mid \alpha \in A, \varepsilon > 0\}$ is a neighborhood base at 0. One can always find directed sets of seminorms:

Proposition Every locally convex space has a directed family of seminorms equivalent to the family defining the space.

Proof If $\{\rho_\alpha\}_{\alpha \in A}$ defines the space, let B be the set of finite subsets of A. If $F \in B$, let $d_F = \sum_{\alpha \in F} \rho_\alpha$. Then $\{d_F\}_{F \in B}$ is directed and equivalent to the initial set. ∎

We consider briefly two examples. In Sections V.3 and V.4 we will discuss several other examples; in particular, the technical appendix of Section V.3 is useful to the reader desiring experience with equivalent semi-norms and directed sets of seminorms.

Example 1 Let X be a vector space and suppose that Y is a set of linear functionals on X separating points. In Section IV.5, we introduced the $\sigma(X,Y)$-topology. It is precisely the locally convex topology generated by the seminorms $\{\rho_\ell \mid \ell \in Y\}$ where $\rho_\ell(x) = |\ell(x)|$. While this topology is given by seminorms, it is never given by norms if Y has infinite algebraic dimension (Problem 2).

Example 2 Let D be a region of the complex plane, that is, D is connected and open. Let $\mathcal{O}_D$ be the vector space of all (single-valued) analytic functions in D. For any compact $C \subset D$, let $\rho_C(f) = \sup_{z \in C} |f(z)|$. $\mathcal{O}_D$, topologized by the seminorms ρ_C, is a locally convex space which is complete. For suppose f_α is a ρ_C-Cauchy net for all C. Then, $f_\alpha(z) \to f(z)$ uniformly on compacts sets. By a classical theorem of Weierstrass, f is analytic (essentially because f is analytic if and only if it obeys the Cauchy integral formula which is preserved by uniform limits). Let

$$\rho_C^{(2)}(f) = \iint_{x+iy \in C} |f(x+iy)|^2 \, dx \, dy$$

The families $\{\rho_C^{(2)}\}$ and $\{\rho_C\}$ are equivalent families (Problem 7).

We are now prepared to discuss the reason for the name locally convex and the associated geometrical ideas and construction. The neighborhoods $N_{\alpha_1,\alpha_2,\ldots,\alpha_n;\varepsilon}$ have special geometric properties:

Definition A set $C \subset V$, a vector space, is called **convex** if x and $y \in V$, $0 \leq t \leq 1$, implies $tx + (1-t)y \in C$. C is called **balanced** (or **circled**) if $x \in C$ and $|\lambda| = 1$ implies $\lambda x \in C$. Finally, C is called **absorbing** (or **absorbent**) if $\bigcup_{t>0} tC = V$, that is, if for every $x \in V$, $sx \in C$ for some $s > 0$.

If C is convex and V is a vector space over the reals, balanced means only that $-x \in C$ whenever $x \in C$; if V is a vector space over the complex numbers, balanced means $e^{i\theta}x \in C$ whenever $\theta \in [0, 2\pi)$ and $x \in C$ (so circled is a more suitable name).

It is an elementary application of the definitions to see that the $N_{\alpha_1,\ldots,\alpha_n;\varepsilon}$ are convex; in fact:

Proposition If $\rho_{\alpha_1}, \ldots, \rho_{\alpha_n}$ are seminorms on a vector space V then $\{x \mid |\rho_{\alpha_1}(x)| < \varepsilon, \ldots, |\rho_{\alpha_n}(x)| < \varepsilon\}$ is a balanced, convex, absorbing set.

This proposition is a one-half of the basic theorem:

Theorem V.1 Let V be a vector space with a Hausdorff topology in which addition and scalar multiplication are separately continuous. Then V is a locally convex space (that is, has a topology given by a family of seminorms) if and only if 0 has a neighborhood base of balanced, convex, absorbing sets.

The proof of the other half of the theorem, that is, that V has a topology generated by seminorms if 0 has a neighborhood base of balanced, convex, absorbing sets relies on the following technical device:

Definition Let C be an absorbing subset of a vector space V with the additional property that if $x \in C$ and $0 \leq t \leq 1$, then $tx \in C$. The **Minkowski functional** or **gauge** of C is the map $\rho: V \to [0, \infty)$ given by

$$\rho(x) = \inf\{\lambda \mid x \in \lambda C\}$$
$$= [\sup\{\mu \mid \mu x \in C\}]^{-1}$$

Lemma
  (a) If $t \geq 0$, then $\rho(tx) = t\rho(x)$ for the gauge of any set C.
  (b) ρ obeys $\rho(x + y) \leq \rho(x) + \rho(y)$ if C is convex.
  (c) ρ obeys $\rho(\lambda x) = |\lambda| \rho(x)$ if C is circled.
  (d) $\{x \mid \rho(x) < 1\} \subset C \subset \{x \mid \rho(x) \leq 1\}$.

The proof of this beautiful lemma is left to the problems.

Proof of Theorem V.1 Let $\mathcal{U}$ be a neighborhood base at 0 containing only convex, balanced, absorbing sets; for each $U \in \mathcal{U}$, let ρ_U be the gauge of U. By (b) and (c) of the lemma, ρ is a seminorm and by (d) the neighborhoods of 0 in the original topology are the same as those in the locally convex topology given by the seminorms $\{\rho_U \mid U \in \mathcal{U}\}$. Since addition is separately continuous in both topologies, the neighborhoods about any point are identical in the two topologies. ∎

In normed linear spaces, a linear map from X to Y is continuous if and only if it is bounded. A similar result holds in locally convex spaces (Problem 9):

Theorem V.2 Let X and Y be locally convex spaces with families of semi-norms $\{\rho_\alpha\}_{\alpha \in A}$ and $\{d_\beta\}_{\beta \in B}$. Then a linear map $T: X \to Y$, is continuous if and only if for all $\beta \in B$, there are $\alpha_1, \ldots, \alpha_n \in A$ and $C > 0$ with

$$d_\beta(Tx) \leq C(\rho_{\alpha_1}(x) + \cdots + \rho_{\alpha_n}(x))$$

If the $\{\rho_\alpha\}_{\alpha \in A}$ are directed, then T is continuous if and only if for all $\beta \in B$

$$d_\beta(Tx) \leq D\rho_\alpha(x)$$

for some $\alpha \in A$, and $D > 0$.

Finally, we conclude this introduction by discussing two applications of the Hahn–Banach theorem (Theorem III.5) to locally convex spaces. First:

Theorem V.3 Let X be a locally convex space and let $Y \subset X$ be a subspace. Let $\ell: Y \to \mathbb{R}$ (or $\mathbb{C}$ if X is a complex space) be linear and continuous. Then, there is a continuous linear map $L: X \to \mathbb{R}$ (or $\mathbb{C}$) with $L \upharpoonright Y = \ell$.

Proof The relative topology on Y is given by the restrictions of the continuous seminorms to Y. Thus, $|\ell(x)| \leq C\rho(x)$ for some continuous seminorm. Applying Theorem III.5 or III.6, we obtain our result. ∎

Thus, locally convex spaces possess many continuous linear functionals; in fact, enough to separate points. We denote by X^* the family of continuous linear functionals on X and call it the **topological dual**.

The second application of the theorem is more geometric in nature and is related to the idea of slipping a closed hyperplane between disjoint convex sets; see Figure V.1. A **hyperplane** is the set of points where $\ell(x) = a$ for some *real-valued* linear functional (even in the complex case).

Definition We say that two sets A and B in a locally convex space are **separated by a hyperplane** if there is a continuous real-valued functional ℓ and an $a \in \mathbb{R}$ with $\ell(x) \leq a$ for $x \in A$ and $\ell(x) \geq a$ for $x \in B$. If $\ell(x) < a$ for $x \in A$ and $\ell(x) > a$ for $x \in B$, we say that A and B are **strictly separated**.

130 V: LOCALLY CONVEX SPACES

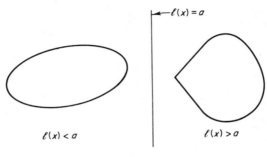

FIGURE V.1

Theorem V.4 (Separating hyperplane theorem) Let A and B be disjoint convex sets in a locally convex space X. Then:

(a) If A is open, they can be separated by a hyperplane.
(b) If A and B are both open, they can be strictly separated by a hyperplane.
(c) If A is compact and B is closed, they can be strictly separated by a hyperplane.

Proof (a) Pick $-x \in A - B = \{y - z \mid y \in A, z \in B\}$. Let $C = A - B + \{x\}$. Then C is open and thus absorbing, convex, $0 \in C$ and $x \notin C$ since A and B are disjoint. Let ρ_C be the Minkowski functional for C. Then $\rho_C(z + y) \leq \rho_C(z) + \rho_C(y)$ and $\rho_C(az) = a\rho_C(z)$ if $a > 0$. Define ℓ on $\{\lambda x \mid \lambda \in \mathbb{R}\}$ by $\ell(\lambda x) = \lambda$. Since $x \notin C$, $\rho_C(x) \geq 1$ so that $\ell(x) \leq \rho_C(x)$. Thus, by Theorem III.5, ℓ has an extension to all of X with $\ell(y) \leq \rho_C(y)$. Since $C \cap (-C) \subset \ell^{-1}[-1, 1]$, ℓ is continuous. By the inequality, $\ell(y) \leq 1$ if $y \in C$. Thus for any $a \in A$ and $b \in B$, $\ell(a) \leq \ell(b) + (1 - \ell(x))$. Since $\ell(x) = 1$,

$$\sup_{a \in A} \ell(a) \leq \inf_{b \in B} \ell(b)$$

so ℓ separates A and B.

(b) It is easy to see that if ℓ is a nonzero linear functional and A is open, then $\ell[A]$ is open. Since $\ell[A]$ and $\ell[B]$ are open and intersect in at most one point, they are automatically disjoint.

(c) Since A and B are disjoint, $0 \notin B - A \equiv S$. Since A is compact, it is easy to see that S is closed. Thus there is an open convex set U disjoint from S with $0 \in U$. Let $A' = A + \frac{1}{2}U$ and $B' = B - \frac{1}{2}U$. Then A' and B' are disjoint, open, convex sets that may be strictly separated by a hyperplane by (b). Since $A \subset A'$ and $B \subset B'$, this hyperplane also separates A and B. ∎

In Chapter XIV we will discuss an "algebraic Hahn–Banach theorem," i.e. a form of the separation theorem which makes no mention of open sets or continuous functions.

V.2 Fréchet spaces

In Section III.5, we saw that complete metric spaces have special properties which imply strong results for Banach spaces. It is thus of interest to single out those locally convex spaces that are also complete metric spaces. First, we must ask which locally convex spaces are metrizable, that is, have a topology generated by a metric. These are not only the spaces whose topology is given by a norm, for if ρ is a metric, $\rho(x, 0)$ need not be a norm since $\rho(\lambda x, 0)$ need not equal $\lambda\rho(x, 0)$.

Theorem V.5 Let X be a locally convex space. The following are equivalent:

(a) X is metrizable.
(b) 0 has a countable neighborhood base.
(c) The topology on X is generated by some countable family of seminorms.

Proof We show (a) $\Rightarrow$ (b) $\Rightarrow$ (c) $\Rightarrow$ (a).
(a) $\Rightarrow$ (b) is a property of any metric space.
(b) $\Rightarrow$ (c) follows from the fact that if $\mathcal{U}$ is any neighborhood base of convex, balanced sets, the gauges of $U \in \mathcal{U}$ generate the topology and the fact that if 0 has a countable neighborhood base, we can find a countable neighborhood base of convex, balanced sets.
(c) $\Rightarrow$ (a). Let $\{\rho_n\}_{n=1,2,...}$ be a family of seminorms generating the topology. Define ρ on $X \times X$ by

$$\rho(x, y) = \sum_{n=1}^{\infty} 2^{-n} \left[\frac{\rho_n(x - y)}{1 + \rho_n(x - y)} \right] \qquad (V.1)$$

Since $a/(1 + a) < 1$ for any $a > 0$, $\rho(x, y) < \infty$. It is easy to see that ρ is a metric and that it generates the same topology as the $\{\rho_n\}_{n=1}^{\infty}$ (Problem 10a). ∎

In addition, the two notions of complete are the same (Problem 10b):

Proposition A net $\{x_\alpha\}$ is Cauchy in the metric ρ of (V.1) if and only if it is Cauchy in each ρ_n. Thus a metrizable locally convex space X is complete as a metric space if and only if it is complete as a locally convex space.

Definition A complete metrizable locally convex space is called a **Fréchet space**.

As complete metric spaces, Fréchet spaces obey the Baire category theorem, and thus one can prove theorems analogous to those found in Section III.5:

Theorem V.6 If X and Y are Fréchet spaces and $f \colon X \to Y$ is a continuous linear surjection, then f is open.

Theorem V.7 Let X and Y be Fréchet spaces; let $\mathscr{F}$ be a family of continuous linear maps from X to Y so that for continuous seminorm ρ on Y and every $x \in X$, $\{\rho(F(x)) \mid F \in \mathscr{F}\}$ is bounded. Then, for each ρ there is a continuous seminorm d on X and a $C > 0$ so that

$$\rho(Fx) \leq C d(x)$$

for all $x \in X$ and $F \in \mathscr{F}$.

For an application of Theorem V.6, see Problem 12. As an application of Theorem V.7, we first notice that the corollary of Theorem III.9 goes through without change:

Corollary If X is a Fréchet space, a separately continuous bilinear functional, B, is jointly continuous, that is, $|B(f, g)| \leq C\rho_1(f)\rho_2(g)$ for some continuous seminorms ρ_1, ρ_2.

We can also prove the following corollary of Theorem V.7 whose proof can be viewed as Theorem I.27 in disguise:

Theorem V.8 Let X be a Fréchet space and let $f_n \in X^*$ be a *sequence* converging to $f \in X^*$ in the $\sigma(X^*, X)$-topology. Then $f_n \to f$ uniformly on compact subsets of X.

Proof Since $f_n(x)$ is convergent, it is bounded, so we can find a continuous seminorm ρ on X so that $|f_n(x)| \leq C\rho(x)$. Given a compact subset D of X, and an ε, pick a finite cover of D by sets $U_1, \ldots, U_m$ so that $x, y \in U_k$ implies

$\rho(x - y) \leq \varepsilon/3C$. Now pick $x_k \in U_k$ and N so $n > N$ implies $|f_n(x_i) - f(x_i)| < \varepsilon/3$ for $i = 1, \ldots, m$. Then an $\varepsilon/3$ argument proves

$$\sup_{x \in D} |f_n(x) - f(x)| < \varepsilon \quad \text{if} \quad n > N \quad \blacksquare$$

V.3 Functions of rapid decrease and the tempered distributions

We want to discuss a very convenient space of functions, the functions of rapid decrease $\mathscr{S}$ and its dual, the tempered distributions. To make the definitions flow more smoothly, we first introduce some notation. We will write functions on $\mathbb{R}^n$ merely as $f(x)$, $x = \langle x_1, \ldots, x_n \rangle$. I_+^n will always denote the set of all n-tuples of nonnegative integers $\alpha = \langle \alpha_1, \ldots, \alpha_n \rangle$ and $|\alpha| = \sum_{i=1}^n \alpha_i$; $I_+^1 = I_+$. Further,

$$D^\alpha \quad \text{will denote} \quad \frac{\partial^{|\alpha|}}{\partial x_1^{\alpha_1} \cdots \partial x_n^{\alpha_n}}$$

and x^α denotes $x_1^{\alpha_1} \cdots x_n^{\alpha_n}$.

Definition The **functions of rapid decrease** $\mathscr{S}(\mathbb{R}^n)$ is the set of infinitely differentiable complex-valued functions $\varphi(x)$ on $\mathbb{R}^n$ for which

$$\|\varphi\|_{\alpha, \beta} \equiv \sup_{x \in \mathbb{R}^n} |x^\alpha D^\beta \varphi(x)| < \infty$$

for all $\alpha, \beta \in I_+^n$.

Thus, the functions in $\mathscr{S}$ are those functions which together with their derivatives fall off more quickly than the inverse of any polynomial.

Theorem V.9 The vector space $\mathscr{S}(\mathbb{R}^n)$ with the natural topology given by the seminorms $\|\cdot\|_{\alpha, \beta}$ is a Fréchet space.

Proof The reader can easily check that $\|\cdot\|_{\alpha, \beta}$ is a seminorm. Since there are countably many of them, $\mathscr{S}(\mathbb{R}^n)$ is metrizable (Theorem V.5). Thus we need only show that $\mathscr{S}(\mathbb{R}^n)$ is complete. Suppose that f_m is Cauchy in each $\|\cdot\|_{\alpha, \beta}$. Then $x^\alpha D^\beta f_m \to g_{\alpha, \beta}$ uniformly as $m \to \infty$ since $C(\mathbb{R}^n)$ is complete. If we can show $g = g_{0,0}$ is C^∞ and $g_{\alpha, \beta} = x^\alpha D^\beta g$, then g will be in $\mathscr{S}$ and $\lim_{m \to \infty} f_m$

will be g in the topology of $\mathscr{S}$. Let us prove that g is C^1 and $dg/dx = g_{0,1}$ in the case of $\mathscr{S}(\mathbb{R}^1)$. The general case is proven by similar means. We know that

$$f_n(x) = f_n(0) + \int_0^x f_n'(t)\,dt$$

Since $f_n' \to g_{0,1}$ uniformly,

$$g(x) = g(0) + \int_0^x g_{0,1}(t)\,dt$$

Thus g is C^1 and $g' = g_{0,1}$. ∎

For technical reasons, we often want a directed family of seminorms, so we define for $k, m \in I_+$ and $f \in \mathscr{S}(\mathbb{R}^n)$,

$$\|f\|_{k,m} = \sum_{\substack{|\alpha| \le k \\ |\beta| \le m}} \|f\|_{\alpha,\beta}$$

Definition The (topological) dual space of $\mathscr{S}(\mathbb{R}^n)$, denoted by $\mathscr{S}'(\mathbb{R}^n)$, is called **the space of tempered distributions**.

For a linear functional T on $\mathscr{S}(\mathbb{R}^n)$ to be in $\mathscr{S}'(\mathbb{R}^n)$, it must be continuous. By Theorem V.2, this is equivalent to the existence of a seminorm $\|\cdot\|_{k,m}$ with $|T(\varphi)| \le C\|\varphi\|_{k,m}$ for all $\varphi \in \mathscr{S}(\mathbb{R}^n)$. We discuss several examples in the case $n = 1$; the reader can easily extend them to the case of general n.

Example 1 ($\mathscr{S}$) Let $g \in \mathscr{S}(\mathbb{R}^1)$ and define a functional $g(\cdot)$ on $\mathscr{S}$ by

$$g(\varphi) = \int_{-\infty}^{\infty} g(x)\varphi(x)\,dx \tag{V.2}$$

$g(\cdot)$ is clearly linear and

$$|g(\varphi)| \le \|g\|_{L^1(\mathbb{R})}\|\varphi\|_\infty$$

and $\|\varphi\|_\infty$ is a continuous seminorm. Moreover, if $g_1 \ne g_2$ as functions in $\mathscr{S}$, $g_1(\cdot) \ne g_2(\cdot)$ as elements in $\mathscr{S}'$. For $\mathscr{S}$ is dense in L^2 so that $g_1 \ne g_2$ in $\mathscr{S}$ implies $g_1 \ne g_2$ in L^2 implies $g_1 \ne g_2$ in $(L^2)^*$ implies $g_1 \ne g_2$ in $\mathscr{S}'$.

We thus see that $\mathscr{S}$ is embedded naturally in $\mathscr{S}'$. This imbedding is continuous when $\mathscr{S}'$ is given the $\sigma(\mathscr{S}', \mathscr{S})$ topology since we will see that $\|\cdot\|_{L^1(\mathbb{R})}$ is a continuous seminorm on $\mathscr{S}$. Moreover, $\mathscr{S}$ is dense in $\mathscr{S}'$ in the

V.3 Functions of rapid decrease and the tempered distributions

$\sigma(\mathscr{S}', \mathscr{S})$ topology (see Problems 16 and 19 or Corollary 1 to Theorem V.14 in the appendix to this section). This suggests extending continuous maps $T: \mathscr{S} \to \mathscr{S}$ to $\mathscr{S}'$ as follows: Since the natural map ι of $\mathscr{S} \to \mathscr{S}'$ is continuous $\iota \circ T: \mathscr{S} \to \mathscr{S}'$ is continuous from $\mathscr{S}$ to $\mathscr{S}'$. Since $\mathscr{S}$ is dense in $\mathscr{S}'$, there is at most one continuous extension of $\iota \circ T: \mathscr{S}' \to \mathscr{S}'$. To find this continuous extension we need some way of constructing continuous maps of $\mathscr{S}' \to \mathscr{S}'$. There is one simple method. Suppose $S: \mathscr{S} \to \mathscr{S}$ is continuous. Define the **adjoint** $S': \mathscr{S}' \to \mathscr{S}'$, $\ell \mapsto S'(\ell)$, by requiring that $[S'(\ell)](g) = \ell(Sg)$ for all g in $\mathscr{S}$. It is evident that if $\ell_\alpha \to \ell$ in the $\sigma(\mathscr{S}', \mathscr{S})$ topology, then $S'(\ell_\alpha) \to S'(\ell)$ in the $\sigma(\mathscr{S}', \mathscr{S})$ topology. Thus, to extend T, we seek a map $S: \mathscr{S} \to \mathscr{S}$ so that $S' \upharpoonright \mathscr{S} = T$ and then extend T by using S'. When no confusion arises, we will denote this extension also by T. Adjoints are further discussed in Section VI.2.

This idea will become clearer as we discuss examples below. We will discuss another topology on $\mathscr{S}'$ in Section V.7 (the $\beta(\mathscr{S}', \mathscr{S}) \equiv \tau(\mathscr{S}', \mathscr{S})$ topology) and we will see that the adjoints S' are continuous when this topology is put on both the domain and range (see Problem 17).

Example 2 (L^p) $\mathscr{S}$ is a subset of each $L^p(\mathbb{R})$ and the identity mapping of $\mathscr{S}$ into L^p is continuous. For $p = 1$, we note that

$$\|f\|_1 = \int_{-\infty}^{\infty} (1 + x^2)^{-1}[(1 + x^2)|f|] \, dx \leq \pi(\|f\|_\infty + \|x^2 f\|_\infty)$$

and for general p, we notice that

$$\|f\|_p \leq \| |f|^{1/p} |f|^{1-1/p} \|_p \leq \|f\|_1^{1/p} \|f\|_\infty^{1-1/p}$$

If $g \in L^q$, $q^{-1} + p^{-1} = 1$ and $\varphi \in \mathscr{S}$ we have

$$\left| \int g(x)\varphi(x) \, dx \right| \leq \|g\|_q \|\varphi\|_p$$

Thus $\varphi \to \int g(x)\varphi(x) \, dx$ is a continuous map of $\mathscr{S}$ into $\mathbb{C}$. This defines a continuous imbedding of L^q into $\mathscr{S}'$.

Thus $\mathscr{S}'$ contains the images of many spaces of functions under natural injections. We usually ignore the injections and refer to the function $g(x) \in \mathscr{S}'$.

Example 3 (the delta function) Let $b \in \mathbb{R}$. Define δ_b to be the linear functional $\delta_b(\varphi) = \varphi(b)$. Since $|\delta_b(\varphi)| \leq \|\varphi\|_{0,0}$, $\delta_b(\cdot) \in \mathscr{S}'(\mathbb{R})$. There is no function $g(x)$ with $\delta_b(\varphi) = \int g(x)\varphi(x)\,dx$ for all $\varphi \in \mathscr{S}$, although there is a measure μ_b (the pure point measure) with

$$\delta_b(\varphi) = \int \varphi(x)\,d\mu_b(x)$$

but the symbolism of (V.2) is so suggestive that one often writes

$$\delta_b(\varphi) = \int \varphi(x)\delta(x-b)\,dx \tag{V.3}$$

$\delta(x-b)$ is not a function; (V.3) should be treated merely as a symbolic expression. $\delta(x-b)$ is called the **delta function at b**.

Example 4 (polynomially bounded measures) Suppose v is any finite measure. Then $f \mapsto \int_{-\infty}^{\infty} f(x)\,dv$ is a linear functional on $\mathscr{S}$ and since $\left|\int f\,dv\right| \leq v(\mathbb{R})\|f\|_\infty$ this functional is in $\mathscr{S}'$. In general, if v is a measure on $\mathbb{R}$ so that $v([-D, D]) \leq C(D^n + 1)$ for some C and n, and all $D \in \mathbb{R}_+$, then $f \mapsto \int_{-\infty}^{\infty} f\,dv$ is in $\mathscr{S}'$.

Example 5 (derivative of $\delta(x)$) To see that not every $g \in \mathscr{S}'$ comes from a linear combination of measures, consider $\delta'(f) = -f'(0)$. This is a continuous linear functional, but it does not comes from a measure (Problem 21).

Example 6 ($\mathscr{P}(1/x)$) The Cauchy principle part integral is given by

$$\mathscr{P}\!\left(\frac{1}{x}\right) : f \mapsto \lim_{\varepsilon \downarrow 0} \int_{|x| \geq \varepsilon} \frac{1}{x} f(x)\,dx$$

To see that this is finite for any $f \in \mathscr{S}(\mathbb{R}^1)$ and is in fact a distribution, we note that

$$\int_{|x| \geq \varepsilon} \frac{1}{x} f(x)\,dx = \int_\varepsilon^\infty \frac{f(x) - f(-x)}{x}\,dx$$

Since $[f(x) - f(-x)]/x \to 2f'(0)$ as $x \to 0$, we can write

$$\mathscr{P}\!\left(\frac{1}{x}\right)(f) = \int_0^\infty \frac{f(x) - f(-x)}{x}\,dx$$

which proves it is finite. Since

$$\left|\frac{1}{x}[f(x) - f(-x)]\right| \leq \frac{1}{x}\int_{-x}^x |f'(t)|\,dt \leq 2\|f'\|_\infty$$

V.3 Functions of rapid decrease and the tempered distributions

we have

$$\left|\mathscr{P}\left(\frac{1}{x}\right)(f)\right| \leq 2\int_0^1 \|f'\|_\infty\, dx + \left|\int_1^\infty (xf(x))\frac{dx}{x^2}\right|$$

$$\leq 2\|f\|_{1,0} + \|f\|_{0,1}$$

Thus $\mathscr{P}(1/x)$ is a distribution. The famous formula

$$\lim_{\varepsilon \downarrow 0} \frac{1}{x - x_0 + i\varepsilon} = \mathscr{P}\left(\frac{1}{x - x_0}\right) - i\pi\delta(x - x_0) \tag{V.4}$$

holds if all the objects are viewed as distributions and $\lim_{\varepsilon \downarrow 0}$ is interpreted in the $\sigma(\mathscr{S}', \mathscr{S})$ topology (Problem 22).

Example 7 (O_M^n) Let O_M^n denote the set of infinitely differentiable functions on $\mathbb{R}^n$ which together with their derivatives are polynomially bounded, that is, $f \in O_M$ means that f is C^∞ and for each $\alpha \in I_+^n$, there is an $N(\alpha)$ and a $C(\alpha)$ with

$$|(D^\alpha f)(x)| \leq C[1 + x^2]^N$$

where $x^2 = \sum_{i=1}^n x_i^2$. $O_M \subset \mathscr{S}'$.

The set O_M is useful for another reason. If $F \in O_M$, it is not hard to see that $Ff \in \mathscr{S}$ if $f \in \mathscr{S}$ and the map $f \mapsto Ff$ is a continuous map of $\mathscr{S} \to \mathscr{S}$ (Problem 23a). In fact, a measurable function F defines a continuous map $f \mapsto Ff$ of $\mathscr{S} \to \mathscr{S}$ if and only if $F \in O_M$ (Problem 23b).

Let $F \in O_M$. Then multiplication by F takes $\mathscr{S}$ into $\mathscr{S}$ continuously. This provides the first test of the philosophy expounded following Example 1. Can we find a map $S: \mathscr{S} \to \mathscr{S}$ so that for any $f, g \in \mathscr{S}$, $(Ff)(g) = (S'f)(g) = f(Sg)$, that is, so that $\int F(x)f(x)g(x)\, dx = \int f(x)(Sg(x))\, dx$. The answer is obvious: take $(Sg)(x) = F(x)g(x)$.

Operation 1 Let $F \in O_M^n$ and let $T \in \mathscr{S}'(\mathbb{R}^n)$. We define $FT \in \mathscr{S}'(\mathbb{R}^n)$ by

$$(FT)(\varphi) = T(F\varphi)$$

$\mathscr{S}$ was chosen partially to ensure that $f \to D^\alpha f$ would be continuous. Notice that $\|D^\alpha f\|_{\gamma,\delta} = \|f\|_{\gamma,\delta+\alpha}$ where $\delta + \alpha = \langle \delta_1 + \alpha_1, \ldots, \delta_n + \alpha_n \rangle$. To extend D^α to $\mathscr{S}'$ we seek $T: \mathscr{S} \to \mathscr{S}$ so that for $f, g \in \mathscr{S}$, $(D^\alpha f)(g) = f(Sg)$, that is, $\int (D^\alpha f)(x)g(x)\, dx = \int f(x)(Sg)(x)\, dx$. At first sight this appears difficult but integration by parts implies that $\int (D^\alpha f)g = (-1)^{|\alpha|} \int f(D^\alpha g)$ so we take $S = (-1)^{|\alpha|} D^\alpha$.

Operation 2 Let $T \in \mathscr{S}'(\mathbb{R}^n)$, $\alpha \in I_+^n$. The **weak derivative**, $D^\alpha T$, or **the derivative in the sense of distributions** is defined by

$$(D^\alpha T)(f) = (-1)^{|\alpha|} T(D^\alpha f)$$

In symbolic notation:

$$\int (D^\alpha T)(x) f(x)\, dx = (-1)^{|\alpha|} \int T(x) \left(\frac{\partial^{|\alpha|}}{\partial x_1^{\alpha_1} \cdots \partial x_n^{\alpha_n}} f \right)(x)\, dx$$

We have thus defined a notion of derivative which coincides with the ordinary derivative on O_M^n and for which integration by parts without boundary terms at ∞ holds by fiat.

Example 8 Let

$$g(x) = \begin{cases} x, & x \geq 0 \\ 0, & x \leq 0 \end{cases} \qquad x \in \mathbb{R}$$

Then g is continuous but is not everywhere differentiable in the classical sense. Since $\int_{-\infty}^{\infty} g(x)\varphi(x)\, dx \leq \|x\varphi\|_{L^1}$, $g \in \mathscr{S}'$ so it has a derivative in $\mathscr{S}'$. By definition

$$\left(\frac{d}{dx} g\right)(\varphi) = -g\left(\frac{d\varphi}{dx}\right) = -\int_0^\infty x\varphi'(x)\, dx$$

$$= \int_0^\infty \varphi(x)\, dx$$

Thus $dg(x)/dx = H(x)$ where H is the **Heaviside function**

$$H(x) = \begin{cases} 1, & x \geq 0 \\ 0, & x < 0 \end{cases}$$

H is not even continuous, but it too has a derivative in $\mathscr{S}'$ given by

$$\left(\frac{d}{dx} H\right)(\varphi) = -H\left(\frac{d\varphi}{dx}\right) = -\int_0^\infty \varphi'(x)\, dx = \varphi(0)$$

so $dH/dx = \delta$. δ, too has a derivative; it is described in Example 5.

This last example shows that even a nonfunction like δ is the second derivative of a continuous function. This is typical of tempered distributions, for

Theorem V.10 (regularity theorem for distributions) Let $T \in \mathcal{S}'(\mathbb{R}^n)$. Then $T = D^\beta g$ for some polynomially bounded continuous function g and some $\beta \in I_+^n$, that is,
$$T(\varphi) = \int (-1)^{|\beta|} g(x)(D^\beta \varphi)(x)\, d^n x$$
for all $\varphi \in \mathcal{S}$.

The proof is given in the appendix to this section (or see Problems 24 and 25).

Another operation is induced by translation. Let $U_a \colon \mathcal{S} \to \mathcal{S}$ by $(U_a f)(x) = f(x - a)$. Then $\int (U_a f)(x) g(x)\, dx = \int f(x)(U_{-a} g)(x)\, dx$ if $f, g \in \mathcal{S}$. Thus

Operation 3 (translation) $U_a T$ is defined for $T \in \mathcal{S}'$ by $(U_a T)(\varphi) = T(U_{-a}\varphi)$.

Similarly, if A is an invertible linear map of $\mathbb{R}^n \to \mathbb{R}^n$, we define $V(A) \colon \mathcal{S} \to \mathcal{S}$ by $(V(A)f)(x) = f(A^{-1}x)$. Then, one has:

Operation 4 (linear coordinates changes)   If $T \in \mathcal{S}'$, $V(A)T$ is given by
$$[V(A)T](\varphi) = |\det A|\, T(V(A^{-1})\varphi)$$
This extends $V(A)$ from $\mathcal{S}$ to $\mathcal{S}'$ (see Problem 28a).

In Chapter IX we discuss two other operations on $\mathcal{S}'$, convolution, and Fourier transform.

It is meaningless to say that a distribution vanishes at a point x, but vanishing in a neighborhood of x makes sense:

Definition   Let Ω be an open set of $\mathbb{R}^n$. We say that $T \in \mathcal{S}'(\mathbb{R}^n)$ **vanishes** in Ω if $T(\varphi) = 0$ whenever φ has support in Ω (that is, whenever φ vanishes outside Ω). The **support of** T, supp T, is the complement of the largest open set on which T vanishes. If $T - S$ vanishes on Ω, we say $T = S$ on Ω.

These notions extend those of ordinary functions $\varphi \in \mathcal{S}$ (Problem 28b). Moreover, one has the following simple and intuitive result (Problem 29):

Theorem V.11   Let $T \in \mathcal{S}'(\mathbb{R}^n)$ and suppose that supp $T = \{0\}$. Then
$$T = \sum_{|\alpha| \leq m} c_\alpha (D^\alpha \delta)$$
for suitable c_α and m.

Example 9 ($(1/x)_+$ renormalized) Consider the function $(1/x)_+ = H(x)x^{-1}$. Since $\int_0^1 (1/x)\, dx = \infty$, $(1/x)_+$ does *not* define a distribution. Let $\mathscr{S}(\mathbb{R}\backslash 0) = \{f \in \mathscr{S} \mid \mathrm{supp}\, f \in \mathbb{R}\backslash 0\}$. If $f \in \mathscr{S}(\mathbb{R}\backslash 0)$, $\int (1/x)_+ f(x)\, dx$ makes sense. Thus $(1/x)_+$ does define a linear functional on $\mathscr{S}(\mathbb{R}\backslash 0)$ which is continous as we shall see. By the Hahn–Banach theorem, this functional on $\mathscr{S}(\mathbb{R}\backslash 0)$ has extensions to all of $\mathscr{S}(\mathbb{R})$ which we call "renormalizations of $(1/x)_+$." As explicit examples, consider

$$\left(\frac{1}{x}\right)_{+,M}(f) = \int_0^M \frac{f(x) - f(0)}{x}\, dx + \int_M^\infty \frac{f(x)}{x}\, dx$$

Since these maps are continuous on $\mathscr{S}$, $(1/x)_+$ is continuous on $\mathscr{S}(\mathbb{R}\backslash 0)$. How much arbitrariness is there in the renormalization? If T and S are two renormalizations of $(1/x)_+$, then $T - S$ vanishes on $\mathscr{S}(\mathbb{R}\backslash 0)$ and so has support $\{0\}$; thus $T - S = \sum_{|\alpha| \leq m} c_\alpha D^\alpha \delta$. For example,

$$\left(\frac{1}{x}\right)_{+,M} - \left(\frac{1}{x}\right)_{+,N} = -\ln\left(\frac{M}{N}\right)\delta(x)$$

With this definition of renormalization, there are an infinity of free constants in the renormalization of $(1/x)_+$. However, one could argue that as long as $f \in \mathscr{S}$ and $f(0) = 0$, $\int_0^\infty [f(x)/x]\, dx < \infty$; so we really want to extend $(1/x)_+$ from $\{f \in \mathscr{S} \mid f(0) = 0\}$ to $\mathscr{S}$. If we adopt this requirement, the only real renormalizations of $(1/x)_+$ are the $(1/x)_{+,M}$ and there is only one free constant in the renormalization (see Problem 32 for a link between these definitions).

Bogoliubov and Hepp have treated the renormalization of x space Feynman graphs in the spirit of Example 9; the renormalization constants, for example, the renormalized mass and charge, enter as the free constants analogous to the $\ln(M/N)$ in Example 9. For more details, see the references in the Notes.

There is one final theorem about $\mathscr{S}$ and $\mathscr{S}'$ which is often useful. To appreciate its significance, let us first consider the case of L^p where the analogous theorem fails. Suppose $p^{-1} + q^{-1} = 1$, $p < \infty$, $q < \infty$, and let $F \in L^q(\mathbb{R}^2) = [L^p(\mathbb{R} \times \mathbb{R})]^*$. Let $f, g \in L^p(\mathbb{R})$; then $f(x)g(y) \in L^p(\mathbb{R}^2)$ so

$$F(f, g) = \int F(x, y) f(x) g(y)\, dx\, dy < \infty$$

Moreover

$$|F(f, g)| \leq \|F\|_{L^q(\mathbb{R}^2)} \|f\|_p \|g\|_p$$

so F defines a continuous bilinear form on L^p. Not every bilinear form is of this type. For example, if $p = 2$, the bilinear form $(\tilde{f}, g) = \int f(x)g(x)\, dx$

cannot be expressed as $(\tilde{f}, g) = \int F(x, y) f(x) g(y) \, dx \, dy$ for some $F \in L^2(\mathbb{R}^2)$. The situation for $\mathscr{S}(\mathbb{R})$ and $\mathscr{S}(\mathbb{R} \times \mathbb{R})^* = \mathscr{S}'(\mathbb{R}^2)$ is very different:

Theorem V.12 (kernel or nuclear theorem) Let $B(f, g)$ be a separately continuous bilinear functional on $\mathscr{S}(\mathbb{R}^n) \times \mathscr{S}(\mathbb{R}^m)$. Then there is a unique tempered distribution $T \in \mathscr{S}'(\mathbb{R}^{n+m})$ with $B(f, g) = T(f \otimes g)$ where

$$(f \otimes g)(x_1, \ldots, x_{n+m}) = f(x_1, \ldots, x_n) g(x_{n+1}, \ldots, x_{n+m})$$

That separate continuity implies joint continuity is a consequence of the fact that $\mathscr{S}$ is Fréchet (Theorem V.9) and the corollary of Theorem V.7. We prove that jointly continuous functionals have the requisite form in the appendix to this section (Corollary 4 to Theorem V.14). Theorem V.12 can be extended to multilinear functionals (Problems 34 and 35).

Appendix to V.3 The N-representation for $\mathscr{S}$ and $\mathscr{S}'$

In this appendix, we will prove some of theorems about $\mathscr{S}$ and $\mathscr{S}'$. These proofs rely on the realization of $\mathscr{S}$ and therefore $\mathscr{S}'$ as sequence spaces (in fact as the space s of Section III.1). This realization depends in turn on two elements. The first element is topologizing $\mathscr{S}$ by an equivalent family of L^2 norms. To forcefully distinguish these norms from the $\|\cdot\|_{\alpha, \beta}$ norms we write

$$\|f\|_{\alpha, \beta, \infty} = \|x^\alpha D^\beta f\|_\infty$$

rather than merely $\|\cdot\|_{\alpha, \beta}$ and define

$$\|f\|_{\alpha, \beta, 2} = \|x^\alpha D^\beta f\|_{L^2(\mathbb{R}^n)}$$

Then:

Lemma 1 The families of seminorms $\{\|\cdot\|_{\alpha, \beta, \infty}\}$ and $\{\|\cdot\|_{\alpha, \beta, 2}\}$ on $\mathscr{S}(\mathbb{R}^n)$ are equivalent.

Proof We provide the proof in the case $n = 1$ for simplicity of notation. Since $(1 + x^2)^{-1} \in L^2$, $\|f\|_2 \leq \|(1 + x^2)^{-1}\|_2 \|(1 + x^2) f\|_\infty$ so

$$\|f\|_{\alpha, \beta, 2} \leq C(\|f\|_{\alpha, \beta, \infty} + \|f\|_{\alpha+2, \beta, \infty})$$

On the other hand, $f(x) = \int_{-\infty}^{x} f'(x)\, dx$, so
$$\|f\|_\infty \leq \|f'\|_1 \leq \|(1+x^2)f'\|_2 \|(1+x^2)^{-1}\|_2$$
Since $(x^\alpha D^\beta f)' = \alpha x^{\alpha-1} D^\beta f + x^\alpha D^{\beta+1} f$ we have
$$\|f\|_{\alpha,\beta,\infty} \leq C(\alpha \|f\|_{\alpha-1,\beta,2} + \|f\|_{\alpha,\beta+1,2} + \alpha\|f\|_{\alpha+1,\beta,2} + \|f\|_{\alpha+2,\beta+1,2}) \quad \blacksquare$$

The second element involves some special properties of Hermite functions (the eigenfunctions of the harmonic oscillator). Consider the maps $A: \mathscr{S}(\mathbb{R}) \to \mathscr{S}(\mathbb{R})$ and $A^\dagger: \mathscr{S}(\mathbb{R}) \to \mathscr{S}(\mathbb{R})$ given by
$$A = \frac{1}{\sqrt{2}}\left(x + \frac{d}{dx}\right) \qquad A^\dagger = \frac{1}{\sqrt{2}}\left(x - \frac{d}{dx}\right)$$
and $N = A^\dagger A$. Let $\|f\|_n \equiv \|(N+1)^n f\|_2$ which is a seminorm on $\mathscr{S}$.

Lemma 2 The seminorms $\{\|\cdot\|_n\}$ are a directed family equivalent to the $\{\|\cdot\|_{\alpha,\beta,2}\}$ family of seminorms on $\mathscr{S}$.

Proof One need only use the inequality $\|A_1^\# \cdots A_m^\# f\|_2 \leq \|(N+m)^{m/2} f\|_2$ where $A^\#$ stands for either A or $A^\dagger$. The details are left to the reader (Problem 36).

Now consider the function ϕ_0 defined by $A\phi_0 = 0$ and $\int_{-\infty}^{\infty} (\phi_0)^2 \, dx = 1$, that is, $\phi_0(x) = \pi^{-\frac{1}{4}} e^{-\frac{1}{2}x^2}$ and let
$$\phi_n = (n!)^{-\frac{1}{2}} (A^\dagger)^n \phi_0 = (2^n n!)^{-\frac{1}{2}} (-1)^n \pi^{-\frac{1}{4}} e^{+\frac{1}{2}x^2} \left(\frac{d}{dx}\right)^n e^{-x^2}$$

The $\{\phi_n\}_{n=0}^\infty$ are called the **Hermite functions** or the harmonic oscillator wave functions since
$$\left(-\frac{d^2}{dx^2} + x^2\right)\phi_n = (2n+1)\phi_n$$
One has:

Lemma 3 The set $\{\phi_n\}_{n=0}^\infty$ is an orthonormal basis for $L^2(\mathbb{R})$.

Proof See Problems 6 and 7 of Chapter IX or Problems 30 and 31 of Chapter X.

Notice that $N\phi_n = n\phi_n$. Suppose $f \in \mathscr{S}$ and consider the L^2-convergent expansion $f = \sum_{n=0}^\infty a_n \phi_n$ where $a_n = (\phi_n, f) \equiv \int_{-\infty}^\infty \overline{\phi_n(x)} f(x)\, dx$. Since

Appendix to V.3 The N-representation for $\mathscr{S}$ and $\mathscr{S}'$

$N^m: \mathscr{S} \to \mathscr{S}$, $N^m f \in \mathscr{S}$ and thus in L^2. But $N^m f = \sum_{n=0}^{\infty} a_n n^m \phi_n$, so $\sum_{n=0}^{\infty} |a_n|^2 n^{2m} < \infty$. In particular $\sup_n |a_n| n^m < \infty$. We have thus proven the first part of:

Theorem V.13 (the N-representation theorem for $\mathscr{S}$) Let s_k be the set of multisequences $\{a_\alpha\}_{\alpha \in I_+^k}$ with the property

$$\sup_{\alpha \in I_+^k} |a_\alpha| |\alpha|^m < \infty$$

for each m. Topologize s_k with the seminorms

$$\|\{a_\alpha\}\|_\beta^2 = \sum_\alpha (\alpha + 1)^{2\beta} |a_\alpha|^2$$

where $\beta \in I_+^k$ and $(\alpha + 1)^{2\beta} = \prod_{i=1}^k (\alpha_i + 1)^{2\beta_i}$. Let $f \in \mathscr{S}(\mathbb{R}^k)$. Then the sequence $\{a_\alpha\}$, $a_\alpha = (\phi_\alpha, f)$ with $\phi_\alpha(x) = \prod_{i=1}^k \phi_{\alpha_i}(x_i)$, is in s_k and the map $f \mapsto \{a_\alpha\}$ is a topological isomorphism. The **Hermite expansion** $f = \sum_\alpha a_\alpha \phi_\alpha$ converges in $\mathscr{S}$. The $\{a_\alpha\}$ are called **Hermite coefficients**.

Proof We give the details in the case $k = 1$. By our previous discussion, if $f \in \mathscr{S}$ and $a_n = (\phi_n, f)$, then $\{a_n\} \in s$. Moreover, $\|\{a_n\}\|_m = \|f\|_m$ in the notation of Lemma 2. Since the $\|\ \|_m$ are norms on $\mathscr{S}$, the map $f \mapsto \{a_n\}$ is injective. Now let $\{a_n\}_{n=0}^\infty \in s$ and let $f_N = \sum_{n=0}^N a_n \phi_n$. A simple computation shows that

$$\|f_N - f_M\|_m^2 = \sum_{n=N+1}^M |a_n|^2 (n+1)^{2m} \to 0$$

as $N, M \to \infty$. Thus f_N is Cauchy in each of $\|\ \|_m$ and thus in $\mathscr{S}$ (by Lemmas 1 and 2). Since $\mathscr{S}$ is complete, $f_N \to f$ for some $f \in \mathscr{S}$. But then $f_N \to f$ in L^2 so $(\phi_n, f) = a_n$. Thus the image of our map of $\mathscr{S} \to s$ is all of s. The equivalence of the topologies follows from the equality of the norms $\|\cdot\|_m$ on $\mathscr{S}$ and s. ∎

We can now identify $\mathscr{S}'$ with a sequence space also:

Theorem V.14 (the N-representation theorem for $\mathscr{S}'$) Let $T \in \mathscr{S}'(\mathbb{R}^k)$. Let $b_\alpha = T(\phi_\alpha)$ for each $\alpha \in I_+^k$. Then for some $\beta \in I_+^k$, $|b_\alpha| \leq C(\alpha+1)^\beta$ for all α. Conversely, if $|b_\alpha| \leq C(\alpha+1)^\beta$ for all α, there is a unique $T \in \mathscr{S}'$ with $T(\phi_\alpha) = b_\alpha$. If $T \in \mathscr{S}'$ and $b_\alpha = T(\phi_\alpha)$ are its **Hermite coefficients**, then $\sum_\alpha b_\alpha \phi_\alpha$ converges in the $\sigma(\mathscr{S}', \mathscr{S})$ topology to T.

Proof Again, we consider only $k = 1$. Let $T \in \mathscr{S}'$. Then $|T(\phi)| \leq C\|\phi\|_m$ for some m and C since $\{\|\cdot\|_m\}$ is a directed set. $\|\phi_n\|_m = (n+1)^m$, so

$|b_n| \leq C(n+1)^m$. Conversely, suppose $|b_n| \leq C(n+1)^m$. For $\{a_n\} \in s$ define $B(\{a_n\}) = \sum_{n=0}^{\infty} b_n a_n$. Then

$$|B(\{a_n\})| \leq \sum_{n=0}^{\infty} |b_n| |a_n| \leq C \sum_{n=0}^{\infty} (n+1)^m |a_n|$$

$$\leq C \left(\sum_{n=0}^{\infty} (n+1)^{2m+2} |a_n|^2 \right)^{1/2} \left(\sum_{n=0}^{\infty} (n+1)^{-2} \right)^{1/2}$$

$$\leq \frac{\pi^2}{6} C \|\{a_n\}\|_{m+1}$$

Thus B defines a continuous linear functional on s. Under the association of $\mathscr{S}$ and s, there is a $T \in \mathscr{S}'$ with $T(\sum_{n=0}^{\infty} a_n \phi_n) = \sum_{n=0}^{\infty} a_n b_n$; in particular, $T(\phi_n) = b_n$. The weak convergence of $\sum_n b_n \phi_n$ to T is easy. ∎

We can now easily prove many interesting theorems about $\mathscr{S}$ with this machinery which has *two* important simplifications: (1) Sequences are easier to deal with than functions. (2) The two conditions in $\mathscr{S}$, fall-off at ∞ and the C^∞ condition, are replaced by a *single* fall-off condition in s.

Corollary 1 $\mathscr{S}$ is dense in $\mathscr{S}'$ in the $\sigma(\mathscr{S}', \mathscr{S})$ topology.

Proof $\sum_{|\alpha| \leq N} b_\alpha \phi_\alpha \in \mathscr{S}$ and converges weakly to $T \in \mathscr{S}'$ as $N \to \infty$ if $b_\alpha = T(\phi_\alpha)$. ∎

Corollary 2 $\mathscr{S}$ is separable in the Fréchet topology. $\mathscr{S}'$ is separable in the $\sigma(\mathscr{S}', \mathscr{S})$ topology (and also in the $\tau(\mathscr{S}', \mathscr{S})$ topology we introduce in Section V.7).

Corollary 3 The regularity theorem for distributions—Theorem V.10.

Proof Again we only consider the case $k = 1$. Since $\|f\|_\infty \leq C \|(1 + x^2) f'\|_2$ we conclude that $\|\phi_n\|_\infty \leq C'(n+1)^{3/2}$, using A and $A^\dagger$ and the estimate in the proof of Lemma 2. (More detailed studies of the ϕ_n show $\|\phi_n\|_\infty \sim D(n+1)^{-1/12}$). Let $T \in \mathscr{S}'$ and let $\{b_n\}$ be its Hermite coefficients. Then $|b_n| \leq E(n+1)^m$ for some m. Let $a_n = (n+1)^{-m-3} b_n$. Then $\sum |a_n| \|\phi_n\|_\infty \leq E \sum (n+1)^{-3/2} < \infty$, so $\sum a_n \phi_n$ converges uniformly to some continuous function F on $\mathbb{R}$. F has Hermite coefficients (as an element of $\mathscr{S}'$), $\{a_n\}$. Extend $A^\dagger$, A and $N = \frac{1}{2}(-d^2/dx^2 + x^2 - 1)$ to $\mathscr{S}'$. Then

$$T = (N+1)^{m+3} F = \frac{1}{2^{m+3}} \left(-\frac{d^2}{dx^2} + x^2 + 1 \right)^{m+3} F$$

Thus T can be written as a sum of polynomials times weak derivatives of polynomially bounded continuous functions. Simple manipulations (Problem 37) now complete the proof. ∎

Corollary 4 (nuclear theorem) Every jointly continuous bilinear functional $B(\cdot, \cdot)$ on $\mathscr{S}(\mathbb{R}^n) \times \mathscr{S}(\mathbb{R}^m)$ is of the form $B(f, g) = T(f \otimes g)$ for some $T \in \mathscr{S}'(\mathbb{R}^{n+m})$.

Proof Since B is jointly continuous, $|B(f, g)| \leq C\|f\|_r \|g\|_s$ for some $r \in I_+^n$, $s \in I_+^m$. Then $|B(\phi_\alpha, \phi_\beta)| \leq C(\alpha + 1)^r (\beta + 1)^s = C[\langle \alpha, \beta \rangle + 1]^{\langle r, s \rangle}$ where

$$\langle \alpha, \beta \rangle = \langle \alpha_1, \ldots, \alpha_n, \beta_1, \ldots, \beta_m \rangle \in I_+^{n+m}$$

As a result $b_{\langle \alpha, \beta \rangle} \equiv B(\phi_\alpha, \phi_\beta)$ are the Hermite coefficients of a distribution $T \in \mathscr{S}'(\mathbb{R}^{n+m})$ with $T(\phi_{\langle \alpha, \beta \rangle}) \equiv T(\phi_\alpha \otimes \phi_\beta) = b_{\langle \alpha, \beta \rangle}$. Let $f = \sum a_\alpha \phi_\alpha$, $g = \sum c_\beta \phi_\beta$. Since these expansions converge in $\mathscr{S}$,

$$T(f \otimes g) = \sum_{\alpha, \beta} a_\alpha c_\beta T(\phi_\alpha \otimes \phi_\beta) = \sum_{\alpha, \beta} a_\alpha c_\beta b_{\langle \alpha, \beta \rangle} = B(f, g) \quad \blacksquare$$

V.4 Inductive limits: generalized functions and weak solutions of partial differential equations

In an intuitive sense, the distributions of the last section had the restriction of being polynomially bounded at infinity. We saw this in Theorem V.10 which told us any $T \in \mathscr{S}'$ is the derivative of a polynomially bounded function. The growth of a tempered distribution $T \in \mathscr{S}'$ is in some sense dual to the decrease restrictions imposed on functions $f \in \mathscr{S}$. This suggests we construct "distributions" without any growth restriction at ∞ as the dual of a space with the severest possible decrease conditions at ∞, that of vanishing outside of a compact set. That is, we want to topologize the C^∞ functions of compact support, $C_0^\infty(\mathbb{R}^n)$, so that it is a complete locally convex space. If K is a compact set in $\mathbb{R}^n$, the functions $C_0^\infty(K)$ which are C^∞ and have support in K have a natural topology given by $\|f\|_{\alpha, \infty} \equiv \sup_{\mathbb{R}^n} |D^\alpha f|$. $C_0^\infty(\mathbb{R}^n)$ is *not* complete when given the $\{\|\cdot\|_{\alpha, \infty}\}_{\alpha \in I_+^n}$ family of norms (see Problem 38), even though $C_0^\infty(K)$ is for each compact set K. In some sense, we want to think of $C_0^\infty(\mathbb{R}^n)$ as $\bigcup_m C_0^\infty(K_m)$ for some family of compact sets $\{K_m\}_{m=1}^\infty$ with $\bigcup_m K_m = \mathbb{R}^n$, and topologize it with a "limit" topology. To do this we describe a general construction.

Theorem V.15 Let X be a complex (or real) vector space. Let X_n be a family of subspaces with $X_n \subseteq X_{n+1}$, $X = \bigcup_{n=1}^{\infty} X_n$. Suppose that each X_n has a locally convex topology so that the restriction of the topology of X_{n+1} to X_n is the given topology on X_n. Let $\mathcal{U}$ be the collection of balanced, absorbing, convex sets $\mathcal{O}$ in X for which $\mathcal{O} \cap X_n$ is open in X_n for each n. Then:

(a) $\mathcal{U}$ is a neighborhood base at 0 for a locally convex topology.

(b) The topology generated by $\mathcal{U}$ is the strongest locally convex topology on X so that the injections $X_n \to X$ are continuous.

(c) The restriction of the topology on X to each X_n is the given topology on X_n.

(d) If each X_n is complete, so is X.

We require a technical lemma for the proof of Theorem V.15:

Lemma Let X be a locally convex space and let X_1 be a subspace with the relative topology (which is automatically locally convex). Let V be an open, convex, balanced subset of X_1. Then there is an open, convex, balanced set $Z \subset X$ so that $Z \cap X_1 = V$.

Proof Since X_1 has the relative topology we can find an open set $\mathcal{O} \subset X$ so $\mathcal{O} \cap X_1 = V$. Since $\mathcal{O}$ is a neighborhood of $0 \in X$ and X is locally compact, we can find $\mathcal{O}_1 \subset \mathcal{O}$ which is balanced, convex, and open in X. Let

$$Z = \{\alpha x + \beta y \mid x \in \mathcal{O}_1, \ y \in V, \ |\alpha| + |\beta| = 1\}$$
$$= \bigcup_{y \in V, \ |\alpha|+|\beta|=1, \ \alpha \neq 0} (\beta y + |\alpha| \mathcal{O}_1)$$

As a union of open sets, Z is open. Since $V \subset Z$, $V \subset (Z \cap X_1)$, but if $\alpha x + \beta y \in X_1 \cap Z$, then $x \in X_1 \cap \mathcal{O}_1 \subset X_1 \cap \mathcal{O} = V$ so $\alpha x + \beta y \in V$; that is, $Z \cap X_1 \subset V$. This proves the lemma. ∎

Proof of Theorem V.15 $\mathcal{U}$ is closed under finite intersections and dilatations, so we need only prove that the generated topology is Hausdorff to conclude (a). But if we can prove (c), it will follow that X is Hausdorff. For if $x \in X_n$ is given, we can find an X_n-open set $\mathcal{O}$ about $0 \in X_n$ with $x \notin \mathcal{O}$ and then, once (c) is proven, we can find U open in X with $U \cap X_n = \mathcal{O}$. Thus (c) implies (a). Given a neighborhood $\mathcal{O}_1$ of $0 \in X_n$ find a balanced, convex open set $N_n \subset \mathcal{O}_1$. Now, using the lemma, find $N_{n+1} \subset X_{n+1}$, convex, balanced, and open in N_{n+1}, so that $N_{n+1} \cap X_n = N_n$ and by induction, N_k ($k > n$) so that N_k is convex, balanced, and open in X_k and $N_k \cap X_{k-1} = N_{k-1}$. Let $N_\infty = \bigcup_{k \geq n} N_k$. It is

V.4 Inductive limits: generalized functions and weak solutions

easy to see $N_\infty \in \mathscr{U}$ and $N_\infty \cap X_k \subset \mathcal{O}_1$. Thus $\mathcal{O}_1$ is a neighborhood in the relative topology. That the relative topology is finer than the given topology follows from the definition of $\mathscr{U}$. This proves (c) and thus (a). (b) is easy. A reference for (d) is given in the notes. ∎

Definition The locally convex space X constructed in Theorem V.15 is called the **strict inductive limit** of the spaces X_n.

We remark that if each X_n is a proper closed subspace of X_{n+1}, then X is not metrizable (Problem 45). One of the nice properties of strict inductive limits is:

Theorem V.16 Let X be the strict inductive limit of the locally convex spaces $\{X_n\}_{n=1}^\infty$. Then a linear map T from X to a locally convex space Y is continuous if and only if each of the restrictions $T \upharpoonright X_n$ is continuous.

Proof If T is continuous, each restriction is continuous. Conversely, suppose each restriction is continuous. Let N be a balanced, convex, open set in Y. Then $T^{-1}[N] \cap X_n = (T \upharpoonright X_n)^{-1}[N]$ is open in X_n since $T \upharpoonright X_n$ is continuous. Since $T^{-1}[N]$ is balanced and convex, it is open, so T is continuous. ∎

Example 1 Let $\kappa(\mathbb{R})$ be the continuous functions on $\mathbb{R}$ which have compact support. Let κ_n be those functions in $\kappa(\mathbb{R})$ with support in $[-n, n]$, normed with $\|\cdot\|_\infty$. Topologize κ with the inductive limit topology. By the last theorem, the dual of κ with this topology is precisely the complex Baire measures on $\mathbb{R}$. This construction works for $\kappa(X)$ when X is any σ-compact locally compact space.

Now let Ω be an open connected set in $\mathbb{R}^n$, $C_0^\infty(\Omega)$ the infinitely differentiable functions with compact support in Ω. Let K_n be an increasing family of compact sets with $\bigcup K_n^\circ = \Omega$. Put the Fréchet topology on $C_0^\infty(K_n)$ generated by the $\|D^\alpha f\|_\infty$ norms. The set $C_0^\infty(\Omega)$ with the inductive limit topology obtained by $C_0^\infty(\Omega) = \bigcup C_0^\infty(K_n)$ is denoted $\mathscr{D}_\Omega$. This topology is independent of the choice of the K_n (Problem 46). *Sequential* convergence in $\mathscr{D}_\Omega$ is fairly simple.

Theorem V.17 Suppose that $X = \bigcup X_n$ has a strict inductive limit topology and that each X_n is a closed proper subspace of X_{n+1}. Then a sequence $f_m \in X$ converges to $f \in X$ if and only if all the f_m are in some X_n

and $f_m \to f$ in the topology of that X_n. In particular, a *sequence* $f \in \mathscr{D}_{\mathbb{R}^n}$ converges to f if and only if all the f_m and f have support inside some fixed compact K and $D^\alpha f_m$ converges uniformly to $D^\alpha f$ for each multiindex α.

Proof Let $f_m \to f$ and suppose for all n, that there is an f_m with $f_m \notin X_n$. Then it is easy to construct subsequences of the f_m, say $g_i = f_{m(i)}$ and the X_n, say $Y_i = X_{n(i)}$ with $g_i \in Y_{i+1} \setminus Y_i$. Since Y_i is closed, we can use the Hahn–Banach theorem to find $\ell_i \in X^*$ so that $\ell_i \equiv 0$ on Y_i and $\ell_n(g_n) = n - \sum_{k=1}^{n-1} \ell_k(g_n)$. Let $\ell = \sum_{n=1}^{\infty} \ell_n$. On any X_n, this sum is effectively finite, so ℓ is continuous on each X_n, and hence, by Theorem V.16, on X. Since $g_m \to f$, and $\ell \in X^*$, $\ell(g_m)$ converges. But $\ell(g_m) = m$; this contradiction proves that all the f_m are in some X_n. ∎

We are now ready to define the distributions on Ω:

Definition A **generalized function** (or **distribution**) is a continuous linear functional on $\mathscr{D}_\Omega$. The space of all continuous linear functionals on $\mathscr{D}_\Omega$ is denoted by $\mathscr{D}'_\Omega$. $\mathscr{D}$ and $\mathscr{D}'$ will denote $\mathscr{D}_{\mathbb{R}^n}$ and $\mathscr{D}'_{\mathbb{R}^n}$ respectively.

Theorem V.16 translates directly into:

Corollary A linear functional T on $\mathscr{D}_{\mathbb{R}^n}$ is continuous if and only if for each compact $K \subset \mathbb{R}^n$, there is a constant C and an integer j so that

$$|T(\varphi)| \leq C \sum_{|\alpha| \leq j} \|D^\alpha \varphi\|_\infty$$

for all $\varphi \in C_0^\infty(K)$.

Example 2 Let f be an arbitrary continuous function on $\mathbb{R}^n$, and define $D^\alpha f \in \mathscr{D}'$ by:

$$(D^\alpha f)(\varphi) = (-1)^{|\alpha|} \int f(x)(D^\alpha \varphi)(x)\, dx$$

Then for each compact set K and $\varphi \in C_0^\infty(K)$

$$|(D^\alpha f)(\varphi)| \leq C \|D^\alpha \varphi\|_\infty \sup_{x \in K} |f(x)|$$

so $D^\alpha f \in \mathscr{D}'_{\mathbb{R}^n}$. Thus $\mathscr{D}'_{\mathbb{R}^n}$ contains the weak derivatives of all continuous functions.

Example 3 Consider $\mathscr{D}$. Let $\delta^n(x-a): \mathscr{D} \to \mathbb{C}$ by $\delta^n(x-a)(f) = D^n f(a)$. Then $\sum_{n=0}^{\infty} \delta^n(x-n) = T$ is in $\mathscr{D}'$. For let $\varphi \in C_0^\infty[-m, m]$. Then

$$|T(\varphi)| = \left| \sum_{n=0}^{m} (D^n \varphi)(n) \right| \leq \sum_{n=0}^{m} \|D^n \varphi\|_\infty$$

V.4 Inductive limits: generalized functions and weak solutions

This last example shows that $\mathscr{D}'_{\mathbb{R}^n}$ contains distributions T which are *not* the αth derivative of any continuous function. Thus, there is no direct analogue of Theorem V.10 for $\mathscr{D}'$ but there is a local regularity theorem (Problem 26). There is also a nuclear theorem for $\mathscr{D}$ (see Problems 59, 60).

One can also carry operations from $\mathscr{D}$ to $\mathscr{D}'$ by the method used in Section V.3. Thus, for example, if $p(x_1, \ldots, x_n)$ is a polynomial of total degree k in n variables, $p(x_1, \ldots, x_n) = \sum_{|\alpha| \leq k} a_\alpha x^\alpha$, the partial differential operator $p(D) = \sum_{|\alpha| \leq k} a_\alpha D^\alpha$ extends to $\mathscr{D}'$ by the formula

$$(p(D)T)(\varphi) = T\left[\sum_{|\alpha| \leq k} (-1)^{|\alpha|} D^\alpha(a_\alpha \varphi)\right] \tag{V.5}$$

Formula (V.5) may also be used if the a_α are x-dependent C^∞ functions.

The extension of partial differential operators to $\mathscr{D}'$ is particularly useful in the theory of partial differential equations. Let f be a continuous function. A k-times continuously differentiable function u (we write $u \in C^k$) for which $p(D)u = f$ is called a **strict solution**. If $T \in \mathscr{D}'$ and $p(D)T = f$ with $p(D)T$ defined by (V.5), then T is called a **weak solution** of the partial differential equation. The difference between strict and weak solutions is only smoothness, for:

Proposition If $u \in C^k$, then $p(D)u$ defined by (V.5) is equal to the classical value of $p(D)u$. In particular, if $u \in C^k$ and f is continuous, u is a weak solution of $p(D)u = f$ if and only if it is a strict solution.

Proof An elementary integration by parts. ∎

The following example shows that not every weak solution is a strict solution:

Example Let $f(x)$ be the characteristic function of $[0, 1]$. We will show that $u(x, t) = f(x - ct)$ is a weak solution of $u_{tt} - c^2 u_{xx} = 0$. Rather than use the definition (V.5) directly (which is a useful exercise), we make use of the fact that the operator $p(D)$ in (V.5) is continuous on $\mathscr{D}'$. Since $f \in L^1(\mathbb{R})$, we can find a sequence f_n in $\mathscr{D}_\mathbb{R}$ with $f_n \to f$ in $L^1(\mathbb{R})$. It is then easy to see that $u_n(x, t) = f_n(x - ct) \to u(x, t)$ in $\sigma(\mathscr{D}', \mathscr{D})$. But $p(D)u_n$ can be computed classically, that is,

$$\frac{\partial^2}{\partial t^2} u_n(x, t) = c^2 f_n''(x - ct) \qquad \frac{\partial^2}{dx^2} u_n(x, t) = f_n''(x - ct)$$

Thus

$$\frac{\partial^2}{\partial t^2} u_n - c^2 \frac{\partial^2}{\partial x^2} u_n = 0 \qquad \text{so} \qquad \frac{\partial^2}{\partial t^2} u - c^2 \frac{\partial^2}{\partial x^2} u = 0$$

In Problem 47, we discuss the notion of a distribution being only a function of $x - ct$ and prove that any such distribution T satisfies

$$\frac{\partial^2}{\partial t^2} T - c^2 \frac{\partial^2}{\partial x^2} T = 0$$

The concept of weak solution is particularly useful because it is often easy to prove weak solutions exist (see Section IX.5 for the constant coefficient case). In the case of elliptic equations (see Section IX.6), one can prove a regularity theorem which assures that under certain conditions, every weak solution is a strict solution. Combining these two techniques one concludes the existence of strong solutions for elliptic partial differential equations. Because of the phenomenon of the last example, the case of hyperbolic equations is not so easy.

V.5 Fixed point theorems †

We wish to consider the solution of equations of the form $x = Tx$ in a variety of types of applications. For example, an inhomogeneous integral equation $f(x) = g(x) + \int K(x, y) f(y)\, dy$ is of the form $f = Tf$ with $Tf = g + Kf$, an affine linear map. The famous "bootstrap" equations proposed in particle physics are of the forms $S = T(S)$ where S is the S matrix and T is a very complicated operator. A Lorentz invariance condition on vacuum expectation values takes the form

$$W_n(x_1, \ldots, x_n) = W_n(\Lambda x_1, \ldots, \Lambda x_n)$$

where Λ is a fixed Lorentz transformation.

We want to discuss a variety of existence theorems for such equations —so-called fixed point theorems—and in Section V.6 we discuss some applications. We study them here because several of them are stated quite naturally in the language of locally convex spaces. We first consider "nonlinear" theorems, that is, theorems that make no assumptions on the linearity of the map T involved, and then one simple theorem which employs linearity.

Definition Let $T: X \to X$ be a map on a set X. A point $x \in X$ for which $Tx = x$ is called a **fixed point** of T.

The first nonlinear theorem is very simple and is probably familiar to the reader:

† A supplement to this section begins on p. 363.

V.5 Fixed point theorems

Definition Let $\langle S, \rho \rangle$ be a metric space. A map $T: S \to S$ for which $\rho(Tx, Ty) \leq \rho(x, y)$ is called a **contraction**. If there is a $K < 1$ for which $\rho(Tx, Ty) \leq K\rho(x, y)$, T is called a **strict contraction**.

Theorem V.18 (contraction mapping principle)   A strict contraction on a complete metric space has a unique fixed point.

Proof First, let us prove uniqueness. If $Tx = x$, $Ty = y$, then $\rho(x, y) = \rho(Tx, Ty) \leq K\rho(x, y)$. Since $K < 1$ and $\rho(x, y) \geq 0$, we conclude $\rho(x, y) = 0$, i.e. $x = y$. To prove existence, we first note that T is automatically continuous since $\rho(x, y) < K^{-1}\varepsilon$ implies $\rho(Tx, Ty) < \varepsilon$. Now, let x_0 be arbitrary and let $T^n x_0 \equiv x_n$. We will show that $\{x_n\}$ is Cauchy.

$$\rho(x_n, x_{n+1}) = \rho(Tx_{n-1}, Tx_{n-2}) \leq K\rho(x_{n-1}, x_{-2})$$
$$\leq K^2 \rho(x_{n-2}, x_{n-3})$$
$$\cdots \leq K^{n-1} \rho(x_1, x_0)$$

Thus if $n > m$,

$$\rho(x_n, x_m) \leq \sum_{j=m+1}^{n} \rho(x_j, x_{j-1}) \leq K^m (1-K)^{-1} \rho(x_0, x_1) \to 0 \text{ as } m \to \infty$$

Thus $\{x_n\}$ is Cauchy, so $x_n \to x$ for some x. Since T is continuous, $Tx = \lim Tx_n = \lim x_{n+1} = x$ which proves the theorem. ∎

The second theorem is much more difficult to prove and we have no intention of proving it here (see the notes); it generalizes the Brouwer fixed-point theorem which says that a continuous map of the closed unit ball in $\mathbb{R}^n$ into itself has at least one fixed point, a theorem which is already quite deep.

Theorem V.19 (Leray–Schauder–Tychonoff theorem)   Let C be a nonempty compact convex subset of a locally convex space X. Let $T: C \to C$ be a continuous map. Then T has a fixed point.

As preparation for our last general fixed point theorem, we will prove Theorem V.19 in one special case (see the lemma below).

Definition Let X and Y be vector spaces, C a convex subset of X. A map $T: C \to Y$ is called an **affine linear map** on C if

$$T(tx + (1-t)y) = tTx + (1-t)Ty$$

for all $x, y \in C$ and all $0 \leq t \leq 1$.

Unlike the case of linear functionals on subspaces, continuous affine functionals on convex sets may not have continuous extensions to all of X (Problem 49).

Lemma Let T be a continuous affine map of C into itself where C is a compact convex subset of a locally convex space X. Then T has a fixed point.

Proof Let

$$x_n = \frac{1}{n} \sum_{i=0}^{n-1} T^i x_0$$

where x_0 is picked in C. Since C is convex, each $x_n \in C$. Since C is compact, some subnet $x_{n(\alpha)}$ converges to a limit x. We wish to show $Tx = x$. By the Hahn–Banach theorem, it is enough to show $\ell(Tx) = \ell(x)$ for any $\ell \in X^*$. Since C is compact, $\sup_{x \in C} |\ell(x)| \equiv M_\ell < \infty$ for any fixed ℓ. Thus

$$|\ell(Tx_n - x_n)| = \left|\ell\left(\frac{1}{n} T^n x_0 - \frac{1}{n} x_0\right)\right| \leq \frac{2}{n} M_\ell \to 0 \quad \text{as} \quad n \to \infty$$

As a result $\ell(Tx - x) = \lim_\alpha \ell(Tx_{n(\alpha)} - x_{n(\alpha)}) = 0$. ∎

The last fixed-point theorem that we consider deals with a whole family of maps:

Definition A family of maps $\mathscr{F}$ from a set X to itself is said to possess a **common fixed point** if there is an $x \in X$ so that $Tx = x$ for all $T \in \mathscr{F}$.

Theorem V.20 (Markov–Kakutani theorem) Let $\mathscr{F}$ be a family of *commuting* continuous affine maps of C into itself where C is a compact convex subset of a locally convex space; that is, $TSx = STx$ for all $S, T \in \mathscr{F}$ and $x \in C$. Then $\mathscr{F}$ has a common fixed point.

Proof For each finite subset $F \subset \mathscr{F}$, let $f_F = \{x \in C \mid Tx = x \text{ for all } T \in F\}$. Since the T are all continuous, each f_F is closed and clearly $f_{F_1} \cap f_{F_2} = f_{F_1 \cup F_2}$. Thus, if we can show each f_F is nonempty, $\bigcap_F f_F \neq \emptyset$ by the f.i.p. criterion, so there is an x with $Tx = x$ for all $T \in \mathscr{F}$. Since the $T \in \mathscr{F}$ are affine linear each f_F is convex. $x \in f_F$ implies $Sx \in f_F$ for each $S \in \mathscr{F}$ because $T \in F$ implies $T(Sx) = S(Tx) = Sx$ when $Tx = x$. Since f_F is convex, compact, and $S: f_F \to f_F$, there is an $x \in f_F$ with $Sx = x$, that is, $f_{F \cup \{S\}} \neq \emptyset$ if $f_F \neq \emptyset$. By induction, each f_F is nonempty so $\bigcap_F f_F \neq \emptyset$. As remarked above, this implies $\mathscr{F}$ has a common fixed point. ∎

V.6 Applications of fixed point theorems

In Chapter XIV, we return to the special properties of compact, convex subsets of locally convex spaces; in particular, in Section XIV.6, we extend Theorem V.20 to a variety of noncommutative cases (see also Problem 50).

V.6 Applications of fixed point theorems

A. Ordinary differential equations

Let F be a continuous function from $\mathbb{R} \times \mathbb{R}^n$ to $\mathbb{R}^n$. We are interested in solving the differential equation $\dot{y} \equiv dy/dt = F(t, y)$ with initial conditions; i.e. given $y_0 \in \mathbb{R}^n$, we want to find a continuously differentiable function, $y(t)$, on $\mathbb{R}$ for which $y(0) = y_0$ and $\dot{y}(t) = F(t, y(t))$ for all $t \in \mathbb{R}$. We will discuss how the fixed point theorems of Section V.5 can be applied to prove the existence of *local* solutions, that is, given y_0, we will find δ and a function, $y(t)$ on $(-\delta, \delta)$ obeying $y(0) = 0$ and $\dot{y}(t) = F(t, y(t))$ for all $|t| < \delta$. We remark that a pth order equation $y^{(p)} = F(t, y, \dot{y}, \ldots, y^{(p-1)})$ on $\mathbb{R}^k$ can be translated into a first-order equation on $\mathbb{R}^{kp}$ by letting Y be the column vector

$$Y = \begin{pmatrix} y \\ \dot{y} \\ \vdots \\ y^{(p-1)} \end{pmatrix} \equiv \begin{pmatrix} y_1 \\ y_2 \\ \vdots \\ y_p \end{pmatrix} \quad \text{and solving} \quad \dot{Y} = \begin{pmatrix} y_2 \\ y_3 \\ \vdots \\ F(t, y_1, \ldots, y_p) \end{pmatrix}$$

Differentiation makes functions less smooth—it cannot usually be defined as a map from a space to itself unless the space contains only C^∞ functions. Integration is a much smoother operation—it takes continuous functions on an interval into themselves. Therefore, it is useful to rewrite the differential equation in integral form:

$$y(t) = y_0 + \int_0^t F(s, y(s))\, ds \tag{V.6}$$

It is easy to see that a continuous function $y(t)$ on $(-\delta, \delta)$ obeys (V.6) if and only if it is a local solution of $\dot{y}(t) = F(t, y(t))$ with the initial condition $y(0) = y_0$.

Thus given y_0 and δ, we consider the map $G: C[-\delta, \delta] \to C[-\delta, \delta]$, on the continuous functions from $[-\delta, \delta]$ to $\mathbb{R}^n$, given by

$$[G(g)](t) = y_0 + \int_0^t F(s, g(s))\, ds$$

Solving (V.6) is equivalent to finding a fixed point of G!

Consider, first the case where F is Lipschitz continuous; that is, given y_0, there is a K, δ, and ε so that $\|y - y_0\| < \varepsilon$, $|t| < \delta$ implies $\|F(y, t) - F(z, t)\| \leq K\|y - z\|$ if $\|z - y_0\| < \varepsilon$. $\|\cdot\|$ denotes the Euclidean norm on $\mathbb{R}^n$. By shrinking δ we can be sure that

$$\delta \max_{\substack{|t|<\delta \\ \|y-y_0\|<\varepsilon}} \|F(t, y)\| < \varepsilon \quad \text{and} \quad \delta K < \tfrac{1}{2}$$

Now, let

$$S = \{g \in C[-\delta, \delta] \mid \|g(t) - y_0\| \leq \tfrac{1}{2}\varepsilon, \forall t \in (-\delta, \delta)\}.$$

S is a complete metric space under $\|\cdot\|_\infty$ and $\delta \max \|F(t, y)\| < \varepsilon$ implies that $G(g) \in S$ if $g \in S$. The conditions $\delta K < \tfrac{1}{2}$ and $\|F(y, t) - F(z, t)\| \leq K\|y - z\|$ imply $\|G(g_1) - G(g_2)\|_\infty < \tfrac{1}{2}\|g_1 - g_2\|_\infty$, if $g_1, g_2 \in S$. Thus G is a strict contraction on S, so there is a *unique* $g \in S$ which satisfies (V.6) by Theorem V.18. It is not hard to see that any solution of (V.6) must obey $\|g(t) - y_0\| \leq \tfrac{1}{2}\varepsilon$ for t small and so must agree with the unique solution in S when t is small. In this case, we thus prove local existence and uniqueness.

Now suppose F is merely continuous. Given y_0, pick δ so that

$$\max_{\substack{|t|<\delta \\ \|y-y_0\|\leq 1}} |F(t, y)| < \delta^{-1}$$

Let

$$C_0 = \{g \in C[-\delta, \delta] \mid \|g(t) - y_0\| \leq 1, \forall t \in (-\delta, \delta)\}$$

Then $G(g) \in C_0$, if $g \in C_0$. Actually, more can be said; let

$$C_1 = \{g \in C_0 \mid \|g(t) - g(s)\| \leq |t - s| \max_{\substack{|t|<\delta \\ \|y-y_0\|\leq 1}} |F(t, y)|, \forall t, s \in (-\delta, \delta)\}$$

Then $g \in C_0$ implies $G(g) \in C_1$. Thus $G: C_1 \to C_1$. G is continuous and C_1 is convex and compact by equicontinuity arguments. Thus G has a fixed point by Theorem V.19.

Notice that the second method gives existence of local solutions, but not uniqueness also. In fact, when F is continuous but not Lipschitz, uniqueness may not hold: for example, if $F(t, y) = 2\sqrt{y}$, the equation $\dot{y} = 2\sqrt{y}$, $y(0) = 0$, has the two solutions, $y(t) \equiv 0$ and $y(t) = t^2$, which differ in any neighborhood of $t = 0$. The compactness of C_1 depends on the finite dimensionality of $\mathbb{R}^n$ where the unit ball is compact. However, the contraction in the Lipschitz case does not use compactness and thus goes through immediately for ordinary differential equations for functions with values in a Banach space.

The question of when local solutions can be continued to global solutions is much touchier.

B. Haar measure on commutative compact groups

One can use the Markov–Kakutani theorem for an easy construction of an invariant measure on a compact abelian group. Let G be a compact topological space which is also an abelian group and suppose that all the group operations are continuous. We wish to construct a measure $\mu \in \mathcal{M}_{+,1}(G)$ so that $\int f \, d\mu = \int f_g \, d\mu$ for any $g \in G$ and $f \in C(G)$ where f_g is the translate of f: $f_g(h) = f(h-g)$. For any $\mu \in \mathcal{M}_{+,1}(G)$, let $T_g \mu$ be defined by $T_g \mu(f) = \mu(f_g)$. Then $T_g: \mathcal{M}_{+,1}(G) \to \mathcal{M}_{+,1}(G)$ and is continuous in the vague (weak-$*$) topology. $\mathcal{M}_{+,1}$ is convex and is compact by the Banach–Alaoglu theorem. The various T_g commute since G is abelian. They are affine, so Theorem V.20 implies the existence of a common fixed point μ_{Haar} with the desired invariance property. In this case the fixed point theorem implies existence; uniqueness must be proven by a distinct argument.

C. Bootstrap equations

Without going into details, either technical or physical, we will describe an application of the Leray–Schauder–Tychonoff theorem to proving the consistency of certain bootstrap schemes.

In order to describe the bootstrap idea, let us fall back on potential scattering where there are fewer complications. Consider the scattering of two equal mass particles. Since the total momentum $\mathbf{P}$ is conserved and we will suppose that all forces depend only on the relative position of the particles, we can describe the scattering in a coordinate frame where $\mathbf{P} = 0$ (see Figure V.2).

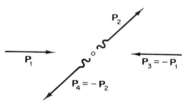

FIGURE V.2 Scattering when $\mathbf{P} = 0$.

For simplicity, we take $m = \tfrac{1}{2}$ where m is the mass of the particles. The natural variables in the nonrelativistic case are the energy $E = P_1^2 = P_2^2$ and the scattering angle θ, but for the proper relativistic correspondence, we take instead $s = 4E$ and the momentum transfer $t = -(\mathbf{P}_1 - \mathbf{P}_2)^2 = 2E(\cos \theta - 1)$.

In quantum mechanics, scattering is basically a wave phenomenon and is thus described by the magnitude of the scattered wave and its phase relative to the incident wave; that is, by a complex number $A_{\text{phys}}(s, t)$, the scattering amplitude discussed in Section XI.6. It is defined in the region $E \geq 0$, $-1 \leq \cos \theta \leq 1$, or equivalently $0 \leq s < \infty$, $-s \leq t \leq 0$. As we shall see in Section XI.7, for a large variety of potentials the amplitude $A_{\text{phys}}(s, 0)$ is the boundary value of a function analytic in the cut s plane (Figure V.3). In

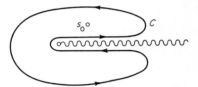

FIGURE V.3 The curve C in the cut s plane.

general there may be poles along the negative axis which we ignore for simplicity. Thus, $A_{\text{phys}}(s, 0) = \lim_{\varepsilon \downarrow 0} A(s + i\varepsilon, 0) \equiv A(s + i0, 0)$ for some analytic function in $\{s \mid \arg s \neq 0\}$. Moreover, $A(s, 0)$ is real for $s < 0$, so the Schwarz reflection principle implies that the physical amplitude obeys

$$\text{Im } A(s + i0, 0) = \frac{1}{2i}[A(s + i0, 0) - A(s - i0, 0)]$$

Suppose, again for simplicity, that $A(s, 0) \to 0$ at infinity. In general, this is false and one must modify the argument below by what is known as making subtractions (see Section XI.7). Then, by the Cauchy integral theorem,

$$A(s_0, 0) = \frac{1}{2\pi i} \oint_C \frac{A(s, 0)}{s - s_0} ds$$

where C is the contour in Figure V.3. If we now make the large circle bigger and bigger, it will make no contribution since we assumed $A(s, 0) \to 0$ as $s \to \infty$. By shrinking the straight sections to the real axis, we find

$$A(s, t_0 = 0) = \frac{1}{\pi} \int_0^\infty \frac{D(s, t_0 = 0)}{s - s_0} ds \qquad (\text{V.7a})$$

where $D(s, t_0 = 0) = (1/2i)[A(s + i0, 0) - A(s - i0, 0)] = \text{Im } A(s + i0, 0)$ is the "discontinuity" of A.

This "dispersion relation" (V.7a) can be proven to hold with $t = t_0$ for all real positive t_0 for a small class of potentials. This class includes sums of Yukawa potentials which are thought to be the nonrelativistic analogues of nuclear forces. Moreover, in that case, for s_0 fixed, $D(s_0, t)$ is the boundary value of an analytic function in a plane cut from $t = \sigma(s)$ to ∞. $\sigma(s)$ is a function depending on the potential which is explicitly known and is called the Mandelstam boundary. We can write a dispersion relation for D

$$D(s, t_0) = \frac{1}{\pi} \int_{\sigma(s)}^{\infty} \frac{\rho(s, t)}{t - t_0} dt$$

where $\rho(s, t) = (2i)^{-1}[D(s, t + i0) - D(s, t - i0)]$ is called the "double spectral function."

Putting our two dispersion relations together, we obtain the "Mandelstam representation" for A

$$A(s, t) = \frac{1}{\pi^2} \int_0^{\infty} ds' \int_{\sigma(s')}^{\infty} dt' \frac{\rho(s', t')}{(s' - s)(t' - t)} \quad \text{(V.7b)}$$

This *linear* relation is essentially a statement of the analyticity properties of A.

The second element of the bootstrap scheme is the "unitarity of the S matrix." Let us temporarily revert to the (E, θ) variables and write $A(E, \theta)$ for the amplitude at $s = 4E$; $t = -2E(1 - \cos \theta)$. Unitarity is an expression of the fact that as many particles should leave the scattering region as enter it. Put differently, the number of particles scattered out of a beam must be the number lost from the beam (for a deeper discussion, see Section XI.4). Quantum mechanically, the decrease in the number of particles in the beam is due to interference between the scattered wave and the unscattered wave. This interference is proportional to Im $A(E, 0)$ while the amount scattered out of the beam is proportional to $|A|^2$. Thus, one finds the *nonlinear* relation

$$\text{Im } A(E, 0) = c \int |A(E, \theta)|^2 \, d\Omega$$

where $c(E)$ is a function of E dependent on normalization of A and $d\Omega$ is the angular measure on the sphere. One can extend this relation to nonzero θ and thus obtain a nonlinear integral relation between $D(s, t)$ and $A(s', t')$ (quadratic in A). By taking the discontinuity of D and using the Mandelstam representation for A one finds the relation

$$\rho = T(\rho)$$

where T is an explicit but complicated function of ρ. If ρ obeys $\rho = T(\rho)$ and has the proper decrease to make certain integrals converge, then one can show that A defined by (V.7b) obeys unitarity. Thus, the existence of $A(s, t)$ with the right analyticity and unitarity properties is equivalent to the existence of fixed points of T. Of course, in the nonrelativistic case, one knows that such ρ exist since one can show that the scattering amplitude for a superposition of Yukawa potentials has the right analyticity and unitarity properties.

In the relativistic case, for example, $\pi^0\pi^0$ scattering, there are two additional complications:

(i) *Crossing symmetry* Replace the momenta p_1, p_2 in Figure V.1 by the energy–momentum four-vectors $p_i = \langle \sqrt{\mu^2 + \mathbf{p}_i^2}, \mathbf{p}_i \rangle$ where μ is the mass of the pion. Given a four-vector $a = \langle a_0, \mathbf{a} \rangle$, we define $a^2 = a_0^2 - \mathbf{a}^2$; for example, $p_i^2 = \mu^2$, an expression of the relativistic energy–momentum relation. Now one defines the Mandelstam variables, $s = (p_1 + p_3)^2 = 4(p^2 + \mu^2)$ if p is the center of mass momentum, $t = (p_1 - p_2)^2 = \frac{1}{2}(s - 4\mu^2)(\cos\theta - 1)$ and $u = (p_1 - p_4)^2 = \frac{1}{2}(s - 4\mu^2)(-\cos\theta - 1)$. Of course, $s, t,$ and u are not independent for $s + t + u = 4\mu^2$. Crossing symmetry expresses a deep fact of relativistic quantum theory, namely that the analytically continued amplitude for say $\pi^0\pi^0 \to \pi^0\pi^0$ is a symmetric function of s and t, and a symmetric function of s and u when the change of variables $\langle s, t \rangle \to \langle s, u \rangle$ is made by using $u = 4\mu^2 - s - t$. This automatically implies additional branch cuts in the domain of the function $A(s, t)$. For example, analogous to the cut in the domain of the nonrelativistic amplitude, running from $E = 0$ to $E = \infty$, is a cut in $A(s, 0)$ running from $s = 4\mu^2$ to $s = \infty$. Crossing symmetry implies that there must also be a cut running from $u = 4\mu^2$ to $u = \infty$, or equivalently, since $t = 0$, from $s = 0$ to $s = -\infty$. The analogue of nonrelativistic Mandelstam analyticity is then expressed by the relativistic Mandelstam relation

$$A(s_0, t_0) = \frac{1}{\pi^2} \int_{4\mu^2}^\infty ds \int_{\sigma(s)}^\infty dt\, \rho(s, t)$$

$$\times \left[\frac{1}{(s - s_0)(t - t_0)} + \frac{1}{(s - s_0)(t + s_0 + t_0 - 4\mu^2)} \right.$$

$$\left. + \frac{1}{(s + s_0 + t_0 - 4\mu^2)(t - t_0)} \right]$$

In this formula

$$\sigma(s) = \min\left\{ \frac{4s}{s - 16\mu^2}, \frac{16s}{s - 4\mu^2} \right\}$$

and ρ must obey $\rho(s, t) = \rho(t, s)$. The last two terms are just

$$\frac{\rho(s, u)}{(s - s_0)(u - u_0)} \quad \text{and} \quad \frac{\rho(t, u)}{(t - t_0)(u - u_0)}$$

after change of variable. This Mandelstam relation is just the expression of crossing symmetry plus certain analyticity properties.

(ii) *Inelastic processes* It is characteristic of relativistic systems that if there is sufficient energy available, large numbers of particles can be produced. For example if $s > 16\mu^2$, the reaction $\pi^0 + \pi^0 \to \pi^0 + \pi^0 + \pi^0 + \pi^0$ is

V.6 Applications of fixed point theorems 159

possible. Unitarity comes from a connection between the interference of the scattered and unscattered waves and the total amount scattered in *all* processes. It thus gives a relation between an $A(s, 0)$ and a sum of terms, one of which comes from $\pi^0 + \pi^0 \to \pi^0 + \pi^0$. Even when $4\mu^2 \leq \alpha \leq 16\mu^2$, an inelastic process $\pi^0 + \pi^0 \to \pi^+ + \pi^-$ is possible. A complete treatment would require consideration of all the $\pi + \pi \to \pi + \pi$ amplitudes. For simplicity we consider a model when there are no π^+ and π^-. Unitarity is thus a nonlinear equality on A only when $4\mu^2 \leq s \leq 16\mu^2$.

The "bootstrap hypothesis" of Chew and Mandelstam is the philosophical idea that there is only one set of amplitudes for all processes with the "usual" analyticity properties and which obey all the unitarity equations (coupling various processes). In practice, one approximates the equations, for example by replacing the coupled unitarity equations with inequalities when $s \geq 16\mu^2$ as we have done above. Whether one accepts the bootstrap philosophy or not, the various bootstrap equations are of interest since they can be viewed as an expression of the constraints placed by unitarity, crossing symmetry, and analyticity on the amplitude. Even if these do not determine the amplitude (and we do not subscribe to the bootstrap philosophy), they do put severe restrictions on the amplitude. In fact, it is not clear, a priori, that *any* functions $A(s, t)$ exist which obey the requisite analyticity *and* crossing symmetry, elastic unitarity if $4\mu \leq s \leq 16\mu^2$, and the unitarity inequalities for $16\mu^2 \leq s$.

The existence of such functions has been established by Atkinson in a beautiful application of the Leray–Schauder–Tychonoff theorem. The basic idea of the proof is the following. One seeks a function $\rho(s, t)$ to put into the Mandelstam representation. If A obeyed elastic unitarity everywhere, one would have

$$\rho(s, t) = (T^{\text{el}}\rho)(s, t)$$

where T^{el} is a complicated nonlinear map. Since elastic unitarity is only obeyed in certain regions, one has this equality only in certain regions of the s-t plane. In general $\rho(s, t) = (T^{\text{el}}\rho)(s, t) + v(s, t)$ where setting $v = 0$ in certain regions is equivalent to elastic unitarity. If v obeys certain other conditions, any solution of $\rho = T^{\text{el}}\rho + v$ obeying certain integrability conditions yields an A obeying elastic unitarity in $4\mu^2 \leq s \leq 16\mu^2$ and the inelastic unitarity inequalities in $s \geq 16\mu^2$. Thus, the existence of solutions of these approximate bootstrap equations would follow if $\rho = T\rho$ had a solution with $T\rho = T^{\text{el}}\rho + v$. What Atkinson does is to construct a convex set S_v, depending on v, of uniformly bounded, equicontinuous functions which is compact in the $\|\cdot\|_\infty$ topology and so that $T: S \to S$ and is continuous. The Leray–Schauder–Tychonoff theorem then provides the existence of solutions of of the approximate bootstrap equations.

D. Determining the phase of the scattering amplitude

According to quantum scattering theory (see Chapter XII), the "differential" scattering cross section at fixed energy is given by a function

$$D(\theta) = |F(\theta)|^2$$

where $F(\theta)$ is a complex-valued function of the scattering angle θ. At energies where there is only elastic scattering, F must obey the nonlinear "unitarity relation"

$$\operatorname{Im} F(\theta) = \int \overline{F(\theta_1)} F(\theta_2) \sin \theta_1 \, d\theta_1 \, d\varphi_1$$

where θ_2 is the function of θ_1, θ, and φ_1 determined by the spherical geometry in Figure V.4. $z_2 = \cos \theta_2$ is given in terms of $z = \cos \theta$, $z_1 = \cos \theta_1$, and φ_1 by

$$z_2 = z z_1 + (1 - z^2)^{1/2}(1 - z_1^2)^{1/2} \cos \varphi_1$$

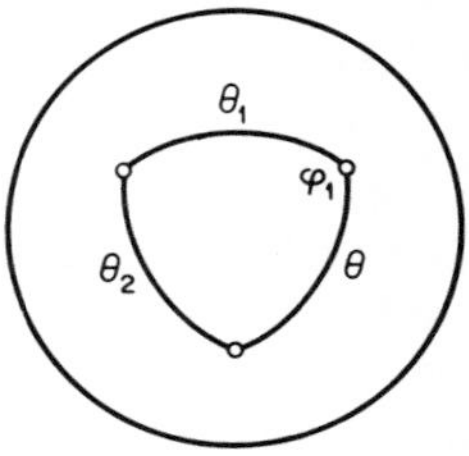

FIGURE V.4   The angle θ_2.

In experiments, one measures $D(\theta)$ while $F(\theta)$ is of great theoretical interest. There are two questions one immediately wants to ask: (1) Does the unitarity relation place any restrictions on the possible functions $D(\theta)$ which come from functions F obeying the unitarity relation. (2) Given $D(\theta)$, is F determined by the condition $|F(\theta)| = |D(\theta)|^{1/2}$ and the unitarity condition. The reader should realize by now that these questions are really the existence and uniqueness aspects of a single question.

Introduce the variables $z_i = \cos \theta_i$. Let $K(z_1, z_2; z)$ be the Jacobian of the transformation from $\langle z_1, \varphi_1 \rangle$ to $\langle z_1, z_2 \rangle$. Let $B(z) = |D(\theta)|^{1/2}$ be given and write the putative $F(\theta) = B(z)e^{i\varphi(z)}$. Then

$$\sin \varphi(z) = \int_{-1}^{1} \int_{-1}^{1} K(z_1, z_2; z) \frac{B(z_1)B(z_2)}{B(z)} e^{-i(\varphi(z_1) - \varphi(z_2))} \, dz_1 \, dz_2$$

or (V.8)

$$\varphi(z) = \sin^{-1}\left[\iint K(z_1, z_2; z) \frac{B(z_1)B(z_2)}{B(z)} \cos[\varphi(z_1) - \varphi(z_2)] \, dz_1 \, dz_2 \right]$$

Let

$$M = \iint |K(z_1, z_2; z)| \frac{B(z_1)B(z_2)}{B(z)} dz_1 dz_2$$

Suppose $M < 1$. Then for any continuous function $\varphi(z)$ on $[-1, 1]$ the function $\mathscr{F}\varphi$ given by

$$(\mathscr{F}\varphi)(z) = \sin^{-1}\left[\iint K(z_1, z_2; z) \frac{B(z_1)B(z_2)}{B(z)} \cos[\varphi(z_1) - \varphi(z_2)] dz_1 dz_2\right]$$

is well defined. Question (1) has a positive answer if and only if $\mathscr{F}$ has a fixed point. (2) is related to uniqueness as follows: Let $\mu = \sin^{-1}(M)$. First note that if the branch of $\sin^{-1}$ with $-\pi/2 < \sin^{-1} x \leq \pi/2$ is chosen, $\mathscr{F}$ takes the functions φ on $[-1, 1]$ with $\|\varphi\|_\infty \leq \mu$ into themselves. Also, if φ satisfies (V.8), is continuous, and $|\varphi(0)| < \pi/2$, then $\|\varphi\|_\infty \leq \mu$. Finally, we remark that $\varphi(z)$ satisfies (V.8) if and only if $\tilde{\varphi}(z) = \pi - \varphi(z)$ satisfies (V.8). Thus $\mathscr{F}\varphi = \varphi$ has exactly two solutions if and only if there is only one solution with $|\varphi(z)| \leq \mu$ for all z.

If $M < 0.79$, Martin has shown that $\mathscr{F}$ is a contraction on

$$\{\varphi \mid \|\varphi\|_\infty \leq \mu; \varphi \text{ continuous}\}$$

with a suitable metric. This implies uniqueness and existence. For the general $M < 1$, existence (but not uniqueness) has been shown by an application of the Leray–Schauder–Tychonoff theorem.

E. Existence of correlation functions at low density

Finally we will briefly discuss the application of a very simple fixed-point theorem to statistical mechanics. The theorem is (Problem 51):

Theorem V.21 Let K be a linear map of a Banach space onto itself with $\|K\| < 1$. Then, for any g, $f = g + Kf$ has a *unique* solution $f = \sum_{n=0}^{\infty} K^n g$. Let K_m be a sequence of such maps so that $\|K_m - K\| \to 0$ and $\|K_m\| < 1$. If $g_m \to g$, $f_m = g_m + K_m f_m$ and $f = g + Kf$, then $f_m \to f$.

In equilibrium statistical mechanics, one introduces "correlation functions," $\rho(x_1), \rho_2(x_1, x_2), \ldots, \rho_n(x_1, \ldots, x_n), \ldots$. In the infinite volume limit, where the boundary effects stop playing a role, $\rho_n(x_1, \ldots, x_n)$ is the probability *density* for finding particles at $x_1, \ldots, x_n$. Thus, for example, $\rho_1(x)$ should be a constant equal to the density.

In a box Λ of finite size, elementary statistical mechanics gives an explicit formula for the correlation functions $\rho_n^{(\Lambda)}(x_1, \ldots, x_n; \beta, z)$ where $\beta = (kT)^{-1}$

is the inverse temperature, $z = e^{\beta\mu}$ is the fugacity, and μ is the chemical potential. z is small when the density is low, i.e. when the system is a gas. One goal of classical statistical mechanics is to prove the existence of

$$\lim_{\Lambda \to \infty} \rho_n^{(\Lambda)}(x_1, \ldots, x_n, \beta, z)$$

at least for small z and/or β, i.e., for high temperature and/or low density where the system should be a gas, so there are no problems associated with the presence of several phases. This can be accomplished by the use of Theorem V.21. In a finite box, one shows that $\{\rho_n^{(\Lambda)}\}_{n=1}^\infty$ obeys a system of coupled integral equations (the Kirkwood–Salzburg equations):

$$\rho_1^{(\Lambda)}(x_1) = g_1^{(\Lambda)}(x_1) + \sum_{n=1}^\infty \int dy_1 \cdots dy_n \, K_{1,n}^{(\Lambda)}(x_1; y_1, \ldots, y_n) \, \rho_n^{(\Lambda)}(y_1, \ldots, y_n)$$

$$\rho_m^{(\Lambda)}(x_1, \ldots, x_m) = g_m^{(\Lambda)}(x_1, \ldots, x_m) \rho_{m-1}^{(\Lambda)}(x_2, \ldots, x_m)$$

$$+ \sum_{n=1}^\infty \int dy_1 \cdots dy_n \, K_{m,n}^{(\Lambda)}(x_1 \cdots x_m; y_1 \cdots y_n)$$

$$\times \rho_{n+m-1}^{(\Lambda)}(x_1, \ldots, x_m; y_1, \ldots, y_n)$$

Introducing a vector $\boldsymbol{\rho} = (\rho_1, \ldots, \rho_n, \ldots)$, these can be written schematically

$$\boldsymbol{\rho}^{(\Lambda)} = \mathbf{g}^{(\Lambda)} + \mathbf{K}^{(\Lambda)} \boldsymbol{\rho}^{(\Lambda)}$$

It turns out that one can introduce a norm on the set of $\boldsymbol{\rho}$ so that $\mathbf{K}^{(\Lambda)}$ is a bounded operator. For small z and/or β, one has $\mathbf{g}^{(\Lambda)} \to \mathbf{g}$ and $\mathbf{K}^{(\Lambda)} \to \mathbf{K}$ in the infinite volume limit, and all $\mathbf{K}^{(\Lambda)}$, $\mathbf{K}$ have norm less than one. By Theorem V.21, $\boldsymbol{\rho}^{(\Lambda)}$ converges in the limit of infinite volume to the unique solution of the infinite volume Kirkwood–Salzburg equations.

With regard to these last three examples, we caution the reader against thinking he has any real understanding of the technical details. We have tried to explain in a vague sense how fixed-point theorems arise—but all the hard work is in the choice of "suitable" norms or spaces and the proofs of bounds or continuity of maps. We have not considered these details at all.

V.7 Topologies on locally convex spaces: duality theory and the strong dual topology

In this section, we want to consider relations among various locally convex topologies on a vector space X. These topologies will not be used until Chapter XIV but their study develops useful intuition and illuminates the choice of topology on $\mathscr{S}$ and $\mathscr{D}$.

V.7 Topologies on locally convex spaces

Let X be a vector space and Y a space of linear functionals on X which separates points; such a pair $\langle X, Y \rangle$ is called a **dual pair**. We already know that X has a locally convex topology, $\sigma(X, Y)$, in which the topological dual is precisely Y (Theorem IV.20). We first want to ask what other locally convex topologies on X produce Y as the topological dual.

Definition Let $\langle X, Y \rangle$ be a dual pair. The **Mackey topology** on X, $\tau(X, Y)$, is the topology of uniform convergence on $\sigma(Y, X)$-compact convex sets of Y; that is, $x_\alpha \xrightarrow{\tau(X,Y)} x$ if and only if $y(x_\alpha) \to y(x)$ uniformly as y runs through any fixed $\sigma(Y, X)$-compact convex subset of Y.

Alternatively, for each $\sigma(Y, X)$-compact convex subset C of Y, define the seminorm ρ_C on X by
$$\rho_C(x) = \sup_{y \in C} |y(x)|$$
The family of seminorms $\{\rho_C \mid C \text{ is } \sigma(Y, X)\text{-compact convex subset of } Y\}$ on X generates the $\tau(X, Y)$ topology. If C is compact, then
$$\tilde{C} = \{\lambda y \mid |\lambda| \leq 1, y \in C\}$$
is also compact (because $\{\lambda \mid |\lambda| \leq 1\}$ is compact) and $\rho_C = \rho_{\tilde{C}}$. Thus, we need only consider balanced, convex, $\sigma(Y, X)$-compact sets, C.

Since single points in Y are compact in the weak topology, if $x_\alpha \xrightarrow{\tau(X,Y)} x$, then $x_\alpha \xrightarrow{\sigma(X,Y)} x$, that is, the weak topology is weaker than the Mackey topology.

Example 1 Let X be a Banach space. Let X^* be its dual space. We claim that the $\tau(X, X^*)$ topology is just the norm topology on X. For the Banach–Alaoglu theorem (Theorem IV.21) tells us that the unit ball, X_1^*, in X^* is $\sigma(X^*, X)$ compact, so
$$\rho_{X_1^*}(x) = \sup_{y \in X_1^*} |y(x)|$$
is a Mackey seminorm. And, the Hahn–Banach theorem implies $\rho_{X_1^*}(\cdot) = \|\cdot\|_X$. On the other hand, if $C \subset X^*$ is $\sigma(X^*, X)$-compact, then for any $x \in X$, $y \mapsto y(x)$ is bounded on C so $C \subset \{y \mid \|y\| \leq m\}$ for some m by the Banach–Steinhaus principle. Thus $\rho_C(x) \leq m\|x\|_X$. This shows that the Mackey topology is generated by $\|\cdot\|_X$.

Example 2 We will see in Problem 52 that any Fréchet space, X, has the Mackey topology, $\tau(X, X^*)$. And, any *strict* inductive limit of Fréchet spaces also has the Mackey topology (Problem 52a,d). In particular, the spaces $\mathscr{S}(\mathbb{R}^n)$ and $\mathscr{D}(\mathbb{R}^n)$ have the Mackey topology.

We return to the question of finding all dual $\langle X, Y \rangle$ topologies. A locally convex topology $\mathscr{T}$ on X is called a **dual-$\langle X, Y \rangle$ topology** if the topological dual of $\langle X, \mathscr{T} \rangle$ is Y.
The basic duality theorem which identifies all dual topologies says:

Theorem V.22 (Mackey–Arens theorem) Let $\langle X, Y \rangle$ be a dual pair. A locally convex topology $\mathscr{T}$ on X is a dual-(X, Y) topology if and only if

$$\sigma(X, Y) \subseteq \mathscr{T} \subseteq \tau(X, Y)$$

Proof For the geometric proof, see the appendix to this section.

Thus the dual topologies are precisely those between the weak topology and the Mackey topology (inclusive).
We have seen how to recover the norm topology on a Banach space X in terms of the dual pair $\langle X, X^* \rangle$. How about the norm topology on X^*? It is not the $\tau(X^*, X)$ topology unless X is reflexive, for the $\tau(X^*, X)$ dual is X. Clearly, the norm topology on X^* is the topology of uniform convergence on the unit ball in X and so we need a locally convex notion that singles out sets contained inside balls. This need is met by the notion of bounded set. Before defining this notion, we note:

Theorem V.23 Let E be a locally convex space with dual F. The following are equivalent for a set $A \subset E$:

(a) For any neighborhood, U, of $0 \in E$, $A \subset nU = \{nx \mid x \in U\}$ for some n.
(b) The polar of A, A° (which we define in the appendix to this section), is absorbing.
(c) For any continuous seminorm ρ on E, $\sup_{x \in A} \rho(x) < \infty$.
(d) For any $\ell \in F$, $\sup_{x \in A} |\ell(x)| < \infty$.

Proof The equivalence of (a) and (c) and of (b) and (d) are essentially matters of definition. Theorem V.4 says that (c) implies (d). So suppose (d) holds and a continuous seminorm ρ is given. Let $K_\rho = \{x \in E \mid \rho(x) = 0\}$ and let E_ρ be the vector space E/K_ρ. Then ρ "lifts" to a norm on E_ρ. Let π be the canonical map of $E \to E_\rho$ and let $A_\rho = \pi[A]$. It is easy to see that

V.7 Topologies on locally convex spaces

$\sup_{x \in A} p(x) < \infty$ if and only if $\sup_{x \in A_p} p(x) < \infty$. Let $\bar{E}_p$ be the completion of E_p. $\bar{E}_p$ is a Banach space. Let $\ell \in (E_p)^*$. Then $\ell \circ \pi \in E^*$ so

$$\sup_{x \in A} |\ell(x)| = \sup_{x \in A} |(\ell \circ \pi)(x)| < \infty$$

Thus by the Banach–Steinhaus principle, $\sup_{x \in A_p} p(x) < \infty$. ∎

Definition A set $A \subset E$, a locally convex space, is called **bounded** if and only if one, and hence all, of the conditions (a)–(d) of Theorem V.23 hold.

Condition (d) makes it clear that the notion of boundedness is the same in all $\langle E, F \rangle$-dual topologies and is thus a notion associated most naturally with a duality on E rather than a single topology on E.

Example 3 If X is a Banach space, $A \subset X$ is bounded if and only if $\sup_{x \in A} \|x\| < \infty$, which holds if and only if A is contained in some multiple of the unit ball. Thus for any bounded linear map on X,

$$\sup_{x \in A} \|Tx\| \leq C_A \|T\|$$

Example 4 Let X be a strict inductive limit of spaces X_n with X_n a proper closed subspace of X_{n+1}. If x_n is a sequence with $x_n \notin X_n$, by using the construction in the proof of Theorem V.17, we can find a linear functional, $\ell \in X^*$ with $\sup_n \ell(x_n) = \infty$. Thus any bounded set $A \subset X$ must actually be a bounded subset of some X_n. Thus, for example, $A \subset \mathscr{D}_\Omega$ is bounded if and only if: (i) there is a compact $K \subset \Omega$ so that $\text{supp} f \subset K$ when $f \in A$; (ii) $\sup_{x \in A} \|D^\alpha f\|_\infty < \infty$ for any $\alpha \in I_+^n$.

The first example tells us how to generalize the norm topology on X^*:

Definition Let E be a locally convex space. Let F be its dual. The **strong topology**, $\beta(F, E)$ on F is the topology of uniform convergence on bounded subsets of E; that is, the topology generated by the seminorms $\{\rho_A | A \subset E \text{ is bounded}\}$ where $\rho_A(f) = \sup_{x \in A} |f(x)|$.

Any $\sigma(E, F)$-compact set C of E is bounded because the $f \in F$ are continuous functions on C. Thus the *strong topology $\beta(F, E)$ is stronger than the Mackey topology*.

We have found a topology $\beta(F, E)$ only dependent on the dual pair $\langle F, E \rangle$ so that the norm topology on X^* is the $\beta(X^*, X)$ topology when X is a

Banach space. Given E, a locally convex space, we can form its dual E^* and put the $\beta(E^*, E)$ topology on it. The dual of E^* in this topology is called the **double dual** of E and is denoted by E^{**} when it is given the $\beta(E^{**}, E^*)$ topology. One can map E into E^{**} by the standard duality map, $\rho \colon E \to E^{**}$ by $\rho(x)(\ell) = \ell(x)$. This map is *not* always continuous (Problem 54). If it is a *topological* isomorphism, we say E is **reflexive**, that is, E is reflexive if:

(i) The $\beta(E^*, E)$ dual of E^* is E.
(ii) The $\beta(E, E^*)$ topology on E is the given topology.

The following criterion for reflexivity is often useful:

Lemma Let E be a locally convex space. Then E is reflexive if and only if all of the following hold:

(a) Every $\sigma(E, E^*)$-closed, bounded set of E is $\sigma(E, E^*)$ compact.
(b) Every $\sigma(E^*, E)$-closed, bounded set of E^* is $\sigma(E^*, E)$ compact.
(c) E has the Mackey (that is, $\tau(E, E^*)$) topology.

Proof It is not hard to see (Problem 55) that (a) holds if and only if the $\beta(E^*, E)$ and $\tau(E^*, E)$ topologies are identical and (b) says the $\beta(E, E^*)$ and $\tau(E, E^*)$ topologies are the same. Now, let E be reflexive. Since $E = E^{**}$, the $\beta(E^*, E)$ topology is a dual topology, so $\beta(E^*, E) \subset \tau(E^*, E)$ by the Mackey–Arens theorem. Since, $\tau \subset \beta$, $\beta(E^*, E) = \tau(E^*, E)$. Similarly, since the dual of E is E^* and E has the $\beta(E, E^*)$ topology, $\beta(E, E^*) = \tau(E, E^*)$ and E has the Mackey topology. Conversely, let (a)–(c) hold. By (a), $E = E^{**}$ as vector spaces since $\beta(E^*, E) = \tau(E^*, E)$ in that case, and the Mackey topology is a dual topology by Theorem V.22. By (b) and (c), E has the $\beta(E, E^*)$ topology. Thus E is reflexive. ∎

Using this lemma, one can prove (Problems 56, 57) that:

Theorem V.24 The spaces $\mathscr{S}(\mathbb{R}^n)$, $\mathscr{D}_\Omega$, and $\mathcal{O}_D$ are reflexive.

In general, the Mackey topology on a space is much stronger than the weak topology, so it is much harder for a set to be Mackey compact than weakly compact. For example, the unit ball in an infinite-dimensional Banach space is never norm compact (Problem 4) but it is weakly compact if X is reflexive. Thus general theorems on compactness in the Mackey topology are particularly strong statements. A most useful one is:

Theorem V.25 In $\mathscr{S}(\mathbb{R}^n)$, $\mathscr{D}_\Omega$, and $\mathcal{O}_D$, any closed bounded set is compact (in the usual Fréchet topology).

Appendix V.7 Polars and the Mackey-Arens theorem

Proof Let $C \subset \mathscr{S}(\mathbb{R})$ be closed and bounded. Since $\sup_{f \in C} \|f'\|_\infty \equiv E < \infty$, $|f(x) - f(y)| \leq E|x - y|$ whenever $f \in C$. As a result, C is a uniformly equicontinuous family of uniformly bounded functions. Similarly,

$$\{x^\alpha D^\beta f \mid f \in C\}$$

are uniformly equicontinuous and uniformly bounded. By the Ascoli theorem and a diagonalization trick, any sequence in C has a convergent subsequence. Since C is closed and the topology is metric, this proves that C is compact. The proofs for $\mathscr{S}(\mathbb{R}^n)$ and $\mathcal{O}_D$ are similar. For $C \subset \mathscr{D}_\Omega$, we note that since C is bounded, $C \subset C^\infty(K)$ for some K and then use the above argument. ∎

A particularly useful consequence of this last theorem and Theorem V.8 is:

Theorem V.26 *A sequence in $\mathscr{S}'$, $\mathscr{D}'$ or $\mathcal{O}'_D$ converges in the weak topology if and only if it converges in the strong topology.*

Proof For $\mathscr{S}'$ and $\mathcal{O}'_D$ this follows directly from Theorems V.8 and Theorem V.25. In the case of $\mathscr{D}'$, we notice that any bounded set $C \subset \mathscr{D}$ lies in $C_0^\infty(K)$, which is a Fréchet space and then apply Theorem V.8 and Theorem V.25. ∎

Even though Theorem V.26 is essentially a corollary to Theorem V.25, we single it out as a theorem because it is very useful in applications. We caution the reader to heed the word *sequence*. The theorem does not hold if "net" replaces "sequence."

Appendix to V.7 Polars and the Mackey–Arens theorem

In this technical appendix, we prove the Mackey–Arens theorem by introducing the machinery of polar sets:

Definition Let $\langle E, F \rangle$ be a dual pair. Let $A \subset E$. The **polar of** A, A°, is $\{f \in F \mid |f(e)| \leq 1 \; \forall e \in A\}$. If we want F to be explicit, we write $(A)_F^\circ$.

Examples (1) Let $\mathscr{H}$ be a Hilbert space in duality with itself. If A is a subspace, $A^\circ \equiv A^\perp$.
(2) If E is a Banach space and F its dual,

$$\{x \mid \|x\|_E \leq k\}^\circ = \{y \mid \|y\|_F \leq k^{-1}\}$$

It is easy to prove the following simple properties of polar sets:

Lemma 1 Let $\langle E, F\rangle$ be a dual pair. Then

(a) A° is convex, balanced, and $\sigma(F, E)$ closed.
(b) If $A \subset B$, then $B^\circ \subset A^\circ$.
(c) If $\lambda \neq 0$, $(\lambda A)^\circ = |\lambda|^{-1} A^\circ$.
(d) $(\bigcup_\alpha A_\alpha)^\circ = \bigcap_\alpha A_\alpha^\circ$.

Polar sets are simply related to duals:

Theorem V.27 Let E be a locally convex space and $\mathscr{U}$ a neighborhood base at 0. Consider the dual pair, $\langle E, E^*_{\text{alg}}\rangle$, where E^*_{alg}, the algebraic dual, is the set of all linear maps of E into $\mathbb{C}$. Then the topological dual of E is $\bigcup_{U \in \mathscr{U}} U^\circ$ where the polars are taken relative to E^*_{alg}.

Proof $\ell \in E^*_{\text{alg}}$ is continuous if and only if $|\ell(x)| \leq 1$ for all x in some $U \in \mathscr{U}$, i.e. if and only if $\ell \in U^\circ$ for some $U \in \mathscr{U}$. ∎

Theorem V.28 (the bipolar theorem)   Let E and F be a dual pair. Then using the $\sigma(E, F)$-topology on E, we have

$$E^{\circ\circ} = \overline{\text{a.c.h. }(E)}$$

where a.c.h.(E), the absolutely convex hull of E, is the smallest balanced convex set containing E, that is,

$$\text{a.c.h.}(E) = \left\{ \sum_{n=1}^N \alpha_n x_n \,\middle|\, x_1, \ldots, x_N \in E, \sum_{n=1}^N |\alpha_n| = 1, N = 1, 2, \ldots \right\}$$

and the closure is in the $\sigma(E, F)$ topology.

Proof Let $E_C = \overline{\text{a.c.h.}(E)}$. Clearly $E \subset E^{\circ\circ}$ and since $(E^\circ)^\circ$ is convex, balanced, and $\sigma(E, F)$-closed, $E_C \subset (E^\circ)^\circ$. On the other hand, if $x \notin E_C$, we can find $\ell \in F$ with $\text{Re }\ell(e) \leq 1$ for $e \in E_C$ and $\text{Re }\ell(x) > 1$ (Theorem V.4). Since E_C is balanced, $\sup_{e \in E_C} |\ell(e)| \leq 1$ so $\ell \in E^\circ$. But then $|\ell(x)| > 1$ implies $x \notin E^{\circ\circ}$. ∎

Lemma 2 The Mackey topology is a dual topology.

Proof We use Theorem V.27 to compute the $\tau(E, F)$-dual of E. The $\{\rho_C\}$ as C runs through all $\sigma(F, E)$-compact, absolutely convex sets of F generate the $\tau(E, F)$-topology. Consider $C \subset F \subset E^*_{\text{alg}}$. Since the restriction of the $\sigma(E^*_{\text{alg}}, E)$ topology to F is $\sigma(F, E)$, C is $\sigma(E^*_{\text{alg}}, E)$-compact and so $\sigma(E^*_{\text{alg}}, E)$-

closed in E^*_{alg}. Thus, by Theorem V.28, $(C^\circ)^\circ_{E^*_{\text{alg}}} = C$. But $C^\circ = \{x \mid |\rho_C(x)| \leq 1\}$. The sets

$\{C^\circ \mid C$ is a convex, balanced $\sigma(F, E)$-compact subset of $F\}$

thus form a neighborhood base at $0 \in E$ for $\tau(E, F)$. Therefore,

$$E^*_\tau = \bigcup_C (C^\circ)^\circ_{E^*_{\text{alg}}} = \bigcup_C C = F \quad \blacksquare$$

Lemma 3 (the Bourbaki–Alaoglu theorem)   Let $U \subset E$ be a balanced, convex neighborhood of 0 in some $\langle E, F \rangle$ dual topology. Then U°_F is a $\sigma(F, E)$-compact set in F.

Proof This is essentially a restatement of the Banach–Alaoglu theorem (Theorem IV.21); see Problem 58.

Lemma 4   Every dual topology is weaker than the Mackey topology.

Proof Let ρ be a seminorm on E in some given dual topology. We will show that $\rho = \rho_C$ for some $\sigma(F, E)$-compact, convex subset, C, in F. Let $U = \{x \mid |\rho(x)| \leq 1\}$. Then U is balanced, convex and $\sigma(E, F)$-closed by an application of Theorem V.4 (see Problem 20c). Thus, $(U^\circ)^\circ = U$ by the double polar theorem. Let $C = U^\circ \subset F$. By Lemma 3, C is $\sigma(F, E)$-compact and it is convex. By definition $(U^\circ)^\circ = \{x \mid |\rho_C(x)| \leq 1\} = U$, so $\rho_C = \rho$. $\blacksquare$

We are now ready for:

Proof of Theorem V.22   Since the $\sigma(E, F)$ and $\tau(E, F)$ topologies are dual topologies (Lemma 2 and Theorem IV.20) any $\mathcal{T}$ in between is also a dual topology. By definition, $\sigma(E, F)$ is the weakest possible dual topology and by Lemma 4, $\tau(E, F)$ is the strongest possible dual topology. $\blacksquare$

NOTES

Section V.1   For general references on locally convex spaces, see: Choquet's book (ref. to Section IV.1); J. Kelley and I. Namioka, *Linear Topological Spaces*, Van Nostrand–Reinhold, Princeton, New Jersey, 1963; G. Köthe, *Topological Vector Spaces*, Springer-Verlag, Berlin and New York, 1969; A. and W. Robertson, *Topological Vector Spaces*, Cambridge Univ. Press, London and New York, 1964; H. Schaeffer, *Topological Vector Spaces*, Macmillan, New York, 1966.

170 V: LOCALLY CONVEX SPACES

The Robertsons' book is a delightful 154 page monograph, and of the remaining books we find the translation of Köthe's German classic the most readable. For those who like problems, the Kelley–Namioka book has many, but it is not as superb as Kelley's topology book.

The first formulation of the Hahn–Banach theorem in terms of separating convex sets is in S. Mazur, "Über konvexe Mengen in linearen normierten Räumen," *Studia Math.* **4** (1933), 70–84. The more modern form (Theorem V.4) is due to M. Eidelheit, "Zur Theorie der konvexen Mengen in linearen normierten Räumen," *Studia Math.* **6** (1936), 104–111, and S. Kakutani, "Ein Beweis der Satzes von M. Eidelheit über konvexe Mengen," *Proc. Imp. Acad. Tokyo* **13** (1937), 93–94.

Locally convex spaces are topological vector spaces in which the Hahn–Banach theorem holds and thus are spaces with large topological duals. The L^p spaces, $0 < p < 1$, serve as examples of spaces without any continuous linear functionals; see Köthe, pp. 156–158.

Section V.2 The term "Fréchet space" was coined by Banach in his classic book. Theorem V.5 is a special case of a general theorem on the metrizability of uniform spaces; see Kelley's *General Topology*, pp. 184–190.

Sections V.3 and V.4. The theory of general and tempered distributions was first formalized by L. Schwartz and is very well described in his classic: *Théorie des distributions*, Vols. I–II, Hermann, Paris, 1957, 1959. The five volumes by Gel'fand and co-workers, *Generalized Functions*, Academic Press, New York, 1964–1967, are also quite readable. Informally, many of the notions were already discussed by Bochner, Friedrichs, and Sobolev in the 1930s.

The treatment of renormalization of Feynman amplitudes mentioned in Section V.3 is discussed in N. N. Bogoliubov and O. S. Parasiuk, "Über die Multiplikation der Kausalfunktionen in der Quantentheorie der Felder," *Acta. Math* **97** (1957), 227–266; K. Hepp, "Proof of the Bogoliubov–Parasiuk Theorem on Renormalization," *Commun. Math. Phys.* **2** (1966), 301–326; and E. Speer, *Generalized Feynman Amplitudes*, Ann. Math. Study #62.

The nuclear theorem (Theorem V.12) is the starting point of a general discussion of spaces in which such theorems hold. The theory of such nuclear spaces was first developed in A. Grothendieck: "Produits tensoriels topologiques et espaces nucléaires," *Mem. Amer. Math. Soc.* No. 16 (1955). See also Schaeffer (ref. to Section V.1), pp. 92–107, Gel'fand, Vol. 4, and F. Trèves, *Topological Vector Spaces, Distributions and Kernels*, Academic Press, New York, 1967.

The discussion in Appendix V.3 follows that in B. Simon, "Distributions and Their Hermite Expansions," *J. Math. Phys.* **12** (1971), 140–148. A representation of $\mathscr{S}$ by entire functions which also allows a proof of the nuclear theorem is discussed in the paper of V. Bargmann: "On a Hilbert Space of Analytic Functions and an Associated Integral Transform; II: A Family of Related Function Spaces. Application to Distribution Theory," *Comm. Pure App. Math.* **20** (1967), 1–102.

The notion of inductive limit was first discussed systematically by the French school of L. Schwartz, J. Dieudonné, and A. Grothendieck. There is a generalization of the notion of strict inductive limit which only requires that the injection $X_n \to X_{n+1}$ be continuous (rather than continuous and open). In addition, in defining this "inductive limit," the indexing set may be any directed set. For additional discussion, see Köthe, pp. 215–233, and Robertson and Robertson, pp. 76–100, 127–130. In particular, (d) of Theorem V.15 is proven in the Robertsons' book, p. 128.

For additional discussion of weak solutions of partial differential equations the reader may consult the following (listed in order of increasing sophistication required): I. Stakgold, *Boundary Value Problems of Mathematical Physics*, Vol. 1 and 2, Macmillan, New York, 1968; A. Friedman, *Partial Differential Equations*, Holt, New York, 1969; S. Agmon, *Lectures on Elliptic Boundary Value Problems*, Van Nostrand–Reinhold, Princeton, New Jersey, 1965; L. Hörmander, *Linear Partial Differential Operators*, Springer–Verlag, Berlin, 1963.

Section V.5   For a general discussion of fixed point theorems in nonlinear contexts, see T. L. Saaty and J. Bram, *Nonlinear Mathematics*, McGraw-Hill, New York, 1964, or M. A. Kransnosel'ski, *Topological Methods in the Theory of Nonlinear Integral Equations*, Pergamon, New York, 1964. Of particular interest are attempts to apply notions from algebraic topology to infinite dimensional spaces; see A. Granas, *Introduction to Topology of Function Spaces*, Univ. of Chicago Math. Notes, 1961, or J. Cronin, *Fixed Points and Topological Degree in Nonlinear Analysis*, Amer. Math. Soc., Providence, Rhode Island, 1964. The proof of Theorem V.19 may be found in N. Dunford and J. Schwartz, *Linear Operators*. Vol. I, pp. 453–457, Wiley (Interscience), 1957. The deepest part of the theorem relies on Brouwer's theorem on the closed unit ball—an "analytic proof" of the theorem can be found on pp. 468–470 of Dunford–Schwartz. For a more "natural" algebraic topological proof, see any text on homology theory, e.g. P. Hilton and S. Wylie, *Homology Theory*, Cambridge Univ. Press, London and New York.

That the Brouwer theorem extends to some infinite dimensional spaces was first noted by G. D. Birkoff and O. D. Kellogg, "Invariant Points in Function Space," *Trans. Amer. Math. Soc.* **23** (1922), 96–115. In two papers, J. Schauder, "Zur Theorie Stetiger Abbildungen in Funktionalräumen," *Math. Z.* **26** (1927), 47–65, 417–431; "Der Fixpunktsatz in Funktionalräumen," *Studia Math.* **2** (1930), 171–180, the case where X is a Banach space was proven. The general theorem was proved by A. Tychonoff, "Ein Fixpunktsatz," *Math. Ann.* **111** (1935), 767–776.

The Markov–Kakutani theorem was first proven in A. Markov, "Quelques théorèmes sur les ensembles abéliens," *Dokl. Akad. Nauk. SSSR* **10** (1936), 311–314, using Tychonoff's theorem on products of compact sets. The proof we give is that given by S. Kakutani, "Two Fixed-Point Theorems Concerning Bicompact Convex Sets," *Proc. Imp. Akad. Tokyo* **14** (1938), 242–245.

There is a rather large literature on additional fixed point theorems. For example E. Begle "A Fixed Point Theorem," *Ann. Math.* **51** (1950), 544–550; H. Bohenhwst and S. Karlin, "On a Theorem of Ville," in *Contributions to the Theory of Games* (H. W. Kuhn and A. W. Tucker, eds.), Princeton Univ. Press, Princeton, New Jersey, 1950; F. Browder, "Asymptotic Fixed Point Theorems," *Math. Ann.* **185** (1970), 38–61; S. Eilenberg and D. Montgomery, "Fixed Point Theorems for Multi-valued Transformations," *Amer. J. Math.* **68**, (1946), 214–222; K. Fan, "A Generalization of Tychonoff's Fixed Point Theorem," *Math. Ann.* **142** (1961), 305–310; I. Glicksberg, "A Further Generalization of the Kakutani Fixed Point theorem with Applications to Nash Equilibrium Points," *Proc. Amer. Math. Soc.* **3**, (1952), 170–174; W. Horn, "Some Fixed Point Theorems for Compact Maps and Flows in Banach Spaces," *Trans. Amer. Math. Soc.* **149** (1970), 391–404; S. Kakutani, "A Generalization of Brouwer's Fixed Point Theorem," *Duke Math. J.* **8** (1941), 457–459; J. Leray, "Théorie des points fixés, indice total et nombre de Lefschetz," *Bull. Soc. Math. France* **87** (1959), 221–233; R. Nussbaum, "Some Fixed Point Theorems," *Bull. Amer. Math. Soc.* **77** (1971), 360–365.

V: LOCALLY CONVEX SPACES

Section V.6   For a discussion of the application of fixed point theorems to the theory of ordinary differential equations, see L. Loomis and S. Sternberg, *Advanced Calculus*, pp. 266–304, Addison–Wesley, Reading, Massachusetts, 1968, who also discuss an application of the contraction mapping theorem to differential calculus (pp. 166–167; 230–234), or C. Goffman, "Preliminaries to Functional Analysis," in *Studies in Modern Analysis*, (R. C. Buck, ed.), pp. 149–150, 153–154, Prentice Hall, Englewood Cliffs, New Jersey, 1962. Goffman gives a proof of our existence theorem employing equicontinuity and the Stone–Weierstrass theorem but avoiding the Leray–Schauder–Tychonoff theorem.

We discuss the Haar integral in more detail in Chapter XIV. For its history, see the notes to that chapter.

For a general discussion of the bootstrap philosophy, see G. Chew, *The Analytic S-Matrix*, Benjamin, New York, 1966.

The Mandelstam representation was first proposed by S. Mandelstam in "Determination of the Pion-Nuclear Scattering Amplitude from Dispersion Relations and Unitarity, General Theory," *Phys. Rev.* **112** (1958), 1344–1360. In potential scattering, it was first proven by R. Blankenbecker, M. L. Goldberger, N. N. Khuri, and S. A. Treiman, "Mandelstam Representation for Potential Scattering," *Ann. Phys.* **10** (1960), 62–93. For additional discussion of the potential scattering case, see V. de Alfaro and T. Regge, *Potential Scattering*, North–Holland Publ., Amsterdam, 1965.

The work of Atkinson which we discuss in Section V.6c can be found in D. Atkinson: "A Proof of the Existence of Functions that Satisfy Exactly Both Crossing and Unitarity," I, II, III, IV, *Nucl. Phys.* **B7** (1968), 375–408; **B8** (1968), 377–390; **B13** (1969), 415–436; **B23** (1970), 397–412. For additional discussion of fixed point theorems applied to integral equations in high energy physics, see C. Lovelace, "Uniqueness and Symmetry Breaking in S-Matrix Theory," *Commun. Math. Phys.* **4** (1967), 261–302; J. Kupsch, "Scattering Amplitudes that Satisfy a Mandelstam Representation with One Subtraction and Unitarity," *Nucl. Phys.* **B11** (1969), 573–587, or R. L. Warnock, "Nonlinear Analysis Applied to S-Matrix Theory," Boulder Lectures in Theoretical Physics, 1968. pp. 72–86, Benjamin, New York, 1969.

The work of Martin on determining the phase of the scattering amplitude from its magnitude and unitarity may be found in A. Martin, "Construction of the Scattering Amplitude from Differential Cross Sections," *Nuovo Cimento* **59A** (1969), 131–151. A discussion of the Martin results and some of the Atkinson bootstrap results using only the contraction mapping theorem may be found in D. Atkinson, "Introduction to the Use of Non-Linear Techniques in S-Matrix Theory," *Acta Phys. Austriaca Suppl.* **7** (1970), 32–70. The first person to use fixed point theorems to study the problem seems to be R. G. Newton in "Determination of the Amplitude from the Differential Cross Section by Unitarity," *J. Math. Phys.* **9** (1968), 2050–2055. Examples of differential cross sections with non-unique associated amplitudes (and $M > 1$ in the notation we use in Example (d)) are constructed in J. Crichton, "Phase-Shift Ambiguities for Spin-Independent Scattering," *Nuovo Cimento* **45A** (1966), 256–258.

Discussion of fixed-point theorems in statistical mechanics can be found in: D. Ruelle, *Statistical Mechanics*, pp. 72–86, Benjamin, New York, 1969; J. Groenveld, "Two Theorems on Classical Many-Particle Systems," *Phys. Lett.* **3** (1962), 50–51; O. Penrose, "Convergence of Fugacity Expansions for Fluids and Lattice Gases," *J. Math. Phys.* **4** (1963), 1312–1320; D. Ruelle, "Correlation Functions of Classical Gases," *Ann. Phys.* **25** (1963), 109–1230; and G. Gallavotti and S. Miracle-Sole, "Correlation Functions of a Lattice System," *Commun. Math. Phys.* **7** (1968), 274–288.

Section V.7

The notion of dual pair is due to J. Dieudonné and G. Mackey. The Mackey–Arens theorem was first proven by G. Mackey, "On Convex Topological Linear Spaces," *Trans. Amer. Math. Soc.* **60** (1946), 519–537, and by R. F. Arens, "Duality in Linear Spaces," *Duke Math. J.* **14** (1947), 787–794.

A space in which every closed convex, balanced, absorbing set is a neighborhood of zero is called *barrelled*. By the Baire theorem, every Fréchet space is barrelled and, by an elementary argument, any strict inductive limit of Fréchet spaces is barrelled. A barrelled space in which any closed bounded set is compact is called a *Montel* space. Thus Theorem V.25 says certain spaces are Montel spaces. Montel spaces are automatically reflexive (so Theorem V.24 follows from Theorem V.25) and their duals are also Montel spaces when given the Mackey (strong) topology. See Köthe, pp. 369–372 for details. In particular, $\mathscr{S}'$ and $\mathscr{D}'$ are also Montel spaces.

Often, the polar, $A°$, is defined to be $A° = \{f \in F \mid \operatorname{Re} f(e) \geq -1 \text{ for all } e \in A\}$. If A is convex and balanced, this agrees with the definition we gave. When A is a cone, this new definition is much more useful. For example, with the definition in the text, the polar of $\{f \in C(X) \mid f \geq 0\}$ is $\{0\}$ while for the new definition, the polar is all positive measures.

PROBLEMS

1. (a) Prove that a locally convex space has a topology given by a single norm if the topology is generated by finitely many seminorms.
   (b) Prove that a locally convex space has a topology generated by a single norm if and only if 0 has a bounded neighborhood.

2. Let X be an infinite dimensional locally convex space and X^* its dual. Prove that no $\sigma(X, X^*)$-continuous semi-norm is actually a norm; so seminorms are essential.

3. (a) Let $\{\rho_\alpha\}_{\alpha \in A}$ be a family of seminorms so that some finite sum $\rho_{\alpha_1} + \cdots + \rho_{\alpha_n}$ is actually a norm. Prove that $\{\rho_\alpha\}_{\alpha \in A}$ is equivalent to a family of norms.
   (b) Prove that any locally convex topology on $\mathbb{R}^n$ is the usual topology. [Hint: Use the equivalence of all norms on $\mathbb{R}^n$ and the following construction: Pick a seminorm $\rho_1 \neq 0$ and let $V_1 = \{x \mid \rho_1(x) = 0\}$; dim $V_1 \leq n - 1$. If $V_1 \neq \{0\}$, pick $x_1 \in V_1$ and ρ_2 so $\rho_2(x_1) \neq 0$. Let $V_2 = \{x \mid (\rho_1 + \rho_2)(x) = 0\}$; dim $V_2 \leq n - 2$; etc.]
   (c) Let X be a locally convex space. Show that any linear functional on a finite dimensional subspace of X has a continuous extension to all of X.
   (d) Prove that any finite dimensional subspace of a locally convex space is closed.

4. The purpose of this exercise is to prove that every locally compact, locally convex space is finite dimensional.
   (a) Let U be a compact neighborhood of 0. Show that one can find $x_1, \ldots, x_n$ so that $U \subset \bigcup_{i=1}^n (x_i + \frac{1}{2}U)$ and thus a finite dimensional space, M, with $U \subset M + \frac{1}{2}U$.
   (b) Prove that $U \subset M + (\frac{1}{2})^m U$ for any m.
   (c) Prove that $U \subset \overline{M}$.
   (d) Conclude that $\overline{M} = X = M$.

5. Let X be a Banach space, X^* and X^{**} its dual and double dual. Let X_1 be the unit ball of X and X_1^{**} the unit ball in X^{**}. Prove that

(a) X_1 is $\sigma(X^{**}, X^*)$ dense in X_1^{**}. (Hint: Use theorem V.6(c) on X^* with the $\sigma(X^{**}, X^*)$ topology.)
(b) X is reflexive if and only if its unit ball is $\sigma(X, X^*)$-compact.

†6. Prove the three propositions at the beginning of Section V.1.

†7. Prove that the families $\{\rho_C^{(2)}\}$ and $\{\rho_C\}$ on $\mathcal{O}_D$ are equivalent. (Hint: Use the Cauchy integral formula integrated over an annular region.)

†8. Let C be an absorbing subset of V with $tx \in C$ if $x \in C$ and $0 \leq t < 1$. Let ρ be the Minkowski functional for C. Prove that
(a) $\rho(tx) = t\rho(x)$ if $t \geq 0$.
(b) $\rho(x + y) \leq \rho(x) + \rho(y)$ if and only if for all $u, v \in C$ and $0 \leq t < 1$, $t(\frac{1}{2}u + \frac{1}{2}v) \in C$.
(c) $\rho(\lambda x) = |\lambda|\rho(x)$ if and only if for all $u \in C$ and λ with $|\lambda| < 1$, $\lambda u \in C$.
(d) $\{x \mid \rho(x) < 1\} \subset C \subset \{x \mid \rho(x) \leq 1\}$.

*†9. Prove Theorem V.2.

†10. (a) Complete the proof of Theorem V.5.
(b) Prove the proposition following Theorem V.5.

11. Let A and B be absorbing sets with the property: If $x \in A$ (respectively B) and $0 \leq t \leq 1$, then $tx \in A$ (respectively B). Let ρ_A, ρ_B be their Minkowski functionals. Prove that
(a) $\rho_B \leq \rho_A$ if and only if $tA \subset B$ for all $0 \leq t < 1$.
(b) $\rho_{A \cap B} = \max(\rho_A, \rho_B)$.
(c) $\rho_{A \cup B} = \min(\rho_A, \rho_B)$.

12. In the theory of functions of several complex variables, one can have open sets $\mathcal{O}$, $\mathcal{O}'$ in $\mathbb{C}^n$ with $\mathcal{O} \subset \mathcal{O}'$, properly contained, but with the property that every function analytic in $\mathcal{O}$ has a continuation to $\mathcal{O}'$. In that case, let $K \subset \mathcal{O}$ be compact and let $\hat{K} = \{z \in \mathcal{O}' \mid |f(z)| \leq \sup_{w \in K} |f(w)| \text{ for all } f \text{ analytic in } \mathcal{O}'\}$. It is a useful theorem in the theory that
$$\bigcup_{\substack{K \subset \mathcal{O} \\ \text{compact}}} \hat{K} = \mathcal{O}'$$
(a) Show that $\bigcup \hat{K} = \mathcal{O}'$ is implied by the equality of the topologies of uniform convergence on compact subsets of $\mathcal{O}$ and compacts of $\mathcal{O}'$.
(b) Use Theorem V.6 to prove the equality of topologies mentioned in (a).

13. Let Z be a metric space. Let X^* be the dual of a Fréchet space. Suppose that $f: Z \to X^*$ is continuous when X^* is given the $\sigma(X^*, X)$ topology. Prove that it is continuous when X^* is given the $\tau(X^*, X)$ topology. (Hint: Use Theorem V.8.)

14. If Z in Problem 13 is replaced with an arbitrary topological space, what happens to the conclusion?

15. The purpose of this problem is to prove that every uniformly convex Banach space (see Problem 25 of Chapter III for the definition of uniformly convex) is reflexive.
(a) Let X be uniformly convex and let $x, y \in X$, $\ell \in X^*$. Prove that, if $\|x\| = \|y\| = \|\ell\| = 1$, $\text{Re}\,\ell(x) > 1 - \delta(\varepsilon)$ and $\text{Re}\,\ell(y) > 1 - \delta(\varepsilon)$, then $\|x - y\| < \varepsilon$.
(b) Let X be uniformly convex, and suppose that $\{x_\alpha\}$ is a net in X so that $x_\alpha \to x \in X^{**}$ in the $\sigma(X^{**}, X^*)$ topology. Suppose that $\|x\| = 1$, $\|x_\alpha\| \leq 1$ for all α. Using (a), prove that x_α is $\|\cdot\|$-Cauchy.
(c) Prove that $X = X^{**}$. (Use Problem 5a.)

Remark: The theorem proven in Problem 15 is due to D. P. Milman, "On some criteria for the regularity of spaces of type (B)", *Dokl. Akad. Nauk SSSR* **20** (1938), 243–246, and B. J. Pettis, "A proof that every uniformly convex space is reflexive," *Duke Math. J.* **5** (1939), 249–253.

16. (a) Let $\varphi \in \mathscr{S}$ and let φ_y be the function in $\mathscr{S}$ defined by $\varphi_y(x) = \varphi(x - y)$. Prove that the map $y \to \varphi_y$ is a C^∞ function from $\mathbb{R}^n$ to $\mathscr{S}(\mathbb{R}^n)$ with $D^\alpha(\varphi_y) = (-1)^\alpha (D^\alpha \varphi)_y$. To say $y \to \varphi_y$ has derivative $\partial \varphi_y / \partial y_j$ as a function with values in $\mathscr{S}$ means

$$\lim_{y \to y_0} |y - y_0|^{-1} \left[\varphi_y - \varphi_{y_0} - \sum_{j=1}^N \frac{\partial}{\partial y_j} (\varphi_y) \cdot (y - y_0)_j \right] = 0$$

in the topology of $\mathscr{S}$.

(b) Let $T \in \mathscr{S}'$. Let $\varphi \in \mathscr{S}$. Define T^φ to be the function, $T^\varphi(y) = T(\varphi_y)$. Prove that $T^\varphi \in C^\infty$.

(c) Let $\varphi_n \in \mathscr{S}$ with $\varphi_n \to \delta$ in the weak topology on $\mathscr{S}'$. Prove that $T^{\varphi_n} \to T$ for all $T \in \mathscr{S}'$ in the weak topology on $\mathscr{S}'$.

(d) Prove that $\mathscr{S}$ is dense in $\mathscr{S}'$.

17. Let X and Y be locally convex spaces and let X^*, Y^* be their topological duals. Suppose that $T: X \to Y$ is linear and continuous. Define the adjoint $T': Y^* \to X^*$ by $[T'(y^*)](x) = y^*(Tx)$. Prove that
   (a) If X^* and Y^* are given the $\sigma(X^*, X)$ and $\sigma(Y^*, Y)$ topology, then T' is continuous.
   (b) If X and Y are given the $\sigma(X, X^*)$ and $\sigma(Y, Y^*)$ topology, then T is continuous.
   (c) If X^* and Y^* are given the $\tau(X^*, X)$ and $\tau(Y^*, Y)$ topology, then T' is continuous. (Hint: use (b).)
   (d) T takes bounded subsets of X into bounded subsets of Y.
   (e) If X^* and Y^* are given the $\beta(X^*, X)$ and $\beta(Y^*, Y)$ topology, then T' is continuous.

18. Let $T: X \to Y$, locally convex spaces, with T linear and continuous. Put the weak topology on both X^* and Y^* so that $(X^*)^* = X$; $(Y^*)^* = Y$.
   (a) Prove that $(T')' = T$ in this case.
   (b) Conclude that T is continuous from X to Y when they are given the $\tau(X, X^*)$ and $\tau(Y, Y^*)$ topologies.
   (c) Conclude that T is continuous from X to Y when they are given the $\beta(X, X^*)$ and $\beta(Y, Y^*)$ topologies.

19. Let $T: X \to Y$ locally convex spaces with T linear and continuous. Let T' be the adjoint map from $Y^* \to X^*$. Prove that
   (a) T' is injective if and only if $\overline{\operatorname{Ran} T} = Y$.
   (b) $\overline{\operatorname{Ran} T'} = X^*$ in the $\sigma(X^*, X)$ or $\tau(X^*, X)$ topology if and only if T is injective.
   (c) Let $\iota: \mathscr{S} \to \mathscr{S}'$ be the natural map of $\mathscr{S}$ into $\mathscr{S}'$, then ι is continuous when $\mathscr{S}$ is given the $\sigma(\mathscr{S}, \mathscr{S}')$ topology and $\mathscr{S}'$ is given the $\sigma(\mathscr{S}', \mathscr{S})$ topology.
   (d) Prove $\iota' = \iota$!
   (e) Conclude $\iota(\mathscr{S})$ is dense in $\mathscr{S}'$ in the $\tau(\mathscr{S}', \mathscr{S})$ and the $\sigma(\mathscr{S}', \mathscr{S})$ topology.

20. Let X be a locally convex space, with dual X^*.
   (a) Prove that any closed subspace in X is $\sigma(X, X^*)$ closed.
   (b) Prove that all dual-$\langle X, Y \rangle$ topologies have the same closed subspaces.
   (c) Let C be a closed convex subset of X. Prove that C is $\sigma(X, X^*)$ closed. (Hint: Use the separation theorem.)

(d) Prove that all dual-$\langle X, Y \rangle$ topologies have the same closed convex sets.
(e) Do all dual-(X, Y) topologies have the same compact convex sets?

†21. Prove directly that δ' (Example 5 of Section V.3) is in $\mathscr{S}'$. Prove that δ' does not come from a measure.

†22. (a) Prove that
$$\lim_{\varepsilon \downarrow 0} \frac{x - x_0}{(x - x_0)^2 + \varepsilon^2} = \mathscr{P}\left(\frac{1}{x - x_0}\right)$$
in the weak topology on $\mathscr{S}'$.
(b) Let φ_n be a sequence of bounded functions on $\mathbb{R}$ so that $\int_{|x - x_0| \geq \varepsilon} \varphi_n(x)\, dx \to 0$ as $n \to \infty$ for each $\varepsilon > 0$, $\varphi_n(x) \geq 0$, and $\int \varphi_n(x)\, dx = c$ independent of n. Prove that $\varphi_n \to c\delta(x - x_0)$ in the topology of $\mathscr{S}'$.
(c) Prove that
$$\lim_{\varepsilon \to 0} \frac{\varepsilon}{(x - x_0)^2 + \varepsilon^2} = \pi\delta(x - x_0)$$
(d) Prove formula (V.4).

†23. (a) Let $F \in O_M$. Prove that $f \to Ff$ is a bounded map of $\mathscr{S}$ into $\mathscr{S}$.
(b) Let F be a measurable function so that $Ff \in \mathscr{S}$ for all $f \in \mathscr{S}$. Prove that F is C^∞.
(c) If $f \to Ff$ is continuous, prove that $F \in O_M$.

24. Let $T \in \mathscr{S}'(\mathbb{R})$ with $|T(f)| \leq C \sum_{\alpha, \beta = 0}^{n} \|x^\alpha (d/dx)^\beta f\|_\infty$. Map $\mathscr{S} \to C_n(\mathbb{R}) \oplus \cdots \oplus C_n(\mathbb{R})$ ($n + 1$ times) by $f \to \langle f, f', \ldots, f^{(n)} \rangle$ where $C_n(\mathbb{R})$ is the Banach space of continuous functions, f, with sup $\|x^\alpha f\|_\infty < \infty$ for $\alpha = 1, \ldots, n$ with norm $\|f\|^{(n)} = \sum_{\alpha = 0}^{n} \|x^\alpha f\|_\infty$. Use the Hahn–Banach and Riesz–Markov theorems to prove that T can be written
$$Tf = \sum_{\beta = 0}^{n} \int D^\beta f\, d\mu_\beta$$
where $\mu_0, \ldots, \mu_n$ are complex measures of polynomial growth.

25. (a) Let μ be a polynomially bounded measure and let $F(x) = \int_0^x d\mu$; $G(x) = \int_0^x F(x)\, dx$. Prove that $\mu = G''$ in the sense of distributions.
(b) Prove the regularity theorem for $\mathscr{S}'(\mathbb{R})$ using Problems 24 and 25(a).

26. By mimicking Problems 24 and 25, prove the local regularity theorem for $\mathscr{D}'$: Given $T \in \mathscr{D}'(\mathbb{R}^n)$ and a compact set $C \subset \mathbb{R}^n$, there is a continuous function F on C and an α so that $Tf = (-1)^\alpha \int F(x)(D^\alpha f)(x)\, dx$ for all $f \in \mathscr{D}(\mathbb{R}^n)$ with supp $f \subset C$.

27. Let U_a be translation by a as an operator on $\mathscr{S}'(\mathbb{R})$. Let d/dx be the derivative operation on $\mathscr{S}'$. Prove that $(U_a - 1)a^{-1}$ converges pointwise in the $\sigma(\mathscr{S}', \mathscr{S})$ topology to d/dx.

†28. (a) Prove that $V(A)$ defined by Operation 4 in Section V.3 agrees with $(V(A)\varphi)(x) = \varphi(A^{-1}x)$ if $\varphi \in \mathscr{S}$ is regarded as an element of $\mathscr{S}'$.
(b) Show that the support of a function $\varphi \in \mathscr{S}$ as a distribution is identical to $\overline{\{x \mid \varphi(x) \neq 0\}}$.

†29. (a) Let $\varphi \in \mathscr{S}(\mathbb{R}^n)$ with $\varphi(0) = 0$. Prove that there exist $\{\varphi_k\} \in \mathscr{S}$ with $0 \notin \operatorname{supp} \varphi_k$ for all k so that $\|x^\alpha(\varphi_k - \varphi)\|_\infty \to 0$ for all $\alpha \in I_+^n$.
(b) Let $\varphi \in \mathscr{S}(\mathbb{R}^n)$ with $(D^\beta \varphi)(0) = 0$ for all $\beta \in I_+^n$ with $|\beta| \leq m$. Prove that there are $\{\varphi_k\} \subset \mathscr{S}$ with $0 \notin \operatorname{supp} \varphi_k$ for all k so that $\|x^\alpha D^\beta(\varphi_k - \varphi)\|_\infty \to 0$ for all $\alpha \in I_+^n$ and β with $|\beta| \leq m$.
(c) Let $T \in \mathscr{S}'(\mathbb{R}^n)$ with

$$|T(f)| \leq C \sum_{\substack{|\beta| \leq m \\ |\alpha| \leq n}} \|x^\alpha D^\beta f\|_\infty$$

Suppose that $\operatorname{supp} T = \{0\}$ and let $\varphi \in \mathscr{S}$ with $D^\beta \varphi(0) = 0$ for $|\beta| \leq m$. Prove that $T(\varphi) = 0$.
(d) Let $T \in \mathscr{S}'(\mathbb{R}^n)$ obey

$$|T(\cdot)| \leq C \sum_{\substack{|\beta| \leq m \\ |\alpha| \leq n}} \|x^\alpha D^\beta \cdot \|_\infty$$

Let $\operatorname{supp} T = \{0\}$. Find constants $\{C_\beta\}_{|\beta| \leq m}$ so that

$$T(\psi) = \sum_{|\beta| \leq m} (-1)^\beta C_\beta (D^\beta \psi)(0)$$

for all $\psi \in \mathscr{S}$. Hint: Pick $\eta \in \mathscr{S}$, identically 1 near 0 and let

$$\varphi = \psi - \eta \sum_{|\beta| \leq m} \frac{x^\beta}{\beta!} (D^\beta \psi)(0)$$

(e) Prove Theorem V.11.

30. Let $F \in \mathcal{O}_M(\mathbb{R})$, $T \in \mathscr{S}'(\mathbb{R})$. Let $'$ denote the derivative on $\mathscr{S}'$. By manipulating the definitions of multiplication and $'$, prove $(FT)' = F'T + FT'$.

31. A map $S: \mathscr{S} \to \mathscr{S}$ is called **local** if $\operatorname{supp} S\varphi \subset \Omega$ whenever $\operatorname{supp} \varphi \subset \Omega$. A map $S: \mathscr{S}' \to \mathscr{S}'$ is called **local** if $\operatorname{supp} ST \subset \Omega$ whenever $\operatorname{supp} T \subset \Omega$.
(a) Let $S: \mathscr{S} \to \mathscr{S}$ be local. Prove that $S': \mathscr{S}' \to \mathscr{S}'$ is local.
(b) Which of the operations 1–4 are local?

*32. A distribution $T \in \mathscr{S}'$ is said to be of order at most n if

$$|T(\varphi)| \leq C \sum_{\substack{|\alpha| \leq k \\ |\beta| \leq n}} \|x^\alpha D^\beta \varphi\|$$

for some C and k. It is said to be of order at most n^- if for some fixed j and for any $D > 0$, there are k and C so that

$$|T(\varphi)| \leq D \sum_{\substack{|\alpha| \leq j \\ |\beta| = n}} \|x^\alpha D^\beta \varphi\| + C \sum_{\substack{|\alpha| \leq k \\ |\beta| \leq n-1}} \|x^\alpha D^\beta \varphi\|$$

We say T is of **order** n, if it is of order at most n, but not of order at most n^-. We say T is of **order** n^-, if it is of order at most n^- but not of order at most $n - 1$.
(a) Prove that the renormalizations $(1/x)_{+,M}$ of Example 9 are of order 1^-.
(b) Prove that any other renormalization of $(1/x)_+$ has order not less than 1.

*33. Prove that not every bilinear form on $L^p(\mathbb{R})$ ($1 < p < \infty$) is of the form $F(f, g) = \int F(x, y) f(x) g(y) \, dx \, dy$ for some $F(x, y) \in L^q(\mathbb{R}^2)$.

[Hint: If $p \geq 2$, let $F(f, g) = \int G(x)f(x)g(x)\,dx$ for some $G \in L^r$; $1/r + 2/p = 1$; if $1 < p \leq 2$, let $F(f, g) = \int G(x - y)f(x)g(y)\,dx\,dy$ with $G \in L^r$, $1/r = 2(1 - 1/p)$.]

Remarks. 1) The $1 < p \leq 2$ example suggested in the hint requires Young's inequality which we prove in Section IX.4.
2) The nuclear theorem is true for $L^1(\mathbb{R})$, that is, every bilinear form on $L^1(\mathbb{R})$ is of the form $F(f, g) = \int F(x, y)f(x)g(y)\,dy$ for some $F \in L^\infty(\mathbb{R}^2)$. It is an interesting exercise to prove this from the Dunford–Pettis theorem which says: Let E be a separable Banach space, $T: E \to L^\infty(\mathbb{R})$. Then there is a measurable function g on $\mathbb{R}$ with values in E^* so that $\sup_{x \in \mathbb{R}} \|g(x)\| = \|T\|$ and $[T(e)](x) = [g(x)](e)$, a.e. x. For a proof of the Dunford-Pettis theorem, see Trèves, (ref. to Section V.3) pp. 469–473.

†34. Prove that a separately continuous multilinear form on $F_1 \times \cdots \times F_n$, is jointly continuous if all the F_i are Fréchet spaces.

†35. Extend the proof of the nuclear theorem in the appendix of V.3 to multilinear functionals; explicitly, if $B(f_1, \ldots, f_k)$ is a separately continuous k-linear functional on $\mathscr{S}(\mathbb{R}^{n_1}) \times \cdots \times \mathscr{S}(\mathbb{R}^{n_k})$, there is a $T \in \mathscr{S}'(\mathbb{R}^{n_1 + \cdots + n_k})$ with

$$B(f_1, \ldots, f_k) = T(f_1 \otimes \cdots \otimes f_k).$$

†36. Provide the details of the proof of Lemma 2 in the appendix to Section V.3.

†37. Complete the proof of Corollary 3 in the appendix to Section V.3.

38. Define $\mathscr{D}_{L^\infty}(\mathbb{R}^n) = \{f \mid f \text{ is a } C^\infty \text{ function on } \mathbb{R}^n \text{ with all its derivatives in } L^\infty(\mathbb{R}^n)\}$. Put the seminorms $\{\|\cdot\|_{\alpha, \infty}\}$ on $\mathscr{D}_{L^\infty}(\mathbb{R}^n)$, where $\|f\|_{\alpha, \infty} = \|D^\alpha f\|_\infty$.
  (a) Prove that $\mathscr{D}_{L^\infty}$ is complete.
  †(b) Prove that $C_0^\infty(\mathbb{R}^n)$ is not closed in $\mathscr{D}_{L^\infty}$ and find its closure.

39. Let $\mathscr{E}(\mathbb{R}^n) = \{f \mid f \text{ a } C^\infty \text{ function on } \mathbb{R}^n\}$. For any integer m and $\alpha \in I_+^n$, let

$$\|f\|_{(m), \alpha} = \sup_{|x| \leq m} |(D^\alpha f)(x)|$$

Norm $\mathscr{E}(\mathbb{R}^n)$ with $\{\|\cdot\|_{(m), \alpha} \mid \alpha \in I_+^n, m \in I_+\}$.
  (a) Prove that $\mathscr{E}$ is complete.
  (b) Prove that the natural injection, $\iota: \mathscr{D} \subset \mathscr{E}$ is continuous, so $\iota': \mathscr{E}' \subset \mathscr{D}'$ in a natural way.
  (c) Prove that $T \in \mathscr{D}'$ is in $\mathscr{E}'$ if and only if T has compact support.

40. Prove that the natural map $\mathscr{D} \subset \mathscr{S}$ is continuous and conclude that $\mathscr{S}' \subset \mathscr{D}'$ in a natural way. Prove that $\mathscr{E}' \subset \mathscr{S}'$.

†41. Prove part (b) of Theorem V.15.

42. Extend Lemma 1 of the appendix to Section V.3 to prove that $\{\|\cdot\|_{\alpha, \beta, \infty}\}$ and $\{\|\cdot\|_{\alpha, \beta, p}\}$ are equivalent families for any fixed $1 \leq p < \infty$ if $\|f\|_{\alpha, \beta, p} = \|x^\alpha D^\beta f\|_{L^p(\mathbb{R}^n)}$.

43. *(a) Prove that s_m and s_n are isomorphic, that is, there is a continuous linear bijection with continuous inverse $T: s_m \to s_n$.
  (b) Prove that $\mathscr{S}(\mathbb{R}^n)$ and $\mathscr{S}(\mathbb{R}^m)$ are topologically isomorphic.

Problems 179

44. Prove that the norms $\{\|\cdot\|_\beta\}$ and $\{\|\cdot\|_{\beta;\,p}\}$ on s_m are equivalent families if $\|a\|_{\beta;\,p}^p = \sum_\alpha (\alpha+1)^\beta |a_\alpha|^p$.

45. Let X be a strict inductive limit of X_n with each X_n a proper closed subspace of X. Suppose that $\{U_n\}$ is a countable decreasing family of neighborhoods of 0. Pick $x_n \in U_n \backslash X_n$.
    (a) Prove that $\{x_n\}$ is not bounded.
    (b) Show that $\{x_n\}$ would be bounded if U_n were a neighborhood base.
    (c) Conclude X is not metrizable.

†46. (a) Suppose X is the strict inductive limit of spaces X_n. Suppose $\{Y_n\}$ is an increasing family of subspaces of X so that for any n, there is an N with $X_n \subset Y_N$. Prove that X is the strict inductive limit of the Y_n.
    (b) Let $K \subset \Omega \subset \mathbb{R}^n$ with K compact and Ω open. Prove that if $\mathscr{D}_\Omega$ has a topology given by some family $\{K_n\}$, then the restriction of this topology to $C_0^\infty(K)$ is the one given by the $\|D^\gamma f\|_\infty$ norms.
    (c) Prove that the topology on $\mathscr{D}_\Omega$ is independent of the choice of the increasing family K_n of compact sets.

†47. Let $y_1, \ldots, y_n$ be coordinates on $\mathbb{R}^n$. We say a distribution $T \in \mathscr{D}'(\mathbb{R}^n)$ is **independent** of $y_{k+1}, \ldots, y_n$ or is a function of $y_1, \ldots, y_k$ if and only if for any translation

$$U_{a_{k+1},\ldots,a_n} : (y_1,\ldots,y_n) \to (y_1,\ldots,y_k, y_{k+1}+a_{k+1},\ldots,y_n+a_n)$$

$$U_{a_{k+1},\ldots,a_n} T = T$$

(a) Let F be a measurable function on $\mathbb{R}^k$ which is locally L^1. Let T be the distribution associated with the function $F(y_1,\ldots,y_k)$ on $\mathbb{R}^n$. Prove that T is independent of $y_{k+1},\ldots,y_n$.
(b) Let T be independent of $y_{k+1},\ldots,y_n$. Prove that $(\partial T/\partial y_i) = 0$ if $i = k+1,\ldots,n$. [Hint: Look at Problem 27.]
(c) Let T be a distribution in $\mathscr{D}(\mathbb{R}^2)$ which is only a function of $x\text{-}ct = y_1$. Prove this notion is independent of what independent coordinate is taken for y_2.
(d) Let T be as in (c). Prove that $(\partial T/\partial t) = -c(\partial T/\partial x)$ and that $\partial T/\partial t$ is also a distribution which is only a function of $x\text{-}ct$.
(e) Conclude that

$$\frac{\partial^2}{\partial t^2} T - c^2 \frac{\partial^2}{\partial x^2} T = 0$$

48. Let T and S be commuting maps of a metric space into itself. Let $f_T = \{x \,|\, Tx = x\}$.
    (a) Prove that $Sx \in f_T$ if $x \in f_T$.
    (b) Suppose that T is a strict contraction. Prove that S has a fixed point.
    (c) Let T^n be a strict contraction for some n. Prove that T has a unique fixed point.

†49. Let $X = \{\{x_n\} \in \ell_2 \,|\, |x_n| \leq 1/3^n\}$.
    (a) Prove that X is a compact convex subset of l_2.
    (b) Let $f: X \to \mathbb{C}$ be given by $f(x) = \sum_{n=1}^\infty 2^n x_n$. Prove that f is a continuous affine linear map on X.
    (c) Prove that f has no continuous extension to all of ℓ_2.

V: LOCALLY CONVEX SPACES

50. Suppose that G is a group with an abelian subgroup N so that G/N is abelian (for example the family of rotations and translations of $\mathbb{R}^2$). Let C be a compact, convex subset of a locally convex space X. For each $g \in G$, let T_g, an affine linear map of C into C, be given so that $T_g T_h = T_{gh}$.
    (a) Let $C_N = \{x \in C \mid T_n x = x \text{ for all } n \in N\}$. Prove that C_N is compact, convex, and nonempty.
    (b) Suppose g_1, g_2 are in the same coset in G/N. Prove that $T_{g_1} \restriction C_N = T_{g_2} \restriction C_N$.
    (c) Prove that there is $x \in C$ with $T_g x = x$ for all $g \in G$.

51. Prove Theorem V.21 directly.

52. A locally convex space X is called a **Mackey space** or a **bornological space** if for any locally convex space Y, any linear map $T: X \to Y$ which takes bounded sets into bounded sets is continuous.
    (a) Let X be a Mackey space. Prove that it has the Mackey topology, $\tau(X, X^*)$. (Hint: Consider $id: X \to \langle X, \tau \rangle$.)
    (b) Let $x_n \to 0$ in a metrizable locally convex space. Prove that there exists $\{\rho_n\} \in \mathbb{R}$ with $\rho_n \to \infty$ so $\rho_n x_n \to 0$. [Hint: Let U_n be a countable neighborhood basis, $U_{n+1} \subset U_n$. Pick n_k so $n \geq n_k$ implies $x_n \in (1/k)U_k$. Pick $\rho_n = k$ if $n_k \leq n < n_{k+1}$.]
    (c) Prove that every metrizable locally convex space is a Mackey space. [Hint: Use (b) and the fact that a convergent sequence is bounded.]
    (d) Prove that a strict inductive limit $X = \bigcup X_n$ with X_n a proper closed subspace of X_{n+1} is Mackey if each X_n is Mackey.
    (e) Prove that a strict inductive limit of Fréchet spaces is a Mackey space.
    (f) Conclude that the natural topologies on $\mathscr{S}$ and $\mathscr{D}$ are the Mackey topologies.

53. Let E be a locally convex space. Define the **natural topology** $\eta(E^{**}, E)$ on E^{**} as follows. Let $\mathscr{U}$ be the family of all balanced convex neighborhoods of $0 \in E$. For $U \in \mathscr{U}$, let $\tilde{U}$ be the E^{**} polar of $U^\circ \subset E^*$. The $\tilde{U}$ generate the natural topology. Prove:
    (a) The natural topology is weaker than the $\beta(E^{**}, E^*)$ topology. (Hint: Every U° is bounded.)
    (b) The restriction of the natural topology on E^{**} to E is the original topology on E, that is, $\rho: E \to \langle E^{**}, \eta \rangle$ is continuous and open.
    (c) The *inverse* of the natural injection $\rho: E \to \langle E^{**}, \beta \rangle$ from Ran ρ to E is always continuous.

54. Let E be a Banach space with the *weak* topology. Prove that the injection $\rho: E \to \langle E^{**}, \beta \rangle$ is *never* continuous if E is infinite dimensional.

†55. Let $\langle E, F \rangle$ be a dual pair. Prove that every $\sigma(E, F)$-closed bounded set of E is $\sigma(E, F)$-compact if and only if the $\tau(F, E)$ and $\beta(F, E)$ topologies on F are identical.

*56. (a) Let E be a Fréchet space. Prove that any $\sigma(E^*, E)$-closed bounded subset of E^* is $\sigma(E^*, E)$-compact. (Hint: Mimic the Banach–Alaoglu theorem, invoking the principle of uniform boundedness at the crucial place.)
    (b) Prove the conclusion of (a) when E is a strict inductive limit of Fréchet spaces.

†57. Combine Problems 52 and 56 with Theorem V.25 to prove Theorem V.24.

†58. Prove Lemma 3 in the appendix to Section V.7.

Problems 181

59. Let $[-a, a] \subset \mathbb{R}$. Let $\varphi_n(x) = (2a)^{-1/2} \exp(\pi i n x/a)$; $n = 0, \pm 1, \pm 2, \ldots$. Given $f \in L^2[-a, a]$, let $a_n = (\varphi_n, f)$. Prove that $f \in L^2[-a, a]$ is in $\overline{C_0^\infty[-a, a]}$ if and only if $n^k a_n \to 0$ for all k (as $n \to \infty$). Develop norm inequalities to prove that s and $\overline{C_0^\infty[-a, a]}$ are isomorphic. Prove that the closure of $C_0^\infty[-a, a]$ is in $\mathcal{D}$.

60. (a) Let $B(\cdot, \cdot)$ be a bilinear functional on $C_0^\infty[-a, a]$ which is separately continuous. Prove that there is a $T \in C_0^\infty([-a, a] \times [-a, a])^*$ with $B(f, g) = T(f \otimes g)$ by mimicking the appendix to Section V.3 and using Problem 59.
    (b) Let $B(\cdot, \cdot)$ be a bilinear functional on $\mathcal{D}_\mathbb{R}$, which is separately continuous. Prove that there is a $T \in \mathcal{D}'_{\mathbb{R}^2}$ with $B(f, g) = T(f \otimes g)$.

61. Prove that $C_0^\infty(\mathbb{R})$ is nonempty, that is, construct an explicit function of compact support which is infinitely differentiable. Hint: First show that the function $f(x) = \chi_{(0, \infty)}(x) e^{-1/x}$ is C^∞ where $\chi_{(0, \infty)}$ is the characteristic function of $(0, \infty)$.

VI: Bounded Operators

I was at the mathematical school, where the master taught his pupils after a method scarce imaginable to us in Europe. The proposition and demonstration were fairly written on a thin wafer, with ink composed of a cephalic tincture. This the student was to swallow upon a fasting stomach, and for three days following eat nothing but bread and water. As the wafer digested the tincture mounted to the brain, bearing the proposition along with it.

Jonathan Swift in **Gulliver's Travels**

VI.1 Topologies on bounded operators

We have already introduced $\mathscr{L}(X, Y)$, the Banach space of operators from one Banach space to another. In this chapter we will study $\mathscr{L}(X, Y)$ more closely. We emphasize the case which will arise most frequently later, namely, $\mathscr{L}(\mathscr{H}, \mathscr{H}) \equiv \mathscr{L}(\mathscr{H})$ where $\mathscr{H}$ is a separable Hilbert space. Theorem III.2 shows that $\mathscr{L}(X, Y)$ is a Banach space with the norm

$$\|T\| = \sup_{x \neq 0} \frac{\|Tx\|_Y}{\|x\|_X}$$

The induced topology on $\mathscr{L}(X, Y)$ is called the **uniform operator topology** (or **norm** topology). In this topology the map $\langle A, B \rangle \to BA$ of $\mathscr{L}(X, Y) \times \mathscr{L}(Y, Z) \to \mathscr{L}(X, Z)$ is jointly continuous.

We now introduce two new topologies on $\mathscr{L}(X, Y)$, the weak and strong operator topologies. There are other interesting and useful topologies on $\mathscr{L}(X, Y)$, but we delay their introduction until we need them in a later volume (see however the discussion at the end of Section 6 and the Notes).

The **strong operator topology** is the weakest topology on $\mathscr{L}(X, Y)$ such that the maps

$$E_x : \mathscr{L}(X, Y) \to Y$$

given by $E_x(T) = Tx$ are continuous for all $x \in X$. A neighborhood basis at the origin is given by sets of the form

$$\{S \mid S \in \mathscr{L}(X, Y), \quad \|Sx_i\|_Y < \varepsilon, \quad i = 1, \ldots, n\}$$

where $\{x_i\}_{i=1}^n$ is a finite collection of elements of X and ε is positive. In this topology a net $\{T_\alpha\}$ of operators converges to an operator T (written $T_\alpha \xrightarrow{s} T$) if and only if $\|T_\alpha x - Tx\| \to 0$ for all $x \in X$. The map $\langle A, B \rangle \to AB$ is separately but not jointly continuous if X, Y, and Z are infinite dimensional (see Problem 6a, b). We sometimes denote strong limits by the symbol s-lim.

The **weak operator topology** on $\mathscr{L}(X, Y)$ is the weakest topology such that the maps

$$E_{x,\ell}: \mathscr{L}(X, Y) \to \mathbb{C}$$

given by $E_{x,\ell}(T) = \ell(Tx)$ are all continuous for all $x \in X$, $\ell \in Y^*$. A basis at the origin is given by sets of the form

$$\{S \mid S \in \mathscr{L}(X, Y), \quad |\ell_i(Tx_j)| < \varepsilon, \quad i = 1, \ldots, n, \quad j = 1, \ldots, m\}$$

where $\{x_i\}_{i=1}^n$ and $\{\ell_j\}_{j=1}^m$ are finite families of elements of X and Y^* respectively. A net of operators $\{T_\alpha\}$ converges to an operator T in the weak operator topology (written $T_\alpha \xrightarrow{w} T$) if and only if $|\ell(T_\alpha x) - \ell(Tx)| \to 0$ for each $\ell \in Y^*$ and $x \in X$. Notice that in the case $\mathscr{L}(\mathscr{H})$, $T_\gamma \to T$ weakly just means that the "matrix elements" $(y, T_\gamma x)$ converge to (y, Tx). In the weak topology the map $\langle A, B \rangle \to AB$ is separately, but not jointly continuous if X, Y, and Z are infinite dimensional (see Problem 6c).

Remark The reader should not confuse the weak operator topology on $\mathscr{L}(X, Y)$ with the weak (Banach space) topology on $\mathscr{L}(X, Y)$. The former is the weakest topology such that the bounded linear functionals on $\mathscr{L}(X, Y)$ of the form $\ell(\cdot x)$ are continuous for all $x \in X$ and $\ell \in Y^*$. The latter is the weakest topology such that *all* bounded linear functionals on $\mathscr{L}(X, Y)$ are continuous (see Section VI.6).

Notice that the weak operator topology is weaker than the strong operator topology which is weaker than the uniform operator topology. In general, the weak and strong operator topologies on $\mathscr{L}(X, Y)$ will not be first countable so that questions of compactness, net convergence, and sequential convergence are complicated. The following simple example illustrates the different topologies on $\mathscr{L}(\ell_2)$.

Example   Consider the bounded operators on ℓ_2.

(i) Let T_n be defined by

$$T_n(\xi_1, \xi_2, \ldots) = \left(\frac{1}{n}\xi_1, \frac{1}{n}\xi_2, \ldots\right)$$

Then $T_n \to 0$ uniformly.

(ii) Let S_n be defined by

$$S_n(\xi_1, \xi_2, \ldots) = (\underbrace{0, 0, \ldots, 0}_{n \text{ places}}, \xi_{n+1}, \xi_{n+2}, \ldots)$$

Then $S_n \to 0$ strongly but not uniformly.

(iii) Let W_n be defined by

$$W_n(\xi_1, \xi_2, \ldots) = (\underbrace{0, 0, \ldots, 0}_{n \text{ places}}, \xi_1, \xi_2, \ldots)$$

Then $W_n \to 0$ in the weak operator topology but not in the strong or uniform topologies.

The following result in the Hilbert space case is sometimes useful and provides a nice application of the uniform boundedness theorem.

Theorem VI.1   Let $\mathscr{L}(\mathscr{H})$ denote the bounded operators on a Hilbert space $\mathscr{H}$. Let T_n be a *sequence* of bounded operators and suppose that $(T_n x, y)$ converges as $n \to \infty$ for each $x, y \in \mathscr{H}$. Then there exists $T \in \mathscr{L}(\mathscr{H})$ such that $T_n \xrightarrow{w} T$.

Proof   We begin by showing that for each x, $\sup_n \|T_n x\| < \infty$. Since for any $x \in \mathscr{H}$, $(x, T_n y)$ converges we have

$$\sup_n |(T_n x, y)| < \infty$$

For each n, $T_n x \in \mathscr{L}(\mathscr{H}, \mathbb{C})$, and since $\sup_n |(T_n x)(y)|_{\mathbb{C}} < \infty$, the uniform boundedness theorem implies that the operator norms of the $T_n x$ in $\mathscr{L}(\mathscr{H}, \mathbb{C})$ are uniformly bounded. But the norm of $T_n x$ as an operator in $\mathscr{L}(\mathscr{H}, \mathbb{C})$ is the same as its norm in $\mathscr{H}$; thus $\|T_n x\|_{\mathscr{H}}$ is uniformly bounded.

Now, we use the uniform boundedness theorem again. Since

$$\sup_n \|T_n x\|_{\mathscr{H}} < \infty,$$

we conclude

$$\sup_n \|T_n\|_{\mathscr{L}(\mathscr{H})} < \infty$$

Define $B(x, y) = \lim_n (T_n x, y)$. Then it is easily verified that $B(x, y)$ is sesquilinear and

$$|B(x, y)| \leq \overline{\lim_n} |(T_n x, y)| \leq \|x\| \|y\| (\sup_n \|T_n\|)$$

Thus $B(x, y)$ is a bounded sesquilinear form on $\mathcal{H}$ and so, by the corollary to the Riesz lemma, there is a bounded operator $T \in \mathcal{L}(\mathcal{H})$ such that $B(x, y) = (Tx, y)$. Clearly $T_n \xrightarrow{w} T$. ∎

If a *sequence* of operators T_n on a Hilbert space has the property that $T_n x$ converges for each $x \in \mathcal{H}$, then there exists $T \in \mathcal{L}(\mathcal{H})$ such that $T_n \xrightarrow{s} T$. The reader is asked to prove this theorem and various generalizations in Problem 3.

Let $T \in \mathcal{L}(X, Y)$. The set of vectors $x \in X$ so that $Tx = 0$ is called the **kernel** of T, written Ker T. The set of vectors $y \in Y$ so that $y = Tx$ for some $x \in X$ is called the **range** of T, written Ran T. Notice that both Ker T and Ran T are subspaces. Ker T is necessarily closed, but Ran T may not be closed (Problem 7).

VI.2 Adjoints

In this section we define adjoints of bounded operators on Banach and Hilbert spaces. The reader should be cautioned at the outset that the Hilbert space adjoint of an operator $T \in \mathcal{L}(\mathcal{H})$ is not equal to the Banach space adjoint although it is closely related to it.

Definition   Let X and Y be Banach spaces, T a bounded linear operator from X to Y. The Banach space **adjoint** of T, denoted by T', is the bounded linear operator from Y^* to X^* defined by

$$(T'\ell)(x) = \ell(Tx)$$

for all $\ell \in Y^*$, $x \in X$.

Example   Let $X = \ell_1 = Y$ and let T be the right shift operator

$$T(\xi_1, \xi_2, \ldots) = (0, \xi_1, \xi_2, \ldots)$$

Then $T' : \ell_\infty \to \ell_\infty$ is the operator

$$T'(\xi_1, \xi_2, \ldots) = (\xi_2, \xi_3, \ldots)$$

In this example, $\|T\| = 1 = \|T'\|$. In fact the norms of T and T' are always equal:

186 VI: BOUNDED OPERATORS

Theorem VI.2   Let X and Y be Banach spaces. The map $T \to T'$ is an isometric isomorphism of $\mathscr{L}(X, Y)$ into $\mathscr{L}(Y^*, X^*)$.

Proof The map $T \to T'$ is linear. The fact that T' is bounded and that the map is an isometry follows from the computation

$$\|T\|_{\mathscr{L}(X,Y)} = \sup_{\|x\| \leq 1} \|Tx\|_Y$$

$$= \sup_{\|x\| \leq 1} \left(\sup_{\|\ell\| \leq 1} |\ell(Tx)| \right) \qquad \ell \in Y^*$$

$$= \sup_{\|\ell\| \leq 1} \left(\sup_{\|x\| \leq 1} |(T'\ell)(x)| \right)$$

$$= \sup_{\|\ell\| \leq 1} \|T'\ell\|$$

$$= \|T'\|_{\mathscr{L}(Y^*, X^*)}$$

The second equality uses a corollary of the Hahn–Banach theorem. ∎

We are mostly interested in the case where T is a bounded linear transformation of a Hilbert space $\mathscr{H}$ to itself. The Banach space adjoint of T is then a mapping of $\mathscr{H}^*$ to $\mathscr{H}^*$. Let $C: \mathscr{H} \to \mathscr{H}^*$ be the map which assigns to each $y \in \mathscr{H}$, the bounded linear functional $(y, \cdot)$ in $\mathscr{H}^*$. C is a *conjugate* linear isometry which is surjective by the Riesz lemma. Now define a map T^*: $\mathscr{H} \to \mathscr{H}$ by

$$T^* = C^{-1} T' C$$

Then T^* satisfies

$$(x, Ty) = (Cx)(Ty) = (T'Cx)(y) = (C^{-1}T'Cx, y) = (T^*x, y)$$

T^* is called the **Hilbert space adjoint** of T, but usually we will just call it the adjoint and let the * distinguish it from T'. Notice that the map $T \to T^*$ is *conjugate* linear, that is, $\alpha T \to \bar{\alpha} T^*$. This is because C is conjugate linear. We summarize the properties of the map $T \to T^*$:

Theorem VI.3   (a) $T \to T^*$ is a conjugate linear isometric isomorphism of $\mathscr{L}(\mathscr{H})$ onto $\mathscr{L}(\mathscr{H})$.
  (b) $(TS)^* = S^* T^*$.
  (c) $(T^*)^* = T$.
  (d)   If T has a bounded inverse, T^{-1}, then T^* has a bounded inverse and $(T^*)^{-1} = (T^{-1})^*$.

(e) The map $T \to T^*$ is always continuous in the weak and uniform operator topologies but is only continuous in the strong operator topology if $\mathcal{H}$ is finite dimensional.

(f) $\|T^*T\| = \|T\|^2$.

Proof (a) follows from Theorem VI.2 and the fact that C is an isometry. (b) and (c) are easily checked. Since $T^{-1}T = I = TT^{-1}$ we have from (b)

$$T^*(T^{-1})^* = I^* = I = I^* = (T^{-1})^*T^*$$

which proves (d).

Continuity of $T \to T^*$ in the weak and uniform operator topologies is trivial. In the case $\mathcal{H} = \ell_2$, here is a counter example which shows that $T \to T^*$ is not continuous in the strong operator topology. The general infinite dimensional case is similar. Let W_n be right shift on ℓ_2 by n places. Then W_n converges weakly but not strongly to zero. However, $W_n^* = V_n$ converges strongly to zero. Thus $V_n \xrightarrow{s} 0$, but $V_n^* = W_n$ does not converge strongly to zero.

(f) Note that $\|T^*T\| \leq \|T\| \|T^*\| = \|T\|^2$ and

$$\|T^*T\| \geq \sup_{\|x\|=1} (x, T^*Tx) = \sup_{\|x\|=1} \|Tx\|^2 = \|T\|^2 \quad \blacksquare$$

Definition A bounded operator T on a Hilbert space is called **self-adjoint** if $T = T^*$.

Self-adjoint operators play a major role in functional analysis and mathematical physics and much of our time is devoted to studying them. Chapter VII is devoted to proving a structure theorem for bounded self-adjoint operators. In Chapter VIII we introduce unbounded self-adjoint operators and continue their study in Chapter X. We remind the reader that on $\mathbb{C}^n$, a linear transformation is self-adjoint if and only if its matrix in any orthonormal basis is invariant under the operation of reflection across the diagonal followed by complex conjugation.

An important class of operators on Hilbert spaces is that of the projections.

Definition If $P \in \mathcal{L}(\mathcal{H})$ and $P^2 = P$, then P is called a **projection**. If in addition $P = P^*$, then P is called an **orthogonal projection**.

Notice that the range of a projection is always a closed subspace on which P acts like the identity. If in addition P is orthogonal, then P acts like the zero operator on $(\text{Ran } P)^\perp$. If $x = y + z$, with $y \in \text{Ran } P$ and $z \in (\text{Ran } P)^\perp$, is the decomposition guaranteed by the projection theorem, then $Px = y$. P is called the orthogonal projection onto Ran P. Thus, the projection theorem sets up a one to one correspondence between orthogonal projections and closed subspaces. Since orthogonal projections arise more frequently than nonorthogonal ones, we normally use the word projection to mean orthogonal projection.

VI.3 The spectrum

If T is a linear transformation on $\mathbb{C}^n$, then the eigenvalues of T are the complex numbers λ such that the determinant of $\lambda I - T$ is equal to zero. The set of such λ is called the spectrum of T. It can consist of at most n points since $\det(\lambda I - T)$ is a polynomial of degree n. If λ is not an eigenvalue, then $\lambda I - T$ has an inverse since $\det(\lambda I - T) \neq 0$.

The spectral theory of operators on infinite-dimensional spaces is more complicated, more interesting, and very important for an understanding of the operators themselves.

Definition Let $T \in \mathscr{L}(X)$. A complex number λ is said to be in the **resolvent set** $\rho(T)$ of T if $\lambda I - T$ is a bijection with a bounded inverse. $R_\lambda(T) = (\lambda I - T)^{-1}$ is called the **resolvent** of T at λ. If $\lambda \notin \rho(T)$, then λ is said to be in the **spectrum** $\sigma(T)$ of T.

We note that by the inverse mapping theorem, $\lambda I - T$ automatically has a bounded inverse if it is bijective. We distinguish two subsets of the spectrum.

Definition Let $T \in \mathscr{L}(X)$.

(a) An $x \neq 0$ which satisfies $Tx = \lambda x$ for some $\lambda \in \mathbb{C}$ is called an **eigenvector** of T; λ is called the corresponding **eigenvalue**. If λ is an eigenvalue, then $\lambda I - T$ is not injective so λ is in the spectrum of T. The set of all eigenvalues is called the **point spectrum** of T.
(b) If λ is not an eigenvalue and if $\operatorname{Ran}(\lambda I - T)$ is not dense, then λ is said to be in the **residual spectrum**.

At the end of this section we present an example which illustrates these kinds of spectra. The reason that we single out the residual spectrum is that it does not occur for a large class of operators, for example, for self-adjoint operators (see Theorem VI.8).

The spectral analysis of operators is very important for mathematical physics. For example, in quantum mechanics the Hamiltonian is an unbounded self-adjoint operator on a Hilbert space. The point spectrum of the Hamiltonian corresponds to the energy levels of bound states of the system. The rest of the spectrum plays an important role in the scattering theory of the system (see Chapter XII).

We will shortly prove that the resolvent set $\rho(T)$ is open and that $R_\lambda(T)$ is an analytic operator-valued function on $\rho(T)$. This fact allows one to use complex analysis to study $R_\lambda(T)$ and thus to obtain information about T. We begin with a brief aside about vector–valued analytic functions.

Let X be a Banach space and let D be a region in the complex plane, i.e., a connected open subset of $\mathbb{C}$. A function, $x(\cdot)$, defined on D with values in X, is said to be **strongly analytic** at $z_0 \in D$ if the limit of $(x(z_0 + h) - x(z_0))/h$ exists in X as h goes to zero in $\mathbb{C}$. Starting from this point one can develop a theory of vector-valued analytic functions which is almost exactly parallel to the usual theory; in particular, a strongly analytic function has a norm-convergent Taylor series. We do not repeat this development here; see the notes for references. We do want to discuss one important point. There is another natural way to define Banach-valued analytic functions. Namely: a function $x(\cdot)$ on D with values in X is said to be **weakly analytic** if $\ell(x(\cdot))$ is a complex valued analytic function on D for each $\ell \in X^*$. Although this second definition of analytic is a priori weaker than the first, the two definitions are equivalent, a fact we will prove in a moment. This is very important, since weak analyticity is often much easier to check.

Lemma Let X be a Banach space. Then a sequence $\{x_n\}$ is Cauchy if and only if $\{\ell(x_n)\}$ is Cauchy, uniformly for $\ell \in X^*$, $\|\ell\| \leq 1$.

Proof If $\{x_n\}$ is Cauchy, then $|\ell(x_n) - \ell(x_m)| \leq \|x_n - x_m\|$ for all ℓ with $\|\ell\| \leq 1$, so $\{\ell(x_n)\}$ is Cauchy uniformly. Conversely,

$$\|x_n - x_m\| = \sup_{\|\ell\| \leq 1} |\ell(x_n - x_m)|$$

Thus, if $\{\ell(x_n)\}$ is Cauchy, uniformly for $\|\ell\| \leq 1$, then $\{x_n\}$ is norm-Cauchy. ∎

Theorem VI.4 Every weakly analytic function is strongly analytic.

Proof Let $x(\cdot)$ be a weakly analytic function on D with values in X. Let

190 VI: BOUNDED OPERATORS

$z_0 \in D$ and suppose that Γ is a circle in D containing z_0 whose interior is contained in D. If $\ell \in X^*$ then $\ell(x(z))$ is analytic and

$$\ell\left(\frac{x(z_0 + h) - x(z_0)}{h}\right) - \frac{d}{dz}\ell(x(z_0))$$

$$= \frac{1}{2\pi i}\oint_\Gamma \left[\frac{1}{h}\left(\frac{1}{z - (z_0 + h)} - \frac{1}{z - z_0}\right) - \frac{1}{(z - z_0)^2}\right]\ell(x(z))\, dz$$

Since $\ell(x(z))$ is continuous on Γ and Γ is compact, $|\ell(x(z))| \leq C_\ell$ for all $z \in \Gamma$. Regarding $x(z)$ as a family of mappings $x(z): X^* \to \mathbb{C}$ we see that $x(z)$ is pointwise bounded at each ℓ so by the uniform boundedness theorem $\sup_{z \in \Gamma} \|x(z)\| \leq C < \infty$. Thus

$$\left|\ell\left(\frac{x(z_0 + h) - x(z_0)}{h}\right) - \frac{d}{dz}\ell(x(z_0))\right|$$

$$\leq (2\pi)^{-1}\|\ell\|\left(\sup_{z \in \Gamma}\|x(z)\|\right)\oint_\Gamma \left|\frac{1}{(z - (z_0 + h))(z - z_0)} - \frac{1}{(z - z_0)^2}\right|\, dz$$

This estimate shows that $[x(z_0 + h) - x(z_0)]/h$ is uniformly Cauchy for $\|\ell\| \leq 1$. By the lemma, $[x(z_0 + h) - x(z_0)]/h$ converges in X, proving that $x(\cdot)$ is strongly analytic. ∎

We now prove the theorem we promised about the resolvent.

Theorem VI.5 Let X be a Banach space and suppose $T \in \mathscr{L}(X)$. Then $\rho(T)$ is an open subset of $\mathbb{C}$ and $R_\lambda(T)$ is an analytic $\mathscr{L}(X)$-valued function on each component (maximal connected subset) of D. For any two points λ, $\mu \in \rho(T)$, $R_\lambda(T)$ and $R_\mu(T)$ commute and

$$R_\lambda(T) - R_\mu(T) = (\mu - \lambda)R_\mu(T)R_\lambda(T) \tag{VI.1}$$

Proof We begin with the following formal computation, temporarily ignoring questions of convergence. Let $\lambda_0 \in \rho(T)$.

$$\frac{1}{\lambda - T} = \frac{1}{\lambda - \lambda_0 + (\lambda_0 - T)} = \left(\frac{1}{\lambda_0 - T}\right)\frac{1}{1 - \left(\dfrac{\lambda_0 - \lambda}{\lambda_0 - T}\right)}$$

$$= \left(\frac{1}{\lambda_0 - T}\right)\left[1 + \sum_{n=1}^\infty \left(\frac{\lambda_0 - \lambda}{\lambda_0 - T}\right)^n\right]$$

This suggests that we define

$$\tilde{R}_\lambda(T) = R_{\lambda_0}(T)\left\{I + \sum_{n=1}^\infty (\lambda_0 - \lambda)^n [R_{\lambda_0}(T)]^n\right\}$$

Since
$$\|[R_{\lambda_0}(T)]^n\| \le \|R_{\lambda_0}(T)\|^n$$
the series on the right converges in the uniform operator topology if
$$|\lambda - \lambda_0| < \|R_{\lambda_0}(T)\|^{-1}$$
For such λ, $\tilde{R}_\lambda(T)$ is well defined, and it is easily checked that
$$(\lambda I - T)\tilde{R}_\lambda(T) = I = \tilde{R}_\lambda(T)(\lambda I - T)$$
This proves that $\lambda \in \rho(T)$ if $|\lambda - \lambda_0| < \|R_{\lambda_0}(T)\|^{-1}$ and that $\tilde{R}_\lambda(T) = R_\lambda(T)$. Thus $\rho(T)$ is open. Since $R_\lambda(T)$ has a power series expansion, it is analytic. The expression
$$R_\lambda(T) - R_\mu(T) = R_\lambda(T)(\mu I - T)R_\mu(T) - R_\lambda(T)(\lambda I - T)R_\mu(T)$$
proves (VI.1). Interchanging μ and λ shows that $R_\lambda(T)$ and $R_\mu(T)$ commute. ∎

Equation (VI.1) is called the **first resolvent formula**. A nice example of the use of complex analytic methods is given by the proof of the following corollary.

Corollary Let X be a Banach space, $T \in \mathscr{L}(X)$. Then the spectrum of T is not empty.

Proof Formally,
$$\frac{1}{\lambda - T} = \left(\frac{1}{\lambda}\right)\frac{1}{1 - T/\lambda} = \frac{1}{\lambda}\left(1 + \sum_{n=1}^{\infty}\left(\frac{T}{\lambda}\right)^n\right)$$
which suggests that for large values of $|\lambda|$,
$$R_\lambda(T) = \frac{1}{\lambda}\left\{I + \sum_{n=1}^{\infty}\left(\frac{T}{\lambda}\right)^n\right\} \qquad (VI.2)$$
If $|\lambda| > \|T\|$, then the series on the right converges in norm and it is easily checked that for such λ, its limit is indeed the inverse of $(\lambda I - T)$. Thus, as $|\lambda| \to \infty$, $\|R_\lambda(T)\| \to 0$. If $\sigma(T)$ were empty, $R_\lambda(T)$ would be an entire bounded analytic function. By Liouville's theorem, $R_\lambda(T)$ would be zero which is a contradiction. Thus, $\sigma(T)$ is not empty. ∎

The series (VI.2) is called the **Neumann series** for $R_\lambda(T)$. The proof of the corollary shows that $\sigma(T)$ is contained in the closed disc of radius $\|T\|$. Actually, we can say more about $\sigma(T)$.

192   VI: BOUNDED OPERATORS

Definition   Let $\displaystyle r(T) = \sup_{\lambda \in \sigma(T)} |\lambda|$

$r(T)$ is called the **spectral radius** of T.

Theorem VI.6   Let X be a Banach space, $T \in \mathscr{L}(X)$. Then $\lim_{n \to \infty} \|T^n\|^{1/n}$ exists and is equal to $r(T)$. If X is a Hilbert space and A is self-adjoint, then $r(A) = \|A\|$.

Proof   The reader can check that $\lim_{n \to \infty} \|T^n\|^{1/n}$ exists by following the clever subadditivity argument outlined in Problem 11. The crux of the proof of the theorem is to establish that the radius of convergence of the Laurent series of $R_\lambda(T)$ about ∞ is just $r(T)^{-1}$. First notice that the radius of convergence cannot be smaller than $r(T)^{-1}$ since we have proven that $R_\lambda(T)$ is analytic on $\rho(T)$ and $\{\lambda \mid |\lambda| > r(T)\} \subset \rho(T)$. On the other hand, (VI.2) is just the Laurent series about ∞ and we have seen that where it converges absolutely, $R_\lambda(T)$ exists. Since a Laurent series converges absolutely inside the circle of convergence, we conclude that the radius of convergence cannot be larger than $r(T)^{-1}$. That $r(T) = \lim_{n \to \infty} \|T^n\|^{1/n}$ follows from the vector-valued version of Hadamard's theorem which says that the radius of convergence of (VI.2) is just the inverse of

$$\overline{\lim_{n}} \|T^n\|^{1/n} = \lim_{n \to \infty} \|T^n\|^{1/n}$$

Finally, if X is a Hilbert space and A is self-adjoint, then $\|A\|^2 = \|A^2\|$ by part (f) of Theorem VI.3. This implies that $\|A^{2^n}\| = \|A\|^{2^n}$ so

$$r(A) = \lim_{k \to \infty} \|A^k\|^{1/k} = \lim_{n \to \infty} \|A^{2^n}\|^{2^{-n}} = \|A\| \quad \blacksquare$$

The following theorem is sometimes useful in determining spectra.

Theorem VI.7   (Phillips)   Let X be a Banach space, $T \in \mathscr{L}(X)$. Then $\sigma(T) = \sigma(T')$ and $R_\lambda(T') = R_\lambda(T)'$. If $\mathscr{H}$ is a Hilbert space, then $\sigma(T^*) = \{\lambda \mid \bar{\lambda} \in \sigma(T)\}$ and $R_{\bar{\lambda}}(T^*) = R_\lambda(T)^*$.

We note that the Hilbert space case follows from (d) of Theorem VI.3.

We now work out in some detail an example which illustrates the various kinds of spectra.

Example   Let T be the operator on ℓ_1 which acts by

$$T(\xi_1, \xi_2, \ldots) = (\xi_2, \xi_3, \ldots)$$

VI.3 The spectrum

The adjoint of T, T', acts on ℓ_∞ by

$$T'(\xi_1, \xi_2, \ldots) = (0, \xi_1, \xi_2, \ldots)$$

We first observe that $\|T\| = \|T'\| = 1$, so that all λ with $|\lambda| > 1$ are in $\rho(T)$ and $\rho(T')$. Suppose $|\lambda| < 1$. Then the vector $x_\lambda = (1, \lambda, \lambda^2, \ldots)$ is in ℓ_1 and satisfies $(\lambda I - T)x_\lambda = 0$. Thus all such λ are in the point spectrum of T. Since the spectrum is closed, $\sigma(T) = \{\lambda \mid |\lambda| \leq 1\}$. By Theorem VI.7 this set is also the spectrum of T'.

We want to show that T' has no point spectrum. Suppose that $\{\xi_n\}_{n=1}^\infty \in \ell_\infty$ and $(\lambda I - T')\{\xi_n\} = 0$. Then

$$\lambda \xi_0 = 0$$
$$\lambda \xi_1 - \xi_0 = 0$$
$$\vdots$$

These equations together imply that $\{\xi_n\}_{n=1}^\infty = 0$ so $\lambda I - T'$ is one to one and T' has no point spectrum. Next, suppose $|\lambda| < 1$. Then for all $L \in \ell_\infty$

$$[(\lambda I - T')L](x_\lambda) = L((\lambda I - T)x_\lambda) = 0.$$

where $x_\lambda \in \ell_1$ is the eigenvector with eigenvalue λ. By the Hahn-Banach theorem we know that there is a linear functional in ℓ_∞ which does not vanish on x_λ so the range of $\lambda I - T'$ is not dense. Thus $\{\lambda; |\lambda| < 1\}$ is in the residual spectrum of T'.

It remains to consider the boundary $|\lambda| = 1$. Suppose that $|\lambda| = 1$ and $(\lambda I - T)\{\xi_n\} = 0$ for some $\{\xi_n\}$ in ℓ_1. Then

$$\xi_1 = \lambda \xi_0$$
$$\xi_2 = \lambda \xi_1$$
$$\vdots$$

so $\{\xi_n\}_{n=1}^\infty = \xi_0(1, \lambda, \lambda^2, \ldots)$ which is not in ℓ_1. Thus λ is not point spectrum. If the range of $\lambda I - T$ were not dense there would be a nonzero $L \in \ell_\infty$ such that $L[(\lambda I - T)x] = 0$ for all $x \in \ell_1$. But then $[(\lambda I - T')L](x) = 0$ which would imply that λ is in the point spectrum of T' which we have proven cannot occur. Thus, $\{\lambda \mid |\lambda| = 1\}$ is neither in the point spectrum of T nor in the residual spectrum of T.

Finally, we prove that $\{\lambda \mid |\lambda| = 1\}$ is in the residual spectrum of T' by explicitly finding an open ball disjoint from $\text{Ran}(\lambda - T')$. If $a = \{a_n\}$ and $b = \{b_n\}$ are in ℓ_∞ and obey $a = (\lambda I - T')b$, then

$$a_0 = \lambda b_0$$
$$\vdots$$
$$a_n = \lambda b_n - b_{n-1}$$

so $b_n = (\bar\lambda)^{n+1} \sum_{m=0}^n \lambda^m a_m$. Let $c = \{c_n\}$ with $c_n = \bar\lambda^n$ and suppose that $d \in \ell_\infty$ and $\|d - c\|_\infty \leq \frac{1}{2}$. Then

$$\operatorname{Re}\{\bar\lambda^n d_n\} \geq \operatorname{Re}\{\bar\lambda^n c_n\} - \|d - c\|_\infty \geq \tfrac{1}{2}$$

Thus, if $(\lambda - T')e = d$ for some $e \in \ell_\infty$, then since

$$e_n = (\bar\lambda)^{n+1} \sum_{m=0}^n \lambda^m d_m$$

$|e_n| \geq n/2$ which is impossible. Therefore, $\operatorname{Ran}(\lambda - T')$ does not intersect the ball of radius $\frac{1}{2}$ about c so λ is in the residual spectrum.

Operator	Spectrum	Point spectrum	Residual spectrum
T	$|\lambda| \leq 1$	$|\lambda| < 1$	$\varnothing$
T'	$|\lambda| \leq 1$	$\varnothing$	$|\lambda| \leq 1$

As in the above example, one can prove in general

Proposition Let X be a Banach space and $T \in \mathscr{L}(X)$. Then,

(a) If λ is in the residual spectrum of T, then λ is in the point spectrum of T'.

(b) If λ is in the point spectrum of T, then λ is in either the point or the residual spectrum of T'.

Finally, we note:

Theorem VI.8 Let T be a self-adjoint operator on a Hilbert space $\mathscr{H}$. Then,

(a) T has no residual spectrum.
(b) $\sigma(T)$ is a subset of $\mathbb{R}$.
(c) Eigenvectors corresponding to distinct eigenvalues of T are orthogonal.

Proof If λ and μ are real, we compute

$$\|[T - (\lambda + i\mu)]x\|^2 = \|(T - \lambda)x\|^2 + \mu^2 \|x\|^2$$

Thus $\|[T - (\lambda + i\mu)]x\|^2 \geq \mu^2 \|x\|^2$, so if $\mu \neq 0$, then $T - (\lambda + i\mu)$ is one to one and has a bounded inverse on its range, which is closed. If $\operatorname{Ran}(T - (\lambda + i\mu)) \neq \mathscr{H}$, then, by the above proposition, $\lambda - i\mu$ would be in the point spectrum of T, which is impossible by the inequality. Thus if $\mu \neq 0$, $\lambda + i\mu$ is in $\rho(T)$. This proves (b). If a real λ were in the residual spectrum of T, then $\bar\lambda = \lambda$ would be in the point spectrum of $T^* = T$, which is impossible since the point and residual spectrum are disjoint by definition. This proves (a). The easy proof of (c) is left as an exercise (Problem 8). ∎

VI.4 Positive operators and the polar decomposition

We want to prove the existence of a special decomposition for operators on a *Hilbert space* which is analogous to the decomposition $z = |z|e^{i\arg z}$ for complex numbers. First we must describe a suitable analogue of the positive numbers.

Definition Let $\mathcal{H}$ be a Hilbert space. An operator $B \in \mathcal{L}(\mathcal{H})$ is called **positive** if $(Bx, x) \geq 0$ for all $x \in \mathcal{H}$. We write $B \geq 0$ if B is positive and $B \leq A$ if $A - B \geq 0$.

Every (bounded) positive operator on a *complex* Hilbert space is self-adjoint. To see this, notice that $(x, Ax) = \overline{(x, Ax)} = (Ax, x)$ if (Ax, x) takes only real values. By the polarization identity (Chapter II, Problem 4), $(Ax, y) = (x, Ay)$ if $(Ax, x) = (x, Ax)$ for all x. Thus, if A is positive, it is self-adjoint. This is false on real Hilbert spaces because it is not possible to recover (x, Ay) by knowing (x, Ax) for all x.

For any $A \in \mathcal{L}(\mathcal{H})$, notice that $A^*A \geq 0$ since $(A^*Ax, x) = \|Ax\|^2 \geq 0$. Just as $|z| = \sqrt{\bar{z}z}$ we would like to define $|A| = \sqrt{A^*A}$. To do this we must show that we can take square roots of positive operators. We begin with a lemma.

Lemma The power series for $\sqrt{1-z}$ about zero converges absolutely for all complex numbers z satisfying $|z| \leq 1$.

Proof Let $\sqrt{1-z} = 1 + c_1 z + c_2 z^2 + \cdots$ be the power series of $\sqrt{1-z}$ about the origin. Since $\sqrt{1-z}$ is analytic for $|z| < 1$, the series converges absolutely there. The derivatives of $\sqrt{1-z}$ at the origin are all negative, so the c_i are negative if $i \geq 1$. Thus

$$\sum_{n=0}^{N} |c_n| = 2 - \sum_{n=0}^{N} c_n$$

$$= 2 - \lim_{x \to 1-} \sum_{n=0}^{N} c_n x^n$$

$$\leq 2 - \lim_{x \to 1-} \sqrt{1-x}$$

$$= 2$$

where $\lim_{x \to 1-}$ means the limit as x approaches one from below. Since this is true for all N, $\sum_{n=0}^{\infty} |c_n| \leq 2$, which implies that the series converges absolutely for $|z| = 1$. ∎

Theorem VI.9 (square root lemma)   Let $A \in \mathcal{L}(\mathcal{H})$ and $A \geq 0$. Then there is a unique $B \in \mathcal{L}(\mathcal{H})$ with $B \geq 0$ and $B^2 = A$. Furthermore, B commutes with every bounded operator which commutes with A.

Proof It is sufficient to consider the case where $\|A\| \leq 1$. Since
$$\|I - A\| = \sup_{\|\varphi\|=1} |((I - A)\varphi, \varphi)| \leq 1$$
the above lemma implies that the series $1 + c_1(I - A) + c_2(I - A)^2 + \cdots$ converges in norm to an operator B. Since the convergence is absolute we can square the series and rearrange terms which proves that $B^2 = A$. Furthermore, since $0 \leq I - A \leq I$ we have $0 \leq (\varphi, (I - A)^n \varphi) \leq 1$ for all $\varphi \in \mathcal{H}$ with $\|\varphi\| = 1$. Thus
$$(\varphi, B\varphi) = 1 + \sum_{n=1}^{\infty} c_n(\varphi, (I - A)^n \varphi)$$
$$\geq 1 + \sum_{n=1}^{\infty} c_n = 0$$
where we have used the fact that $c_n < 0$ and the estimate in the lemma. Thus, $B \geq 0$. Since the series for B converges absolutely, it commutes with any operator that commutes with A.

Suppose there is a B', with $B' \geq 0$ and $(B')^2 = A$. Then since
$$B'A = (B')^3 = AB'$$
B' commutes with A and thus with B. Therefore
$$(B - B')B(B - B') + (B - B')B'(B - B') = (B^2 - B'^2)(B - B') = 0 \quad (\text{VI}.3)$$
Since both terms in (VI.3) are positive, they must both be zero, so their difference $(B - B')^3 = 0$. Since $B - B'$ is self-adjoint, $\|B - B'\|^4 = \|(B - B')^4\| = 0$, so $B - B' = 0$. ∎

We are now ready to define $|A|$.

Definition   Let $A \in \mathcal{L}(\mathcal{H})$. Then $|A| = \sqrt{A^*A}$.

The reader should be wary of the emotional connotations of the symbol $|\cdot|$. While it is true that $|\lambda A| = |\lambda| \, |A|$ for $\lambda \in \mathbb{C}$, it is in general *false* that

VI.4 Positive operators and the polar decomposition

$|AB| = |A| \, |B|$ or that $|A| = |A^*|$. Furthermore it is not true in general that $|A + B| \le |A| + |B|$ (Problem 16). In fact, while it is known that $|\cdot|$ is norm continuous (see Problem 15), it is not known whether it is Lipschitz, that is, whether $\| \, |A| - |B| \, \| \le c \, \|A - B\|$ for some constant c (however, see Problem 17).

The analogue of the complex numbers of modulus one is a little more complicated. At first one might expect that the unitary operators would be sufficient, but the following example shows that this is not the case.

Example Let A be the right shift operator on ℓ_2. Then $|A| = \sqrt{A^*A} = I$ so if we write $A = U|A|$ we must have $U = A$. However, A is not unitary since $(1, 0, 0, \ldots)$ is not in its range.

Definition An operator $U \in \mathcal{L}(\mathcal{H})$ is called an **isometry** if $\|Ux\| = \|x\|$ for all $x \in \mathcal{H}$. U is called a **partial isometry** if U is an isometry when restricted to the closed subspace $(\operatorname{Ker} U)^\perp$.

Thus, if U is a partial isometry, $\mathcal{H}$ can be written as $\mathcal{H} = \operatorname{Ker} U \oplus (\operatorname{Ker} U)^\perp$ and $\mathcal{H} = \operatorname{Ran} U \oplus (\operatorname{Ran} U)^\perp$ and U is a unitary operator between $(\operatorname{Ker} U)^\perp$, the **initial subspace** of U, and $\operatorname{Ran} U$, the **final subspace** of U. It is not hard to see that U^* is a partial isometry from $\operatorname{Ran} U$ to $(\operatorname{Ker} U)^\perp$ which acts as the inverse of the map $U \colon (\operatorname{Ker} U)^\perp \to \operatorname{Ran} U$.

Proposition Let U be a partial isometry. Then $P_i = U^*U$ and $P_f = UU^*$ are respectively the projections onto the initial and final subspaces of U. Conversely, if $U \in \mathcal{L}(\mathcal{H})$ with U^*U and UU^* projections, then U is a partial isometry.

The proof of the proposition is left to Problem 18. We are now ready to prove the analogue of the decomposition $z = |z| e^{i \arg z}$.

Theorem VI.10 (Polar decomposition) Let A be a bounded linear operator on a Hilbert space $\mathcal{H}$. Then there is a partial isometry U such that $A = U|A|$. U is uniquely determined by the condition that $\operatorname{Ker} U = \operatorname{Ker} A$. Moreover, $\operatorname{Ran} U = \overline{\operatorname{Ran} A}$.

Proof Define $U \colon \operatorname{Ran} |A| \to \operatorname{Ran} A$ by $U(|A| \psi) = A\psi$. Since

$$\| \, |A| \psi \|^2 = (\psi, |A|^2 \psi) = (\psi, A^*A\psi) = \|A\psi\|^2$$

VI: BOUNDED OPERATORS

U is well-defined, that is, if $|A|\psi = |A|\phi$ then $A\psi = A\phi$. U is isometric and so extends to an isometry of $\overline{\operatorname{Ran} |A|}$ to $\overline{\operatorname{Ran} A}$. Extend U to all of $\mathcal{H}$ by defining it to be zero on $(\operatorname{Ran} |A|)^\perp$. Since $|A|$ is self-adjoint, $(\operatorname{Ran} |A|)^\perp = \operatorname{Ker} |A|$. Furthermore, $|A|\psi = 0$ if and only if $A\psi = 0$ so that $\operatorname{Ker} |A| = \operatorname{Ker} A$. Thus $\operatorname{Ker} U = \operatorname{Ker} A$. Uniqueness is left to the reader. ∎

In Problem 20 of Chapter VII, the reader will prove that U is a strong limit of polynomials in A and A^* so that U is in the "von Neumann algebra" generated by A.

VI.5 Compact operators †

Many problems in classical mathematical physics can be handled by reformulating them in terms of integral equations. A famous example is the Dirichlet problem discussed at the end of this section. Consider the simple operator K, defined in $C[0, 1]$ by

$$(K\varphi)(x) = \int_0^1 K(x, y)\varphi(y)\, dy \qquad (\text{VI.4})$$

where the function $K(x, y)$ is continuous on the square $0 \le x, y \le 1$. $K(x, y)$ is called the **kernel** of the **integral operator** K. Since

$$|(K\varphi)(x)| \le \left(\sup_{0 \le x, y \le 1} |K(x, y)|\right)\left(\sup_{0 \le y \le 1} |\varphi(y)|\right)$$

we see that

$$\|K\varphi\|_\infty \le \left(\sup_{0 \le x, y \le 1} |K(x, y)|\right) \|\varphi\|_\infty$$

so K is a bounded operator on $C[0, 1]$. K has another property which is very important. Let B_M denote the functions φ in $C[0, 1]$ such that $\|\varphi\|_\infty \le M$. Since $K(x, y)$ is continuous on the square $0 \le x, y \le 1$ and since the square is compact, $K(x, y)$ is uniformly continuous. Thus, given an $\varepsilon > 0$, we can find $\delta > 0$ such that $|x - x'| < \delta$ implies $|K(x, y) - K(x', y)| < \varepsilon$ for all $y \in [0, 1]$. Thus, if $\varphi \in B_M$,

$$|(K\varphi)(x) - (K\varphi)(x')| \le \left(\sup_{y \in [0, 1]} |K(x, y) - K(x', y)|\right) \|\varphi\|_\infty$$

$$\le \varepsilon M$$

† A supplement to this section begins on p. 368.

Therefore the functions $K[B_M]$ are equicontinuous. Since they are also uniformly bounded by $\|K\|M$, we can use the Ascoli theorem (Theorem I.28) to conclude that for every sequence $\varphi_n \in B_M$, the sequence $K\varphi_n$ has a convergent subsequence (the limit may not be in $K[B_M]$). Another way of saying this is that the set $K[B_M]$ is **precompact**; that is, its closure is compact in $C[0, 1]$. It is clear that the choice of M was not important so what we have shown is that K takes bounded sets into precompact sets. It is this property which makes the so called "Fredholm alternative" hold for nice integral equations like (VI.4). This section is devoted to studying such operators.

Definition Let X and Y be Banach spaces. An operator $T \in \mathscr{L}(X, Y)$ is called **compact** (or completely continuous) if T takes bounded sets in X into precompact sets in Y. Equivalently, T is compact if and only if for every bounded sequence $\{x_n\} \subset X$, $\{Tx_n\}$ has a subsequence convergent in Y.

The integral operator (VI.4) is one example of a compact operator. Another class of examples is:

Example (finite rank operators) Suppose that the range of T is finite dimensional. That is, every vector in the range of T can be written $Tx = \sum_{i=1}^{N} \alpha_i y_i$, for some fixed family $\{y_i\}_{i=1}^{N}$ in Y. If x_n is any bounded sequence in X, the corresponding α_i^n are bounded since T is bounded. The usual subsequence trick allows one to extract a convergent subsequence from $\{Tx_n\}$ which proves that T is compact.

An important property of compact operators is given by (compare Problem 34):

Theorem VI.11 A compact operator maps weakly convergent *sequences* into norm convergent sequences.

Proof Suppose $x_n \xrightarrow{w} x$. By the uniform boundedness theorem, the $\|x_n\|$ are bounded. Let $y_n = Tx_n$. Then $\ell(y_n) - \ell(y) = (T'\ell)(x_n - x)$ for any $\ell \in Y^*$. Thus, y_n converges weakly to $y = Tx$ in Y. Suppose that y_n does not converge to y in norm. Then, there is an $\varepsilon > 0$ and a subsequence $\{y_{n_k}\}$ of $\{y_n\}$ so that $\|y_{n_k} - y\| \geq \varepsilon$. Since the sequence $\{x_{n_k}\}$ is bounded and T is compact $\{y_{n_k}\}$ has a subsequence which converges to a $\tilde{y} \neq y$. This subsequence must then also converge weakly to $\tilde{y}$, but this is impossible since y_n converges weakly to y. Thus y_n converges to y in norm. ∎

We note that if X is reflexive then the converse of Theorem VI.11 holds (Problem 20). The following theorem is important since one can use it to prove that an operator is compact by exhibiting it as a norm limit of compact operators or as an adjoint of a compact operator.

Theorem VI.12 Let X and Y be Banach spaces, $T \in \mathscr{L}(X, Y)$.

(a) If $\{T_n\}$ are compact and $T_n \to T$ in the norm topology, then T is compact.

(b) T is compact if and only if T' is compact.

(c) If $S \in \mathscr{L}(Y, Z)$ with Z a Banach space and if T or S is compact, then ST is compact.

Proof (a) Let $\{x_m\}$ be a sequence in the unit ball of X. Since T_n is compact for each n, we can use the diagonalization trick of I.5 to find a subsequence of $\{x_m\}$, call it $\{x_{m_k}\}$, so that $T_n x_{m_k} \to y_n$ for each n as $k \to \infty$. Since $\|x_{m_k}\| \leq 1$ and $\|T_n - T\| \to 0$, an $\varepsilon/3$-argument shows that the sequence $\{y_n\}$ is Cauchy, so $y_n \to y$. It is not difficult to show using an $\varepsilon/3$ argument that $T x_{m_k} \to y$. Thus T is compact.

(b) See the Notes and Problem 36.

(c) The proof is elementary (Problem 37). ∎

We are mostly interested in the case where T is a compact operator from a separable Hilbert space to itself, so we will not pursue the general case any further (however, see the discussion in the Notes). We denote the Banach space of compact operators on a separable Hilbert space by $\text{Com}(\mathscr{H})$. By the first example and Theorem VI.12 the norm limit of a sequence of finite rank operators is compact. The converse is also true in the Hilbert space case.

Theorem VI.13 Let $\mathscr{H}$ be a separable Hilbert space. Then every compact operator on $\mathscr{H}$ is the norm limit of a sequence of operators of finite rank.

Proof Let $\{\varphi_j\}_{j=1}^\infty$ be an orthonormal set in $\mathscr{H}$. Define

$$\lambda_n = \sup_{\substack{\psi \in [\varphi_1, \ldots, \varphi_n]^\perp \\ \|\psi\| = 1}} \|T\psi\|$$

Clearly, $\{\lambda_n\}$ is monotone decreasing so it converges to a limit $\lambda \geq 0$. We first show that $\lambda = 0$. Choose a sequence $\psi_n \in [\varphi_1, \ldots, \varphi_n]^\perp$, $\|\psi_n\| = 1$, with

$\|T\psi_n\| \geq \lambda/2$. Since $\psi_n \xrightarrow{w} 0$, $T\psi_n \to 0$ by Theorem VI.11. Thus, $\lambda = 0$. As a result

$$\sum_{j=1}^{n} (\varphi_j, \cdot) T\varphi_j \to T$$

in norm since λ_n is just the norm of the difference. ∎

We have discussed a wide variety of properties of compact operators but we have not yet described any property which explains our special interest in them. The basic principle which makes compact operators important is the Fredholm alternative: If A is compact, then either $A\psi = \psi$ has a solution or $(I - A)^{-1}$ exists. This is not a property shared by all bounded linear transformations. For example, if A is the operator $(A\varphi)(x) = x\varphi(x)$ on $L^2[0, 2]$, then $A\varphi = \varphi$ has no solutions but $(I - A)^{-1}$ does not exist (as a bounded operator). In terms of "solving equations" the Fredholm alternative is especially nice: It tells us that if for any φ there is at most one ψ with $\psi = \varphi + A\psi$, then there is always exactly one. That is, compactness and uniqueness together imply existence; for an example, see the discussion of the Dirichlet problem at the end of the section.

As one might expect, since the Fredholm alternative holds for finite-dimensional matrices, it is possible to prove the Fredholm alternative for compact operators (in the Hilbert space case) by using the fact that any compact operator A can be written as $A = F + R$ where F has finite rank and R has small norm. Compactness combines very nicely with analyticity so we first prove an elegant result which is of great use in itself (see Sections XI.6, XI.7, XIII.4, and XIII.5).

Theorem VI.14 (analytic Fredholm theorem)   Let D be an open connected subset of $\mathbb{C}$. Let $f: D \to \mathscr{L}(\mathscr{H})$ be an analytic operator-valued function such that $f(z)$ is compact for each $z \in D$. Then, either

(a) $(I - f(z))^{-1}$ exists for no $z \in D$.

(b) $(I - f(z))^{-1}$ exists for all $z \in D\backslash S$ where S is a discrete subset of D (i.e. a set which has no limit points in D). In this case, $(I - f(z))^{-1}$ is meromorphic in D, analytic in $D\backslash S$, the residues at the poles are finite rank operators, and if $z \in S$ then $f(z)\psi = \psi$ has a nonzero solution in $\mathscr{H}$.

Proof   We will prove that near any z_0 either (a) or (b) holds. A simple connectedness argument allows one to convert this into a statement about all of D

(Problem 21). Given $z_0 \in D$, choose an r so that $|z - z_0| < r$ implies $\|f(z) - f(z_0)\| < \frac{1}{2}$ and pick F, an operator with finite rank so that

$$\|f(z_0) - F\| < \frac{1}{2}$$

Then, for $z \in D_r$, the disc of radius r about z_0, $\|f(z) - F\| < 1$. By expanding in a geometric series we see that $(I - f(z) + F)^{-1}$ exists and is analytic.

Since F has finite rank, there are independent vectors $\psi_1, \ldots, \psi_N$ so that $F(\varphi) = \sum_{i=1}^{N} \alpha_i(\varphi)\psi_i$. The $\alpha_i(\cdot)$ are bounded linear functionals on $\mathcal{H}$ so by the Riesz lemma there are vectors $\phi_1, \ldots, \phi_N$ so that $F(\varphi) = \sum_{i=1}^{N} (\phi_i, \varphi)\psi_i$ for all $\varphi \in \mathcal{H}$. Let $\phi_n(z) = ((I - f(z) + F)^{-1})^*\phi_n$ and

$$g(z) = F(I - f(z) + F)^{-1} = \sum_{n=1}^{N} (\phi_n(z), \cdot)\psi_n$$

By writing

$$(I - f(z)) = (I - g(z))(I - f(z) + F)$$

we see that $I - f(z)$ is invertible for $z \in D_r$ if and only if $I - g(z)$ is invertible and that $\psi = f(z)\psi$ has a nonzero solution if and only if $\varphi = g(z)\varphi$ has a solution.

If $g(z)\varphi = \varphi$, then $\varphi = \sum_{n=1}^{N} \beta_n \psi_n$ and the β_n satisfy

$$\beta_n = \sum_{m=1}^{N} (\phi_n(z), \psi_m)\beta_m \tag{VI.5a}$$

Conversely, if (VI.5a) has a solution $\langle \beta_1, \ldots, \beta_N \rangle$, then $\varphi = \sum_{n=1}^{N} \beta_n \psi_n$ is a solution of $g(z)\varphi = \varphi$. Thus $g(z)\varphi = \varphi$ has a solution if and only if the determinant

$$d(z) = \det\{\delta_{nm} - (\phi_n(z), \psi_m)\} = 0$$

Since $(\phi_n(z), \psi_m)$ is analytic in D_r so is $d(z)$ which means that either $S_r = \{z \mid z \in D_r, d(z) = 0\}$ is a discrete set in D_r or $S_r = D_r$. Now, suppose $d(z) \neq 0$. Then, given ψ, we can solve $(I - g(z))\varphi = \psi$ by setting $\varphi = \psi + \sum_{n=1}^{N} \beta_n \psi_n$ if we can find β_n satisfying

$$\beta_n = (\phi_n(z), \psi) + \sum_{m=1}^{N} (\phi_n(z), \psi_m)\beta_m \tag{VI.5b}$$

But, since $d(z) \neq 0$, this equation has a solution. Thus $(I - g(z))^{-1}$ exists if and only if $z \notin S_r$.

The meromorphic nature of $(I - f(z))^{-1}$ and the finite rank residues follow from the fact that there is an explicit formula for the β_n in (VI.5b) in terms of cofactor matrices. ∎

This theorem has four important consequences:

Corollary (the Fredholm alternative) If A is a compact operator on $\mathscr{H}$, then either $(I - A)^{-1}$ exists or $A\psi = \psi$ has a solution.

Proof Take $f(z) = zA$ and apply the last theorem at $z = 1$. ∎

Theorem VI.15 (Riesz–Schauder theorem) Let A be a compact operator on $\mathscr{H}$, then $\sigma(A)$ is a discrete set having no limit points except perhaps $\lambda = 0$. Further, any nonzero $\lambda \in \sigma(A)$ is an eigenvalue of finite multiplicity (i.e. the corresponding space of eigenvectors is finite dimensional).

Proof Let $f(z) = zA$. Then $f(z)$ is an analytic compact operator-valued function on the entire plane. Thus $\{z \mid zA\psi = \psi \text{ has a solution } \psi \neq 0\}$ is a discrete set (it is not the entire plane since it does not contain $z = 0$) and if $1/\lambda$ is not in this discrete set then

$$(\lambda - A)^{-1} = \frac{1}{\lambda}\left(I - \frac{1}{\lambda}A\right)^{-1}$$

exists. The fact that the nonzero eigenvalues have finite multiplicity follows immediately from the compactness of A. ∎

Theorem VI.16 (the Hilbert–Schmidt theorem) Let A be a self-adjoint compact operator on $\mathscr{H}$. Then, there is a complete orthonormal basis, $\{\phi_n\}$, for $\mathscr{H}$ so that $A\phi_n = \lambda_n \phi_n$ and $\lambda_n \to 0$ as $n \to \infty$.

Proof For each eigenvalue of A choose an orthonormal basis for the set of eigenvectors corresponding to the eigenvalue. The collection of all these vectors, $\{\phi_n\}$, is an orthonormal set since eigenvectors corresponding to distinct eigenvalues are orthogonal. Let $\mathscr{M}$ be the closure of the span of $\{\phi_n\}$. Since A is self-adjoint and $A: \mathscr{M} \to \mathscr{M}$, $A: \mathscr{M}^\perp \to \mathscr{M}^\perp$. Let $\tilde{A}$ be the restriction of A to $\mathscr{M}^\perp$. Then $\tilde{A}$ is self-adjoint and compact since A is. By the Riesz–Schauder theorem, if any $\lambda \neq 0$ is in $\sigma(\tilde{A})$, it is an eigenvalue of $\tilde{A}$ and thus of A. Therefore the spectral radius of $\tilde{A}$ is zero since the eigenvectors of A are in $\mathscr{M}$. Because $\tilde{A}$ is self-adjoint, it is the zero operator on $\mathscr{M}^\perp$ by Theorem VI.6. Thus, $\mathscr{M}^\perp = \{0\}$ since if $\varphi \in \mathscr{M}^\perp$, then $A\varphi = 0$ which implies that $\varphi \in \mathscr{M}$. Therefore, $\mathscr{M} = \mathscr{H}$.

The fact that $\lambda_n \to 0$ is a consequence of the first part of the Riesz–Schauder theorem which says that each nonzero eigenvalue has finite multiplicity and the only possible limit point of the λ_n is zero. ∎

Theorem VI.17 (canonical form for compact operators) Let A be a compact operator on $\mathscr{H}$. Then there exist (not necessarily complete) orthonormal sets $\{\psi_n\}_{n=1}^N$ and $\{\phi_n\}_{n=1}^N$ and positive real numbers $\{\lambda_n\}_{n=1}^N$ with $\lambda_n \to 0$ so that

$$A = \sum_{n=1}^{N} \lambda_n (\psi_n, \cdot) \phi_n \qquad (VI.6)$$

The sum in (VI.6), which may be finite or infinite, converges in norm. The numbers, $\{\lambda_n\}$, are called the **singular values** of A.

Proof Since A is compact, so is A^*A (Theorem VI.12). Thus A^*A is compact and self-adjoint. By the Hilbert–Schmidt theorem, there is an orthonormal set $\{\psi_n\}_{n=1}^N$ so that $A^*A\psi_n = \mu_n \psi_n$ with $\mu_n \neq 0$ and so that A^*A is the zero operator on the subspace orthogonal to $\{\psi_n\}_{n=1}^N$. Since A^*A is positive, each $\mu_n > 0$. Let λ_n be the positive square root of μ_n and set $\phi_n = A\psi_n / \lambda_n$. A short calculation shows that the ϕ_n are orthonormal and that

$$A\psi = \sum_{n=1}^{N} \lambda_n (\psi_n, \psi) \phi_n \quad \blacksquare$$

The proof shows that the singular values of A are precisely the eigenvalues of $|A|$.

We conclude with a classical example.

Example (Dirichlet problem)   The main impetus for the study of compact operators arose from the use of integral equations in attempting to solve the classical boundary value problems of mathematical physics. We briefly describe this method. Let D be an open bounded region in $\mathbb{R}^3$ with a smooth boundary surface ∂D. The Dirichlet problem for Laplace's equation is: given a continuous function f on ∂D, find a function u, twice differentiable in D and continuous on $\bar{D}$, which satisfies

$$\Delta u(x) = 0 \qquad x \in D$$
$$u(x) = f(x) \qquad x \in \partial D$$

Let $K(x, y) = (x - y, n_y)/2\pi |x - y|^3$ where n_y is the outer normal to ∂D at the point $y \in \partial D$. Then, as a function of x, $K(x, y)$ satisfies $\Delta_x K(x, y) = 0$ in the interior which suggests that we try to write u as a superposition

$$u(x) = \int_{\partial D} K(x, y) \varphi(y) \, dS(y) \qquad (VI.6a)$$

where $\varphi(y)$ is some continuous function on ∂D and dS is the usual surface measure. Indeed, for $x \in D$, the integral makes perfectly good sense and

$\Delta u(x) = 0$ in D. Furthermore, if x_0 is any point in ∂D and $x \to x_0$ from inside D, it can be proven that

$$u(x) \to -\varphi(x_0) + \int_{\partial D} K(x_0, y)\varphi(y)\, dS(y) \qquad \text{(VI.6b)}$$

If $x \to x_0$ from outside D, the minus is replaced by a plus. Also,

$$\int_{\partial D} K(x_0, y)\varphi(y)\, dS(y)$$

exists and is a continuous function on ∂D if φ is a continuous function on ∂D. The proof depends on the fact that the boundary of D is smooth which implies that for $x, y \in \partial D$, $(x - y, n_y) \approx c|x - y|^2$ as $x \to y$.

Since we wish $u(x) = f(x)$ on ∂D, the whole question reduces to whether we can find φ so that

$$f(x) = -\varphi(x) + \int_{\partial D} K(x, y)\varphi(y)\, dS(y), \qquad x \in \partial D$$

Let $T: C(\partial D) \to C(\partial D)$ be defined by

$$T\varphi = \int_{\partial D} K(x, y)\varphi(y)\, dS(y)$$

Not only is T bounded but (as we will shortly see) T is also compact. Thus, by the Fredholm alternative, either $\lambda = 1$ is in the point spectrum of T in which case there is a $\psi \in C(\partial D)$ such that $(I - T)\psi = 0$, or $-f = (I - T)\varphi$ has a unique solution for each $f \in C(\partial D)$. If u is defined by (VI.6a) with ψ replacing φ, then $u \equiv 0$ in D by the maximum principle. Further, $\partial u/\partial n$ is continuous across ∂D and therefore equals zero on ∂D. By an integration by parts this implies that $u \equiv 0$ outside ∂D. Therefore, by (VI.6b), $2\psi(x) \equiv 0$ on ∂D, so the first alternative does not hold.

The idea of the compactness proof is the following. Let

$$K_\delta(x, z) = \frac{(x - z, n_z)}{|x - z|^3 + \delta}$$

If $\delta > 0$, the kernel K_δ is continuous, so, by the discussion at the beginning of this section, the corresponding integral operators T_δ, are compact. To prove that T is compact, we need only show that $\|T - T_\delta\| \to 0$ as $\delta \to 0$. By the estimate

$$|(T_\delta f)(x) - (Tf)(x)| \leq \|f\|_\infty \int_{\partial D} |K(x, z) - K_\delta(x, z)|\, dS(z)$$

we must only show that the integral converges to zero uniformly in x as $\delta \to 0$. To prove this, divide the integration region into the set where $|x-z| \geq \varepsilon$ and its complement. For fixed ε, the kernels converge uniformly on the first region. By using the fact that K is integrable, the contribution from the second region can be made arbitrarily small for ε sufficiently small.

VI.6 The trace class and Hilbert–Schmidt ideals

In the last section we saw that compact operators have many nice properties and are useful for applications. It is therefore important to have effective criteria for determining when a given operator is compact or, better yet, general statements about whole classes of operators. In this section we will prove that the integral operator

$$(Tf)(x) = \int_M K(x, y) f(y) \, d\mu(y)$$

on $L^2(M, d\mu)$ is compact if $K(\cdot, \cdot) \in L^2(M \times M, d\mu \otimes d\mu)$. First we will develop the trace, a tool which is of great interest in itself. Theorem VI.1 shows that $\text{Com}(\mathcal{H})$, the compact operators on a separable Hilbert space $\mathcal{H}$ form a Banach space. At the conclusion of the section, we will compute the dual and double dual of $\text{Com}(\mathcal{H})$. These calculations illustrate the difference between the weak Banach space topology on $\mathcal{L}(\mathcal{H})$ and the weak operator topology and give a foretaste of the structure of abstract von Neumann algebras which we will study later.

The trace is a generalization of the usual notion of the sum of the diagonal elements of a matrix, but because infinite sums are involved, not all operators will have a trace. The construction of the trace is analogous to the construction of the Lebesgue integral where one first defines $\int f \, d\mu$ for $f \geq 0$; it has values in $[0, \infty]$, including ∞. Then $\mathcal{L}^1$ is defined as those f so that $\int |f| \, d\mu < \infty$. $\mathcal{L}^1$ is a vector space and $f \mapsto \int f \, d\mu$ a linear functional. Similarly we first define the trace, $\text{tr}(\cdot)$, on the positive operators; $A \to \text{tr } A$ has values in $[0, \infty]$. We then define the **trace class**, $\mathcal{I}_1$, to be all $A \in \mathcal{L}(\mathcal{H})$ such that $\text{tr } |A| < \infty$. We will then show that $\text{tr}(\cdot)$ is a linear functional on $\mathcal{I}_1$ with the right properties.

Theorem VI.18 Let $\mathcal{H}$ be a separable Hilbert space, $\{\varphi_n\}_{n=1}^\infty$ an orthonormal basis. Then for any positive operator $A \in \mathcal{L}(\mathcal{H})$ we define $\text{tr } A = \sum_{n=1}^\infty (\varphi_n, A\varphi_n)$. The number $\text{tr } A$ is called the **trace of** A and is independent of the orthonormal basis chosen. The trace has the following properties:

(a) $\operatorname{tr}(A + B) = \operatorname{tr} A + \operatorname{tr} B$.
(b) $\operatorname{tr}(\lambda A) = \lambda \operatorname{tr} A$ for all $\lambda \geq 0$.
(c) $\operatorname{tr}(UAU^{-1}) = \operatorname{tr} A$ for any unitary operator U.
(d) If $0 \leq A \leq B$, then $\operatorname{tr} A \leq \operatorname{tr} B$.

Proof Given an orthonormal basis $\{\varphi_n\}_{n=1}^\infty$, define $\operatorname{tr}_\varphi(A) = \sum_{n=1}^\infty (\varphi_n, A\varphi_n)$. If $\{\psi_m\}_{m=1}^\infty$ is another orthonormal basis then

$$\operatorname{tr}_\varphi(A) = \sum_{n=1}^\infty (\varphi_n, A\varphi_n) = \sum_{n=1}^\infty \|A^{1/2}\varphi_n\|^2$$

$$= \sum_{n=1}^\infty \left(\sum_{m=1}^\infty |(\psi_m, A^{1/2}\varphi_n)|^2 \right)$$

$$= \sum_{m=1}^\infty \left(\sum_{n=1}^\infty |(A^{1/2}\psi_m, \varphi_n)|^2 \right)$$

$$= \sum_{m=1}^\infty \|A^{1/2}\psi_m\|^2$$

$$= \sum_{m=1}^\infty (\psi_m, A\psi_m)$$

$$= \operatorname{tr}_\psi(A)$$

Since all the terms are positive, interchanging the sums is allowed.

Properties (a), (b), and (d) are obvious. To prove (c) we note that if $\{\varphi_n\}$ is an orthonormal basis, then so is $\{U\varphi_n\}$. Thus,

$$\operatorname{tr}(UAU^{-1}) = \operatorname{tr}_{(U\varphi)}(UAU^{-1}) = \operatorname{tr}_\varphi(A) = \operatorname{tr}(A). \blacksquare$$

Definition An operator $A \in \mathscr{L}(\mathscr{H})$ is called **trace class** if and only if $\operatorname{tr}|A| < \infty$. The family of all trace class operators is denoted by $\mathscr{I}_1$.

The basic properties of $\mathscr{I}_1$ are given in the following:

Theorem VI.19 $\mathscr{I}_1$ is a $*$-ideal in $\mathscr{L}(\mathscr{H})$, that is,

(a) $\mathscr{I}_1$ is a vector space.
(b) If $A \in \mathscr{I}_1$ and $B \in \mathscr{L}(\mathscr{H})$, then $AB \in \mathscr{I}_1$ and $BA \in \mathscr{I}_1$.
(c) If $A \in \mathscr{I}_1$, then $A^* \in \mathscr{I}_1$.

Proof (a) Since $|\lambda A| = |\lambda| |A|$ for $\lambda \in \mathbb{C}$, $\mathscr{I}_1$ is closed under scalar multiplication. Now, suppose that A and B are in $\mathscr{I}_1$, we wish to prove that

VI: BOUNDED OPERATORS

$A + B \in \mathcal{I}_1$. Let U, V, and W be the partial isometries arising from the polar decompositions

$$A + B = U|A + B|$$
$$A = V|A|$$
$$B = W|B|$$

Then

$$\sum_{n=1}^{N} (\varphi_n, |A + B|\varphi_n) = \sum_{n=1}^{N} (\varphi_n, U^*(A + B)\varphi_n)$$
$$\leq \sum_{n=1}^{N} |(\varphi_n, U^*V|A|\varphi_n)| + \sum_{n=1}^{N} |(\varphi_n, U^*W|B|\varphi_n)|$$

However,

$$\sum_{n=1}^{N} |(\varphi_n, U^*V|A|\varphi_n)| \leq \sum_{n=1}^{N} \| \, |A|^{1/2} V^* U \varphi_n \| \, \| \, |A|^{1/2} \varphi_n \|$$
$$\leq \left(\sum_{n=1}^{N} \| \, |A|^{1/2} V^* U \varphi_n \|^2 \right)^{1/2} \left(\sum_{n=1}^{N} \| \, |A|^{1/2} \varphi_n \|^2 \right)^{1/2}$$

Thus, if we can show

$$\sum_{n=1}^{N} \| \, |A|^{1/2} V^* U \varphi_n \|^2 \leq \text{tr} \, |A| \tag{VI.7}$$

we can conclude that

$$\sum_{n=1}^{N} (\varphi_n, |A + B|\varphi_n) \leq \text{tr}\,|A| + \text{tr}\,|B| < \infty$$

and thus $A + B \in \mathcal{I}_1$. To show (VI.7), we need only prove that

$$\text{tr}(U^*V|A|V^* U) \leq \text{tr}|A|$$

Picking an orthonormal basis, $\{\varphi_n\}$ with each φ_n in Ker U or (Ker $U)^\perp$ we see that $\text{tr}(U^*(V|A|V^*)U) \leq \text{tr}(V|A|V^*)$. Similarly, picking an orthonormal basis, $\{\psi_m\}$, with each ψ_m in Ker V^* or (Ker $V^*)^\perp$ we find $\text{tr}(V|A|V^*) \leq \text{tr}\,|A|$.

 (b) By the lemma proven below, each $B \in \mathcal{L}(\mathcal{H})$ can be written as a linear combination of four unitary operators so by (a) we need only show that $A \in \mathcal{I}_1$ implies $UA \in \mathcal{I}_1$ and $AU \in \mathcal{I}_1$ if U is unitary. But $|UA| = |A|$ and $|AU| = U^{-1}|A|U$, so by part (c) of Theorem VI.18, AU and UA are in $\mathcal{I}_1$.

 (c) Let $A = U|A|$ and $A^* = V|A^*|$ be the polar decompositions of A and A^*. Then $|A^*| = V^*|A|U^*$. If $A \in \mathcal{I}_1$, then $|A| \in \mathcal{I}_1$, so by part (b) $|A^*| \in \mathcal{I}_1$ and $A^* = V|A^*| \in \mathcal{I}_1$. ∎

VI.6 The trace class and Hilbert–Schmidt ideals

To complete the proof of part (b) above we need the following lemma which we will use in other contexts later.

Lemma Every $B \in \mathcal{L}(\mathcal{H})$ can be written as a linear combination of four unitary operators.

Proof Since $B = \frac{1}{2}(B + B^*) - \frac{i}{2}[i(B - B^*)]$, B can be written as a linear combination of two self-adjoint operators. So, suppose A is self-adjoint and without loss of generality assume $\|A\| \leq 1$. Then $A \pm i\sqrt{I - A^2}$ are unitary and $A = \frac{1}{2}(A + i\sqrt{I - A^2}) + \frac{1}{2}(A - i\sqrt{I - A^2})$. ∎

The proof of the following theorem is left to the reader (Problem 23).

Theorem VI.20 Let $\|\cdot\|_1$ be defined in $\mathcal{I}_1$ by $\|A\|_1 = \text{tr}\,|A|$. Then $\mathcal{I}_1$ is a Banach space with norm $\|\cdot\|_1$ and $\|A\| \leq \|A\|_1$.

We note that $\mathcal{I}_1$ is *not* closed under the operator norm $\|\cdot\|$. The connection between the trace class operators and the compact operators is simple:

Theorem VI.21 Every $A \in \mathcal{I}_1$ is compact. A compact operator A is in $\mathcal{I}_1$ if and only if $\sum_{n=1}^{\infty} \lambda_n < \infty$ where $\{\lambda_n\}_{n=1}^{\infty}$ are the singular values of A.

Proof Since $A \in \mathcal{I}_1$, $|A|^2 \in \mathcal{I}_1$, so $\text{tr}(|A|^2) = \sum_{n=1}^{\infty} \|A\varphi_n\|^2 < \infty$ for any orthonormal basis $\{\varphi_n\}_{n=1}^{\infty}$. Suppose $\psi \in [\varphi_1, \ldots, \varphi_N]^{\perp}$ and $\|\psi\| = 1$, then we have

$$\|A\psi\|^2 \leq \text{tr}(|A|^2) - \sum_{n=1}^{N} \|A\varphi_n\|^2$$

since $\{\varphi_1, \varphi_2, \ldots, \varphi_N, \psi\}$ can always be completed to an orthonormal basis. Thus

$$\sup\{\|A\psi\| \,|\, \psi \in [\varphi_1, \ldots, \varphi_N]^{\perp}, \|\psi\| = 1\} \to 0 \quad \text{as} \quad N \to \infty$$

Therefore $\sum_{n=1}^{N} (\varphi_n, \cdot)A\varphi_n$ is norm convergent to A. Thus A is compact. The second part of the theorem follows easily from the canonical form derived in Theorem VI.17 (Problem 24). ∎

Corollary The finite rank operators are $\|\cdot\|_1$-dense in $\mathcal{I}_1$.

The second class of operators which we will discuss are the Hilbert–Schmidt operators, the analogue of $\mathcal{L}^2$.

Definition An operator $T \in \mathscr{L}(\mathscr{H})$ is called **Hilbert–Schmidt** if and only if tr $T^*T < \infty$. The family of all Hilbert–Schmidt operators is denoted by $\mathscr{I}_2$.

By arguments analogous to those we used for $\mathscr{I}_1$, one can prove.

Theorem VI.22 (a) $\mathscr{I}_2$ is a $*$-ideal.
 (b) If $A, B \in \mathscr{I}_2$, then for any orthonormal basis $\{\varphi_n\}$,

$$\sum_{n=1}^{\infty} (\varphi_n, A^*B\varphi_n)$$

is absolutely summable, and its limit, denoted by $(A, B)_2$, is independent of the orthonormal basis chosen.
 (c) $\mathscr{I}_2$ with inner product $(\cdot, \cdot)_2$ is a Hilbert space.
 (d) If $\|A\|_2 = \sqrt{(A, A)_2} = (\operatorname{tr}(A^*A))^{1/2}$, then

$$\|A\| \leq \|A\|_2 \leq \|A\|_1 \quad \text{and} \quad \|A\|_2 = \|A^*\|_2$$

 (e) Every $A \in \mathscr{I}_2$ is compact and a compact operator, A, is in $\mathscr{I}_2$ if and only if $\sum_{n=1}^{\infty} \lambda_n^2 < \infty$ where λ_n are the singular values of A.
 (f) The finite rank operators are $\|\cdot\|_2$-dense in $\mathscr{I}_2$.
 (g) $A \in \mathscr{I}_2$ if and only if $\{\|A\varphi_n\|\} \in \ell_2$ for some orthonormal basis $\{\varphi_n\}$.
 (h) $A \in \mathscr{I}_1$ if and only if $A = BC$ with B, C in $\mathscr{I}_2$.

We note that $\mathscr{I}_2$ is not $\|\cdot\|$-closed. The important fact about $\mathscr{I}_2$ is that when $\mathscr{H} = L^2(M, d\mu)$, $\mathscr{I}_2$ has a concrete realization.

Theorem VI.23 Let $\langle M, \mu \rangle$ be a measure space and $\mathscr{H} = L^2(M, d\mu)$. Then $A \in \mathscr{L}(\mathscr{H})$ is Hilbert–Schmidt if and only if there is a function

$$K \in L^2(M \times M, d\mu \otimes d\mu)$$

with

$$(Af)(x) = \int K(x, y) f(y) \, d\mu(y)$$

Moreover,

$$\|A\|_2^2 = \int |K(x, y)|^2 \, d\mu(x) \, d\mu(y)$$

Proof Let $K \in L^2(M \times M, d\mu \otimes d\mu)$ and let A_K be the associated integral operator. It is easy to see (Problem 25) that A_K is a well-defined operator on $\mathscr{H}$ and that

$$\|A_K\| \leq \|K\|_{L^2} \tag{VI.8}$$

VI.6 The trace class and Hilbert–Schmidt ideals

Let $\{\varphi_n\}_{n=1}^\infty$ be an orthonormal basis for $L^2(M, d\mu)$. Then $\{\varphi_n(x)\overline{\varphi_m(y)}\}_{n,m=1}^\infty$ is an orthonormal base for $L^2(M \times M, d\mu \otimes d\mu)$ so

$$K = \sum_{n,m=1}^\infty \alpha_{nm} \varphi_n(x)\overline{\varphi_m(y)}$$

Let

$$K_N = \sum_{n,m=1}^N \alpha_{n,m} \varphi_n(x)\overline{\varphi_m(y)}$$

Then each K_N is the integral kernel of a finite rank operator. In fact, $A_{K_N} = \sum_{n,m=1}^N \alpha_{nm}(\varphi_m, \cdot)\varphi_n$. Since $\|K_N - K\|_{L^2} \to 0$ we have $\|A_K - A_{K_N}\| \to 0$ as $N \to \infty$ by (VI.8). Thus A_K is compact and in fact

$$\operatorname{tr}(A_K^* A_K) = \sum_{n=1}^\infty \|A_K \varphi_n\|^2 = \sum_{n=1}^\infty \sum_{m=1}^\infty |\alpha_{nm}|^2 = \|K\|_{L^2}$$

Thus $A_K \in \mathcal{J}_2$ and $\|A_K\|_2 = \|K\|_{L^2}$.

We have shown that the map $K \to A_K$ is an isometry of $L^2(M \times M, d\mu \otimes d\mu)$ into $\mathcal{J}_2$, so its range is closed. But the finite rank operators clearly come from kernels and since they are dense in $\mathcal{J}_2$ the range of $K \mapsto A_K$ is all of $\mathcal{J}_2$. ∎

This theorem provides a simple sufficient condition for an operator to be compact and is therefore very useful. Notice that the condition is not necessary. Also, we have a sufficient condition for an operator on $\mathcal{H} = L^2(M, d\mu)$ to be an integral operator. This condition is also not necessary. Now, we return to defining the trace on $\mathcal{J}_1$.

Theorem VI.24 If $A \in \mathcal{J}_1$ and $\{\varphi_n\}_{n=1}^\infty$ is any orthonormal basis, then $\sum_{n=1}^\infty (\varphi_n, A\varphi_n)$ converges absolutely and the limit is independent of the choice of basis.

Proof We write $A = U|A|^{1/2}|A|^{1/2}$. Then

$$|(\varphi_n, A\varphi_n)| \leq \| |A|^{1/2} U^* \varphi_n\| \, \| |A|^{1/2}\varphi_n\|$$

Thus

$$\sum_{n=1}^\infty |(\varphi_n, A\varphi_n)| \leq \left(\sum_{n=1}^\infty \| |A|^{1/2} U^* \varphi_n\|^2\right)^{1/2} \left(\sum_{n=1}^\infty \| |A|^{1/2}\varphi_n\|^2\right)^{1/2}$$

so since $|A|^{1/2} U^*$ and $|A|^{1/2}$ are in $\mathcal{J}_2$, the sum converges. The proof of the independence of basis is identical to that for $\operatorname{tr} A$ when $A \geq 0$. ∎

Definition The map $\operatorname{tr}: \mathcal{J}_1 \to \mathbb{C}$ given by $\operatorname{tr} A = \sum_{n=1}^\infty (\varphi_n, A\varphi_n)$ where $\{\varphi_n\}$ is any orthonormal basis is called the **trace**.

We remark that it is not true that $\sum_{n=1}^{\infty} |(\varphi_n, A\varphi_n)| < \infty$ for some orthonormal basis implies $A \in \mathscr{I}_1$. For A to be in $\mathscr{I}_1$ the sum must be finite for all orthonormal bases. The spectral theorem which we will prove in the next chapter will tell us that any self-adjoint A can be written $A = A_+ - A_-$ where both A_+ and A_- are positive and $A_+ A_- = 0$. Not surprisingly, $A \in \mathscr{I}_1$ if and only if $\text{tr}(A_+) < \infty$, $\text{tr}(A_-) < \infty$ and in this case $\text{tr } A = \text{tr } A_+ - \text{tr } A_-$. We collect the properties of the trace.

Theorem VI.25 (a) $\text{tr}(\cdot)$ is linear.
(b) $\text{tr } A^* = \overline{\text{tr } A}$.
(c) $\text{tr } AB = \text{tr } BA$ if $A \in \mathscr{I}_1$ and $B \in \mathscr{L}(\mathscr{H})$.

Proof (a) and (b) are obvious. To prove (c) it is sufficient to consider the case where B is unitary since any bounded operator is the sum of four unitaries. In that case

$$\text{tr } AB = \sum_{n=1}^{\infty} (\varphi_n, AB\varphi_n)$$
$$= \sum_{n=1}^{\infty} (B^*\psi_n, A\psi_n)$$
$$= \sum_{n=1}^{\infty} (\psi_n, BA\psi_n)$$
$$= \text{tr } BA$$

where $\psi_n = B\varphi_n$ for all n. ∎

If $A \in \mathscr{I}_1$, the map $B \mapsto \text{tr } AB$ is a linear functional on $\mathscr{L}(\mathscr{H})$. These are not all the continuous linear functionals on $\mathscr{L}(\mathscr{H})$ but such functionals do yield the entire dual of $\text{Com}(\mathscr{H})$, the compact operators. We can also hold $B \in \mathscr{L}(\mathscr{H})$ fixed and obtain a linear functional on $\mathscr{I}_1$ given by the map $A \mapsto \text{tr } BA$. The set of these functionals is just the dual of $\mathscr{I}_1$ (with the operator norm topology). We state this as a theorem; the interested reader can follow the outline of the proof given in Problem 30.

Theorem VI.26 (a) $\mathscr{I}_1 = [\text{Com}(\mathscr{H})]^*$. That is, the map $A \mapsto \text{tr}(A \cdot)$ is an isometric isomorphism of $\mathscr{I}_1$ onto $[\text{Com }(\mathscr{H})]^*$.
(b) $\mathscr{L}(\mathscr{H}) = \mathscr{I}_1^*$. That is, the map $B \mapsto \text{tr}(B \cdot)$ is an isometric isomorphism of $\mathscr{L}(\mathscr{H})$ onto $\mathscr{I}_1^*$.

We now return to the distinction between the weak operator topology on $\mathscr{L}(\mathscr{H})$ (see Section VI.1) and the weak Banach space topology, i.e. the $\sigma(\mathscr{L}(\mathscr{H}), \mathscr{L}(\mathscr{H})^*)$ topology. If $\mathscr{F}$ is the family of finite rank operators, then $\mathscr{F} \subset \mathscr{I}_1$ and each $F \in \mathscr{F}$ can be realized as linear functional on $\mathscr{L}(\mathscr{H})$ via the dual action of $\mathscr{I}_1$ on $\mathscr{L}(\mathscr{H})$. The topology on $\mathscr{L}(\mathscr{H})$ generated by these functionals, that is, $\sigma(\mathscr{L}(\mathscr{H}), \mathscr{F})$, is just the weak operator topology. The set $\mathscr{F}$ is not closed in the $\mathscr{L}(\mathscr{H})^*$-norm. As a matter of fact, the $\mathscr{L}(\mathscr{H})^*$ norm on $\mathscr{F}$ is just $\|\cdot\|_1$ so the closure of $\mathscr{F}$ in this norm is just $\mathscr{I}_1$. The weak topology on $\mathscr{L}(\mathscr{H})$ generated by the functionals in $\mathscr{I}_1$, that is, $\sigma(\mathscr{L}(\mathscr{H}), \mathscr{I}_1)$, is called the **ultraweak** topology on $\mathscr{L}(\mathscr{H})$. Notice that it is stronger than the weak operator topology, since more functionals are required to be continuous, but weaker than the weak Banach space topology on $\mathscr{L}(\mathscr{H})$, since $\mathscr{I}_1$ is not the entire dual of $\mathscr{L}(\mathscr{H})$. In fact, since $\mathscr{L}(\mathscr{H}) = \mathscr{I}_1^*$, the ultraweak topology on $\mathscr{L}(\mathscr{H})$ is just the weak-$*$ topology. This realization of $\mathscr{L}(\mathscr{H})$ as the dual of the Banach space of linear functionals continuous in the $\sigma(\mathscr{L}(\mathscr{H}), \overline{\mathscr{F}})$ topology is valid for a larger class of algebras than just $\mathscr{L}(\mathscr{H})$. Problem 31 gives another example: the multiplication algebra L_∞ on L_2. We will study such algebras in detail in Chapter XVIII. We study the $\mathscr{I}_p$ spaces for $p \neq 1, 2, \infty$ in Sections IX.4 and XIII.17.

NOTES

Section VI.1 The reader may be bewildered by the many topologies we have introduced on $\mathscr{L}(\mathscr{H})$: the weak, strong, and uniform operator topologies, the weak Banach space topology, the ultraweak topology (Section VI.6). Later on we will even encounter the ultrastrong topology. Why is it necessary to introduce all these topologies? The answer is that many of the operators we are interested in are given as some sort of limit of simpler operators. It is important to know exactly what one means by "some sort" and to know what properties of the limiting operator follow from properties in the sequence, for example, the uniform limit of compact operators is compact. Furthermore, when one begins a problem one doesn't always know in what sense limits will exist, so it is useful to have a wide range of topologies at hand. In general it is the weak, strong, and uniform operator topologies which are important in Volumes I and II. The ultraweak and ultrastrong topologies will play a role when we deal with von Neumann algebras. The weak, strong, and ultrastrong operator topologies were introduced in J. von Neumann, "Zur Algebra der Functionaloperationen und Theorie der Normalen Operatoren," *Math. Ann.* **102** (1929–1930), 370–427.

Section VI.2 The spectral theorem for self-adjoint operators on finite dimensional vector spaces is nicely described in P. R. Halmos, *Finite Dimensional Vector Spaces*, Van Nostrand–Reinhold, Princeton, New Jersey, 1958.

Section VI.3 The definitions of various kinds of spectra will also be used for unbounded operators. Theorem VI.5 holds as long as we require that T be closed. If T is bounded it is, of course, automatically closed.

The theory of Banach space-valued analytic functions is described in great detail in *Functional Analysis and Semi-groups*, Amer. Math. Soc., Providence, Rhode Island, 1957, by E. Hille and R. S. Phillips. They also discuss the more difficult notion of analytic functions from one Banach space to another. A proof of Theorem VI.7 can be found in *Functional Analysis*, Academic Press, New York, 1965, by K. Yosida.

Some authors (for example: Yosida or Hille, Phillips) use the term "continuous spectrum" to denote any $\lambda \in \sigma(T)$ which is neither in the point spectrum, nor in the residual spectrum. Other authors (such as Kato or Riesz, Nagy) use the definition that we give in Section VII.2. One important distinction is that with our definition the continuous spectrum and the point spectrum need not be disjoint.

Section VI.4 The polar decomposition has a simple geometric meaning for linear transformations on $\mathbb{R}^n$. Any linear transformation A on $\mathbb{R}^n$ can be written as $A = OS$ where O is orthogonal and S is self-adjoint. By the spectral theorem, S can be thought of as a dilation, contraction, or annihilation in certain preferred orthogonal directions.

The notion of positivity has a natural generalization to operator algebras and will play an important role in our investigations in Volume III.

The statement that the triangle inequality fails for $|\cdot|$, that is, $|A + B|$ may *not* be less than or equal to $|A| + |B|$ (see Problem 16) is a statement that $f(x) = |x|$ is not a convex *operator-valued* function, that is, for $0 \leq t \leq 1$, $f(tA + (1-t)B) \leq tf(A) + (1-t)f(B)$ can be false for general operators A and B despite the fact that $f(tx - (1-t)y) \leq tf(x) + (1-t)f(y)$ is true for x and y real and $0 \leq t \leq 1$. Exactly which matrix and operator valued functions are convex has been studied in: F. Krauss, "Über konvexe Matrixfunktionen," *Math. Z.* **41** (1936) 18–42, and J. Bendat and S. Sherman, "Monotone and Convex Operator Functions," *Trans. Amer. Math. Soc.* **79** (1955), 58–71.

Section VI.5 The proof of the second part of Theorem VI.12 can be found in Yosida's book; it is a nice application of the Ascoli–Arzela and Alaoglu theorems (see also Problem 36).

In a very real sense, the theory of compact operators goes back to Fredholm's great paper on integral operators, "Sur une class d'équations fonctionnelles," *Acta Math.* **27** (1903), 365–390. Fredholm considered solving equations of the form

$$f(x) = g(x) + \lambda \int_a^b K(x, y) f(y) \, dy$$

where g and K are given continuous functions and $-\infty < a < b < \infty$. Fredholm showed that there exists an explicit entire function $d(\lambda)$, not identically zero, and an explicit function $D_\lambda(x, y)$, entire in λ and continuous in x and y, so that if $d(\lambda) \neq 0$, then $f(x) = g(x) + d(\lambda)^{-1} \int_a^b D_\lambda(x, y) g(y) \, dy$ solves the equation. Moreover, he showed that when $d(\lambda) = 0$, then

$$f(x) = \lambda \int_a^b K(x, y) f(y) \, dy$$

has a solution $f \neq 0$. Fredholm thus had Theorem VI.15 and the preceding corollary in this special case. Readable expositions of the Fredholm theory may be found in W. Lovitt, *Linear Integral Equations*, Dover, New York (reprinted 1950; original edition, McGraw-Hill,

New York, 1926), and F. Smithies, *Integral Equations*, Cambridge Univ. Press, London and New York, 1958.

Fredholm's work produced considerable interest among Hilbert and his school and led to the abstraction of many notions we now associate with Hilbert space theory. Hilbert first defined completely continuous operators in a manner whose modern form would be the criterion of Theorem VI.11: D. Hilbert, "Grundzüge einer allgemeinen Theorie der linearen Integralgleichungen, I–VI," *Nachr. Akad. Wiss. Göttingen Math.-Phys. Kl.* **49**–91 (1904), 213–259, 307–388 (1905); 157–222, 439–480 (1906); 355–417 (1910); esp. IV. The extension of the notion of compact operator to arbitrary Banach spaces by the precompactness criteria is due to F. Riesz "Über lineare Functionalgleichungen," *Acta. Math.* **41** (1918), 71–98.

Theorem VI.12b is due to J. Schauder: " Über lineare, vollstetige Functionaloperationen," *Studia Math.* **2** (1930), 183–196.

The idea of using Theorem IV.13 to develop the general theory is due to E. Schmidt, "Auflösung der allgemeinen linearen Integralgleichung," *Math. Ann.* **64** (1907), 161–174. While it is true that compact operators in most explicit Banach spaces are norm limits of finite rank operators, there are Banach spaces where this is false. The earliest examples were constructed by P. Enflo. For extensive discussion, see M. M. Day, *Normed Linear Spaces*, Springer, Berlin, 1973, and J. Lindenstrauss and L. Tzafriri, *Classical Banach Spaces*, Springer Lecture Notes in Math **388**, Springer-Verlag, 1973.

Theorem VI.14, its corollary, and Theorem VI.15 hold in an arbitrary Banach space. For their proof in that case, see N. Dunford and J. Schwartz, *Linear Operators*, Vol. I. Wiley (Interscience), 1958. Our technique of proof for Theorem VI.14 is taken from a technical appendix in W. Hunziker, "On the Spectra of Schrödinger Multiparticle Hamiltonians," *Helv. Phys. Acta.* **39** (1966), 451–462. A similar approach can be found in an appendix of G. Tiktopolous, "Analytic Continuation in Complex Angular Momentum and Integral Equations," *Phys. Rev.* **133B** (1964), 1231–1238. One part of Theorem VI.14 is not proven in the general case in Dunford-Schwartz; a discussion of this extra point can be found in S. Steinberg, " Meromorphic Families of Compact Operators," *Arch. Rat. Mech. Anal.* **31** (1968), 372–379. For extensions to locally convex spaces, see J. Leray, " Valeurs propres et vecteurs propres d'un endomorphisme conplètement continu d'un espace vectoriel à voisinages convexes," *Acta Sci. Math. Szcg.* **12**, Part B, (1950), 177–186. Theorem VI.15 was first proven by Riesz and Schauder in the above cited works (Schauder filled in some details for the general case) and Theorem VI.16 is due to Hilbert and Schmidt in the aforementioned papers.

For a discussion of the use of integral equations in the solution of Dirichlet problem, see *Boundary Value Problems of Mathematical Physics*, Vol. 2, (especially sections 6.4 and 6.5), Macmillan, New York, 1968, by Ivor Stakgold and Volume II of R. Courant and D. Hilbert, *Methods of Mathematical Physics*, Wiley (Interscience).

Section VI.6   For a discussion on $\mathscr{I}_1$, $\mathscr{I}_2$, and the $\mathscr{I}_p$ analogues, see R. Schatten, *Norm Ideals of Completely Continuous Operators*, Springer-Verlag, Berlin and New York, 1960. $\mathscr{I}_p$ is defined as those A with $\text{Tr}(|A|^p) < \infty$ and is equivalently those compact operators with $\sum |\lambda_n|^p < \infty$. For further discussion, see Sections IX.4 and XIII.17.

These norm ideals have been extended to other situations with traces (von Neumann algebras) and more general settings in a manner emphasizing the analogy with L^p by I. Segal: "A Non-Commutative Extension of Abstract Integration," *Ann. Math.* **57** (1953), 401–457; **58** (1953), 595–596, and R. A. Kunze, "L_p Fourier Transforms on Locally Compact Unimodular Groups," *Trans. Amer. Math. Soc.* **89** (1958), 519.

PROBLEMS

†*1.* Prove that the weak operator topology is weaker than the strong operator topology which is weaker than the uniform operator topology.

†*2.* Prove the statements in the example in Section VI.1.

3. (a) Let X and Y be Banach spaces. Prove that if $T_n \in \mathscr{L}(X, Y)$ and $\{T_n x\}$ is a Cauchy sequence for each $x \in X$, then there exists a $T \in \mathscr{L}(X, Y)$ so that $T_n \to T$ strongly.
  *(b) Is the theorem in (a) true if T_n is replaced by a net T_α?

4. (a) Let X and Y be Banach spaces. Prove that a theorem for $\mathscr{L}(X, Y)$ analogous to Theorem VI.1 holds if Y is weakly sequentially complete (which means that every weakly Cauchy sequence has a weak limit.)
  (b) Prove that if a Banach space is reflexive, then it is weakly sequentially complete.

5. (a) Let T_t be the operator $T_t: \varphi(x) \to \varphi(x + t)$ on $L^2(\mathbb{R})$. What is the norm of T_t? To what operator does T_t converge as $t \to \infty$ and in what topology?
  (b) Answer the same question for T_t if the Hilbert space is $L^2(\mathbb{R}, e^{-x^2}\, dx)$.

6. (a) Let $\mathscr{H}$ be an infinite dimensional Hilbert space. Suppose $\psi_1, \ldots, \psi_n$ orthonormal are given and that ε, ψ are given. Show there are A and B with $\|A\psi_i\| < \varepsilon$, $\|B\psi_i\| = \varepsilon$; $i = 1, \ldots, n$, but that $\|AB\psi\| > 1$.
  (b) Prove that multiplication from $\mathscr{L}(\mathscr{H}) \times \mathscr{L}(\mathscr{H}) \to \mathscr{L}(\mathscr{H})$ is not jointly continuous when $\mathscr{L}(\mathscr{H})$ is given the strong topology.
  (c) Suppose $\{A_\alpha\}_{\alpha \in I}$ and $\{B_\alpha\}_{\alpha \in I}$ are *nets*. Let $A_\alpha^* \xrightarrow{s} A^*$, $B_\alpha \xrightarrow{s} B$. Prove that $A_\alpha B_\alpha \xrightarrow{w} AB$.
  (d) Let A_n, B_n be *sequences* so that $A_n \xrightarrow{s} A$, $B_n \xrightarrow{s} B$. Prove that $A_n B_n \xrightarrow{s} AB$.
  (e) Let A_n, B_n be *sequences* so that $A_n \xrightarrow{w} A$, $B_n \xrightarrow{w} B$. Give an example where $A_n B_n \xrightarrow{w} AB$ is *false*.

7. Give an example to show that the range of a bounded operator need not be closed. Prove that if T is bounded, everywhere defined, and an isometry, then Ran T is closed.

†*8.* (a) Let A be a self-adjoint bounded operator on a Hilbert space. Prove that the eigenvalues of A are real and that the eigenvectors corresponding to distinct eigenvalues are orthogonal.
  (b) From the proof of Theorem VI.8 derive a universal (but λ-dependent) bound for the norm of the resolvent of a self-adjoint operator at a nonreal $\lambda \in \mathbb{C}$.

9. (a) Let A be a self-adjoint operator on a Hilbert space, $\mathscr{H}$. Prove that

$$\|A\| = \sup_{\|x\| = 1} |(Ax, x)|$$

Hint: First note that

$$\operatorname{Re}(\psi, A\phi) = \tfrac{1}{4}[(\psi + \phi, A(\psi + \phi)) - (\psi - \phi, A(\psi - \phi))]$$

Then using

$$|(\eta, A\eta)| \leq \|\eta\|^2 \sup_{\|\eta\| = 1} |(\eta, A\eta)|$$

and the parallelogram law, prove that

$$|(\psi, A\phi)| \leq \sup_{\|\eta\| = 1} |(\eta, A\eta)|$$

if $\|\varphi\| = \|\psi\| = 1$.

(b) Find an example which shows that the conclusion of (a) need not be true if A is not self-adjoint.

10. Show that the spectral radius of the Volterra integral operator

$$(Tf)(x) = \int_0^x f(y)\, dy$$

as a map on $C[0, 1]$ is equal to zero. What is the norm of T?

†*11*. Let $T \in \mathscr{L}(X)$. Prove that $\lim_{n \to \infty} \|T^n\|^{1/n}$ exists and is equal to $\inf_n \|T^n\|^{1/n}$ as follows:
   (a) Set $a_n = \log \|T^n\|$ and prove that $a_{m+n} \leq a_m + a_n$.
   (b) For a fixed positive integer m set $n = mq + r$ where q and r are positive integers and $0 \leq r \leq m - 1$. Using (a) conclude that

$$\varlimsup_n \frac{a_n}{n} \leq \frac{a_m}{m}$$

   (c) Prove that $\lim_{n \to \infty} a_n/n = \inf_n a_n/n$ and thus the desired equality.

†*12*. Prove the proposition at the end of Section VI.3.

13. (a) Give an example which shows that a linear transformation on $\mathbb{C}^n$ can be positive without all the entries in a given matrix representation being positive.
   *(b) Derive a necessary and sufficient condition for a $n \times n$ matrix to be positive.

14. (a) Prove that if $A_n \geq 0$, $A_n \to A$ in norm, then $\sqrt{A_n} \to \sqrt{A}$ norm.
   (b) Suppose $A_n \to A$ strongly for a sequence $\{A_n\}$. Prove that $\sqrt{A_n} \to \sqrt{A}$ strongly.

15. (a) Let $A_n \to A$ in norm. Prove that $|A_n| \to |A|$ in norm.
   (b) Suppose $A_n \to A$ and $A_n^* \to A^*$ strongly where A_n is a sequence. Prove that $|A_n| \to |A|$ strongly.
   (c) Find an example which shows that $|\cdot|$ is not weakly continuous on $\mathscr{L}(\mathscr{H})$.

16. Let $\sigma_3 = \begin{pmatrix} 1 & 0 \\ 0 & -1 \end{pmatrix}$, $\sigma_1 = \begin{pmatrix} 0 & 1 \\ 1 & 0 \end{pmatrix}$. Prove that it is false that

$$|(\sigma_3 + 1) + (\sigma_1 - 1)| \leq |(\sigma_3 + 1)| + |(\sigma_1 - 1)|$$

Remark: This example is due to E. Nelson.

17. Show that it is not necessarily true that

$$\| |A| - |B| \| \leq \|A - B\|$$

(*Hint*: See Problem 16.)

†*18*. (a) Prove the proposition preceding Theorem VI.10.
   (b) Prove the uniqueness in Theorem VI.10.

19. Write the matrix $\begin{pmatrix} -1 & -2 \\ 2 & 1 \end{pmatrix}$ as the product of a rotation and a positive symmetric matrix.

*20. Suppose that X is a reflexive Banach space and that $T: X \to X$ a bounded linear operator. Prove that if T takes weakly convergent sequences into norm convergent sequences, then T is compact.

†*21*. Complete the proof of Theorem VI.14 by extending the result to all of D.

218   VI: BOUNDED OPERATORS

22. Using the Stone–Weierstrass theorem prove that every Fredholm integral operator on $C[a, b]$

$$(Tf)(x) = \int_a^b K(x, y) f(y) \, dy$$

where K is continuous, is a norm limit of operators of finite rank.

†23. (a) Prove that $\|A\| \leq \|A\|_1$.
   (b) Suppose $\{A_n\}$ is an $\|\cdot\|_1$-Cauchy sequence. Show that $\{A_n\}$ has a $\|\cdot\|$-limit A and that $\text{tr}|A| < \infty$. Then conclude the proof of Theorem VI.20 by showing that A is the $\|\cdot\|_1$-limit of $\{A_n\}$.

†24. (a) Use the canonical form given by Theorem VI.17 to prove the second statement in Theorem VI.21.
   (b) Prove the corollary to Theorem VI.21.

†25. Let $K \in L^2(M \times M, d\mu \otimes d\mu)$ and let A_K be the integral operator

$$(A_K \varphi)(x) = \int_M K(x, y) \varphi(y) \, d\mu(y)$$

Prove that A_K is well defined and $\|A_K\| \leq \|K\|_{L^2}$.

26. (a) Prove that if $\sum_{n=1}^\infty |(A\varphi_n, \varphi_n)| < \infty$ for all orthonormal bases, then $A \in \mathscr{I}_1$.
    (b) Find an $A \notin \mathscr{I}_1$ so that $\sum_{n=1}^\infty |(A\varphi_n, \varphi_n)| < \infty$ for some fixed orthonormal basis.

27. Prove that $\text{tr}(AB) = \text{tr}(BA)$ if $A, B \in \mathscr{I}_2$.

28. Prove that (a) $\|AB\|_1 \leq \|A\| \, \|B\|_1$
                (b) $\|AB\|_2 \leq \|A\| \, \|B\|_2$
                (c) $\|AB\|_1 \leq \|A\|_2 \|B\|_2$

·29. Prove that $A \in \mathscr{I}_1$ if and only if $A = BC$ with B and C in $\mathscr{I}_2$.

·30. The goal of this problem is to prove Theorem VI.26.
   (a) Let f be a bounded linear functional on $\text{Com}(\mathscr{H})$. Let $(\psi, \cdot)\phi$ be the operator on $\mathscr{H}$ which takes η into $(\psi, \eta)\phi$. Show that there is a unique bounded linear operator, B, with

$$(\psi, B\phi) = f[(\psi, \cdot)\phi]$$

   (b) Using the fact that

$$\sum_{n=1}^N (\phi_n, |B|\phi_n) = f\left[\sum_{n=1}^N (U\phi_n, \cdot)\phi_n\right]$$

   prove that $B \in \mathscr{I}_1$ and $\|B\|_1 \leq \|f\|_{\text{Com}(\mathscr{H})^*}$.
   (c) Prove that $A \mapsto \text{tr}(BA)$ is a bounded linear functional on $\text{Com}(\mathscr{H})$ which is in fact equal to $f(\cdot)$.
   (d) Prove that $\|B\|_1 = \|f\|_{\text{Com}(\mathscr{H})^*}$.
   (e) Let g be a bounded linear functional on $\mathscr{I}_1$. Show that there is a unique bounded linear operator, B, with

$$(\psi, B\phi) = g[(\psi, \cdot)\phi]$$

   (f) Prove that $A \mapsto \text{tr}(BA)$ is a bounded linear functional on $\mathscr{I}_1$ which agrees with g and that $\|g\|_{\mathscr{I}_1^*} = \|B\|$.

Problems 219

31. Let $\langle M, \mu \rangle$ be a measure space and let $L^\infty(M, d\mu)$ act on $\mathscr{H} = L^2(M, d\mu)$ by
$$(T_f \varphi)(x) = f(x)\varphi(x)$$
Prove that the topology on L^∞ induced by the weak operator topology on $\mathscr{L}(\mathscr{H})$ is identical to the weak-∗ topology induced on L^∞ by L^1.

32. Let $C[0, 1]$ act on $L^2(0, 1)$ as in Problem 31. Find a sequence in $C[0, 1]$ convergent in the weak operator topology on $C[0, 1]$ to $f \in C[0, 1]$ which is not convergent in the weak Banach space topology on $C[0, 1]$.

33. Consider $\mathscr{I}_2$ as a Hilbert space with inner product $(A, B)_2 = \text{tr}(A^*B)$. Let $A \mapsto L_A$ and $A \mapsto R_A$ be the maps of $\mathscr{L}(\mathscr{H})$ into $\mathscr{L}(\mathscr{I}_2)$ given by
$$L_A(B) = AB, \qquad R_A(B) = BA^*$$
(a) Prove that $A \mapsto L_A$ is a homomorphism of $\mathscr{L}(\mathscr{H})$ into $\mathscr{L}(\mathscr{I}_2)$.
(b) Prove that $A \mapsto R_A$ is a conjugate linear homomorphism of $\mathscr{L}(\mathscr{H})$ into $\mathscr{L}(\mathscr{I}_2)$.
(c) Suppose that $C \in \mathscr{L}(\mathscr{I}_2)$ and obeys $CL_A = L_A C$ for all $A \in \mathscr{L}(\mathscr{H})$. Prove that $C = R_B$ for some $B \in \mathscr{L}(\mathscr{H})$.

*34. Show that in a Hilbert space, a map $T: \mathscr{H} \to \mathscr{H}$ is continuous if the domain is given the weak topology and the range the norm topology (that is, $x_\alpha \xrightarrow{w} x$ implies $Tx_\alpha \xrightarrow{\|\cdot\|} Tx$ for arbitrary *nets*) if and only if T has finite rank! (Compare with Theorem VI.11.)

35. (a) Suppose T is an operator in $\mathscr{L}(\mathscr{H})$ so that $x_n \xrightarrow{\|\cdot\|} x$ implies $Tx_n \xrightarrow{w} Tx$. Prove T is bounded (so $Tx_n \xrightarrow{\|\cdot\|} Tx$).
(b) Identify the continuous linear maps of $\mathscr{L}(\mathscr{H})$ into itself if both the domain and range are given the weak topology.

36. Use (c) of Theorem VI.12 and the polar decomposition to prove (b) of Theorem VI.12 when $X = Y$ is a Hilbert space.

†*37.* Prove part (c) of Theorem VI.12.

38. Let P and Q be orthogonal projections onto subspaces $\mathscr{M}$ and $\mathscr{N}$ in a Hilbert space $\mathscr{H}$. Suppose that $PQ = QP$.
(a) Prove $1 - P, 1 - Q, PQ, P + Q - PQ$ and $P + Q - 2PQ$ are orthogonal projections.
(b) How are the ranges of the projections in (a) related to $\mathscr{M}$ and $\mathscr{N}$.

*39. Let P and Q be orthogonal projections onto subspaces $\mathscr{M}$ and $\mathscr{N}$ in a Hilbert space $\mathscr{H}$. Prove that s-lim$_{n \to \infty} (PQ)^n$ exists and is the orthogonal projection onto $\mathscr{M} \cap \mathscr{N}$.

*40. Let $\mathscr{I}$ be a norm closed ideal in $\mathscr{L}(\mathscr{H})$, $\mathscr{I} \neq 0$. Prove Com($\mathscr{H}$) $\subset \mathscr{I}$ by proving that any finite rank operator is in $\mathscr{I}$.

Remark: We will see (Chapter VII, Problem 31) that the only norm closed ideals when $\mathscr{H}$ is separable are $\{0\}$, Com($\mathscr{H}$), $\mathscr{L}(\mathscr{H})$.

41. Find a projection on $\mathbb{R}^2$ which is not an orthogonal projection.

42. Let $A \in \mathscr{L}(X)$. Prove that the set of λ such that λ is in $\sigma(A)$ but not an eigenvalue and Ran($\lambda I - A$) is closed but not all of X is a open subset of $\mathbb{C}$.

43. Let M and N be subspaces of a Banach space X such that $M + N = X$ and $M \cap N = \{0\}$. Let P be the projection of X onto M. Prove that P is bounded if and only if *both* M and N are closed.

44. (a) Define the numerical range, $N(T)$, of a bounded operator, T, on a Hilbert space, $\mathscr{H}$, by $N(T) = \{(\psi, T\psi) \mid \psi \in \mathscr{H}, \ \|\psi\| = 1\}$. Prove that $\sigma(T) \subset \overline{N(T)}$. (Hint: First show that if λ is an eigenvalue of T or T^*, then $\lambda \in N(T)$; then show that if $\lambda \in \sigma(T)$ and λ is not an eigenvalue of T or T^*, we can find $\psi_n \in \mathscr{H}$ so that $\|(T - \lambda)\psi_n\| \to 0$.)
   (b) Find an example where $N(T)$ is not closed and $\sigma(T) \not\subset N(T)$.
   (c) Find an example where $\sigma(T) \neq N(T) = \overline{N(T)}$.

Remark: There is a deep result of Hausdorff that $N(T)$ is convex.

45. (a) Let $\{\phi_n\}_{n=1}^\infty$ be an orthonormal basis for a Hilbert space $\mathscr{H}$. Let A be an operator with
$$\sup_{\substack{\psi \in [\phi_1, \ldots, \phi_n]^\perp \\ \|\psi\| = 1}} \|A\psi\| \to 0 \quad \text{as} \quad n \to \infty.$$
Prove that A is compact.
   (b) Let $\{\phi_n\}_{n=1}^\infty$ be any orthonormal basis for a Hilbert space $\mathscr{H}$ and let A be compact. Prove that
$$\sup_{\psi \in [\phi_1, \ldots, \phi_n]^\perp} \|A\psi\| \to 0 \quad \text{as} \quad n \to \infty.$$

46. (a) Let $A \geq 0$ with A compact. Prove that $A^{1/2}$ is compact. (Hint: Use Problem 45.)
   (b) Let $0 \leq A \leq B$. Prove that A is compact if B is compact. (Hint: Prove that $A^{1/2}$ is compact using Problem 45 and part (a).)

47. Let $\mathscr{H}$ and $\mathscr{H}'$ be two Hilbert spaces. If T is a bounded linear map from $\mathscr{H}$ to $\mathscr{H}'$ we define $T^* : \mathscr{H}' \to \mathscr{H}$ by $(T^*\psi, \phi)_{\mathscr{H}} = (\psi, T\phi)$. T is called *Hilbert–Schmidt* if and only if $T^*T : \mathscr{H} \to \mathscr{H}$ is trace class. Let T be Hilbert-Schmidt. Prove that there are real numbers, $\lambda_n > 0$, and orthonormal sets $\{\phi_n\}_{n=1}^N \subset \mathscr{H}$, $\{\psi_n\}_{n=1}^N \in \mathscr{H}'$ so that
$$T\phi = \sum_{n=1}^N \lambda_n(\phi_n, \phi)\psi_n$$

48. Let $\mathscr{H}$ and $\mathscr{H}'$ be the two Hilbert spaces and let $\mathscr{I}_2(\mathscr{H}, \mathscr{H}')$ denote the Hilbert–Schmidt operators from $\mathscr{H}$ to $\mathscr{H}'$.
   (a) Prove that $\mathscr{I}_2(\mathscr{H}, \mathscr{H}')$ with the inner product
$$(S, T) = \mathrm{Tr}_{\mathscr{H}}(S^*T)$$
is a Hilbert space.
   (b) Given $\psi \in \mathscr{H}$, $\phi \in \mathscr{H}'$ define $I(\psi, \phi) \in \mathscr{I}_2(\mathscr{H}^*, \mathscr{H}')$ by $I(\psi, \phi)\ell = \ell(\psi)\phi$ for any $\ell \in \mathscr{H}^*$. Prove that the map J, taking $\psi \otimes \phi$ into $I(\psi, \phi)$, is well defined and extends to an isometry of $\mathscr{H} \otimes \mathscr{H}'$ and $\mathscr{I}_2(\mathscr{H}^*, \mathscr{H}')$.
   (c) Given $\eta \in \mathscr{H} \otimes \mathscr{H}'$ show that there exist reals, $\lambda_n > 0$, and orthonormal sets $\{\phi_n\}_{n=1}^N \subset \mathscr{H}$, $\{\psi_n\}_{n=1}^N \subset \mathscr{H}'$ with N finite or infinite, so that
$$\sum_{n=1}^N |\lambda_n|^2 = \|\eta\|^2 \quad \text{and} \quad \sum_{n=1}^N \lambda_n \phi_n \otimes \psi_n = \eta.$$

VII: The Spectral Theorem

Mathematical proofs, like diamonds, are hard as well as clear, and will be touched with nothing but strict reasoning.
John Locke in **Second Reply to the Bishop of Worcester**

VII.1 The continuous functional calculus

In this chapter, we will discuss the spectral theorem in its many guises. This structure theorem is a concrete description of all self-adjoint operators. There are several apparently distinct formulations of the spectral theorem. In some sense they are all equivalent.

The form we prefer says that every bounded self-adjoint operator is a multiplication operator. (We emphasize the word bounded since we will deal extensively with unbounded self-adjoint operators in the next chapter; there is a spectral theorem for unbounded operators which we discuss in Section VIII.3.) This means that given a bounded self-adjoint operator on a Hilbert space $\mathscr{H}$, we can always find a measure μ on a measure space M and a unitary operator $U \colon \mathscr{H} \to L^2(M, d\mu)$ so that

$$(UAU^{-1}f)(x) = F(x)f(x)$$

for some bounded real-valued measurable function F on M.

This is clearly a generalization of the finite-dimensional theorem, which says any self-adjoint $n \times n$ matrix can be diagonalized, or in an abstract form: Given self-adjoint operator A on an n-dimensional complex space V, there is a unitary operator $U: V \to \mathbb{C}^n$ and real numbers $\lambda_1, \ldots, \lambda_n$ so that

$$(UAU^{-1}f)_i = \lambda_i f_i$$

for each $f = \langle f_1, \ldots, f_n \rangle$ in $\mathbb{C}^n$.

In practice, M will be a union of copies of $\mathbb{R}$ and F will be x, so the core of the proof of the theorem will be the construction of certain measures. This will be done in Section VII.2 by using the Riesz–Markov theorem. Our goal in this section will be to make sense out of $f(A)$, for f a continuous function. In the next section, we will consider the measures defined by the functionals $f \mapsto \langle \psi, f(A)\psi \rangle$ for fixed $\psi \in \mathcal{H}$.

Given a fixed operator A, for which f can we define $f(A)$? First, suppose that A is an arbitrary bounded operator. If $f(x) = \sum_{n=1}^N a_n x^n$ is a polynomial, we want $f(A) = \sum_{n=1}^N a_n A^n$. Suppose that $f(x) = \sum_{n=0}^\infty c_n x^n$ is a power series with radius of convergence R. If $\|A\| < R$, then $\sum_{n=0}^\infty c_n A^n$ converges in $\mathcal{L}(\mathcal{H})$ so it is natural to set $f(A) = \sum_{n=0}^\infty c_n A^n$. In this last case, f was a function analytic in a domain including all of $\sigma(A)$. In general, one can make a reasonable definition for $f(A)$ if f is analytic in a neighborhood of $\sigma(A)$ (see the Notes).

The functional calculus we have talked about thus far works for any operator in any Banach space. The special property of self-adjoint operators (or more generally normal operators; see Problems 3, 5) is that $\|P(A)\| = \sup_{\lambda \in \sigma(A)} |P(\lambda)|$ for any polynomial P, so that one can use the B.L.T. theorem to extend the functional calculus to continuous functions. Our major goal in this section is the proof of:

Theorem VII.1 (continuous functional calculus) Let A be a self-adjoint operator on a Hilbert space $\mathcal{H}$. Then there is a *unique* map $\phi: C(\sigma(A)) \to \mathcal{L}(\mathcal{H})$ with the following properties:

(a) ϕ is an algebraic $*$-homomorphism, that is,

$$\phi(fg) = \phi(f)\phi(g) \qquad \phi(\lambda f) = \lambda \phi(f)$$
$$\phi(1) = I \qquad \phi(\bar{f}) = \phi(f)^*$$

(b) ϕ is continuous, that is, $\|\phi(f)\|_{\mathcal{L}(\mathcal{H})} \leq C \|f\|_\infty$.
(c) Let f be the function $f(x) = x$; then $\phi(f) = A$.

Moreover, ϕ has the additional properties:

(d) If $A\psi = \lambda\psi$, then $\phi(f)\psi = f(\lambda)\psi$.
(e) $\sigma[\phi(f)] = \{f(\lambda) \mid \lambda \in \sigma(A)\}$ [spectral mapping theorem].
(f) If $f \geq 0$, then $\phi(f) \geq 0$.
(g) $\|\phi(f)\| = \|f\|_\infty$ [this strengthens (b)].

We sometimes write $\phi_A(f)$ or $f(A)$ for $\phi(f)$ to emphasize the dependence on A.

The idea of the proof which we give below is quite simple. (a) and (c) uniquely determine $\phi(P)$ for any polynomial $P(x)$. By the Weierstrass theorem, the set of polynomials is dense in $C(\sigma(A))$ so the heart of the proof is showing that

$$\|P(A)\|_{\mathscr{L}(\mathscr{H})} = \|P(x)\|_{C(\sigma(A))} \equiv \sup_{\lambda \in \sigma(A)} |P(\lambda)|$$

The existence and uniqueness of ϕ then follow from the B.L.T. theorem.

To prove the crucial equality, we first prove a special case of (e) (which holds for arbitrary bounded operators):

Lemma 1 Let $P(x) = \sum_{n=0}^{N} a_n x^n$. Let $P(A) = \sum_{n=0}^{N} a_n A^n$. Then

$$\sigma(P(A)) = \{P(\lambda) \mid \lambda \in \sigma(A)\}$$

Proof Let $\lambda \in \sigma(A)$. Since $x = \lambda$ is a root of $P(x) - P(\lambda)$, we have $P(x) - P(\lambda) = (x - \lambda)Q(x)$, so $P(A) - P(\lambda) = (A - \lambda)Q(A)$. Since $(A - \lambda)$ has no inverse neither does $P(A) - P(\lambda)$, that is, $P(\lambda) \in \sigma(P(A))$.

Conversely, let $\mu \in \sigma(P(A))$ and let $\lambda_1, \ldots, \lambda_n$ be the roots of $P(x) - \mu$, that is, $P(x) - \mu = a(x - \lambda_1) \cdots (x - \lambda_n)$. If $\lambda_1, \ldots, \lambda_n \notin \sigma(A)$, then

$$(P(A) - \mu)^{-1} = a^{-1}(A - \lambda_1)^{-1} \cdots (A - \lambda_n)^{-1}$$

so we conclude that some $\lambda_i \in \sigma(A)$, that is, $\mu = P(\lambda)$ for some $\lambda \in \sigma(A)$. ∎

Lemma 2 Let A be a bounded self-adjoint operator. Then

$$\|P(A)\| = \sup_{\lambda \in \sigma(A)} |P(\lambda)|$$

Proof
$$\begin{aligned}
\|P(A)\|^2 &= \|P(A)^* P(A)\| \\
&= \|(\bar{P}P)(A)\| \\
&= \sup_{\lambda \in \sigma(PP(A))} |\lambda| \quad \text{(by Theorem VI.6)} \\
&= \sup_{\lambda \in \sigma(A)} |\bar{P}P(\lambda)| \quad \text{(by Lemma 1)} \\
&= \left(\sup_{\lambda \in \sigma(A)} |P(\lambda)| \right)^2 \quad \blacksquare
\end{aligned}$$

Proof of Theorem VII.1 Let $\phi(P) = P(A)$. Then $\|\phi(P)\|_{\mathscr{L}(\mathscr{H})} = \|P\|_{C(\sigma(A))}$ so ϕ has a unique linear extension to the closure of the polynomials in $C(\sigma(A))$. Since the polynomials are an algebra containing 1, containing complex conjugates, and separating points, this closure is all of $C(\sigma(A))$. Properties (a), (b), (c), (g) are obvious and if $\tilde{\phi}$ obeys (a), (b), (c) it agrees with ϕ on polynomials and thus by continuity on $C(\sigma(A))$. To prove (d), note that $\phi(P)\psi = P(\lambda)\psi$ and apply continuity. To prove (f), notice that if $f \geq 0$, then $f = g^2$ with g real and $g \in C(\sigma(A))$. Thus $\phi(f) = \phi(g)^2$ with $\phi(g)$ self-adjoint, so $\phi(f) \geq 0$. (e) is left for the reader (Problem 8). ∎

Before turning to some examples, we make several remarks:

(1) $\phi(f) \geq 0$ if and only if $f \geq 0$ (Problem 9).

(2) Since $fg = gf$ for all f, g, $\{f(A) \mid f \in C(\sigma(A))\}$ forms an abelian algebra closed under adjoints. Since $\|f(A)\| = \|f\|_\infty$ and $C(\sigma(A))$ is complete, $\{f(A) \mid f \in C(\sigma(A))\}$ is norm-closed. It is thus an **abelian** C^* **algebra** of operators.

(3) Ran ϕ is actually the C^* **algebra generated by** A, that is, the smallest C^*-algebra containing A (Problem 10).

(4) This result, that $C(\sigma(A))$ and the C^*-algebra generated by A are isometrically isomorphic, is actually a special case of the "commutative Gelfand–Naimark theorem" which we discuss in Chapter XV.

(5) (b) actually follows from (a) and abstract nonsense (Problem 11). Thus (a) and (c) alone determine ϕ uniquely.

Finally, we consider two specific examples of $\phi(f)$:

Example 1 As a corollary, we have a new proof of the existence half of the square-root lemma (Theorem VI.9) for if $A \geq 0$, then $\sigma(A) \subset [0, \infty)$ (Problem 12). If $f(x) = x^{1/2}$, then $f(A)^2 = A$.

Example 2 From (g) of Theorem VII.1 we see that $\|(A - \lambda)^{-1}\| = [\text{dist}(\lambda, \sigma(A))]^{-1}$ if A is bounded, self-adjoint, and $\lambda \notin \sigma(A)$.

VII.2 The spectral measures

We are now ready to introduce the measures we have anticipated so often before. Let us fix A, a bounded self-adjoint operator. Let $\psi \in \mathcal{H}$. Then $f \mapsto (\psi, f(A)\psi)$ is a positive linear functional on $C(\sigma(A))$. Thus, by the Riesz–Markov theorem (Theorem IV.14), there is a unique measure μ_ψ on the compact set $\sigma(A)$ with $(\psi, f(A)\psi) = \int_{\sigma(A)} f(\lambda) \, d\mu_\psi$.

VII.2 The spectral measures

Definition The measure μ_ψ is called the **spectral measure associated with the vector** ψ.

The first and simplest application of the μ_ψ is to allow us to extend the functional calculus to $\mathcal{B}(\mathbb{R})$, the bounded Borel functions on $\mathbb{R}$. Let $g \in \mathcal{B}(\mathbb{R})$. It is natural to define $g(A)$ so that $(\psi, g(A)\psi) = \int_{\sigma(A)} g(\lambda) \, d\mu_\psi(\lambda)$. The polarization identity lets us recover $(\psi, g(A)\phi)$ from the proposed $(\psi, g(A)\psi)$ and then the Riesz lemma lets us construct $g(A)$. The properties of this "measurable functional calculus" are given in (Problem 13):

Theorem VII.2 (spectral theorem—functional calculus form) Let A be a bounded self-adjoint operator on $\mathcal{H}$. There is a unique map $\hat\phi \colon \mathcal{B}(\mathbb{R}) \to \mathcal{L}(\mathcal{H})$ so that

(a) $\hat\phi$ is an algebraic $*$-homomorphism.
(b) $\hat\phi$ is norm continuous: $\|\hat\phi(f)\|_{\mathcal{L}(\mathcal{H})} \leq \|f\|_\infty$.
(c) Let f be the function $f(x) = x$; then $\hat\phi(f) = A$.
(d) Suppose $f_n(x) \to f(x)$ for each x and $\|f_n\|_\infty$ is bounded. Then $\hat\phi(f_n) \to \hat\phi(f)$ strongly.

Moreover $\hat\phi$ has the properties:

(e) If $A\psi = \lambda\psi$, then $\hat\phi(f)\psi = f(\lambda)\psi$.
(f) If $f \geq 0$, then $\hat\phi(f) \geq 0$.
(g) If $BA = AB$, then $\hat\phi(f)B = B\hat\phi(f)$.

Theorem VII.2 can be proven directly by extending Theorem VII.1; part (d) requires the dominated convergence theorem. Or, Theorem VII.2 can be proven by an easy corollary of Theorem VII.3 below. The proof of Theorem VII.3 uses only the *continuous* functional calculus. $\hat\phi$ extends ϕ and as before we write $\hat\phi(f) = f(A)$. As in the continuous functional calculus, one has $f(A)g(A) = g(A)f(A)$.

Since $\mathcal{B}(\mathbb{R})$ is the smallest family closed under limits of form (d) containing all of $C(\mathbb{R})$, we know that any $\hat\phi(f)$ is in the smallest C^*-algebra containing A which is also strongly closed; such an algebra is called a von Neumann or W^*-algebra. When we study von Neumann algebras in Chapter XVIII we will see that this follows from (g).

The norm equality of Theorem VII.1 carries over if we define $\|f\|'_\infty$ to be the L^∞-norm with respect to a suitable notion of "almost everywhere." Namely, pick an orthonormal basis $\{\psi_n\}$ and say that a property is true a.e. if it is true a.e. with respect to *each* μ_{ψ_i}. Then $\|\hat\phi(f)\|_{\mathcal{L}(\mathcal{H})} = \|f\|'_\infty$.

In the next section, we will return to the operators $\chi_\Omega(A)$ where χ_Ω is a characteristic function; this is the most important set of operators in the measurable but not in the continuous functional calculus. For the time being, we turn to using the spectral measures to form L^2 spaces. We first define:

Definition A vector $\psi \in \mathcal{H}$ is called a **cyclic vector** for A if finite linear combinations of the elements $\{A^n\psi\}_{n=0}^\infty$ are dense in $\mathcal{H}$.

Not all operators have cyclic vectors (Problem 14), but if they do:

Lemma 1 Let A be a bounded self-adjoint operator with cyclic vector ψ. Then, there is a unitary operator $U: \mathcal{H} \to L^2(\sigma(A), d\mu_\psi)$ with

$$(UAU^{-1}f)(\lambda) = \lambda f(\lambda)$$

Equality is in the sense of elements of $L^2(\sigma(A), d\mu_\psi)$.

Proof Define U by $U\phi(f)\psi \equiv f$ where f is continuous. U is essentially the inverse of the map ϕ of Theorem VII.1. To show that U is well defined we compute

$$\|\phi(f)\psi\|^2 = (\psi, \phi^*(f)\phi(f)\psi) = (\psi, \phi(\bar{f}f)\psi)$$
$$= \int |f(\lambda)|^2 \, d\mu_\psi$$

Therefore, if $f = g$ a.e. with respect to μ_ψ, then $\phi(f)\psi = \phi(g)\psi$. Thus U is well defined on $\{\phi(f)\psi \mid f \in C(\sigma(A))\}$ and is norm preserving. Since ψ is cyclic $\overline{\{\phi(f)\psi \mid f \in C(\sigma(A))\}} = \mathcal{H}$, so by the B.L.T. theorem U extends to an isometric map of $\mathcal{H}$ into $L^2(\sigma(A), d\mu_\psi)$. Since $C(\sigma(A))$ is dense in L^2, Ran $U = L^2(\sigma(A), d\mu_\psi)$. Finally, if $f \in C(\sigma(A))$,

$$(UAU^{-1}f)(\lambda) = [UA\phi(f)](\lambda)$$
$$= [U\phi(xf)](\lambda)$$
$$= \lambda f(\lambda)$$

By continuity, this extends from $f \in C(\sigma(A))$ to $f \in L^2$. ∎

To extend this lemma to arbitrary A, we need to know that A has a family of invariant subspaces spanning $\mathcal{H}$ so that A is cyclic on each subspace:

Lemma 2 Let A be a self-adjoint operator on a separable Hilbert space $\mathcal{H}$. Then there is a direct sum decomposition $\mathcal{H} = \bigoplus_{n=1}^N \mathcal{H}_n$ with $N = 1, 2, \ldots,$ or ∞ so that:

(a) A leaves each $\mathcal{H}_n$ **invariant**, that is, $\psi \in \mathcal{H}_n$ implies $A\psi \in \mathcal{H}_n$.
(b) For each n, there is a $\phi_n \in \mathcal{H}_n$ which is cyclic for $A \upharpoonright \mathcal{H}_n$, i.e. $\mathcal{H}_n = \overline{\{f(A)\phi_n \mid f \in C(\sigma(A))\}}$.

VII.2 The spectral measures 227

Proof A simple Zornication (Problem 15). ∎

We can now combine Lemmas 1 and 2 to prove the form of the spectral theorem which we regard as the most transparent:

Theorem VII.3 (spectral theorem—multiplication operator form) Let A be a bounded self-adjoint operator on $\mathcal{H}$, a separable Hilbert space. Then, there exist measures $\{\mu_n\}_{n=1}^N$ ($N = 1, 2, \ldots$ or ∞) on $\sigma(A)$ and a unitary operator

$$U: \mathcal{H} \to \bigoplus_{n=1}^N L^2(\mathbb{R}, d\mu_n)$$

so that

$$(UAU^{-1}\psi)_n(\lambda) = \lambda \psi_n(\lambda)$$

where we write an element $\psi \in \bigoplus_{n=1}^N L^2(\mathbb{R}, d\mu_n)$ as an N-tuple $\langle \psi_1(\lambda), \ldots, \psi_N(\lambda) \rangle$. This realization of A is called a **spectral representation**.

Proof Use Lemma 2 to find the decomposition and then use Lemma 1 on each component. ∎

This theorem tells us that every bounded self-adjoint operator is a multiplication operator on a suitable measure space; what changes as the operator changes are the underlying measures. Explicitly:

Corollary Let A be a bounded self-adjoint operator on a separable Hilbert space $\mathcal{H}$. Then there exists a *finite* measure space $\langle M, \mu \rangle$, a bounded function F on M, and a unitary map, $U: \mathcal{H} \to L^2(M, d\mu)$, so that

$$(UAU^{-1}f)(m) = F(m)f(m)$$

Proof Choose the cyclic vectors ϕ_n so that $\|\phi_n\| = 2^{-n}$. Let $M = \bigcup_{n=1}^N \mathbb{R}$, i.e. the union of N copies of $\mathbb{R}$. Define μ by requiring that its restriction to the nth copy of $\mathbb{R}$ be μ_n. Since $\mu(M) = \sum_{n=1}^N \mu_n(\mathbb{R}) < \infty$, μ is finite. ∎

We also notice that this last theorem is essentially a rigorous form of the physicists' Dirac notation. If we write $\psi_n(x) = \psi(x; n)$, we see that in the "new representation defined by U" one has

$$(\psi, \phi) = \sum_n \int d\mu_n \overline{\psi(\lambda; n)} \phi(\lambda; n)$$

$$(\psi, A\phi) = \sum_n \int d\mu_n \overline{\psi(\lambda; n)} \lambda \phi(\lambda; n)$$

These are the Dirac type formulas familiar to physicists except that the formal sums of Dirac are replaced with integrals over spectral measures, where we define:

Definition The measures $d\mu_n$ are called **spectral measures**; they are just $d\mu_\psi$ for suitable ψ.

These measures are *not* uniquely determined and we will eventually discuss this nonuniqueness question. First, let us consider a few examples:

Example 1 Let A be an $n \times n$ self-adjoint matrix. The "usual" finite-dimensional spectral theorem says that A has a complete orthonormal set of eigenvectors, $\psi_1, \ldots, \psi_n$, with $A\psi_i = \lambda_i \psi_i$. Suppose first that the eigenvalues are distinct. Consider the sum of Dirac measures, $\mu = \sum_{i=1}^{n} \delta(x - \lambda_i)$. $L^2(\mathbb{R}, d\mu)$ is just $\mathbb{C}^n$ since $f \in L^2$ is determined by $f = \langle f(\lambda_1), \ldots, f(\lambda_n) \rangle$. Clearly, the function λf corresponds to the n-tuple $\langle \lambda_1 f(\lambda_1), \ldots, \lambda_n f(\lambda_n) \rangle$, so A is multiplication by λ on $L^2(\mathbb{R}, d\mu)$. If we take $\tilde{\mu} = \sum_{i=1}^{n} a_i \delta(x - \lambda_i)$ with $a_1, \ldots, a_n > 0$, A can also be represented as multiplication by λ on $L^2(\mathbb{R}, d\tilde{\mu})$. Thus, we explicitly see the nonuniqueness of the measure in this case. We can also see when more than one measure is needed: one can represent a finite-dimensional self-adjoint operator as multiplication on $L^2(\mathbb{R}, d\mu)$ with only one measure if and only if A has no repeated eigenvalues.

Example 2 Let A be compact and self-adjoint. The Hilbert–Schmidt theorem tells us there is a complete orthonormal set of eigenvectors $\{\psi_n\}_{n=1}^{\infty}$ with $A\psi_n = \lambda_n \psi_n$. If there is no repeated eigenvalue, $\sum_{n=1}^{\infty} 2^{-n} \delta(x - \lambda_n)$ works as a spectral measure.

Example 3 Let $\mathcal{H} = \ell^2(-\infty, \infty)$, that is, the set of sequences, $\{a_n\}_{n=-\infty}^{\infty}$ with $\sum_{n=-\infty}^{\infty} |a_n|^2 < \infty$. Let $L: \mathcal{H} \to \mathcal{H}$ by $(La)_n = a_{n+1}$, that is, L shifts to the left. $L^* = R$ with $(Ra)_n = a_{n-1}$. Let $A = R + L$ which is self-adjoint. Can we represent A as a multiplication operator? Map $\mathcal{H}$ into $L^2[0, 1]$ by $U: \{a_n\} \to \sum_{n=-\infty}^{\infty} a_n e^{2\pi i n x}$. Then ULU^{-1} is multiplication by $e^{-2\pi i x}$ and URU^{-1} is multiplication by $e^{+2\pi i x}$ so UAU^{-1} is multiplication by $2\cos(2\pi x)$. The necessary transformations needed to represent A as multiplication by x on $L^2(\mathbb{R}, d\mu_1) \oplus L^2(\mathbb{R}, d\mu_2)$ are left for the problems. μ_1 and μ_2 have support in $[-2, 2]$.

Example 4 Consider $i^{-1}d/dx$ on $L^2(\mathbb{R}, dx)$. This is an unbounded operator and thus not strictly within the context of this section, but we will prove an analogue of Theorem VII.3 in Section VIII.3. We thus seek an operator U and a measure $d\mu$ (it turns out that only one μ is needed) with $U: L^2(\mathbb{R}, dx) \to L^2(\mathbb{R}, d\mu(k))$ so that

$$U\left(\frac{1}{i}\frac{d}{dx}f\right)(k) = k\,Uf(k)$$

The Fourier transform $(Uf)(k) = (2\pi)^{-1/2} \int f(x)e^{-ikx}\,dx$ which we study in Chapter IX precisely does the trick. Thus, the Fourier transform is one example of a spectral representation.

We now investigate the connection between spectral measures and the spectrum.

Definition If $\{\mu_n\}_{n=1}^N$ is a family of measures, the **support** of $\{\mu_n\}$ is the complement of the largest open set B with $\mu_n(B) = 0$ for all n; so

$$\operatorname{supp}\{\mu_n\} = \overline{\bigcup_{n=1}^N \operatorname{supp} \mu_n}$$

Proposition Let A be a self-adjoint operator and $\{\mu_n\}_{n=1}^N$ a family of spectral measures. Then

$$\sigma(A) = \operatorname{supp}\{\mu_n\}_{n=1}^N$$

There is also a simple description of $\sigma(A)$ in terms of the more general multiplication operators discussed after Theorem VII.3:

Definition Let F be a real-valued function on a measure space $\langle M, \mu \rangle$. We say λ is in the **essential range** of F if and only if

$$\mu\{m \mid \lambda - \varepsilon < F(m) < \lambda + \varepsilon\} > 0$$

for all $\varepsilon > 0$.

Proposition Let F be a bounded real-valued function on a measure space $\langle M, \mu \rangle$. Let T_F be the operator on $L^2(M, d\mu)$ given by

$$(T_F g)(m) = F(m)g(m)$$

Then $\sigma(T_F)$ is the essential range of F.

Proof See Problem 17b.

We can now see exactly what information is contained in the spectrum. A unitary invariant of a self-adjoint operator A is a property P so that $P(A) = P(UAU^{-1})$ for all unitary operators U. Thus, unitary invariants are "intrinsic" properties of self-adjoint operators, that is, properties independent of "representation." An example of such a **unitary invariant** is the spectrum $\sigma(A)$. However, the spectrum is a poor invariant: for example, multiplication by x on $L^2([0, 1], dx)$ and an operator with a complete set of eigenfunctions having all rationals in $[0, 1]$ as eigenvalues are very different even though both have spectrum $[0, 1]$.

At the conclusion of this section, we will see that there is a canonical choice of "spectral measures" which forms a complete set of unitary invariants, that is, a set of properties which distinguish two self-adjoint operators A and B unless $A = UBU^{-1}$ for some unitary operator U. This explains why $\sigma(A)$ is such a bad invariant for different sorts of measures can have the same support. If we wish to find better invariants which are, however, simpler than measures, it is reasonable to first decompose spectral measures in some natural way and then pass to supports. Recall Theorem I.13 which says that any measure μ on $\mathbb{R}$ has a unique decomposition into $\mu = \mu_{pp} + \mu_{ac} + \mu_{sing}$ where μ_{pp} is a pure point measure, μ_{ac} is absolutely continuous with respect to Lebesgue measure, and μ_{sing} is continuous and singular with respect to Lebesgue measure. These three pieces are mutually singular so

$$L^2(\mathbb{R}, d\mu) = L^2(\mathbb{R}, d\mu_{pp}) \oplus L^2(\mathbb{R}, d\mu_{ac}) \oplus L^2(\mathbb{R}, d\mu_{sing})$$

It is easy to see (Problem 18) that any $\psi \in L^2(\mathbb{R}, d\mu)$ has an absolutely continuous spectral measure $d\mu_\psi$ if and only if $\psi \in L^2(\mathbb{R}, d\mu_{ac})$, and similarly for pure point and singular measures. If $\{\mu_n\}_{n=1}^N$ is a family of spectral measures we can sum $\bigoplus_{n=1}^N L^2(\mathbb{R}, d\mu_{n;ac})$ by defining:

Definition Let A be a bounded self-adjoint operator on $\mathcal{H}$. Let $\mathcal{H}_{pp} = \{\psi \mid \mu_\psi \text{ is pure point}\}$, $\mathcal{H}_{ac} = \{\psi \mid \mu_\psi \text{ is absolutely continuous}\}$, $\mathcal{H}_{sing} = \{\psi \mid \mu_\psi \text{ is continuous singular}\}$.

We have thus proven:

Theorem VII.4 $\mathcal{H} = \mathcal{H}_{pp} \oplus \mathcal{H}_{ac} \oplus \mathcal{H}_{sing}$. Each of these subspaces is invariant under A. $A \upharpoonright \mathcal{H}_{pp}$ has a complete set of eigenvectors, $A \upharpoonright \mathcal{H}_{ac}$ has only absolutely continuous spectral measures and $A \upharpoonright \mathcal{H}_{sing}$ has only continuous singular spectral measures.

Definition $\sigma_{pp}(A) = \{\lambda \mid \lambda \text{ is an eigenvalue of } A\}$
$\sigma_{cont}(A) = \sigma(A \upharpoonright \mathscr{H}_{cont} \equiv \mathscr{H}_{sing} \oplus \mathscr{H}_{ac})$
$\sigma_{ac}(A) = \sigma(A \upharpoonright \mathscr{H}_{ac})$
$\sigma_{sing}(A) = \sigma(A \upharpoonright \mathscr{H}_{sing})$

These sets are called the **pure point, continuous, absolutely continuous**, and **singular** (or **continuous singular**) **spectrum** respectively.

While it may happen that $\sigma_{ac} \cup \sigma_{sing} \cup \sigma_{pp} \neq \sigma$, this is only true because we did not define σ_{pp} as $\sigma(A \upharpoonright \mathscr{H}_{pp})$ but rather as the actual set of eigenvalues. One always has

Proposition $\sigma_{cont}(A) = \sigma_{ac}(A) \cup \sigma_{sing}(A)$
$\sigma(A) = \overline{\sigma_{pp}(A)} \cup \sigma_{cont}(A)$

The sets need not be disjoint, however. The reader should be warned that $\sigma_{sing}(A)$ may have nonzero Lebesgue measure (Problem 7). For many purposes, breaking up the spectrum in this way gives useful information. In Section VII.3, we introduce another breakup which is also useful.

As we discussed in the notes to Section VI.3, some authors use a notion of "continuous spectrum" which is distinct from the above, namely they define the continuous spectrum to be the set of $\lambda \in \sigma(T)$ which are neither in the point spectrum nor in the residual spectrum. To illustrate the difference between the two definitions we let $\mathscr{H} = \mathbb{C} \oplus L^2[0, 1]$ and define $A: \langle \alpha, f(x) \rangle \rightarrow \langle \frac{1}{2}\alpha, xf(x) \rangle$. With our definition, the point $\lambda = \frac{1}{2}$ is in both the pure point and the continuous spectrum. The other authors assign $\lambda = \frac{1}{2}$ to the point spectrum and their continuous spectrum is $[0, \frac{1}{2}) \cup (\frac{1}{2}, 1]$.

Finally, we turn to the question of making canonical choices for the spectral measures, a subject which goes under the title of "multiplicity theory." We will describe the basic results without proof:

1. Multiplicity free operators

We must first ask when A is unitarily equivalent to multiplication by x on $L^2(\mathbb{R}, d\mu)$, that is, when only one spectral measure is needed. A look at Example 1 tells us this happens in the finite-dimensional case only when A has no repeated eigenvalues, so we define:

Definition A bounded self-adjoint operator A is called **multiplicity free** if and only if A is unitarily equivalent to multiplication by λ on $L^2(\mathbb{R}, d\mu)$ for some measure μ.

One is interested in intrinsic characterizations of "multiplicity free" and there are several:

Theorem VII.5 The following are equivalent:
   (a) A is multiplicity free.
   (b) A has a cyclic vector.
   (c) $\{B \mid AB = BA\}$ is an abelian algebra.

2. Measure classes

Next we must ask about the nonuniqueness of the measure in the multiplicity free case. The situation in the finite-dimensional multiplicity free case was seen in Example 1: the "acceptable" measures were $\sum_{n=1}^{N} \alpha_n \delta(\lambda - \lambda_n)$ with each $\alpha_n \neq 0$. There is a natural generalization. Suppose $d\mu$ on $\mathbb{R}$ is given and let F be a measurable function which is positive and nonzero a.e. with respect to μ and locally $L^1(\mathbb{R}, d\mu)$, that is, $\int_C |F| \, d\mu < \infty$ for every compact set $C \subset \mathbb{R}$. Then $dv = F \, d\mu$ is a Borel measure and the map, U,

$$U: L^2(\mathbb{R}, dv) \to L^2(\mathbb{R}, d\mu)$$

given by $(Uf)(\lambda) = \sqrt{F(\lambda)} f(\lambda)$ is unitary (onto since $F \neq 0$ a.e.) and $\lambda(Uf) = U(\lambda f)$. Thus, an operator A with a spectral representation in terms of μ could just as well be represented in terms of v. By the Radon–Nikodym theorem, $dv = F \, d\mu$ with F a.e. nonzero if and only if v and μ have the same sets of measure zero. This suggests the definition:

Definition Two Borel measures μ and v are called **equivalent** if and only if they have the same sets of measure zero. An equivalence class $\langle \mu \rangle$ is called a **measure class**.

Then, the nonuniqueness question is answered by:

Proposition Let μ and v be Borel measures on $\mathbb{R}$ with bounded support. Let A_μ be the operator on $L^2(\mathbb{R}, d\mu)$ given by $(A_\mu f)(\lambda) = \lambda f(\lambda)$ and similarly for A_v on $L^2(\mathbb{R}, dv)$. Then A_μ and A_v are unitarily equivalent if and only if μ and v are equivalent measures.

3. Operators of uniform multiplicity

If one wants a canonical listing of the eigenvalues of a matrix, it is natural to list all eigenvalues of multiplicity one, all eigenvalues of multiplicity two, etc. We thus need a way of saying that A is an operator of uniform multiplicity two, three, etc. It is natural to take:

Definition A bounded self-adjoint operator A is said to be of **uniform multiplicity** m if A is unitarily equivalent to multiplication by λ on $L^2(\mathbb{R}, d\mu) \oplus \cdots \oplus L^2(\mathbb{R}, d\mu)$ where there are m terms in the sum and μ is a fixed Borel measure.

That this is a good definition is shown by

Proposition If A is unitarily equivalent to multiplication by λ on $L^2(\mathbb{R}, d\mu) \oplus \cdots \oplus L^2(\mathbb{R}, d\mu)$ (m times) and on $L^2(\mathbb{R}, d\nu) \oplus \cdots \oplus L^2(\mathbb{R}, d\nu)$ (n times), then $m = n$ and μ and ν are equivalent measures.

4. Disjoint measure classes

In listing eigenvalues of multiplicity one, two, three, etc. in the finite-dimensional case, we must add a requirement that prevents us from counting an eigenvalue of multiplicity three once as an eigenvalue of multiplicity one and once as an eigenvalue of multiplicity two. In the finite-dimensional case, we avoid this "error" by requiring the lists to be disjoint. The analogous notion for measures is:

Definition Two measure classes $\langle \mu \rangle$ and $\langle \nu \rangle$ are called **disjoint** if any $\mu_1 \in \langle \mu \rangle$ and $\nu_1 \in \langle \nu \rangle$ are mutually singular.

5. The multiplicity theorem

We can now state the basic theorem:

Theorem VII. 6 (commutative multiplicity theorem) Let A be a bounded self-adjoint operator on a Hilbert space $\mathcal{H}$. Then there is a decomposition $\mathcal{H} = \mathcal{H}_1 \oplus \mathcal{H}_2 \oplus \cdots \oplus \mathcal{H}_\infty$ so that
(a) A leaves each $\mathcal{H}_m$ invariant.
(b) $A \upharpoonright \mathcal{H}_m$ has uniform multiplicity m.

(c) The measure classes $\langle \mu_m \rangle$ associated with the spectral representation of $A \upharpoonright \mathscr{H}_m$ are mutually disjoint.

Moreover, the subspaces $\mathscr{H}_1, \ldots, \mathscr{H}_m, \ldots, \mathscr{H}_\infty$ (some of which may be zero) and the *measure classes* $\langle \mu_1 \rangle, \ldots, \langle \mu_m \rangle, \ldots, \langle \mu_\infty \rangle$ are uniquely determined by (a)–(c).

The spectral theorem with the multiplicity theory just described is thus one of those gems of mathematics: a structure theorem, that is, a theorem that describes all objects of a certain sort up to a natural equivalence. Each bounded self-adjoint operator A is described by a family of mutually disjoint measure classes on $[-\|A\|, \|A\|]$; two operators are unitarily equivalent if and only if their spectral multiplicity measure classes are *identical*.

VII.3 Spectral projections

In the last section, we constructed a functional calculus, $f \mapsto f(A)$ for any Borel function f and any bounded self-adjoint operator A. The most important functions gained in passing from the continuous functional calculus to the Borel functional calculus are the characteristic functions of sets.

Definition Let A be a bounded self-adjoint operator and Ω a Borel set of $\mathbb{R}$. $P_\Omega \equiv \chi_\Omega(A)$ is called a **spectral projection** of A.

As the definition suggests, P_Ω is an orthogonal projection since $\chi_\Omega^2 = \chi_\Omega = \bar{\chi}_\Omega$ pointwise. The properties of the family of projections $\{P_\Omega \mid \Omega$ an arbitrary Borel set$\}$ is given by the following elementary translation of the functional calculus (Problem 22).

Proposition The family $\{P_\Omega\}$ of spectral projections of a bounded self-adjoint operator, A, has the following properties:

(a) Each P_Ω is an orthogonal projection.
(b) $P_\varnothing = 0;\ P_{(-a,a)} = I$ for some a.
(c) If $\Omega = \bigcup_{n=1}^\infty \Omega_n$ with $\Omega_n \cap \Omega_m = \varnothing$ for all $n \neq m$, then

$$P_\Omega = \text{s-lim}_{N \to \infty} \left(\sum_{n=1}^N P_{\Omega_n} \right)$$

(d) $P_{\Omega_1} P_{\Omega_2} = P_{\Omega_1 \cap \Omega_2}$

Condition (c) is very reminiscent of the condition defining a measure and in fact one defines:

Definition A family of projections obeying (a)–(c) is called a (bounded) **projection-valued measure (p.v.m.)**.

We remark that (d) follows from (a) and (c) by abstract considerations (Problem 22).

As one might guess, one can integrate with respect to a p.v.m. If P_Ω is a p.v.m., then $(\phi, P_\Omega \phi)$ is an ordinary measure for any ϕ. We will use the symbol $d(\phi, P_\lambda \phi)$ to mean integration with respect to this measure. By standard Riesz lemma methods, there is a unique operator B with $(\phi, B\phi) = \int f(\lambda) d(\phi, P_\lambda \phi)$. Thus:

Theorem VII.7 If P_Ω is a p.v.m. and f a bounded Borel function on supp P_Ω, then there is a unique operator B which we denote $\int f(\lambda) dP_\lambda$ so that

$$(\phi, B\phi) = \int f(\lambda) \, d(\phi, P_\lambda \phi), \qquad \forall \phi \in \mathcal{H}$$

Example If A is a bounded self-adjoint operator and $\{P_\Omega\}$ its associated p.v.m., it is easy to see (Problem 23) that $f(A) = \int f(\lambda) dP_\lambda$. In particular $A = \int \lambda \, dP_\lambda$.

Now, suppose a bounded p.v.m. P_Ω is given and we form $A = \int \lambda \, dP_\lambda$. Not surprisingly (Problem 23), P_Ω is just the p.v.m. associated with A. Summarizing:

Theorem VII.8 (spectral theorem—p.v.m. form) There is a one–one correspondence between (bounded) self-adjoint operators A and (bounded) projection valued measures $\{P_\Omega\}$ given by:

$$A \mapsto \{P_\Omega\} = \{\chi_\Omega(A)\}$$

$$\{P_\Omega\} \mapsto A = \int \lambda \, dP_\lambda$$

It is through this theorem and its generalization to unbounded operators that self-adjoint operators arise in quantum mechanics, for the observables occur most naturally as projection-valued measures (see Section VIII.3 for the generalization and the notes to Section VIII.11 for the quantum-mechanical explanation).

Spectral projections can be used to investigate the spectrum of A:

Proposition $\lambda \in \sigma(A)$ if and only if $P_{(\lambda-\varepsilon, \lambda+\varepsilon)}(A) \neq 0$ for any $\varepsilon > 0$.

The essential element of the proof is that $\|(A - \lambda)^{-1}\| = [\text{dist}(\lambda, \sigma(A))]^{-1}$. The details are left to Problem 24.

This suggests that we distinguish between two types of spectrum:

Definition We say $\lambda \in \sigma_{\text{ess}}(A)$, the **essential spectrum** of A, if and only if $P_{(\lambda-\varepsilon, \lambda+\varepsilon)}(A)$ is infinite dimensional for all $\varepsilon > 0$. If $\lambda \in \sigma(A)$, but $P_{(\lambda-\varepsilon, \lambda+\varepsilon)}(A)$ is finite dimensional for some $\varepsilon > 0$, we say $\lambda \in \sigma_{\text{disc}}(A)$, the **discrete spectrum** of A. P is infinite dimensional means Ran P is infinite dimensional.

Thus, we have a second decomposition of $\sigma(A)$. Unlike the first, it is a decomposition into two necessarily disjoint subsets. We note that σ_{disc} is not necessarily closed, but:

Theorem VII.9 $\sigma_{\text{ess}}(A)$ is always closed.

Proof Let $\lambda_n \to \lambda$ with each $\lambda_n \in \sigma_{\text{ess}}(A)$. Since any open interval I about λ contains an interval about some λ_n, $P_I(A)$ is infinite dimensional. ∎

The following three theorems give alternative descriptions of σ_{disc} and σ_{ess}; their proofs are left to the reader (Problem 26).

Theorem VII.10 $\lambda \in \sigma_{\text{disc}}$ if and only if *both* the following hold:

(a) λ is an isolated point of $\sigma(A)$, that is, for some ε, $(\lambda - \varepsilon, \lambda + \varepsilon) \cap \sigma(A) = \{\lambda\}$.
(b) λ is an eigenvalue of finite multiplicity, i.e., $\{\psi \mid A\psi = \lambda\psi\}$ is finite dimensional.

Theorem VII.11 $\lambda \in \sigma_{\text{ess}}$ if and only if *one* or more of the following holds:

(a) $\lambda \in \sigma_{\text{cont}}(A) \equiv \sigma_{\text{ac}}(A) \cup \sigma_{\text{sing}}(A)$.
(b) λ is a limit point of $\sigma_{\text{pp}}(A)$.
(c) λ is an eigenvalue of infinite multiplicity.

Theorem VII.12 (Weyl's criterion) Let A be a bounded self-adjoint operator. Then $\lambda \in \sigma(A)$ if and only if there exists $\{\psi_n\}_{n=1}^{\infty}$ so that $\|\psi_n\| = 1$ and $\lim_{n\to\infty} \|(A - \lambda)\psi_n\| = 0$. $\lambda \in \sigma_{\text{ess}}(A)$ if and only if the above $\{\psi_n\}$ can be chosen to be orthogonal.

As one might guess, the essential spectrum cannot be removed by essentially finite dimensional perturbations. In Section XIII.4, we will prove a general theorem which implies that $\sigma_{\text{ess}}(A) = \sigma_{\text{ess}}(B)$ if $A - B$ is compact.

Finally, we discuss one useful formula relating the resolvent and spectral projections. It is a matter of computation to see that

$$f_\varepsilon(x) \to \begin{cases} 0 & x \notin [a, b] \\ \tfrac{1}{2} & x = a \text{ or } x = b \\ 1 & x \in (a, b) \end{cases}$$

as $\varepsilon \downarrow 0$ where

$$f_\varepsilon(x) = \frac{1}{2\pi i} \int_a^b \left(\frac{1}{x - \lambda - i\varepsilon} - \frac{1}{x - \lambda + i\varepsilon} \right) d\lambda$$

Moreover, $|f_\varepsilon(x)|$ is bounded uniformly in ε, so by the functional calculus, one has:

Theorem VII.13 (Stone's formula) Let A be a bounded self-adjoint operator. Then

$$\operatorname*{s-lim}_{\varepsilon \downarrow 0} (2\pi i)^{-1} \int_a^b [(A - \lambda - i\varepsilon)^{-1} - (A - \lambda + i\varepsilon)^{-1}] \, d\lambda = \tfrac{1}{2}[P_{[a,b]} + P_{(a,b)}]$$

VII.4 Ergodic theory revisited: Koopmanism

In Section II.4 we defined ergodicity for a measure preserving bijective map, $T: \Omega \to \Omega$ where Ω is a measure space with a finite measure μ, and $\mu(T^{-1}(M)) = \mu(M)$ for any measurable set $M \subset \Omega$. Koopman's lemma told us that the map U defined by $(Uf)(w) = f(Tw)$, is a unitary operator on $L^2(\Omega, d\mu)$. T was called ergodic if and only if 1 was a simple eigenvalue of U (that is, an eigenvalue of multiplicity one). In this section, we wish to examine in detail the idea of Koopman that interesting properties of T can be described in terms of spectral properties of U.

238   VII: THE SPECTRAL THEOREM

To appreciate the notion of mixing which we will shortly introduce, let us first consider an example:

Example 1   Let Ω be the surface of a torus: We can think of Ω as all pairs of numbers $\langle x, y \rangle$ with $0 \leq x < 1$; $0 \leq y < 1$ topologized so neighborhoods of 0 include points near 1. We consider two pairs $\langle x, y \rangle$ and $\langle z, w \rangle$ of reals as equivalent if $x - z$ and $y - w$ are integral. Then Ω is also all equivalence classes of pairs. Let us define a two-parameter family of maps $T_{a,b}$ $\Omega \to \Omega$ by $T_{a,b}\langle x, y \rangle = \langle x + a, y + b \rangle$. $T_{a,b}$ is Lebesgue measure preserving. When is $T_{a,b}$ ergodic relative to Lebesgue measure? If one uses the definition of ergodic which requires that there be no invariant sets of measure different from 1 or 0, it is not clear which $T_{a,b}$ are ergodic. However, if one looks at $(U_{a,b}f)(x, y) = f(x + a, y + b)$, one notices that it has a complete family of eigenvectors, $\varphi_{n,m}(x, y) = \exp[2\pi i(nx + my)]$. $U_{a,b}\varphi_{n,m} = \exp[2\pi i(na + mb)] \times \varphi_{n,m}$. When is 1 a simple eigenvalue? Obviously, if and only if $na + mb = k$ has no solution with n, m and k integral, except for $n = m = 0$ (e.g. when $a = \pi, b = \sqrt{2}$).

Thus $T_{\pi, \sqrt{2}}$ is ergodic, so space averages equal time averages in this case. This happens because the images $\{T^n w \mid w \in \Omega\}$ are a set which is dense and fairly uniform rather than because T^n takes a small neighborhood of w and

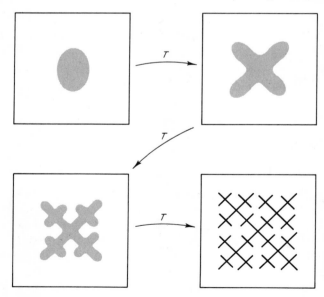

FIGURE VII.1   The intuitive idea of thermodynamic behavior in phase space.

spreads it all out as in Figure VII.1. In fact T^n is "shape preserving in this case." We expect "irreversibility" to mean that nearby points w and u are not nearby after many iterates of T.

How can we describe what it means to have a set M spread out "uniformly" upon iteration? One should expect a point to "forget" at what time it started in A, that is, the probability of being in $T^n A$ and another set B should approach an n-independent limit.

Definition A measure preserving transformation $T: \Omega \to \Omega$ on a measure space $\langle \Omega, \mu \rangle$ with $\mu(\Omega) = 1$ is called **mixing** if and only if for all measurable sets A and B in Ω:

$$\lim_{n \to \infty} \mu(T^n A \cap B) = \mu(A)\mu(B)$$

If T is ergodic, then by the mean ergodic theorem:

$$\lim_{n \to \infty} \frac{1}{n} \sum_{m=0}^{n-1} \mu(T^m A \cap B) = \lim_{n \to \infty} \frac{1}{n} \sum_{m=0}^{n-1} (U^{-m} \chi_A, \chi_B)$$

$$= (\chi_A, 1)(1, \chi_B)$$

$$= \mu(A)\mu(B)$$

Thus, ergodicity is equivalent to the mixing limit existing in a Cesaro sense, so that mixing implies ergodicity. Alternatively, we can see this directly. If $T[A] = A$, then $\lim_{n \to \infty} \mu(T^n A \cap A) = \mu(A)$ so mixing implies that $\mu(A)^2 = \mu(A)$, i.e. $\mu(A) = 0$ or 1. Before stating that mixing implies ergodicity as a formal proposition we introduce an intermediate notion:

Definition A measure preserving transformation $T: \Omega \to \Omega$ on a measure space $\langle \Omega, \mu \rangle$ with $\mu(\Omega) = 1$ is called **weakly mixing** if and only if for all measurable sets A and B in Ω:

$$\lim_{n \to \infty} \frac{1}{n} \sum_{m=0}^{n-1} |\mu(T^m A \cap B) - \mu(A)\mu(B)| = 0$$

Clearly:

Proposition Mixing $\Rightarrow$ weakly mixing $\Rightarrow$ ergodicity.

Let us first consider an example which we will shortly prove is mixing. We will see that Example 1 is *not* mixing.

Example 2 (the Baker's transformation)   Let Ω be the torus. Let

$$T\langle x, y\rangle = \begin{cases} \langle 2x, \tfrac{1}{2}y\rangle & \text{if } 0 \leq x < \tfrac{1}{2} \\ \langle 2x - 1, \tfrac{1}{2} + \tfrac{1}{2}y\rangle & \text{if } \tfrac{1}{2} \leq x < 1 \end{cases}$$

(see Figure VII.2). One can explicitly see the sets being "ripped apart" by T and "spread around."

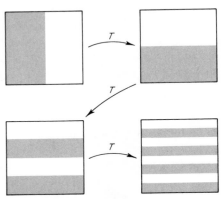

FIGURE VII.2   The Baker's transformation.

Weakly mixing and mixing have simple descriptions in terms of the associated unitary U:

Theorem VII.14   Let T be a measurable transformation and U the associated unitary operator. Then:

(a) U is mixing if and only if

$$w\text{-}\lim_{n\to\infty} U^n = P_1$$

where P_1 is the projection onto the constants, i.e. if and only if

$$\lim_{n\to\infty} (f, U^n g) = (f, 1)(1, g)$$

(b) U is weakly mixing if and only if U has no eigenvalue other than one and one is a simple eigenvalue.

Proof   (a) Mixing is clearly implied by $(f, U^n g) \to (f, 1)(1, g)$ for take $f = \chi_A$ and $g = \chi_B$. Conversely, if T is mixing, the limit statement is true if f and g are characteristic functions, and thus also if they are finite linear combinations of characteristic functions. Since these are dense, $\|U^n\| = 1$, and $\|P_1\| = 1$, the result follows.

(b)   See references in the Notes. ∎

Example 1 is *not* mixing since $U\psi = \lambda\psi$ implies that $w\text{-lim } U^n\psi \neq P_1\psi$ if $\lambda \neq 1$.

There is a spectral condition on U which often is useful in proving mixing. One can prove a spectral theorem for normal (and thus unitary) operators (Problems 3 and 5) so the notion of absolutely continuous spectrum makes good sense.

Theorem VII.15 Let T be a measure preserving transformation and U the associated unitary operator. Then

(a) If U has purely absolutely continuous spectrum on $\{1\}^\perp$, that is, if $\mathscr{H}_{ac} = \{f \mid (f, 1) = \int f \, d\mu = 0\}$, then T is mixing.

(b) If $\{1\}^\perp$ has an orthonormal basis $\{\varphi_{n,m}\}$, $-\infty < n < \infty$, $1 \leq m < N+1$, where N may be finite or infinite, so that $U\varphi_{n,m} = \varphi_{n+1,m}$, then U has purely absolutely continuous spectrum on $\{1\}^\perp$, and T is mixing.

Proof (a) It is quite easy to see that $U^n \overset{w}{\to} P_1$ if and only if $(f, U^n g) \to 0$ for all $f, g \in \{1\}^\perp$. Suppose that U has purely absolutely continuous spectrum. Then we can find functions $\{F_m\}_{m=1}^N$ and realizations of $f \in \{1\}^\perp$ as $f = \langle f_1(\theta), \ldots, f_m(\theta), \ldots \rangle$ so that

$$(f, U^n g) = \sum_{m=1}^N \int_0^{2\pi} e^{in\theta} \overline{f_m(\theta)} g_m(\theta) F_m(\theta) \, d\theta$$

$$= \int_0^{2\pi} e^{in\theta} Q(\theta) \, d\theta$$

where $\int_0^{2\pi} |Q(\theta)| \, d\theta < \infty$. By the Riemann–Lebesgue lemma which we prove in Section IX.2, $(f, U^n g) \to 0$.

(b) On $\{1\}^\perp$, U is just N copies of the right shift on $\ell_2(-\infty, \infty)$. We have analyzed this shift in Example 4 of Section VII.2 and seen that it has absolutely continuous spectrum. ∎

Example 2 (*revisited*) It is intuitively reasonable that the Baker's transformation is mixing and we now have the tools to prove that it is. Let us reexpress T. Write $\langle x, y \rangle \in \Omega$ in a base two decimal expansion $x = .x_1 x_2 \cdots$; $= .y_1 y_2 \cdots$ with each $x_i = 0$ or 1 and each $y_i = 0$ or 1. Then

$$T: (.x_1 x_2 \cdots, .y_1 y_2 \cdots) \to (.x_2 x_3 \cdots, .x_1 y_1 y_2 \cdots),$$

i.e. if we write a point in Ω as $(\ldots, y_3, y_2, y_1, x_1, x_2, \ldots)$ T is just a left shift. Warning: this is not the same as saying U is a left shift! This suggests what we should do. Define the functions $\chi_n(x, y)$ on Ω as follows: If $n > 0$

and $x_n = 0$, let $\chi_n(x, y) = 1$ and if $x_n = 1$, let $\chi_n(x, y) = -1$. If $n \leq 0$ and $y_{-n+1} = 0$, let $\chi_n(x, y) = 1$ and if $y_{-n+1} = 1$, let $\chi_n(x, y) = -1$. If $\{n_1, \ldots, n_m\}$ is a finite set of integers, define

$$\chi_{\{n_1, \ldots, n_m\}}(x, y) = \prod_{k=1}^{m} \chi_{n_k}(x, y)$$

Define $\chi_\emptyset = 1$. Then:

(a) $\chi_A \chi_B = \chi_{A \triangle B}$ where $A \triangle B = (A \backslash B) \cup (B \backslash A)$

(b) $\int \chi_A \, dx \, dy = \begin{cases} 0 & \text{if } A \neq \emptyset \\ 1 & \text{if } A = \emptyset \end{cases}$

(c) The χ_A are an orthonormal set, by (a) and (b).

(d) If m, n, k and j are integers with $0 < m \leq 2^n$, $0 < k \leq 2^j$, then the characteristic function of

$$\left(\frac{m-1}{2^j}, \frac{m}{2^j}\right] \times \left(\frac{k-1}{2^j}, \frac{k}{2^j}\right]$$

can be written as a finite product $\prod_{m=1}^{N} (1 \pm \chi_{n_m})$ for suitable integers, $n_1, \ldots, n_m$.

(e) The χ_A are an orthonormal basis for $L^2(\Omega, dx \otimes dy)$ for the linear combinations of the characteristic functions of (d) are dense.

(f) $U\chi_{\{n_1, \ldots, n_m\}} = \chi_{\{n_1+1, \ldots, n_m+1\}}$. Thus $\{1\}^\perp$ has an orthonormal basis $\psi_{n,m}$ with $U\psi_{n,m} = \psi_{n+1,m}$. m runs through a countably infinite sequence.

Thus the map in Example 2 is mixing.

Ergodicity, mixing and spectral notions are important not only in statistical mechanical contexts but also in studying the following problem: Let $\langle M, \mu \rangle$ and $\langle N, \nu \rangle$ be measure spaces and let $T: M \to M$ and $S: N \to N$ be measurable transformations which are invertible and measure preserving. When are they equivalent? That is, when is there a $R: M \to N$ so that $T = R^{-1}SR$? R is required to be bijective almost everywhere (i.e. $\mu\{x \mid Ry = Rx$ for some $y \neq x\} = 0$ and $\mu\{N \backslash \text{Ran } R\} = 0$) and measure preserving. This is analogous to the unitary equivalence problem for self-adjoint operators which is solved by the multiplicity theorems, but it has *not* been completely solved. In Problem 2 we construct various maps equivalent to the Baker's transformation.

The unitary equivalence problem for self-adjoint operators was solved by finding a *complete* set of invariants. Koopmanism immediately gives whole family of invariants of measure preserving maps, for if $T: M \to M$ and $S: N \to N$ are equivalent, the induced unitaries are unitarily equivalent. Given a measure preserving transformation, T, the multiplicity measure classes of the associated unitary operator are invariants for T, i.e. if S and are related by $T = R^{-1}SR$, then their classes are the same. Since ergodicity and mixing are expressible in terms of the induced Koopman unitaries they are not additional invariants. Are these invariants associated with the induced unitaries complete? That is, if the induced unitaries are unitarily equivalent, is it necessarily true that there exists an R so that $T = R^{-1}SR$? Under some special additional assumptions, the answer is yes.

Theorem VII.16 (Halmos, von Neumann)   Let $T: M \to M$ and $S: N \to N$ be measure-preserving ergodic transformations, U_S and V_T the induced unitaries. Suppose U_S and V_T have only pure point spectrum. Then T and S are measure theoretically equivalent if and only if U_S and V_T are unitarily equivalent.

On the other hand, Kolmogorov and Sinai have constructed an invariant called the entropy) for a class of mixing measurable transformations called K-systems. On $\{1\}^\perp$, the unitary induced by a K-system has an orthonormal basis $\{\psi_{n,m}\}_{n,m=-\infty}^{\infty}$ with $U\psi_{n,m} = \psi_{n+1,m}$. Thus, the unitaries induced by K-systems are all unitarily equivalent to one another. But, there exist K-systems with different entropy so the invariants of the induced unitaries do not distinguish all measure-preserving transformations.

NOTES

Section VII.1   For a proof and discussion of the finite dimensional spectral theorem e P. R. Halmos, *Finite Dimensional Vector Spaces*, Van Nostrand–Reinhold, Princeton ew Jersey, 1958.
For a related but slightly different proof of Theorem VII.1, see E. Nelson, *Topics in ynamics*, Vol. I, Princeton Univ. Press, Princeton, New Jersey, 1969.
The general functional calculus, $A \mapsto f(A)$ for functions f analytic in a neighborhood of A) which we alluded to is often called the **Dunford functional calculus** after its appearance N. Dunford, "Spectral Theory I, Convergence to Projections," *Trans. Amer. Math. Soc.* (1943), 185–217. The basic idea is to pick a contour C in the domain of f with $\sigma(A)$ contained within C. Then one takes $f(A) = (2\pi i)^{-1} \oint_C f(z)(z - A)^{-1} dz$. Thus, for example, e resolvent identity $(z - A)^{-1}(w - A)^{-1} = (w - z)^{-1}[(z - A)^{-1} - (w - A)^{-1}]$ implies)$(A) = f(A)g(A)$. For further discussion, see N. Dunford and J. Schwartz, *Linear erators*, Part I, pp. 556–577, Wiley (Interscience), New York, 1958 (also Problem 1).

Section VII.2   For a summary of the history of the spectral theorem, see the article E. Hellinger and O. Toeplitz, in *Encyklop. Math. Wiss. IIC* **13** (1928), 1335–1616.
The derivation of the functional calculus form of the spectral theorem is also discussed in

J. Diximier, *Les Algèbres d'Opérateurs dans l'Espace Hilbertien*, Gauthier, Paris, 1957, Appendix. The multiplication operator form is discussed in Nelson (ref. to Section VII.1) pp. 66–74.

There has arisen an extensive literature on a "rigorous" Dirac notation which attempts to capture the flavor of bras and kets more fully. For the original Dirac notation, see P. A. M. Dirac, *The Principles of Quantum Mechanics*, Oxford Univ. Press (Clarendon), London and New York, 1935. For the rigorous forms in terms of "rigged Hilbert spaces" see K. Maurin, *General Eigenfunction Expansions and Unitary Representations of Topological Groups*, Polish Scientific Publ., 1968; J. Roberts," The Dirac Bra and Ket Formalism," *J. Math. Phys.* **7** (1966), 1097–1104; "Rigged Hilbert Spaces in Quantum Mechanics," *Commun. Math. Phys.* **3** (1966), 98–119; or J. P. Antoine, "Dirac Formalism and Symmetry Problems in Quantum Mechanics I, II," *J. Math. Phys.* **10** (1969), 53–69, 2277–2290. We must emphasize that we regard the spectral theorem as sufficient for any argument where a nonrigorous approach might rely on Dirac notation; thus, we only recommend the abstract rigged space approach to readers with a strong emotional attachment to the Dirac formalism.

For an additional discussion of σ_{ac}, σ_{sing}, σ_{pp}, see T. Kato, *Perturbation Theory for Linear Operators*, pp. 516–519, Springer–Verlag, Berlin and New York, 1966.

The multiplicity theory for self-adjoint operators dates back at least as far as: H. Hahn "Uber die Integrale des Herrn Hellinger und die orthogonalinvarianten der quadratischer Formen von unendlichen Veränderlichen," *Monatsh. Math. Phy.* **23** (1912), 161–224 and E Hellinger, "Neue Begründung der Theorie quadratischen Formen von unendlichvieler Veranderlichen," *J. Reine Angew. Math.* **136** (1907), 210–271. For a modern readable ap proach, see Nelson (ref. to VII.1) pp. 77–97. The notion of measure class is often the correc continuous analogue to the notion of subset of a discrete set. This idea has been emphasize especially by G. W. Mackey, see *Group Representations and Applications*, pp. 48–80, Oxfor Lectures.

Almost all the spectral theory we discuss has a suitable generalization to the nonseparable case. It is for convenience and brevity that we only discuss the separable case.

Section VII.3 For a discussion of the spectral theorem in p.v.m. form and the relate "resolution of the identity form," see M. Naimark, *Normed Rings*, Nordhoff, New York 1964 or E. Lorch, *Spectral Theory*, Oxford Univ. Press, London and New York. Lorch' method of proof is closely linked to the Dunford functional calculus and Stone's formula See Kato's book (ref. to Section VII.2) for a related point of view.

Stone's formula goes back at least as far as M. Stone's classic book, *Linear Transformation in Hilbert Space and Their Applications to Analysis*, Amer. Math. Soc., Providence, Rhod Island, 1932.

The term essential spectrum goes back to Weyl's famous analysis of singular differentia operators, "Uber gewohnliche Differential-gleichungen mit Singularitaten und die zugehor igen Entwicklungen Willkurlicher Functionen," *Math. Ann.* **68** (1910), 220–269; for example if $H = -d^2/dx^2 + V$ is in the limit point case at infinity and the operators H_λ are the variou extensions of H on $L^2(0, \infty)$ with different boundary conditions, Weyl called $\bigcap \sigma(H$ the essential spectrum, that is, the spectrum independent of boundary conditions. It turn out that $\sigma_{ess}(H_\lambda)$ is the same for éach λ and is just the Weyl essential spectrum. For di cussion of this phenomenon, see Section XIII.4.

Section VII.4 Koopmanism dates back to the fundamental paper of Koopman (se notes to Section II.4). Much of what we discuss may be found in the review article by Wigh man and the books of Avez-Arnold and Halmos (notes to Section II.4). In particula proofs of Theorems VII.14(b) and VII.16 may be found in Halmos' book.

Mixing was introduced by E. Hopf in his note: "Complete Transitivity and the Ergodic Principle," *Proc. Nat. Acad. Sci.* **18** (1932), 204–209.

The theorem of Halmos and von Neumann (Theorem VII.16) was first proven in J. von Neumann, "Zur Operatoren methode in der Klassichen Mechanik," *Ann. Math.* **33** (1932), 587–642. Simplifications and additions may be found in P. R. Halmos and J. von Neumann, "Operator Methods in Classical Mechanics, II," *Ann. Math.* **43** (1942), 332–350. Halmos and von Neumann also prove that the possible discrete spectra of ergodic transformations are all countable subgroups of the circle.

Entropy for K-systems was introduced in A. N. Kolmogorov, "On the Entropy per Time Unit as a Metric Invariant of Automorphisms," *Dokl. Akad. Nauk.* **124** (1959), 754–755. It generalizes the idea of C. Shannon, "A Mathematical Theory of Communication," *Bell. System Tech. J.* **27** (1948), 379–423, 623–656 and is further discussed in J. Sinai, "Dynamical Systems with Countably Multiple Lebesgue Spectrum, I," *Izv. Akad. Nauk. SSSR Mat.* **25** (1961), 899–924 [*Engl. Transl.: Amer. Math. Soc. Transl.* (2) **39** (1964), 83–110]. For a readable introduction to the theory, see P. Billingsley, *Ergodic Theory and Information*, Wiley, New York, 1965.

In an important series of papers, D. Ornstein has clarified the extent to which entropy is a distinguishing invariant. The basic paper in the series is "Bernoulli Shifts with the Same Entropy are Isomorphic," *Advan. Math.* **4** (1970), 337–352.

PROBLEMS

*1. Let f be analytic in a neighborhood of $\sigma(A)$ where A is a bounded operator and let C be a contour as shown in Figure VII.3. Let $f(A)$ be defined by

$$f(A) = \frac{1}{2\pi i} \oint_C f(z)(z - A)^{-1} \, dz$$

Prove that $fg(A) = f(A)g(A)$.

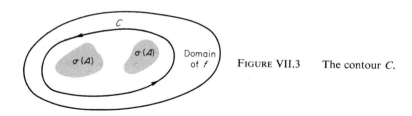

FIGURE VII.3   The contour C.

2. Suppose $\sigma(A)$ is not connected, say $\sigma(A) = \sigma_1 \cup \sigma_2$ where σ_1 and σ_2 are disjoint and closed. Consider the function f which is 1 in a neighborhood of σ_1 and 0 in a neighborhood of σ_2. Prove that $P = f(A)$ defined as in Problem 1 is a projection and that $PA = AP$. Prove that $\mathcal{H}_1 = \operatorname{Ran} P$ is an invariant subspace for A and that $\sigma(A \upharpoonright \mathcal{H}_1) = \sigma_1$.

3. (a) Prove that if A is **normal**, that is, $AA^* = A^*A$, then
$$\|A\| = \sup_{\lambda \in \sigma(A)} |\lambda| \equiv r(A)$$
Hint: Use $\|A\|^2 = \|A^*A\|$ and the formula $r(A) = \lim \|A^n\|^{1/n}$.
(b) Prove that for any polynomial P and any normal operator A, $\|P(A)\| = \sup_{\lambda \in \sigma(A)} |P(\lambda)|$.

*4. Let $A_1, \ldots, A_n$ be *commuting* bounded self-adjoint operators on $\mathscr{H}$, a separable Hilbert space.
(a) Let $\Omega_1, \ldots, \Omega_n$ be Borel sets on $\mathbb{R}$. Prove that $P_{\Omega_1}(A_1), P_{\Omega_2}(A_2), \ldots, P_{\Omega_n}(A_n)$ all commute.
(b) Let f be a function on $\mathbb{R}^n$ which is a linear combination of characteristic functions of rectangles (that is, sets of the form $\Omega = \Omega_1 \times \cdots \times \Omega_n$). Show that f can be written
$$f = \sum_{i=1}^{n} c_i \chi_{\Omega^{(i)}} \quad \text{with} \quad \Omega^{(i)} \cap \Omega^{(j)} = \varnothing \quad \text{if} \quad i \neq j$$
(c) For f of the above form, define
$$f(A_1, \ldots, A_n) = \sum_{i=1}^{n} c_i P_{\Omega_1^{(i)}}(A_1) \cdots P_{\Omega_n^{(i)}}(A_n)$$
where $\Omega^{(i)} = \Omega_1^{(i)} \times \cdots \times \Omega_n^{(i)}$.
(d) By using the B.L.T. theorem, construct a continuous functional calculus, $f(A_1, \ldots, A_n)$ for continuous functions on $[-\|A_1\|, \|A_1\|] \times \cdots \times [-\|A_n\|, \|A_n\|]$.
(e) Construct a Borel functional calculus.
(f) Show there is a $U: \mathscr{H} \to L^2(M, d\mu)$ for some finite measure space $\langle M, \mu \rangle$ and bounded real-valued Borel functions $F_1(m), \ldots, F_n(m)$ on M with
$$[(UA_i U)^{-1} f](m) = F_i(m) f(m)$$

5. Let A be a normal operator. Prove:
(a) $A = B + iC$ where B and C are commuting self-adjoint operators.
(b) There is a $U: \mathscr{H} \to L^2(M, d\mu)$ for some finite measure μ on some space M and a bounded Borel function, $F(m)$ (in general, complex valued) so that
$$(UAU^{-1}f)(m) = F(m)f(m)$$

Reference for Problems 4, 5: Nelson (see the Notes to Section VII.1).

*6. Extend Problem 4 to the case of countably many A_n. *Hint*: Use the product topology on $\mathsf{X}[-\|A_n\|, \|A_n\|]$.

†7. Find a self-adjoint operator A for which $[0, 1] \subset \sigma_{\text{sing}}(A)$ (*Hint*: Take A multiplicity free with a spectral measure which is an infinite weighted sum of translates of Cantor measures.)

†8. Let A be a bounded self-adjoint operator and let f be a continuous function on $\sigma(A)$.
(a) If $\lambda \notin \operatorname{Ran} f$, let $g = (f - \lambda)^{-1}$. Prove that $\phi(g) = (\phi(f) - \lambda)^{-1}$.
(b) Let $\lambda \in \operatorname{Ran} f(A)$. Prove that there are $\psi \in \mathscr{H}$, with $\|\psi\| = 1$ and $\|(\phi(f) - \lambda)\psi\|$ arbitrarily small so that $\lambda \in \sigma(\phi(f))$.
(c) Conclude (e) of Theorem VII.1.

Problems

†9. Suppose that f is continuous and f is not nonnegative on $\sigma(A)$ where A is a bounded self-adjoint operator. Show that there is a $\psi \in \mathcal{H}$ with $(\psi, \phi(f)\psi) < 0$.

†10. Prove that the range of the continuous functional calculus is the C^* algebra generated by A if A is self-adjoint (Theorem VII.1) or normal (Problem 5).

11. Suppose that $\phi: C(X) \to \mathcal{L}(\mathcal{H})$ is an algebraic $*$-homomorphism, X a compact Hausdorff space:
    (a) Prove that $\phi(f) \geq 0$ if $f \geq 0$.
    (b) Prove $\|\phi(f)\| \leq \|f\|_\infty$.

†12. Let $A \geq 0$. Prove that $(A - \lambda)^{-1}$ exists if $\lambda < 0$.

†13. Fill in the details of the proof of Theorem VII.2.

14. Prove that a self-adjoint operator on a finite-dimensional space has a cyclic vector if and only if it has no repeated eigenvalue.

†15. Prove Lemma 2 needed for Theorem VII.3.

16. Complete the reduction of Example 3 of Section VII.2 to a multiplication operator. Does this operator have uniform multiplicity? What is it?

†17. (a) Prove that $\sigma(A) = \text{supp } \{\mu_n\}_{n=1}^N$ if $\{\mu_n\}_{n=1}^N$ are the spectral measures. (The first proposition after Theorem VII.3.)
    (b) Let T_F be the operator of multiplications by F, a real-valued bounded measurable function. Prove that $\sigma(T_F)$ is the essential range of F.

†18. Let A be multiplication by x on $L^2(\mathbb{R}, d\mu) = L^2(\mathbb{R}, d\mu_{ac}) \oplus L^2(\mathbb{R}, d\mu_{pp}) \oplus L^2(\mathbb{R}, d\mu_{sing})$. Let $\psi \in L^2(\mathbb{R}, d\mu)$. Prove that $d\mu_\psi$ is absolutely continuous if and only if $\psi \in L^2(\mathbb{R}, d\mu_{ac})$.

*19. Is it possible to have a measure on $[0, 1]$ absolutely continuous w.r.t. dx with support $[0, 1]$ but which is not equivalent to dx, that is, is supp $\langle \mu \rangle$ a distinguishing invariant for measure classes absolutely continuous w.r.t. dx?

20. Let $A = U|A|$ be the polar decomposition of A. Let f_n be defined by $f_n(x) = 1/x$ if $x \geq 1/n$ and $f_n(x) = 1/n$ if $x \leq 1/n$. Prove that $U = \text{s-lim } Af_n(|A|)$. Conclude that U is in the von Neumann algebra generated by A, that is, the smallest strongly closed $*$-algebra containing A.

†21. (a) Prove that conditions (a) and (b) of Theorem VII.5 are equivalent.
    *(b) Prove condition (a) of Theorem VII.5 implies condition (c) in general. (*Hint*: Prove that $\{B \mid AB = BA\} = L^\infty(M, d\mu)$.)
    (c) Prove (c) implies (a) in the finite-dimensional case.

22. (a) Prove that the properties (a)–(d) of a p.v.m. hold for the spectral projections of an operator A.
    (b) Prove that condition (d) for a p.v.m. follows from (a) and (c).

23. (a) Supply the details of the proof of Theorem VII.7.
    (b) Prove that $f(A) = \int f(\lambda) dP_\lambda$ if $P_\Omega = \chi_\Omega(A)$.
    (c) If $A = \int \lambda \, dP_\lambda$, prove that $P_\Omega = \chi_\Omega(A)$.

24. Prove $\lambda \in \sigma(A)$ if and only if $P_{(\lambda-\varepsilon, \lambda+\varepsilon)}(A) \neq 0$ for all ε.

25. By considering compact operators, prove that σ_{disc} is not always closed.

†26. (a) Prove Theorem VII.10.
   (b) Prove Theorem VII.11.

27. Let C be self-adjoint and compact. What is $\sigma_{\text{ess}}(C)$? How is this related to the invariance property of σ_{ess} described at the end of Section VIII.3?

28. (a) Suppose that μ is a measure of unit mass with the property: Given $0 < x < 1$, there is a set $N \subset M$ with $\mu(N) = x$. Let $T: M \to M$ be measure preserving with $T^k = I$ for some k. Prove that T is not ergodic.
   (b) Prove directly that $T_{a,b}$ of Example 1 of Section VII.4 is not ergodic if $na + mb = r$ has a solution with r integral and $\langle n, m \rangle \neq \langle 0, 0 \rangle$ (that is, without recourse to the unitary U).

29. Show that the following measurable transformations are all equivalent to the Baker's transformation:
   (a) $M = \times_{n=-\infty}^{\infty} A_n$ where each A_n is the two-point set $\{H_n, T_n\} = A_n$[Heads, Tails]. μ is the product measure $\mu = \otimes_{n=-\infty}^{\infty} \mu_n$ where $\mu_n(\{H_n\}) = \frac{1}{2} = \mu_n(\{T_n\})$. Let $T: M \to M$ by shifting to the right. (This is called the honest coin toss space.)
   (b) $M = \times_{n=-\infty}^{\infty} B_n$ where each B_n is the three-point space, $B = \{0, 1, 2\}$. μ is the product measure $\mu = \otimes_{n=-\infty}^{\infty} \nu_n$ where $\nu_v(\{0\}) = \frac{1}{2} = \nu_n(\{2\})$; $\nu_n(\{1\}) = 0$. T is a right shift.
   (c) M is the square with the product of the Cantor measure with itself. T is given by
   $$T\langle x, y \rangle = \begin{cases} \langle 3x, \tfrac{1}{3}y \rangle & \text{if } 0 \leq x < \tfrac{1}{3} \\ \langle 3x - 1, \tfrac{1}{3}y + \tfrac{1}{3} \rangle & \text{if } \tfrac{1}{3} \leq x < \tfrac{2}{3} \\ \langle 3x - 2, \tfrac{1}{3}y + \tfrac{2}{3} \rangle & \text{if } \tfrac{2}{3} \leq x < 1 \end{cases}$$

30. Let $\langle M, \mu \rangle$ be a measure space with $T: M \to M$. Define $T \otimes T: M \times M \to M \times M$ by $(T \otimes T)\langle x, y \rangle = \langle Tx, Ty \rangle$.
   (a) Show that $\mu \otimes \mu$ is an invariant measure for $T \otimes T$ if μ is an invariant measure for T.
   (b) Find an example which shows that $T \otimes T$ may be not ergodic for $\langle M \times M, \mu \otimes \mu \rangle$ even if T is ergodic for $\langle M, \mu \rangle$. [Hint: Look at Example 1 of Section VII.4.]
   (c) Show that $T \otimes T$ is mixing for $\langle M \times M, \mu \otimes \mu \rangle$ if T is mixing for $\langle M, \mu \rangle$.
   (d) Show, once more, that Example 1 is not mixing by using (c).

   Remark: It is known that $T \otimes T$ is ergodic if and only if T is weakly mixing.

31. Prove that the only closed $*$-ideals in $\mathscr{L}(\mathscr{H})$, where $\mathscr{H}$ is separable, are $\{0\}$, $\text{Com}(\mathscr{H})$, and $\mathscr{L}(\mathscr{H})$. *Hint*: If the ideal $\mathscr{I}$ strictly contains $\text{Com}(\mathscr{H})$, find a self-adjoint, noncompact, operator $A \in \mathscr{I}$. Show that for any interval (a, b), with $0 \notin (a, b)$, $P_{(a,b)}(A) \in \mathscr{I}$. Conclude that $\mathscr{I}$ contains an infinite-dimensional projection and thus that $I \in \mathscr{I}$.

32. (a) Let A be self-adjoint and let U_λ be the partial isometry in $U_\lambda |A - \lambda| = A - \lambda$. Prove that $U_\lambda = P_{(\lambda, \infty)} - P_{(-\infty, \lambda)}$ and that $P_{(-\infty, \lambda)} = \lim_{\mu \uparrow \lambda} \tfrac{1}{2}(1 - U_\mu)$ and $P_{(-\infty, \lambda]} = \lim_{\mu \downarrow \lambda} \tfrac{1}{2}(1 - U_\mu)$.
   (b) Given a self-adjoint operator, use the polar decomposition and the formula of (a) to prove the spectral theorem without recourse to the functional calculus.

VIII: Unbounded Operators

I tell them that if they will occupy themselves with the study of mathematics they will find in it the best remedy against the lusts of the flesh.
                                    Thomas Mann in **The Magic Mountain**

VIII.1 Domains, graphs, adjoints, and spectrum

It is a fact of life that many of the most important operators which occur in mathematical physics are not bounded. In this chapter we will introduce some of the basic definitions and theorems necessary for dealing with unbounded operators on Hilbert spaces. The Hellinger–Toeplitz theorem (see Section III.5) says that an everywhere-defined operator A which satisfies $(A\varphi, \psi) = (\varphi, A\psi)$ is necessarily a bounded operator suggesting that a general unbounded operator T will only be defined on a dense linear subset of the Hilbert space $\mathscr{H}$. Thus an **operator** on a Hilbert space $\mathscr{H}$ is a linear map from its domain, a linear subspace of $\mathscr{H}$, into $\mathscr{H}$. Unless we specify otherwise, we will always suppose that the domain is dense. This subspace, which we denote by $D(T)$, is called the **domain** of the operator T. So, to identify an unbounded operator on a Hilbert space one must first give the domain on which it acts and then specify how it acts on that subspace.

Example 1 (the position operator)   Let $\mathscr{H} = L^2(\mathbb{R})$ and let $D(T)$ be the set of functions φ in $L^2(\mathbb{R})$ which satisfy $\int_{\mathbb{R}} x^2 |\varphi(x)|^2 \, dx < \infty$. For $\varphi \in D(T)$ define $(T\varphi)(x) = x\varphi(x)$. It is clear that T is unbounded since if we choose φ to have support near plus or minus infinity, we can make $\|T\varphi\|$ as large as we like while keeping $\|\varphi\| = 1$. Of course, even if $\varphi \notin D(T)$, $x\varphi(x)$ has a well-defined meaning as a function, but it is not in $L^2(\mathbb{R})$. Thus, if we want to deal only with the Hilbert space $L^2(\mathbb{R})$ we must restrict the domain of T. The domain we have chosen is the largest one for which the range is in $L^2(\mathbb{R})$.

Example 2 Let $\mathcal{H} = L^2(\mathbb{R})$ and $D(T) = \mathscr{S}(\mathbb{R})$. On $D(T)$ define $T\psi = -\psi''(x) + x^2\psi(x)$. If $\varphi_n(x)$ is the nth Hermite function (see the appendix to Section V.3), then $\varphi_n(x) \in D(T)$ and $T\varphi_n(x) = (2n + 1)\varphi_n(x)$. Thus T must be unbounded since it has arbitrarily large eigenvalues.

The notion of the graph of a linear transformation, introduced by von Neumann, is very useful for studying unbounded operators.

Definition The **graph** of the linear transformation T is the set of pairs
$$\{\langle \varphi, T\varphi \rangle \mid \varphi \in D(T)\}$$
The graph of T, denoted by $\Gamma(T)$, is thus a subset of $\mathcal{H} \times \mathcal{H}$ which is a Hilbert space with inner product
$$(\langle \varphi_1, \psi_1 \rangle, \langle \varphi_2, \psi_2 \rangle) = (\varphi_1, \varphi_2) + (\psi_1, \psi_2)$$
T is called a **closed** operator if $\Gamma(T)$ is a closed subset of $\mathcal{H} \times \mathcal{H}$.

Definition Let T_1 and T be operators on $\mathcal{H}$. If $\Gamma(T_1) \supset \Gamma(T)$, then T_1 is said to be an **extension** of T and we write $T_1 \supset T$. Equivalently, $T_1 \supset T$ if and only if $D(T_1) \supset D(T)$ and $T_1\varphi = T\varphi$ for all $\varphi \in D(T)$.

Definition An operator T is **closable** if it has a closed extension. Every closable operator has a smallest closed extension, called its **closure**, which we denote by $\overline{T}$.

A natural way to try to obtain a closed extension of an operator, T, is to take the closure of its graph in $\mathcal{H} \times \mathcal{H}$. The trouble with this is that $\overline{\Gamma(T)}$ may not be the graph of an operator (for example, see Problem 1). However, most operators which we deal with will be symmetric operators (introduced in Section VIII.2) and we will see that they always have closed extensions.

Proposition If T is closable, then $\Gamma(\overline{T}) = \overline{\Gamma(T)}$.

Proof Suppose that S is a closed extension of T. Then $\overline{\Gamma(T)} \subset \Gamma(S)$ so if $\langle 0, \psi \rangle \in \overline{\Gamma(T)}$ then $\psi = 0$. Define R with $D(R) = \{\psi \mid \langle \psi, \phi \rangle \in \overline{\Gamma(T)}$ for some

VIII.1 Domains, graphs, adjoints, and spectrum

ϕ} by $R\psi = \phi$ where $\phi \in \mathcal{H}$ is the unique vector so that $\langle \psi, \phi \rangle \in \overline{\Gamma(T)}$. Then $\Gamma(R) = \overline{\Gamma(T)}$ so R is a closed extension of T. But $R \subset S$ which is an arbitrary closed extension, so $R = \overline{T}$. ∎

The following example illustrates the concepts we have just introduced.

Example 3 Let $\mathcal{H} = L^2(\mathbb{R})$, $D(T) = C_0^\infty(\mathbb{R})$, and $D(T_1) = C_0^1(\mathbb{R})$, the once continuously differentiable functions with compact support. Let $Tf = if'(x)$ if $f \in D(T)$ and $T_1 f = if'(x)$ if $f \in D(T_1)$. T_1 is an extension of T. We will show that $\overline{\Gamma(T)} \supset \Gamma(T_1)$. When we prove that T is symmetric and therefore closable, it will follow that $\overline{T}$ extends T_1. First we introduce the approximate identity, $\{j_\varepsilon(x)\}$. Let $j(x)$ be any positive, infinitely differentiable function with support in $(-1, 1)$ so that $\int_{-\infty}^\infty j(x)\, dx = 1$. Define $j_\varepsilon(x) = \varepsilon^{-1} j(x/\varepsilon)$. If $\varphi \in D(T_1)$, set

$$\varphi_\varepsilon(x) = \int_{-\infty}^\infty j_\varepsilon(x - t)\varphi(t)\, dt$$

Then

$$|\varphi_\varepsilon(x) - \varphi(x)| \leq \int j_\varepsilon(x-t)|\varphi(t) - \varphi(x)|\, dt$$

$$\leq \left(\sup_{\{t\mid |x-t|\leq \varepsilon\}} |\varphi(t) - \varphi(x)| \right) \int_{\mathbb{R}} j_\varepsilon(x-t)\, dt$$

$$= \sup_{\{t\mid |x-t|\leq \varepsilon\}} |\varphi(t) - \varphi(x)|$$

Since φ has compact support, it is uniformly continuous which implies that $\varphi_\varepsilon \to \varphi$ uniformly. Since the φ_ε have support in a fixed compact set, $\varphi_\varepsilon \to \varphi$ in $L^2(\mathbb{R})$. Similarly,

$$i \frac{d}{dx} \varphi_\varepsilon(x) = \int_{-\infty}^\infty i \frac{d}{dx} j_\varepsilon(x - t)\varphi(t)\, dt$$

$$= \int_{-\infty}^\infty -i\left(\frac{d}{dt} j_\varepsilon(x - t) \right) \varphi(t)\, dt$$

$$= \int_{-\infty}^\infty j_\varepsilon(x - t) i \frac{d}{dt} \varphi(t)\, dt$$

$$\xrightarrow{L^2(\mathbb{R})} i \frac{d}{dx} \varphi(x)$$

Since $j_\varepsilon(x)$ has compact support and is infinitely differentiable, $\varphi_\varepsilon \in C_0^\infty(\mathbb{R})$. Thus, $\varphi_\varepsilon \in D(T)$ for each $\varepsilon > 0$. What we have shown above is that $\varphi_\varepsilon \xrightarrow{L^2(\mathbb{R})} \varphi$ and $T\varphi_\varepsilon \xrightarrow{L^2(\mathbb{R})} T_1\varphi$ for any $\varphi \in D(T_1)$. Thus, the closure of the graph of T contains the graph of T_1.

The notion of adjoint operator can be extended to the unbounded case.

Definition Let T be a densely defined linear operator on a Hilbert space $\mathcal{H}$. Let $D(T^*)$ be the set of $\varphi \in \mathcal{H}$ for which there is an $\eta \in \mathcal{H}$ with

$$(T\psi, \varphi) = (\psi, \eta) \quad \text{for all} \quad \psi \in D(T) \tag{VIII.1}$$

For each such $\varphi \in D(T^*)$, we define $T^*\varphi = \eta$. T^* is called the **adjoint** of T. By the Riesz lemma, $\varphi \in D(T^*)$ if and only if $|(T\psi, \varphi)| \leq C \|\psi\|$ for all $\psi \in D(T)$.

We note that $S \subset T$ implies $T^* \subset S^*$.

Notice that for η to be uniquely determined by (VIII.1) we need the fact that $D(T)$ is dense. Unlike the case of bounded operators, the domain of T^* may not be dense as the following example shows. As a matter of fact it is possible to have $D(T^*) = \{0\}$.

Example 4 Suppose that f is a bounded measurable function, but that $f \notin L^2(\mathbb{R})$. Let $D(T) = \{\psi \in L^2(\mathbb{R}) \mid \int |f(x)\psi(x)| \, dx < \infty\}$. $D(T)$ certainly contains all the L^2 functions with compact support so $D(T)$ is dense in $L^2(\mathbb{R})$. Let ψ_0 be some fixed vector in $L^2(\mathbb{R})$ and define $T\psi = (f, \psi)\psi_0$ for $\psi \in D(T)$. Suppose that $\varphi \in D(T^*)$, then

$$(\psi, T^*\varphi) = (T\psi, \varphi) = ((f, \psi)\psi_0, \varphi)$$
$$= \overline{(f, \psi)}(\psi_0, \varphi)$$
$$= (\psi, (\psi_0, \varphi)f)$$

for all $\psi \in D(T)$. Thus $T^*\varphi = (\psi_0, \varphi)f$. Since $f \notin L^2(\mathbb{R})$, $(\psi_0, \varphi) = 0$. Thus any $\varphi \in D(T^*)$, is orthogonal to ψ_0, so $D(T^*)$ is not dense. In fact, $D(T^*)$ is just the vectors perpendicular to ψ_0, and on that domain T^* is the zero operator.

If the domain of T^* is dense, then we can define $T^{**} = (T^*)^*$. There is a simple relationship between the notions of adjoint and closure.

Theorem VIII.1 Let T be a densely defined operator on a Hilbert space $\mathcal{H}$. Then:

(a) T^* is closed.
(b) T is closable if and only if $D(T^*)$ is dense in which case $\overline{T} = T^{**}$.
(c) If T is closable, then $(\overline{T})^* = T^*$.

Proof We define a unitary operator V on $\mathcal{H} \times \mathcal{H}$ by
$$V\langle \phi, \psi \rangle = \langle -\psi, \phi \rangle$$
Since V is unitary, $V[E^\perp] = (V[E])^\perp$ for any subspace E. Let T be a linear operator on $\mathcal{H}$ and suppose $\langle \phi, \eta \rangle \in \mathcal{H} \times \mathcal{H}$. Then $\langle \phi, \eta \rangle \in V[\Gamma(T)]^\perp$ if and only if $(\langle \phi, \eta \rangle, \langle -T\psi, \psi \rangle) = 0$ for all $\psi \in D(T)$ which holds if and only if $(\phi, T\psi) = (\eta, \psi)$ for all $\psi \in D(T)$, that is, if and only if $\langle \phi, \eta \rangle \in \Gamma(T^*)$. Thus $\Gamma(T^*) = V[\Gamma(T)]^\perp$. Since $V[\Gamma(T)]^\perp$ is always a closed subspace of $\mathcal{H} \times \mathcal{H}$, this proves (a).

To prove (b), observe that $\Gamma(T)$ is a linear subset of $\mathcal{H} \times \mathcal{H}$ so
$$\begin{aligned}\overline{\Gamma(T)} &= (\Gamma(T)^\perp)^\perp \\ &= (V^2 \Gamma(T)^\perp)^\perp \\ &= (V(V\Gamma(T))^\perp)^\perp \\ &= (V\Gamma(T^*))^\perp\end{aligned}$$

Thus, by the proof of (a), if T^* is densely defined, $\overline{\Gamma(T)}$ is the graph of T^{**}.

Conversely, suppose that $D(T^*)$ is not dense and that $\psi \in D(T^*)^\perp$. A simple computation shows that $\langle \psi, 0 \rangle \in [\Gamma(T^*)]^\perp$ so $V[\Gamma(T^*)]^\perp$ is not the graph of a (single-valued) operator. Since $\overline{\Gamma(T)} = (V\Gamma(T^*))^\perp$, we see that T is not closable.

To prove (c), notice that if T is closable,
$$T^* = \overline{(T^*)} = T^{***} = (\overline{T})^* \quad \blacksquare$$

Definition Let T be a closed operator on a Hilbert space $\mathcal{H}$. A complex number λ is in the **resolvent set**, $\rho(T)$, if $\lambda I - T$ is a bijection of $D(T)$ onto $\mathcal{H}$ with a bounded inverse. If $\lambda \in \rho(T)$, $R_\lambda(T) = (\lambda I - T)^{-1}$ is called the **resolvent** of T at λ.

For a point to be in the resolvent set of T, several conditions must be satisfied. These conditions are not all independent. For example, if $\lambda I - T$ is a bijection of $D(T)$ onto $\mathcal{H}$, by the closed-graph theorem, its inverse is automatically bounded. For other relationships, see Problem 2.

The definitions of **spectrum**, **point spectrum**, and **residual spectrum** are the same for unbounded operators as they are for bounded operators. We will sometimes refer to the spectrum of nonclosed, but closable operators. In this case we always mean the spectrum of the closure.

Theorem VIII.2 Let T be a closed densely defined linear operator. Then the resolvent set of T is an open subset of the complex plane on which the resolvent is an analytic operator-valued function. Furthermore,

$$\{R_\lambda(T) \mid \lambda \in \rho(T)\}$$

is a commuting family of bounded operators satisfying

$$R_\lambda(T) - R_\mu(T) = (\mu - \lambda) R_\mu(T) R_\lambda(T) \quad \text{(VIII.2)}$$

The proof of this theorem is exactly the same as the proof of the bounded case (Theorem VI.5).

It may seem to the reader that many of the questions about domains and closures of unbounded operators are just a technical inconvenience; that one need only choose any dense domain which is small enough so that the unbounded operator makes sense and that is good enough. However, the choice of an appropriate domain is often intimately connected with the physics of the situation being described; see, for example, the discussion in Section X.1. Further, many of the properties of operators which are important are very sensitive to the choice of domain. The following example shows that the spectrum is such a property. In the example we use the notion of "absolutely continuous function" and the corresponding fundamental theorem of calculus. The reader who is unfamiliar with the definition and theorem can find them in the notes.

Example 5 We denote by $AC[0,1]$ the set of absolutely continuous functions on $[0,1]$ whose derivatives are in $L^2[0,1]$. Let T_1 and T_2 be the operation $i\, d/dx$ with domains

$$D(T_1) = \{\varphi \mid \varphi \in AC[0,1]\}$$
$$D(T_2) = \{\varphi \mid \varphi \in AC[0,1] \text{ and } \varphi(0) = 0\}$$

Both $D(T_1)$ and $D(T_2)$ are dense in $L^2[0,1]$ and both of the operators are closed. But:

(a) The spectrum of T_1 is $\mathbb{C}$.
(b) The spectrum of T_2 is empty.

The proof that T_1 and T_2 are closed is left as an exercise (Problem 3). To see that the spectrum of T_1 is the whole plane we observe that

$$(\lambda I - T_1) e^{-i\lambda x} = 0 \quad \text{and} \quad e^{-i\lambda x} \in D(T_1)$$

for all $\lambda \in \mathbb{C}$. As for T_2, the operator

$$(S_\lambda g)(x) = i \int_0^x e^{-i\lambda(x-s)} g(s)\, ds$$

satisfies $(\lambda I - T_2)S_\lambda = I$ and $S_\lambda(\lambda I - T_2)$ is the identity on $D(T_2)$. Moreover,

$$\|S_\lambda g\|_2^2 = \int_0^1 |(S_\lambda g)(x)|^2 \, dx$$

$$\leq \left(\sup_{x \in [0,1]} |(S_\lambda g)(x)| \right)^2$$

$$\leq \left(\sup_{x \in [0,1]} \int_0^x |e^{-i\lambda(x-s)} g(s)| \, ds \right)^2$$

$$\leq \left(\sup_{x \in [0,1]} \left(\int_0^x |e^{-i\lambda(x-s)}|^2 \, ds \right) \right) \left(\sup_{x \in [0,1]} \int_0^x |g(s)|^2 \, ds \right)$$

$$\leq C(\lambda) \|g\|_2^2$$

so S_λ is bounded. By the remark immediately after the definition of resolvent set, we need only have shown that $\lambda I - T_2$ is a bijection to conclude that S_λ is bounded. So, we could have avoided the above computation.

VIII.2 Symmetric and self-adjoint operators: the basic criterion for self-adjointness

Definition A densely defined operator T on a Hilbert space is called **symmetric** (or **Hermitian**) if $T \subset T^*$, that is, if $D(T) \subset D(T^*)$ and $T\varphi = T^*\varphi$ for all $\varphi \in D(T)$. Equivalently, T is symmetric if and only if

$$(T\varphi, \psi) = (\varphi, T\psi) \quad \text{for all} \quad \varphi, \psi \in D(T)$$

Definition T is called **self-adjoint** if $T = T^*$, that is, if and only if T is symmetric and $D(T) = D(T^*)$.

A symmetric operator is always closable, since $D(T^*) \supset D(T)$ is dense in $\mathscr{H}$. If T is symmetric, T^* is a closed extension of T, so the smallest closed extension T^{**} of T must be contained in T^*. Thus for symmetric operators, we have

$$T \subset T^{**} \subset T^*$$

For closed symmetric operators,

$$T = T^{**} \subset T^*$$

And, for self-adjoint operators,

$$T = T^{**} = T^*$$

From this one can easily see that a closed symmetric operator T is self-adjoint if and only if T^* is symmetric.

The distinction between closed symmetric operators and self-adjoint operators is very important. It is only for *self-adjoint* operators that the spectral theorem holds (see Section VIII.3) and it is only *self-adjoint* operators that may be exponentiated to give the one-parameter unitary groups (see Section VIII.4) which give the dynamics in quantum mechanics. Chapter X is mainly devoted to studying methods for proving that operators are self-adjoint. We content ourselves here with proving the basic criterion for self-adjointness. First, we introduce the useful notion of essential self-adjointness.

Definition A symmetric operator T is called **essentially self-adjoint** if its closure $\bar{T}$ is self-adjoint. If T is closed, a subset $D \subset D(T)$ is called a **core** for T if $\overline{T \restriction D} = T$.

If T is essentially self-adjoint, then it has one and only one self-adjoint extension, for suppose that S is a self-adjoint extension of T. Then, S is closed and thereby, since $S \supset T$, $S \supset T^{**}$. Thus, $S = S^* \subset (T^{**})^* = T^{**}$, and so $S = T^{**}$. The converse is also true; namely, if T has one and only one self-adjoint extension, then T is essentially self-adjoint (see Section X.1). Since $T^* = \bar{T}^* = T^{***}$, T is essentially self-adjoint if and only if

$$T \subset T^{**} = T^*$$

The importance of essential self-adjointness is that one is often given a nonclosed symmetric operator T. If T can be shown to be essentially self-adjoint, then there is uniquely associated to T a self-adjoint operator $\bar{T} = T^{**}$. Another way of saying this is that if A is a self-adjoint operator, then to specify A uniquely one need not give the exact domain of A (which is often difficult), but just some core for A.

Now, suppose that T is a self-adjoint operator and that there is a $\varphi \in D(T^*) = D(T)$ so that $T^*\varphi = i\varphi$. Then $T\varphi = i\varphi$ and

$$-i(\varphi, \varphi) = (i\varphi, \varphi) = (T\varphi, \varphi) = (\varphi, T^*\varphi) = (\varphi, T\varphi) = i(\varphi, \varphi)$$

so $\varphi = 0$. A similar proof shows that $T^*\varphi = -i\varphi$ can have no solutions. The converse statement, that if T is a closed symmetric operator and $T^*\varphi = \pm i\varphi$ has no solutions, then T is self-adjoint, is the basic criterion of self-adjointness.

Theorem VIII.3 (the basic criterion for self-adjointness) Let T be a symmetric operator on a Hilbert space $\mathscr{H}$. Then the following three statements are equivalent:

(a) T is self-adjoint.
(b) T is closed and $\text{Ker}(T^* \pm i) = \{0\}$.
(c) $\text{Ran}(T \pm i) = \mathcal{H}$.

Proof We have just seen that (a) implies (b). Suppose that (b) holds; we will prove (c). Since $T^*\varphi = -i\varphi$ has no solutions, $\text{Ran}(T-i)$ must be dense. Otherwise, if $\psi \in \text{Ran}(T-i)^\perp$, we would have $((T-i)\varphi, \psi) = 0$ for all $\varphi \in D(T)$, so $\psi \in D(T^*)$ and $(T-i)^*\psi = (T^* + i)\psi = 0$ which is impossible since $T^*\psi = -i\psi$ has no solutions. (Reversing this last argument we can show that if $\text{Ran}(T-i)$ is dense, the kernel of $T^* + i$ is $\{0\}$.) Since $\text{Ran}(T-i)$ is dense, we need only prove it is closed to conclude that $\text{Ran}(T-i) = \mathcal{H}$. But for all $\varphi \in D(T)$

$$\|(T-i)\varphi\|^2 = \|T\varphi\|^2 + \|\varphi\|^2$$

Thus if $\varphi_n \in D(T)$ and $(T-i)\varphi_n \to \psi_0$, we conclude that φ_n converges to some vector φ_0, and $T\varphi_n$ converges too. Since T is closed, $\varphi_0 \in D(T)$ and $(T-i)\varphi_0 = \psi_0$. Thus, $\text{Ran}(T-i)$ is closed, so $\text{Ran}(T-i) = \mathcal{H}$. Similarly, $\text{Ran}(T+i) = \mathcal{H}$.

Finally, we will show that (c) implies (a). Let $\varphi \in D(T^*)$. Since $\text{Ran}(T-i) = \mathcal{H}$, there is an $\eta \in D(T)$ so that $(T-i)\eta = (T^* - i)\varphi$. $D(T) \subset D(T^*)$, so $\varphi - \eta \in D(T^*)$ and

$$(T^* - i)(\varphi - \eta) = 0$$

Since $\text{Ran}(T+i) = \mathcal{H}$, $\text{Ker}(T^* - i) = \{0\}$, so $\varphi = \eta \in D(T)$. This proves that $D(T^*) = D(T)$, so T is self-adjoint. ∎

Corollary Let T be a symmetric operator on a Hilbert space. Then the following are equivalent:

(a) T is essentially self-adjoint.
(b) $\text{Ker}(T^* \pm i) = \{0\}$.
(c) $\text{Ran}(T \pm i)$ are dense.

We conclude with an example which shows that a symmetric operator may have many self-adjoint extensions. Lest the reader be misled we remark that a symmetric operator may have *no* self-adjoint extensions (see Problem 4 and Section X.1).

Example Let $T = i\, d/dx$ with

$$D(T) = \{\varphi \mid \varphi \in AC[0, 1],\ \varphi(0) = 0 = \varphi(1)\}$$

A simple integration by parts shows that as an operator on $L^2[0, 1]$, T is symmetric. We begin by determining T^*. Let j_ε be the function defined in

Example 3 of Section VIII.1. Fix $0 < \alpha < \beta < 1$ and define

$$f_\varepsilon^{\alpha,\beta}(x) = j_\varepsilon(x-\beta) - j_\varepsilon(x-\alpha)$$

$$g_\varepsilon^{\alpha,\beta}(x) = \int_0^x f_\varepsilon^{\alpha,\beta}(t)\, dt$$

Let $\psi \in D(T^*)$. For ε small enough, $g_\varepsilon^{\alpha,\beta} \in D(T)$ so

$$(Tg_\varepsilon^{\alpha,\beta}, \psi) = (g_\varepsilon^{\alpha,\beta}, T^*\psi) \tag{VIII.3}$$

As $\varepsilon \to 0, -g_\varepsilon^{\alpha,\beta}$ converges to the characteristic function of (α, β) in $L^2(0, 1)$ so

$$(g_\varepsilon^{\alpha,\beta}, T^*\psi) \to -\int_\alpha^\beta (T^*\psi)(x)\, dx$$

The basic estimate of Example 3 shows that

$$J_\varepsilon \varphi = \int_0^1 j_\varepsilon(x-t)\varphi(t)\, dt$$

converges in L^2 to φ if φ is continuous. Moreover, each J_ε is a bounded operator of norm not greater than one. For if $\psi \in L^2(0, 1)$, then

$$|(\psi, J_\varepsilon \varphi)| \leq \iint j_\varepsilon(x-t)|\varphi(t)||\psi(x)|\, dx\, dt$$

$$= \iint j_\varepsilon(y)|\varphi(t)||\psi(y+t)|\, dy\, dt$$

$$\leq \|\varphi\|\,\|\psi\| \int j_\varepsilon(y)\, dy$$

$$= \|\varphi\|\,\|\psi\|$$

By an $\varepsilon/3$ argument, $J_\varepsilon \varphi \xrightarrow{L^2} \varphi$ for all $\varphi \in L^2[0, 1]$. Thus, the left side of (VIII.3) converges to $-i(\psi(\beta) - \psi(\alpha))$ in mean square as $\varepsilon \to 0$. So, for almost all α, β

$$i(\psi(\beta) - \psi(\alpha)) = \int_\alpha^\beta (T^*\psi)(x)\, dx$$

This means that ψ is absolutely continuous (see the Notes for Section VIII.1) and

$$i\frac{d}{dx}\psi(x) = (T^*\psi)(x)$$

Thus, $\psi \in AC[0, 1]$ and $T^*\psi = i\, d\psi(x)/dx$. Conversely, integration by parts shows that any $\psi \in AC[0, 1]$ is in the domain of T^* and $T^*\psi = i\, d\psi/dx$. Therefore $T^* = i\, d/dx$ on $D(T^*) = AC[0, 1]$.

It is easy to see that T is not essentially self-adjoint since

$$e^{\pm x} \in D(T^*) \quad \text{and} \quad i\frac{d}{dx}e^{\pm x} = \pm i e^{\pm x}$$

In fact, T is closed (Problem 6) so T is a closed, symmetric but not self-adjoint operator.

Does T have any self-adjoint extensions? Yes, uncountably many different ones! Let $\alpha \in \mathbb{C}$, $|\alpha| = 1$, and define $T_\alpha = i\, d/dx$ on

$$D(T_\alpha) = \{\varphi \mid \varphi \in AC[0, 1], \quad \varphi(0) = \alpha \varphi(1)\}$$

Each of these operators T_α is a different self-adjoint extension of T (Problem 7). Of course, each T_α is in turn extended by T^*. These extensions are depicted in Figure VIII.1. The reason why there is exactly a circle of different possible extensions will be made clear in Section X.1.

FIGURE VIII.1   The self-adjoint extensions of T.

VIII.3   The spectral theorem

A good definition should be the hypothesis of a theorem.

J. Glimm

In this section we will show how the spectral theorem for bounded self-adjoint operators which we developed in Chapter VII can be extended to unbounded self-adjoint operators. To indicate what we are aiming for, we first prove the following:

Proposition 1   Let $\langle M, \mu \rangle$ be a measure space with μ a finite measure. Suppose that f is a measurable, real-valued function on M which is finite a.e.$[\mu]$. Then the operator $\varphi \xrightarrow{T_f} f\varphi$ on $L^2(M, d\mu)$ with domain

$$D(T_f) = \{\varphi \mid f\varphi \in L^2(M, \mu)\}$$

is self-adjoint and $\sigma(T_f)$ is the essential range of f.

Proof T_f is clearly symmetric. Suppose that $\psi \in D(T_f^*)$ and let

$$\chi_N(m) = \begin{cases} 1 & \text{if } |f(m)| \leq N \\ 0 & \text{otherwise} \end{cases}$$

Then, using the monotone convergence theorem,

$$\|T_f^*\psi\| = \lim_{N\to\infty} \|\chi_N T_f^*\psi\|$$

$$= \lim_{N\to\infty} \left(\sup_{\|\varphi\|=1} |(\varphi, \chi_N T_f^*\psi)|\right)$$

$$= \lim_{N\to\infty} \left(\sup_{\|\varphi\|=1} |(T_f \chi_N \varphi, \psi)|\right)$$

$$= \lim_{N\to\infty} \left(\sup_{\|\varphi\|=1} |(\varphi, \chi_N f\psi)|\right)$$

$$= \lim_{N\to\infty} \|\chi_N f\psi\|$$

Thus, $f\psi \in L^2(M, \mu)$, so $\psi \in D(T_f)$ and therefore T_f is self-adjoint. That $\sigma(T_f)$ is the essential range of f follows as in the bounded case (Problem 17 of Chapter VII). ∎

With more information about f, one can say something about the domains on which T_f is essentially self-adjoint:

Proposition 2 Let f and T_f obey the conditions in Proposition 1. Suppose in addition that $f \in L^p(M, d\mu)$ for $2 < p < \infty$. Let D be any dense set in $L^q(M, d\mu)$ where $q^{-1} + p^{-1} = 1/2$. Then D is a core for T_f.

Proof Let us first show that L^q is a core for T_f. By Hölder's inequality $\|g\|_2 \leq \|1\|_p \|g\|_q$ and $\|fg\|_2 \leq \|f\|_p \|g\|_q$ so $L^q \subset D(T_f)$. Moreover, if $g \in D(T_f)$, let g_n be that function which is zero where $|g(m)| > n$ and equal to g otherwise. By the dominated convergence theorem, $g_n \to g$ and $fg_n \to fg$ in L^2. Since each g_n is in L^q, we conclude that L^q is a core for T_f.

Now let D be dense in L^q and let $g \in L^q$. Find $g_n \in D$ with $g_n \to g$ in L^q. Since $\|g_n - g\|_2 \leq \|1\|_p \|g_n - g\|_q$ and $\|T_f(g_n - g)\|_2 \leq \|f\|_p \|g_n - g\|_q$, $g \in D(\overline{T_f \restriction D})$. Thus $L^q \subset D(\overline{T_f \restriction D})$ so D is a core. ∎

Unless $f \in L^\infty(M, \mu)$ the operator T_f described in Propositions 1 and 2 will be unbounded. Thus, we have found a large class of unbounded self-adjoint operators. In fact, we have found them all.

Theorem VIII.4 (spectral theorem—multiplication operator form) Let A be a self-adjoint operator on a separable Hilbert space $\mathcal{H}$ with domain $D(A)$. Then there is a measure space $\langle M, \mu \rangle$ with μ a finite measure, a unitary operator $U: \mathcal{H} \to L^2(M, d\mu)$, and a real-valued function f on M which is finite a.e. so that

(a) $\psi \in D(A)$ if and only if $f(\cdot)(U\psi)(\cdot) \in L^2(M, d\mu)$.
(b) If $\varphi \in U[D(A)]$, then $(UAU^{-1}\varphi)(m) = f(m)\varphi(m)$.

Proof In the proof of Theorem VIII.3 it was shown that $A + i$ and $A - i$ are one to one and $\text{Ran}(A \pm i) = \mathcal{H}$. Since $A \pm i$ are closed, $(A \pm i)^{-1}$ are closed and therefore bounded (Theorem III.12). By Theorem VIII.2, $(A + i)^{-1}$ and $(A - i)^{-1}$ commute. The equality

$$((A - i)\psi, (A + i)^{-1}(A + i)\varphi) = ((A - i)^{-1}(A - i)\psi, (A + i)\varphi)$$

and the fact that $\text{Ran}(A \pm i) = \mathcal{H}$ shows that $((A + i)^{-1})^* = (A - i)^{-1}$. Thus $(A + i)^{-1}$ is normal.

We now use the easy extension of the spectral theorem for bounded self-adjoint operators to bounded normal operators. The proof of this extension is outlined in Problems 3, 4, and 5 of Chapter VII. We conclude that there is a measure space $\langle M, \mu \rangle$ with μ a finite measure, a unitary operator $U: \mathcal{H} \to L^2(M, \mu)$, and a measurable, bounded, complex-valued function $g(m)$ so that $U(A + i)^{-1}U^{-1}\varphi(m) = g(m)\varphi(m)$ for all $\varphi \in L^2(M, d\mu)$.

Since $\text{Ker}(A + i)^{-1}$ is empty, $g(m) \neq 0$ a.e.$[\mu]$, so the function $f(m) = g(m)^{-1} - i$ is finite a.e.$[\mu]$. Now, suppose $\psi \in D(A)$. Then $\psi = (A + i)^{-1}\varphi$ for some $\varphi \in \mathcal{H}$ and $U\psi = gU\varphi$. Since fg is bounded, we conclude that $f(U\psi) \in L^2(M, d\mu)$. Conversely, if $f(U\psi) \in L^2(M, d\mu)$, then there is a $\varphi \in \mathcal{H}$ so that $U\varphi = (f + i)U\psi$. Thus, $gU\varphi = g(f + i)U\psi = U\psi$, so $\psi = (A + i)^{-1}\varphi$, which shows that $\psi \in D(A)$. This proves (a).

To prove (b) notice that if $\psi \in D(A)$, then $\psi = (A + i)^{-1}\varphi$ for some $\varphi \in \mathcal{H}$ and $A\psi = \varphi - i\psi$. Therefore,

$$\begin{aligned}(UA\psi)(m) &= (U\varphi)(m) - i(U\psi)(m) \\ &= (g(m)^{-1} - i)(U\psi)(m) \\ &= f(m)(U\psi)(m)\end{aligned}$$

Finally, if $\text{Im}(f) > 0$ on a set of nonzero measure, there is a bounded set B in the upper half plane so that $S = \{x \mid f(x) \in B\}$ has nonzero measure. If χ is the characteristic function of S, then $f\chi \in L^2(M, d\mu)$ and $\text{Im}(\chi, f\chi) > 0$. This contradicts the fact that multiplication by f is self-adjoint (since it is unitarily equivalent to A). Thus f is real-valued. ∎

There is a natural way to define functions of a self-adjoint operator by using the above theorem. Given a bounded Borel function h on $\mathbb{R}$ we define

$$h(A) = U^{-1}T_{h(f)}U$$

where $T_{h(f)}$ is the operator on $L^2(M, d\mu)$ which acts by multiplication by the function $h(f(m))$. Using this definition the following theorem follows easily from Theorem VIII.4.

Theorem VIII.5 (spectral theorem—functional calculus form)   Let A be a self-adjoint operator on $\mathcal{H}$. Then there is a unique map $\hat{\phi}$ from the bounded Borel functions on $\mathbb{R}$ into $\mathcal{L}(\mathcal{H})$ so that

(a) $\hat{\phi}$ is an algebraic *-homomorphism.
(b) $\hat{\phi}$ is norm continuous, that is, $\|\hat{\phi}(h)\|_{\mathcal{L}(\mathcal{H})} \leq \|h\|_\infty$.
(c)   Let $h_n(x)$ be a sequence of bounded Borel functions with $h_n(x) \xrightarrow[n \to \infty]{} x$ for each x and $|h_n(x)| \leq |x|$ for all x and n. Then, for any $\psi \in D(A)$, $\lim_{n \to \infty} \hat{\phi}(h_n)\psi = A\psi$.
(d)   If $h_n(x) \to h(x)$ pointwise and if the sequence $\|h_n\|_\infty$ is bounded, then $\hat{\phi}(h_n) \to \hat{\phi}(h)$ strongly.

In addition:
(e)   If $A\psi = \lambda\psi$, $\hat{\phi}(h)\psi = h(\lambda)\psi$.
(f)   If $h \geq 0$, then $\hat{\phi}(h) \geq 0$.

The functional calculus is very useful. For example, it allows us to define the exponential e^{itA} and prove easily many of its properties as a function of t (see the next section). In the case where A is bounded we do not need the functional calculus to define the exponential since we can define e^{itA} by the power series which converges in norm.

The functional calculus is also used to construct spectral measures and can be used to develop a multiplicity theory similar to that for bounded self-adjoint operators. A vector ψ is said to be **cyclic** for A if $\{g(A)\psi \,|\, g \in C_\infty(\mathbb{R})\}$ is dense in $\mathcal{H}$. If ψ is a cyclic vector, then it is possible to represent $\mathcal{H}$ as $L^2(\mathbb{R}, d\mu_\psi)$ where μ_ψ is the measure satisfying

$$\int_\mathbb{R} g(x) \, d\mu_\psi(x) = (\psi, g(A)\psi)$$

in such a way that A becomes multiplication by x. In general, $\mathcal{H}$ decomposes into a direct sum of cyclic subspaces so the measure space, M, in Theorem VIII.4 can be realized as a union of copies of $\mathbb{R}$. As in the case of bounded operators we can define $\sigma_{ac}(A)$, $\sigma_{pp}(A)$, $\sigma_{sing}(A)$, and decompose $\mathcal{H}$ accordingly.

Finally, the spectral theorem in its projection-valued measure form follows easily from the functional calculus. Let P_Ω be the operator $\chi_\Omega(A)$ where χ_Ω is the characteristic function of the measurable set $\Omega \subset \mathbb{R}$. The family of operators $\{P_\Omega\}$ has the following properties:

(a) Each P_Ω is an orthogonal projection.
(b) $P_\varnothing = 0$, $P_{(-\infty, \infty)} = I$
(c) If $\Omega = \bigcup_{n=1}^\infty \Omega_n$ with $\Omega_n \cap \Omega_m = \varnothing$ if $n \neq m$, then $P_\Omega = \text{s-lim}_{N\to\infty} \sum_{n=1}^N P_{\Omega_n}$
(d) $P_{\Omega_1} P_{\Omega_2} = P_{\Omega_1 \cap \Omega_2}$

Such a family is called a **projection-valued measure** (p.v.m.). This is a generalization of the notion of bounded projection-valued measure introduced in Chapter VII in that we only require $P_{(-\infty, \infty)} = I$ rather than $P_{(-a, a)} = I$ for some a. For $\varphi \in \mathcal{H}$, $(\varphi, P_\Omega \varphi)$ is a well-defined Borel measure on $\mathbb{R}$ which we denote by $d(\varphi, P_\lambda \varphi)$ as in Chapter VII. The complex measure $d(\varphi, P_\lambda \psi)$ is defined by polarization. Thus, given a bounded Borel function g we can define $g(A)$ by

$$(\varphi, g(A)\varphi) = \int_{-\infty}^\infty g(\lambda)\, d(\varphi, P_\lambda \varphi) \qquad \text{(VIII.4)}$$

It is not difficult to show that this map $g \mapsto g(A)$ has the properties (a)–(d) of Theorem VIII.5, so $g(A)$ as defined by (VIII.4) coincides with the definition of $g(A)$ given by Theorem VIII.4. Now, suppose g is an unbounded complex-valued Borel function and let

$$D_g = \{\varphi \mid \int_{-\infty}^\infty |g(\lambda)|^2\, d(\varphi, P_\lambda \varphi) < \infty\} \qquad \text{(VIII.5)}$$

Then, D_g is dense in $\mathcal{H}$ and an operator $g(A)$ is defined on D_g by

$$(\varphi, g(A)\varphi) = \int_{-\infty}^\infty g(\lambda)\, d(\varphi, P_\lambda \varphi)$$

As in Chapter VII, we write symbolically

$$g(A) = \int g(\lambda)\, dP_\lambda$$

In particular, for $\varphi, \psi \in D(A)$,

$$(\varphi, A\psi) = \int_{-\infty}^\infty \lambda\, d(\varphi, P_\lambda \psi)$$

If g is real-valued, then $g(A)$ is self-adjoint on D_g. We summarize:

Theorem VIII.6 (spectral theorem—projection valued measure form) There is a one-to-one correspondence between self-adjoint operators A and projection-valued measures $\{P_\Omega\}$ on $\mathcal{H}$, the correspondence being given by

$$A = \int_{-\infty}^\infty \lambda\, dP_\lambda$$

If $g(\cdot)$ is a real-valued Borel function on $\mathbb{R}$, then

$$g(A) = \int_{-\infty}^{\infty} g(\lambda)\, dP_\lambda$$

defined on D_g (VIII.5) is self-adjoint. If g is bounded, $g(A)$ coincides with $\hat{\phi}(g)$ in Theorem VIII.5.

We conclude with several remarks. First, Stone's formula, given in Theorem VII.13, relates the resolvent and the projection-valued measure associated with any self-adjoint operator. The proof is the same as in the bounded case.

The spectrum of an unbounded self-adjoint operator is an unbounded subset of the real axis. One can define discrete and essential spectrum; they are still characterized by Theorems VII.9, VII.10, and VII.11. Theorem VII.12 (Weyl's criterion) still holds if one adds the criterion that the vectors $\{\psi_n\}$ must be in the domain of A.

Finally, we note that the measure space of Theorem VIII.4 can always be chosen so that Proposition 2 is applicable:

Proposition 3 Let A be a self-adjoint operator on a separable Hilbert space $\mathscr{H}$. Then the measure space $\langle M, \mu \rangle$ and the function f of Theorem VIII.4 can be chosen so that $f \in L^p(M, d\mu)$ for all p with $1 \leq p < \infty$.

Proof By Theorem VIII.4, we know that A is unitarily equivalent to T_f on some measure space $\langle M, \nu \rangle$ with $\nu(M) < \infty$. Let μ be the measure given by

$$d\mu = e^{-f^2}\, d\nu$$

Then T_f on $L^2(M, d\mu)$ is unitarily equivalent to T_f on $L^2(M, d\nu)$ under $V: L^2(M, d\nu) \to L^2(M, d\mu)$ given by $Vg(m) = (e^{+f^2/2}g)(m)$. Moreover, $f \in L^p(M, d\mu)$ for any $1 \leq p < \infty$. ∎

VIII.4 Stone's theorem

In this section we prove a theorem due to Stone which, like the spectral theorem, is fundamental for quantum mechanics. Suppose that A is a self-adjoint operator on $\mathscr{H}$. If A is bounded, we can define the exponential of A by

$$e^{itA} = \sum_{n=0}^{\infty} \frac{(it)^n A^n}{n!}$$

since the series converges in norm. If A unbounded *and* self-adjoint, we cannot use the power series directly, but we can use the functional calculus developed in the last section to define e^{itA}.

Theorem VIII.7 Let A be a self-adjoint operator and define $U(t) = e^{itA}$. Then

(a) For each $t \in \mathbb{R}$, $U(t)$ is a unitary operator and $U(t + s) = U(t)U(s)$ for all $s, t \in \mathbb{R}$.

(b) If $\varphi \in \mathscr{H}$ and $t \to t_0$, then $U(t)\varphi \to U(t_0)\varphi$.

(c) For $\psi \in D(A)$, $\dfrac{U(t)\psi - \psi}{t} \to iA\psi$ as $t \to 0$.

(d) If $\lim\limits_{t \to 0} \dfrac{U(t)\psi - \psi}{t}$ exists, then $\psi \in D(A)$.

Proof (a) follows immediately from the functional calculus and the corresponding statements for the complex-valued function $e^{it\lambda}$. To prove (b) observe that

$$\|e^{itA}\varphi - \varphi\|^2 = \int_R |e^{it\lambda} - 1|^2 \, d(P_\lambda \varphi, \varphi)$$

Since $|e^{it\lambda} - 1|^2$ is dominated by the integrable function $g(\lambda) = 2$ and since

$$|e^{it\lambda} - 1|^2 \xrightarrow[t \to 0]{} 0 \quad \text{for each } \lambda \in \mathbb{R}$$

we conclude that $\|U(t)\varphi - \varphi\| \to 0$ by the Lebesgue dominated-convergence theorem. Thus $t \mapsto U(t)$ is strongly continuous at $t = 0$, which by the group property proves $t \mapsto U(t)$ is strongly continuous everywhere. The proof of (c), which again uses the dominated convergence theorem and the estimate $|e^{ix} - 1| \leq |x|$, is left to the reader (Problem 11). To prove (d), we define

$$D(B) = \{\psi \mid \lim_{t \to 0} \dfrac{U(t)\psi - \psi}{t} \text{ exists}\}$$

and let $iB\psi = \lim_{t \to 0} [U(t)\psi - \psi]/t$. A simple computation shows that B is symmetric. By (c), $B \supset A$, so $B = A$. ∎

Definition An operator-valued function $U(t)$ satisfying (a) and (b) is called **a strongly continuous one-parameter unitary group**.

The following theorem says that every strongly continuous unitary group arises as the exponential of a self-adjoint operator.

Theorem VIII.8 (Stone's theorem)   Let $U(t)$ be a strongly continuous one-parameter unitary group on a Hilbert space $\mathscr{H}$. Then, there is a self-adjoint operator A on $\mathscr{H}$ so that $U(t) = e^{itA}$.

Proof   Part (d) of Theorem VIII.7 suggests that we obtain A by differentiating $U(t)$ at $t = 0$. We will show that this can be done on a dense set of especially nice vectors and then show that the limiting operator is essentially self-adjoint by using the basic criterion. Finally, we show that the exponential of this limiting operator is just $U(t)$.

Let $f \in C_0^\infty(\mathbb{R})$ and for each $\varphi \in \mathscr{H}$ define

$$\varphi_f = \int_{-\infty}^{\infty} f(t) U(t) \varphi \, dt$$

Since $U(t)$ is strongly continuous the integral can be taken to be a Riemann integral. Let D be the set of finite linear combinations of all such φ_f for $\varphi \in \mathscr{H}$ and $f \in C_0^\infty(\mathbb{R})$. If $j_\varepsilon(x)$ is the approximate identity introduced in Section VIII.1, then

$$\|\varphi_{j_\varepsilon} - \varphi\| = \left\| \int_{-\infty}^{\infty} j_\varepsilon(t)(U(t)\varphi - \varphi) \, dt \right\|$$

$$\leq \left(\int_{-\infty}^{\infty} j_\varepsilon(t) \, dt \right) \sup_{t \in [-\varepsilon, \varepsilon]} \|U(t)\varphi - \varphi\|$$

Since $U(t)$ is strongly continuous, D is dense in $\mathscr{H}$. We have used the inequality $\|\int h(t) \, dt\| \leq \int \|h(t)\| \, dt$ for Banach space-valued continuous functions on the real line (which can be proven using the approximate partial sums as in the real-valued case).

For $\varphi_f \in D$, we compute

$$\left(\frac{U(s) - I}{s} \right) \varphi_f = \int_{-\infty}^{\infty} f(t) \left(\frac{U(s+t) - U(t)}{s} \right) \varphi \, dt$$

$$= \int_{-\infty}^{\infty} \frac{f(\tau - s) - f(\tau)}{s} U(\tau) \varphi \, d\tau$$

$$\to -\int f'(\tau) U(\tau) \varphi \, d\tau$$

$$= \varphi_{-f'}$$

since $[f(\tau - s) - f(\tau)]/s$ converges to $-f'(\tau)$ uniformly. For $\varphi_f \in D$, we define $A\varphi_f = i^{-1}\varphi_{-f'}$. Notice that $U(t): D \to D$, $A: D \to D$, and $U(t)A\varphi_f =$

$AU(t)\varphi_f$ for $\varphi_f \in D$. Furthermore, if $\varphi_f, \varphi_g \in D$, we have

$$(A\varphi_f, \varphi_g) = \lim_{s \to 0} \left(\left(\frac{U(s) - I}{is} \right) \varphi_f, \varphi_g \right)$$

$$= \lim_{s \to 0} \left(\varphi_f, \left(\frac{I - U(-s)}{is} \right) \varphi_g \right)$$

$$= \left(\varphi_f, \frac{1}{i} \varphi_{-g'} \right)$$

$$= (\varphi_f, A\varphi_g)$$

so A is symmetric.

Now we show that A is essentially self-adjoint. Suppose that there is a $u \in D(A^*)$ so that $A^*u = iu$. Then for each $\varphi \in D(A) = D$,

$$\frac{d}{dt}(U(t)\varphi, u) = (iAU(t)\varphi, u)$$

$$= -i(U(t)\varphi, A^*u)$$

$$= -i(U(t)\varphi, iu)$$

$$= (U(t)\varphi, u)$$

Thus, the complex-valued function $f(t) = (U(t)\varphi, u)$ satisfies the ordinary differential equation $f' = f$ so $f(t) = f(0)e^t$. Since $U(t)$ has norm one, $|f(t)|$ is bounded, which implies that $f(0) = (\varphi, u) = 0$. Since D is dense, $u = 0$. A similar proof shows that $A^*u = -iu$ can have no nonzero solutions. Therefore, by the corollary to Theorem VIII.3, A is essentially self-adjoint on D.

Let $V(t) = e^{it\bar{A}}$. It remains to show that $U(t) = V(t)$. Let $\varphi \in D$. Since $\varphi \in D(\bar{A})$, $V(t)\varphi \in D(\bar{A})$ and $V'(t)\varphi = i\bar{A}V(t)\varphi$ by (c) of Theorem VIII.7. We already know that $U(t)\varphi \in D \subset D(\bar{A})$ for all t. Let $w(t) = U(t)\varphi - V(t)\varphi$. Then $w(t)$ is a strongly differentiable vector-valued function and

$$w'(t) = iAU(t)\varphi - i\bar{A}V(t)\varphi$$

$$= i\bar{A}w(t)$$

Thus

$$\frac{d}{dt}\|w(t)\|^2 = -i(\bar{A}w(t), w(t)) + i(w(t), \bar{A}w(t))$$

$$= 0$$

Therefore $w(t) = 0$ for all t since $w(0) = 0$. This implies that $U(t)\varphi = V(t)\varphi$ for all $t \in \mathbb{R}$, $\varphi \in D$. Since D is dense, $U(t) = V(t)$. ∎

268　VIII: UNBOUNDED OPERATORS

Definition　If $U(t)$ is a strongly continuous one-parameter unitary group, then the self-adjoint operator A with $U(t) = e^{itA}$ is called the **infinitesimal generator** of $U(t)$.

Suppose that $U(t)$ is a *weakly* continuous one-parameter unitary group. Then

$$\|U(t)\varphi - \varphi\|^2 = \|U(t)\varphi\|^2 - (U(t)\varphi, \varphi) - (\varphi, U(t)\varphi) + \|\varphi\|^2$$
$$\to 2\|\varphi\|^2 - 2\|\varphi\|^2 = 0$$

as $t \to 0$. Thus $U(t)$ is actually *strongly* continuous. As a matter of fact, to conclude that $U(t)$ is strongly continuous one need only show that $U(t)$ is weakly measurable, that is, that $(U(t)\varphi, \psi)$ is measurable for each φ and ψ. This startling result, proven by von Neumann, is sometimes useful since in applications one can often show that $(U(t)\varphi, \psi)$ is the limit of a sequence of continuous functions; $(U(t)\varphi, \psi)$ is therefore measurable and by von Neumann's theorem $U(t)$ is then strongly continuous.

Theorem VIII.9　(von Neumann)　Let $U(t)$ be a one-parameter group of unitary operators on a *separable* Hilbert space $\mathscr{H}$. Suppose that for all φ, $\psi \in \mathscr{H}$, $(U(t)\psi, \varphi)$ is measurable. Then $U(t)$ is strongly continuous.

Proof　Let $\psi \in \mathscr{H}$. Then for all $\varphi \in \mathscr{H}$, $(U(t)\psi, \varphi)$ is a bounded measurable function and

$$\varphi \mapsto \int_0^a (U(t)\psi, \varphi)\, dt$$

is a linear functional on $\mathscr{H}$ of norm less than or equal to $a\|\psi\|$. Thus, by the Riesz lemma there is a $\psi_a \in \mathscr{H}$ so that

$$(\psi_a, \varphi) = \int_0^a (U(t)\psi, \varphi)\, dt$$

Now,

$$(U(b)\psi_a, \varphi) = (\psi_a, U(-b)\varphi)$$
$$= \int_0^a (U(t)\psi, U(-b)\varphi)\, dt$$
$$= \int_0^a (U(t+b)\psi, \varphi)\, dt$$
$$= \int_b^{a+b} (U(t)\psi, \varphi)\, dt$$

Thus

$$|(U(b)\psi_a, \varphi) - (\psi_a, \varphi)| = \left|\int_0^b (U(t)\psi, \varphi)\, dt\right| + \left|\int_a^{a+b} (U(t)\psi, \varphi)\, dt\right|$$
$$\leq 2b\|\varphi\|\,\|\psi\|$$

Therefore,

$$\lim_{b \to 0} (U(b)\psi_a, \varphi) = (\psi_a, \varphi)$$

so that $U(b)$ is weakly and therefore strongly continuous on the set of vectors of the form $\{\psi_a \mid \psi \in \mathcal{H}\}$. It remains only to show that this set is dense, since by an $\varepsilon/3$ argument we can then conclude that $t \mapsto U(t)$ is strongly continuous on $\mathcal{H}$. Suppose that $\varphi \in \{\psi_a \mid \psi \in \mathcal{H}, a \in \mathbb{R}\}^\perp$ and let $\{\psi^{(n)}\}$ be an orthonormal basis for $\mathcal{H}$. Then for each n,

$$0 = (\psi_a^{(n)}, \varphi) = \int_0^a (U(t)\psi^{(n)}, \varphi)\, dt$$

for all a which implies that $(U(t)\psi^{(n)}, \varphi) = 0$ except for $t \in S_n$, a set of measure zero. Choose $t_0 \notin \bigcup_{n=1}^\infty S_n$. Then $(U(t_0)\psi^{(n)}, \varphi) = 0$ for all n which implies that $\varphi = 0$, since $U(t_0)$ is unitary. ∎

The proof of essential self-adjointness in Theorem VIII.8 directly implies the following self-adjointness criterion.

Theorem VIII.10 Suppose that $U(t)$ is a strongly continuous one-parameter unitary group. Let D be a dense domain which is invariant under $U(t)$ and on which $U(t)$ is strongly differentiable. Then i^{-1} times the strong derivative of $U(t)$ is essentially self-adjoint on D and its closure is the infinitesimal generator of $U(t)$.

This theorem has a reformulation which is sufficiently important that we state it as a theorem.

Theorem VIII.11 Let A be a self-adjoint operator on $\mathcal{H}$ and D be a dense linear set contained in $D(A)$. If for all t, $e^{itA}: D \to D$, then D is a core for A.

Finally, we have the following generalization of Stone's theorem.

Theorem VIII.12 Let $t \to U(t) = U(t_1, \ldots, t_n)$ be a strongly continuous map of $\mathbb{R}^n$ into the unitary operators on a separable Hilbert space $\mathcal{H}$ satisfying $U(t + s) = U(t)U(s)$ and $U(0) = I$. Let D be the set of finite linear combinations of vectors of the form

$$\varphi_f = \int_{\mathbb{R}^n} f(t)U(t)\varphi \, dt \qquad \varphi \in \mathcal{H}, \quad f \in C_0^\infty(\mathbb{R}^n)$$

Then D is a domain of essential self-adjointness for each of the generators A_j of the one-parameter subgroups $U(0, 0, \ldots, t_j, \ldots, 0)$, each $A_j: D \to D$ and the A_j commute, $j = 1, \ldots, n$. Furthermore, there is a projection-valued measure P_Ω on $\mathbb{R}^n$ so that

$$(\varphi, U(t)\psi) = \int_{\mathbb{R}^n} e^{it \cdot \lambda} d(\varphi, P_\lambda \psi)$$

for all $\varphi, \psi \in \mathcal{H}$.

Proof Let A_j be the infinitesimal generator of $U_j(t_j) = U(0, \ldots, t_j, \ldots, 0)$. The procedure used in the proof of Theorem VIII.8 shows that $D \subset D(A_j)$, $A_j: D \to D$, and $U_j(t_j): D \to D$. Theorem VIII.11 shows that A_j is essentially self-adjoint on D. Because of the relation $U(t + s) = U(t)U(s)$, $U_j(t_j)$ commutes with $U_i(t_i)$ for all $t_i, t_j \in \mathbb{R}$. Therefore, it follows from Theorem VIII.13, that A_i and A_j commute in the sense defined in the next section; that is, their spectral projections commute.

Let P_Ω^j be the projection-valued measure on $\mathbb{R}$ corresponding to A_j. Define a projection valued measure P_Ω on $\mathbb{R}^n$ by defining it first on rectangles $r = (a_1, b_1) \times \cdots \times (a_n, b_n)$ by $P_r = P_{(a_1, b_1)}^1 P_{(a_2, b_2)}^2 \cdots P_{(a_n, b_n)}^n$ and then letting P_Ω be the unique extension to the smallest σ-algebra containing the rectangles, namely the Borel sets. Notice that, by Theorem VIII.13, the $P_{\Omega_j}^j$ commute since the groups U_j commute. For each $\varphi, \psi \in \mathcal{H}$, $(\varphi, P_\Omega \psi)$ is a complex-valued measure of finite mass which we denote by $d(\varphi, P_\lambda \psi)$. Applying Fubini's theorem we easily conclude that

$$\int_{\mathbb{R}^n} e^{it \cdot \lambda} d(\varphi, P_\lambda \psi) = (\varphi, U_1(t_1) \ldots, U_n(t_n)\psi)$$
$$= (\varphi, U(t)\psi) \quad \blacksquare$$

VIII.5 Formal manipulation is a touchy business: Nelson's example

From the theorems proven in the last two sections it may seem to the reader that unbounded operators are just like bounded operators except that one needs to be a little careful about domains. First, of all, it is sometimes

VIII.5 Formal manipulation is a touchy business

difficult to determine the domain of a self-adjoint operator exactly, and it is not always enough to check statements on a core. Furthermore, formal calculations can be misleading. These statements are illustrated by the question of commutativity and the surprising example of Nelson which show how difficult it is to deal with unbounded operators.

Suppose that A and B are two unbounded self-adjoint operators on a Hilbert space $\mathcal{H}$. We would like to find a reasonable meaning for the statement "A and B commute." This cannot be done in the straightforward way since $AB - BA$ may not make sense on any vector in $\mathcal{H}$; for example, one might have $(\operatorname{Ran} A) \cap D(B) = \{0\}$ in which case BA does not have a meaning. This suggests that we find an equivalent formulation of commutativity for bounded self-adjoint operators. The spectral theorem for bounded self-adjoint operators A and B shows that in that case $AB - BA = 0$ if and only if all their projections, $\{P_\Omega^A\}$ and $\{P_\Omega^B\}$, commute. We take this as our definition in the unbounded case.

Definition Two (possibly unbounded) self-adjoint operators A and B are said to **commute** if and only if all the projections in their associated projection-valued measures commute.

The spectral theorem shows that if A and B commute, then all the bounded Borel functions of A and B also commute. In particular, the resolvents $R_\lambda(A)$ and $R_\mu(B)$ commute and the unitary groups e^{itA} and e^{isB} commute. The converse statement is also true and this shows that the above definition of "commute" is reasonable:

Theorem VIII.13 Let A and B be self-adjoint operators on a Hilbert space $\mathcal{H}$. Then the following three statements are equivalent:

(a) A and B commute.
(b) If $\operatorname{Im} \lambda$ and $\operatorname{Im} \mu$ are nonzero, then $R_\lambda(A)R_\mu(B) = R_\mu(B)R_\lambda(A)$.
(c) For all $s, t \in \mathbb{R}$, $e^{itA}e^{isB} = e^{isB}e^{itA}$.

Proof The fact that (a) implies (b) and (c) follows from the functional calculus. The fact that (b) implies (a) easily follows from the formula which expresses the spectral projections of A and B as strong limits of the resolvents (Stone's formula) together with the fact that

$$\operatorname*{s-lim}_{\varepsilon \downarrow 0} i\varepsilon R_{a+i\varepsilon}(A) = P_{\{a\}}^A$$

To prove that (c) implies (a), we use some simple facts about the Fourier transform proven in Section IX.1. Let $f \in \mathscr{S}(\mathbb{R})$. Then, by Fubini's theorem,

$$\int_{-\infty}^{\infty} f(t)(e^{itA}\varphi, \psi)\, dt = \int_{-\infty}^{\infty} f(t)\left(\int_{-\infty}^{\infty} e^{-it\lambda} d_\lambda(P_\lambda^A \varphi, \psi)\right) dt$$

$$= \sqrt{2\pi} \int_{-\infty}^{\infty} \hat{f}(\lambda)\, d_\lambda(P_\lambda^A \varphi, \psi)$$

$$= \sqrt{2\pi}\,(\varphi, \hat{f}(A)\psi)$$

Thus, using (c) and Fubini's theorem again,

$$(\varphi, \hat{f}(A)\hat{g}(B)\psi) = \int_{-\infty}^{\infty}\int_{-\infty}^{\infty} f(t)g(s)(\varphi, e^{-itA}e^{-isB}\psi)\, ds\, dt$$

$$= (\varphi, \hat{g}(B)\hat{f}(A)\psi)$$

so, for all $f, g \in \mathscr{S}(\mathbb{R})$, $\hat{f}(A)\hat{g}(B) - \hat{g}(B)\hat{f}(A) = 0$. Since the Fourier transform maps $\mathscr{S}(\mathbb{R})$ onto $\mathscr{S}(\mathbb{R})$ we conclude that $f(A)g(B) = g(B)f(A)$ for all $f, g \in \mathscr{S}(\mathbb{R})$. But, the characteristic function, $\chi_{(a,\,b)}$ can be expressed as the pointwise limit of a sequence f_n of uniformly bounded functions in $\mathscr{S}$. By the functional calculus,

$$f_n(A) \xrightarrow{s} P^A_{(a,\,b)}$$

Similarly, we find uniformly bounded $g_n \in A$ converging pointwise to $\chi_{(c,\,d)}$ and

$$g_n(B) \xrightarrow{s} P^B_{(c,\,d)}$$

Since the f_n and g_n are uniformly bounded and

$$f_n(A)g_n(B) = g_n(B)f_n(A)$$

for each n, we conclude that $P^A_{(a,\,b)}$ and $P^B_{(c,\,d)}$ commute which proves (a). ∎

Although the above theorem shows that the definition of "commute" is reasonable, it is not always easy to deal with. In practice, one is usually given A and B on sets of essential self-adjointness, $D_0(A)$ and $D_0(B)$, and it may be very difficult to construct the spectral projections, the resolvents, or the groups corresponding to $\bar{A}$ and $\bar{B}$. Thus, one would like to have a criterion in terms of the operators themselves. In Problem 13, the reader is asked to find disjoint domains of essential self-adjointness for the operators x and x^2 on $L^2(\mathbb{R})$, so such a commutativity criterion could never be necessary, but with enough restrictions might be sufficient. Here are two conjectures which seem reasonable but which are *false*:

1. Let D be a dense subspace in $\mathscr{H}$ which is contained in the domains of A and B. Suppose further that $A: D \to D$ and $B: D \to D$. Then, if $AB\varphi - BA\varphi = 0$ for each $\varphi \in D$, A and B commute (*FALSE!*).

VIII.5 Formal manipulation is a touchy business

2. Let D be a dense domain of essential self-adjointness for A and B. Suppose further that $A: D \to D$ and $B: D \to D$. Then, if $AB\varphi - BA\varphi = 0$ for all $\varphi \in D$, A commutes with B (*FALSE!*).

Both of these statements are *false*, the hypotheses are *not* sufficient to guarantee commutativity. This is surprising for several reasons. First, the conditions seem reasonable. Secondly, in condition (2), D is assumed to be a domain of essential self-adjointness for both A and B, so the action of A and B on D should provide enough information to determine whether A and B commute. Finally, in a formal sense

$$e^{itA} \approx I + \sum_{n=1}^{\infty} \frac{(itA)^n}{n!}$$

$$e^{isB} \approx I + \sum_{n=1}^{\infty} \frac{(isB)^n}{n!}$$
(VIII.6)

Since it follows from the conditions in (1) and (2) that $A^n B^m \varphi - B^m A^n \varphi = 0$ for all $\varphi \in D$, one might expect that e^{itA} and e^{isB} commute for all s and t. By Theorem VIII.13 this would imply that A and B commute. The trouble with this argument is that the expressions in (VIII.6) are only formal and may have meaning on no vectors in D since A and B are unbounded. The finite sums make sense on D and

$$\left(I + \sum_{n=1}^{N} \frac{(itA)^n}{n!}\right)\left(I + \sum_{m=1}^{M} \frac{(isB)^m}{m!}\right)\varphi = \left(I + \sum_{m=1}^{M} \frac{(isB)^m}{m!}\right)\left(I + \sum_{n=1}^{N} \frac{(itA)^n}{n!}\right)\varphi$$

but we cannot conclude from this that e^{itA} and e^{isB} commute. The following example is due to Nelson.

Example 1 Suppose that M is the Riemann surface of the $\sqrt{z}$ and $\mathscr{H} = L^2(M)$ with Lebesgue measure (locally). Let $A = -i\partial/\partial x$ and $B = -i\partial/\partial y$ on the domain D which consists of all infinitely differentiable functions with compact support not containing the origin. Then

(a) A and B are essentially self-adjoint on D.
(b) $A: D \to D$, $B: D \to D$
(c) $AB\varphi = BA\varphi$ for $\varphi \in D$
(d) $e^{it\bar{A}}$ and $e^{is\bar{B}}$ do not commute.

The proofs of (b) and (c) are obvious. To prove (a), first observe that integration by parts shows that A and B are symmetric. Let $D_x \subset D$ be the functions in D whose support does not contain the x axis on either sheet. D_x is also dense in $L^2(M)$. On D_x define $(U(t)\varphi)(x, y) = \varphi(x + t, y)$. Then $U(t)$ is a norm-preserving map with dense range and so extends to a unitary operator on

$L^2(M)$. Since $U(t)$ is strongly continuous on D_x, $U(t)$ is strongly continuous on $L^2(M)$. Now, on D_x, $U(t)$ is strongly differentiable and i^{-1} times its strong derivative is A. Thus, by Theorem VIII.10, A is essentially self-adjoint on D_x and therefore on D (Problem 14) and its closure generates $U(t)$. A similar proof shows that the closure of B is the infinitesimal generator of translation in the y direction defined by extension from a domain D_y. This proves (a).

To prove (d), let φ be an infinitely differentiable function with support contained in a small circle about the point $(-\frac{1}{2}, -\frac{1}{2})$ on the first sheet. Then

$$U(1)V(1)\varphi \neq V(1)U(1)\varphi$$

since the functions will have their support around $(\frac{1}{2}, \frac{1}{2})$ on different sheets. ∎

Example 2 (canonical commutation relations) A pair P, Q of self-adjoint operators is said to "satisfy" the **canonical commutation relations** if

$$PQ - QP = -iI \tag{VIII.7}$$

P and Q cannot both be bounded. For, if they were, the relation $PQ^n - Q^nP = -inQ^{n-1}$ [which follows directly from (VIII.7)] would imply

$$n\|Q\|^{n-1} = n\|Q^{n-1}\| \leq 2\|P\| \|Q\|^n$$

So for all n, $2\|P\| \|Q\| \geq n$, which is a contradiction. Thus, either P or Q or both must be unbounded so we cannot discuss the relation (VIII.7) without worrying about domains. The standard realization or "representation" used in quantum mechanics is the **Schrödinger representation** where $\mathcal{H} = L^2(\mathbb{R})$ and P and Q are the closures of $i^{-1}d/dx$ and multiplication by x on $\mathcal{S}(\mathbb{R})$. $\mathcal{S}(\mathbb{R})$ is a domain of essential self-adjointness for $i^{-1}d/dx$ and x,

$$\frac{1}{i}\frac{d}{dx} : \mathcal{S}(\mathbb{R}) \to \mathcal{S}(\mathbb{R}), \qquad x : \mathcal{S}(\mathbb{R}) \to \mathcal{S}(\mathbb{R})$$

and for $\varphi \in \mathcal{S}(\mathbb{R})$,

$$\frac{1}{i}\frac{d}{dx}(x\varphi) - x\left(\frac{1}{i}\frac{d}{dx}\varphi\right) = -i\varphi$$

The question is: in what sense is the Schrödinger representation the "only" representation of the relation (VIII.7). One method of dealing with (VIII.7) is with the unitary groups. If $U(t) = e^{itP}$ and $V(s) = e^{isQ}$, then a *formal* calculation using (VIII.7) and the formal power series expansion for e^{itP} and e^{isQ} yields

$$U(t)V(s) = e^{its}V(s)U(t) \tag{VIII.8}$$

VIII.5 Formal manipulation is a touchy business

The reader should not be surprised to learn that the two problems, solving the relation (VIII.7) and solving the relation (VIII.8), which are related by a formal calculation are *not* equivalent. Let us first discuss (VIII.8), which is an easier problem because we need deal only with bounded operators. Two continuous one-parameter groups satisfying (VIII.8) are said to satisfy the **Weyl relations**. The reader can easily check that in the Schrödinger representation, the groups e^{itP} and e^{isQ} do satisfy the Weyl relations; e^{itP} is just translation to the left by t, and e^{isQ} is multiplication by e^{isx}. The following theorem states that up to multiplicity and unitary equivalence, the Weyl relations have only one solution (for a proof see Theorem XI.84 or Problem 30 in Chapter X).

Theorem VIII.14 (von Neumann) Let $U(t)$ and $V(s)$ be one-parameter, continuous, unitary groups on a separable Hilbert space $\mathscr{H}$ satisfying the Weyl relations. Then, there are closed subspaces $\mathscr{H}_\ell$ so that

(a) $\mathscr{H} = \oplus_{\ell=1}^{N} \mathscr{H}_\ell$ (N a positive integer or ∞)
(b) $U(t): \mathscr{H}_\ell \to \mathscr{H}_\ell$, $V(s): \mathscr{H}_\ell \to \mathscr{H}_\ell$ for all $s, t \in \mathbb{R}$.
(c) For each ℓ, there is a unitary operator $T_\ell: \mathscr{H}_\ell \to L^2(\mathbb{R})$ such that $T_\ell U(t) T_\ell^{-1}$ is translation to the left by t and $T_\ell V(s) T_\ell^{-1}$ is multiplication by e^{isx}.

Corollary Let $U(t)$ and $V(s)$ be continuous one-parameter unitary groups satisfying the Weyl relations on a separable Hilbert space $\mathscr{H}$. Let P be the generator of $U(t)$, Q the generator of $V(s)$. Then there is a dense domain $D \subset \mathscr{H}$ so that

(a) $P: D \to D$, $Q: D \to D$
(b) $PQ\varphi - QP\varphi = -i\varphi$ for all $\varphi \in D$
(c) P and Q are essentially self-adjoint on D.

This corollary (whose easy proof is given as Problem 36) shows that any solution of the Weyl relations has infinitesimal generators which satisfy the canonical commutation relations in the sense given by (a), (b), and (c). The converse of this statement is not true as is shown by the following slight perturbation of Nelson's example. Let $\mathscr{H} = L^2(M)$ as in Example 1,

$$P = \frac{1}{i}\frac{\partial}{\partial x} \qquad Q = x + \frac{1}{i}\frac{\partial}{\partial y}$$

on the domain D given there. P and Q satisfy properties (a), (b), and (c). The proof of self-adjointness is similar to the proof in Example 1. But the groups they generate do not satisfy the Weyl relations.

One point of this section was to show that formal calculations with the formal power series expansion for e^{itA} can lead to false conclusions. This does not imply that one can never make sense out of the formal series for e^{itA} if A is unbounded. In fact, suppose that A is an unbounded self-adjoint operator and P_Ω is the corresponding projection-valued measure. Then the set D_c of vectors of the form $P_{[-M, M]}\varphi$, where $\varphi \in \mathcal{H}$ and M is arbitrary but finite, is a dense set contained in $D(A^n)$ for all n and A is essentially self-adjoint on D_c. Furthermore, if $\psi = P_{[-M, M]}\varphi$, then $\|A^n\psi\| \leq M^n\|\psi\|$, so

$$\sum_{n=0}^{\infty} \frac{t^n \|A^n\psi\|}{n!} < \infty \qquad (\text{VIII.9})$$

for all t. Therefore, if $\psi \in D_c$ the series $\sum_{n=0}^{\infty} (it)^n A^n\psi/n!$ converges. Vectors $\psi \in \bigcap_{n=1}^{\infty} D(A^n)$ which satisfy (VIII.9) for some $t > 0$ are called **analytic vectors** for A. On such vectors the power series for $e^{itA}\psi$ makes sense and converges to $e^{itA}\psi$ as long as t is sufficiently small. We will return to analytic vectors in Section X.6 where we prove a theorem of Nelson that if a symmetric operator A has a dense set of analytic vectors in its domain D then A is essentially self-adjoint.

VIII.6 Quadratic forms

One consequence of the Riesz lemma is that there is a one-to-one correspondence between bounded quadratic forms and bounded operators; that is, any sesquilinear map $q: \mathcal{H} \times \mathcal{H} \to \mathbb{C}$ which satisfies $|q(\varphi, \psi)| \leq M \|\varphi\| \|\psi\|$ is of the form $q(\varphi, \psi) = (\varphi, A\psi)$ for some bounded operator A. As one might expect, the situation is more complicated if one removes the boundedness restriction. It is the relationship between unbounded forms and unbounded operators which we study briefly in this section.

Definition A **quadratic form** is a map $q: Q(q) \times Q(q) \to \mathbb{C}$, where $Q(q)$ is a dense linear subset of $\mathcal{H}$ called the **form domain**, such that $q(\cdot, \psi)$ is conjugate linear and $q(\varphi, \cdot)$ is linear for $\varphi, \psi \in Q(q)$. If $q(\varphi, \psi) = \overline{q(\psi, \varphi)}$ we say that q is **symmetric**. If $q(\varphi, \varphi) \geq 0$ for all $\varphi \in Q$, q is called **positive**, and if $q(\varphi, \varphi) \geq -M \|\varphi\|^2$ for some M, we say that q is **semibounded**.

Notice that if q is semibounded, then it is automatically symmetric if $\mathcal{H}$ is complex.

Example 1 Let $\mathscr{H} = L^2(\mathbb{R})$ and $Q(q) = C_0^\infty(\mathbb{R})$ with $q(f, g) = \bar{f}(0)g(0)$. Then q is a positive quadratic form. Since $q(f, g) = \delta(\bar{f}g)$ one could *formally* write $q(f, g) = (f, Ag)$ where $A: g \mapsto \delta(x)g(x)$. Since multiplication by $\delta(x)$ is not an operator, q is an example of a quadratic form not likely to be associated with an operator.

Example 2 Let A be a self-adjoint operator on $\mathscr{H}$. Let us pass to a spectral representation of A, so that A is multiplication by x on $\bigoplus_{n=1}^N L^2(\mathbb{R}, \mu_n)$. Let

$$Q(q) = \left\{ \{\psi_n(x)\}_{n=1}^N \ \Big| \ \sum_{n=1}^N \int_{-\infty}^\infty |x| \, |\psi_n(x)|^2 \, d\mu_n < \infty \right\}$$

and for $\psi, \varphi \in Q(q)$ define

$$q(\varphi, \psi) = \sum_{n=1}^N \int_{-\infty}^\infty x \overline{\varphi_n(x)} \psi_n(x) \, d\mu_n$$

We call q the **quadratic form** associated with A and write $Q(q) = Q(A)$; $Q(A)$ is called the **form domain of the operator** A. For $\psi, \varphi \in Q(A)$, we will often write $q(\varphi, \psi) = (\varphi, A\psi)$ although A does not make sense on all $\psi \in Q(A)$. $Q(A)$ is in some sense the largest domain on which q can be defined.

To investigate the deep connection between self-adjointness and semi-bounded quadratic forms we need to extend the notion of "closed" from operators to forms. An operator A is closed if and only if its graph is closed which is the same as saying that $D(A)$ is complete under the norm $\|\psi\|_A = \|A\psi\| + \|\psi\|$ (Problem 15). Analogously, we define:

Definition Let q be a semibounded quadratic form, $q(\psi, \psi) \geq -M \|\psi\|^2$. q is called **closed** if $Q(q)$ is complete under the norm

$$\|\psi\|_{+1} = \sqrt{q(\psi, \psi) + (M+1)\|\psi\|^2}$$

If q is closed and $D \subset Q(q)$ is dense in $Q(q)$ in the $\|\cdot\|_{+1}$ norm, then D is called a **form core** for q.

Notice that $\|\psi\|_{+1}$ comes from the inner product

$$(\psi, \varphi)_{+1} = q(\psi, \varphi) + (M+1)(\psi, \varphi)$$

It is not hard to see (Problem 15) that q is closed if and only if whenever $\varphi_n \in Q(q)$, $\varphi_n \xrightarrow{\mathscr{H}} \varphi$ and $q(\varphi_n - \varphi_m, \varphi_n - \varphi_m) \to 0$, as $n, m \to \infty$, then $\varphi \in Q(q)$ and $q(\varphi_n - \varphi, \varphi_n - \varphi) \to 0$. This criterion and the dominated convergence theorem show that the form q associated with a semibounded self-adjoint operator (Example 2) is closed. Furthermore, any operator core for A is a form core for q (Problem 16).

Now, let $q(f, g) = \bar{f}(0)g(0)$ as in Example 1 and φ_n be the C_0^∞ function shown in Figure VIII.2. Then $\varphi_n \xrightarrow{\mathscr{H}} 0$, and $q(\varphi_n - \varphi_m, \varphi_n - \varphi_m) \to 0$, but $q(\varphi_n, \varphi_n) \to 1 \neq q(0, 0)$ which proves that q has no closed extensions. Therefore, even though q is positive (and therefore symmetric) there is no semibounded self-adjoint operator A so that $q(f, g) = (f, Ag)$ for all $f, g \in C_0^\infty$.

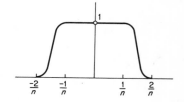

FIGURE VIII.2   The graph of φ_n.

The deep fact about semibounded quadratic forms is that *unlike* the case for operators, they cannot be closed and symmetric, yet not self-adjoint.

Theorem VIII.15   If q is a closed semibounded quadratic form, then q is the quadratic form of a unique self-adjoint operator.

Proof   We may assume without loss of generality that q is positive. Then, since q is closed and symmetric, $Q(q)$ is a Hilbert space, which we denote by $\mathscr{H}_{+1}$, under the inner product

$$(\varphi, \psi)_{+1} = q(\varphi, \psi) + (\varphi, \psi)$$

We denote by $\mathscr{H}_{-1}$ the space of bounded conjugate linear functionals on $\mathscr{H}_{+1}$. Let j, given by $\psi \xmapsto{j} (\cdot, \psi)$, be the linear imbedding of $\mathscr{H}$ into $\mathscr{H}_{-1}$. $j(\psi)$ is bounded because

$$|[j(\psi)](\varphi)| \leq \|\varphi\| \|\psi\| \leq \|\varphi\|_{+1} \|\psi\|$$

Since the identity map i embeds $\mathscr{H}_{+1}$ in $\mathscr{H}$ we have a "scale of spaces"

$$\mathscr{H}_{+1} \xrightarrow{i} \mathscr{H} \xrightarrow{j} \mathscr{H}_{-1}$$

We now exploit the Riesz lemma. Given $\Phi \in \mathscr{H}_{+1}$, let $\hat{B}\Phi$ be the element of $\mathscr{H}_{-1}$ which acts by

$$[\hat{B}\Phi](\varphi) = q(\varphi, \Phi) + (\varphi, \Phi)$$

By the Riesz lemma, $\hat{B}$ is an isometric isomorphism of $\mathcal{H}_{+1}$ onto $\mathcal{H}_{-1}$. Let $D(B) = \{\psi \in \mathcal{H}_{+1} \mid \hat{B}\psi \in \operatorname{Ran} j\}$. Define B on $D(B)$ by $B = j^{-1}\hat{B}$.

$$\mathcal{H} \supset \mathcal{H}_{+1} \xrightarrow{\hat{B}} \mathcal{H}_{-1} \xleftarrow{j} \mathcal{H}$$

First, we prove that the range of j is dense in $\mathcal{H}_{-1}$. If it were not, there would be a $\lambda \in \mathcal{H}_{-1}^*$ so that $\lambda \neq 0$, and $\lambda[j(\psi)] = 0$ for each $\psi \in \mathcal{H}$. By the Riesz Lemma, there is a $\varphi_\lambda \neq 0$ in $\mathcal{H}_{+1}$ so that $0 = \lambda[j(\psi)] = [j(\psi)](\varphi_\lambda) = (\varphi_\lambda, \psi)$ for all $\psi \in \mathcal{H}$. Since $\varphi_\lambda \neq 0$, this is impossible. Therefore $\operatorname{Ran} j$ is dense in $\mathcal{H}_{-1}$. Since B is an isometric isomorphism we conclude that $D(B)$ is $\|\cdot\|_{+1}$ dense in $\mathcal{H}_{+1}$. Further, since $\|\cdot\| \leq \|\cdot\|_{+1}$ and $\mathcal{H}_{+1}$ is norm dense in $\mathcal{H}$, $D(B)$ is norm dense in $\mathcal{H}$.

Suppose $\varphi, \psi \in D(B)$. Then

$$\begin{aligned}(\varphi, B\psi) &= q(\varphi, \psi) + (\varphi, \psi) \\ &= \overline{q(\psi, \varphi) + (\psi, \varphi)} \\ &= \overline{(\psi, B\varphi)} \\ &= (B\varphi, \psi)\end{aligned}$$

Thus, B is a densely defined symmetric operator.

We will prove that B is self-adjoint. Let $C = \hat{B}^{-1}j$. C takes $\mathcal{H}$ into $\mathcal{H}$ and is an everywhere defined symmetric operator. By the Hellinger–Toeplitz theorem, C is a bounded self-adjoint operator. Moreover, C is injective. A simple application of the spectral theorem in multiplication operator form shows that $C^{-1}: \operatorname{Ran} C \to \mathcal{H}$ is a self-adjoint operator. But $C^{-1} = B$.

We now define $A = B - I$. Then A is also self-adjoint on $D(A) = D(B)$ and for $\varphi, \psi \in D(A)$, $(\varphi, A\psi) = q(\varphi, \psi)$. Since $D(A)$ is $\|\cdot\|_{+1}$-dense in $\mathcal{H}_{+1}$, q is the quadratic form associated to A. Uniqueness is left as a problem. ∎

Thus, there is an interesting distinction between semi-bounded symmetric operators and semi-bounded quadratic forms. For symmetric operators, there is never any problem finding closed extensions. A smallest closed extension always exists (the double adjoint), but it is possible that none of these closed extensions is self-adjoint. On the other hand, semibounded forms need not have any closed extensions, but when such extensions exist and are semi-bounded, they are the quadratic forms associated with self-adjoint operators. We caution the reader about several pitfalls.

(1) If A and B are self-adjoint operators and $D(A) \subset D(B)$ with $B \restriction D(A) = A$, then $A = B$. But it can happen that a and b are closed semi-bounded quadratic forms and $b \restriction Q(a) \times Q(a) = a$ without having $a = b$.

(2) Let A be a symmetric operator that is semibounded. Let q be the quadratic form $q(\varphi, \psi) = (\varphi, A\psi)$ with $Q(a) = D(A)$. Suppose that q has a closure (it always will, see Section X.3) $\bar{q}$, that is, a smallest closed form which extends it. Then the self-adjoint operator $\hat{A}$ which corresponds to $\bar{q}$ (by Theorem VIII.15) may be bigger than the operator closure of A.

(3) While a general quadratic form may have no closed extensions, forms that come directly from semibounded operators always have closures and thus semibounded operators always have self-adjoint extensions: see Section X.3.

The following example illustrates the first two of these phenomena.

Example 3 Let $AC^2[0, 1]$ denote the functions $f \in L^2[0, 1]$ such that f is differentiable, f' is absolutely continuous and $f'' \in L^2[0, 1]$. We define

$$D_0 = \{f \mid f \in AC^2[0, 1], \ f(0) = f(1) = 0 = f'(0) = f'(1)\}$$
$$D_{a,b} = \{f \mid f \in AC^2[0, 1], \ af(0) + f'(0) = 0 = bf(1) + f'(1)\}$$
$$D_{\infty, \infty} = \{f \mid f \in AC^2[0, 1], \ f(0) = 0 = f(1)\}$$
$$D = \{f \mid f \in AC^2[0, 1]\}$$

and let T_0, $T_{a,b}$, $T_{\infty, \infty}$, and T be the operation $-d^2/dx^2$ with domains D_0, $D_{a,b}$, $D_{\infty, \infty}$, and D respectively. Then

(a) The operators T_0, $T_{a,b}$, $T_{\infty, \infty}$, and T are closed. T_0 is symmetric but not self-adjoint; its adjoint is T.

(b) $T_{a,b}(-\infty < a < \infty, \ -\infty < b < \infty)$ and $T_{\infty, \infty}$ are all distinct self-adjoint extensions of T_0 (there are others!).

(c) If $t_0(\varphi, \psi) = (\varphi, T_0 \psi)$ for $\varphi, \psi \in D_0$, the form t_0 has a smallest closed extension $\hat{t}_0$. This extension is the form associated with $T_{\infty, \infty}$ which illustrates the remark (2) above.

(d) The form $t_{a,b}$ associated with $T_{a,b}$ has a form domain $Q(t_{a,b})$ which contains the form domain $Q(t_{\infty, \infty})$ of $t_{\infty, \infty}$, the form associated with $T_{\infty, \infty}$, and $t_{a,b} \upharpoonright Q(t_{\infty, \infty}) = t_{\infty, \infty}$. This illustrates the remark (1) above.

Finally, we extend some of the ideas we have considered to nonsymmetric forms. Our use of the terms "sectorial" and "accretive" is not precisely the standard one (see the Notes).

VIII.6 Quadratic forms

Definition A quadratic form q is called **strictly m-accretive** if
(i) q is closed in the sense that if $\varphi_n \in Q(q)$, $\varphi_n \to \varphi$, and
$$\lim_{n,m \to \infty} q(\varphi_n - \varphi_m, \varphi_n - \varphi_m) = 0$$
then $\varphi \in Q(q)$ and $q(\varphi_n - \varphi, \varphi_n - \varphi) \to 0$.
(ii) there is a θ, $0 < \theta < \pi/2$, with $|\arg[q(\varphi, \varphi)]| \leq \theta$ for all $\varphi \in Q(q)$.

Now, suppose that q is strictly m-accretive. Define a new quadratic form R_q by

$$R_q(\varphi, \psi) = \frac{1}{4}\left\{ \operatorname{Re}[q(\varphi + \psi, \varphi + \psi)] - \operatorname{Re}[q(\varphi - \psi, \varphi - \psi)] \right.$$
$$\left. + \frac{1}{i}\operatorname{Re}[q(\varphi + i\psi, \varphi + i\psi)] - \frac{1}{i}\operatorname{Re}[q(\varphi - i\psi, \varphi - i\psi)] \right\}$$

Notice that $R_q(\varphi, \varphi) = \operatorname{Re}[q(\varphi, \varphi)]$ so R_q is a closed positive form. We can now use R_q to set up a scale of spaces $\mathcal{H}_{+1} \subset \mathcal{H} \subset \mathcal{H}_{-1}$ as was done in the proof of Theorem VIII.15, and obtain a map $\hat{T}: \mathcal{H}_{+1} \to \mathcal{H}_{-1}$ so that $[\hat{T}\Phi](\varphi) = \overline{q(\Phi, \varphi)}$. By using the proof of Theorem VIII.15, and by taking T to be a suitable restriction of $\hat{T}$, one can prove:

Theorem VIII.16 Let q be a strictly m-accretive quadratic form. Then there is a unique operator T on $\mathcal{H}$ such that

(a) T is closed.
(b) $D(T) \subset Q(q)$, and if $\varphi, \psi \in D(T)$, then $q(\varphi, \psi) = (\varphi, T\psi)$. Further, $D(T)$ is a form core for q.
(c) $D(T^*) \subset Q(q)$ and if $\varphi, \psi \in D(T^*)$, then $q(\varphi, \psi) = (T^*\varphi, \psi)$. Further, $D(T^*)$ is a form core for q.

The unique operator T given by the above theorem is called the **operator associated with the form** q. Of course, T is called a **strictly m-accretive operator**. Such operators have spectral properties which result from the properties of the associated forms.

Lemma Let T be a strictly m-accretive operator. Then any λ with $\operatorname{Re} \lambda < 0$ is in $\rho(T)$ and $\|(T - \lambda)^{-1}\| \leq (-\operatorname{Re} \lambda)^{-1}$.

Proof Let $\lambda = \mu + i\nu$ with $\mu < 0$. Then

$$\|(T-\lambda)\varphi\|^2 = ((T-\lambda)\varphi, (T-\lambda)\varphi)$$
$$= (\|T\varphi\|^2 - 2\nu\,\mathrm{Im}(\varphi, T\varphi) + \nu^2\|\varphi\|^2) - 2\mu\,\mathrm{Re}(\varphi, T\varphi) + \mu^2\|\varphi\|^2$$
$$\geq \mu^2\|\varphi\|^2$$

since $\mathrm{Re}(\varphi, T\varphi) \geq 0$ and

$$\|T\varphi\|^2 - 2\nu\,\mathrm{Im}(\varphi, T\varphi) + \nu^2\|\varphi\|^2$$
$$\geq \|T\varphi\|^2 - 2|\nu|\,\|T\varphi\|\,\|\varphi\| + \nu^2\|\varphi\|^2 \geq 0$$

As a result, $T - \lambda$ is injective and $\mathrm{Ran}(T - \lambda)$ is closed. Similarly,

$$\|(T-\lambda)^*\varphi\| \geq \mu^2\|\varphi\|^2$$

so $(\mathrm{Ran}(T-\lambda))^\perp = \mathrm{Ker}(T-\lambda)^* = 0$. Thus, $T - \lambda$ is invertible and

$$\|(T-\lambda)^{-1}\| \leq (-\mu)^{-1} \quad\blacksquare$$

Before stating this lemma in slightly extended form as a theorem, we extend the notion of accretive.

Definition A form q is called **strictly m-sectorial** if there are complex numbers z and $e^{i\alpha}$, with α real, so that $e^{i\alpha}q + z$ is strictly m-accretive. The operator T associated with q is also called **strictly m-sectorial**.

Notice that if q is strictly m-sectorial, then the values of $q(\varphi, \varphi)$ lie in a sector

$$S_q = \{\omega \mid \theta_0 \leq \arg(\omega - z) \leq \theta_1, \quad \text{with} \quad |\theta_1 - \theta_0| < \pi\}$$

S_q is called a **sector** for q.

Theorem VIII.17 Let q be a strictly m-sectorial form, S_q a sector for q, and T the associated operator. If $\lambda \notin S_q$, then $\lambda \in \rho(T)$ and $\|(T-\lambda)^{-1}\| \leq [\mathrm{dist}(\lambda, S_q)]^{-1}$.

The idea of the proof is to translate and rotate S_q so that $\mathrm{dist}(\lambda, S_q)$ is arbitrarily close to the real part of the translate of λ (see Figure VIII.3).

Example 4 Let H_0 and V be positive self-adjoint operators with $Q(H_0) \cap Q(V)$ dense. Let $Q(h) = Q(H_0) \cap Q(V)$. Given $\beta \in \mathbb{C} \backslash (-\infty, 0)$, define $h(\varphi, \psi) = (\varphi, H_0\psi) + \beta(\varphi, V\psi)$. Then h is closed and is a strictly m-sectorial form so we can use Theorem VIII.17 to gain information about $H = H_0 + \beta V$.

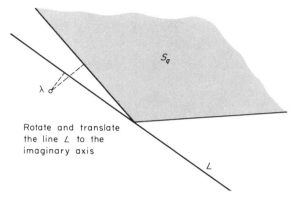

FIGURE VIII.3

In the past, the symbol $A + B$ meant the operator sum defined on $D(A) \cap D(B)$, or perhaps its operator closure. In Example 4, the plus in $H_0 + \beta V$ indicates the operator associated with the sum of *forms* defined on $Q(H_0) \cap Q(V)$. In the future in cases where no confusion can arise we will write $A + B$ without being explicit about the meaning of $+$.

VIII.7 Convergence of unbounded operators†

One of the main difficulties with unbounded operators is that they are only densely defined. This difficulty is especially troublesome when one wants to find a notion of convergence for a sequence $A_n \to A$ of unbounded operators since the domains of the operators A_n may have no vector in common. For example, if $A_n = (1 - 1/n)x$ on $L^2(\mathbb{R})$, it is clear that in some sense $A_n \to A = x$; yet we could have been given domains $D(A_n)$ and $D(A)$ of essential self-adjointness for these operators which have no nonzero vector in common (Problem 19). Of course, in this simple case the closures of A_n and A all have the same domain, but in general this will not be true, and in any case, one is often forced to deal with domains of essential self-adjointness since closures of operators are sometimes difficult to compute. It is very natural to say that self-adjoint operators are "close" if certain bounded functions of them are "close." Most of this section is devoted to this approach. However, we also introduce graph limits, a topic which will be explored further in Section X.8.

† A supplement to this section begins on p. 372.

Definition   Let A_n, $n = 1, 2, \ldots$ and A be self-adjoint operators. Then A_n is said to converge to A in the **norm resolvent sense** (or **norm generalized sense**) if $R_\lambda(A_n) \to R_\lambda(A)$ in norm for all λ with $\operatorname{Im} \lambda \neq 0$. A_n is said to converge to A in the **strong resolvent sense** (or **strong generalized sense**) if $R_\lambda(A_n) \to R_\lambda(A)$ strongly for all λ with $\operatorname{Im} \lambda \neq 0$.

We have not introduced the notion of weak resolvent convergence since weak resolvent convergence implies strong resolvent convergence (Problem 20). The following theorem shows that norm resolvent convergence is the right generalization of norm convergence for bounded self-adjoint operators. A similar result holds for strong resolvent convergence (see Problem 28), but the analogue for weak convergence is not true (Problem 30).

Theorem VIII.18   Let $\{A_n\}_{n=1}^\infty$ and A be a family of uniformly bounded self-adjoint operators. Then $A_n \to A$ in the norm resolvent sense if and only if $A_n \to A$ in norm.

Proof   Let $A_n \to A$ in norm. Then if $\operatorname{Im} \lambda \neq 0$, $(A_n - A)(A - \lambda)^{-1} \to 0$ in norm. Thus, using

$$(A_n - \lambda)^{-1} = (A - \lambda)^{-1}(I + (A_n - A)(A - \lambda)^{-1})^{-1}$$

we see that $(A_n - \lambda)^{-1} \to (A - \lambda)^{-1}$ in norm.

Conversely, suppose $A_n \to A$ in the norm resolvent sense. Then, since

$$A_n - A = (A_n - i)[(A - i)^{-1} - (A_n - i)^{-1}](A - i)$$

and $\sup \|A_n\| < \infty$, we conclude that

$$\|A_n - A\| \leq (\sup \|A_n\| + 1)\|(A - i)^{-1} - (A_n - i)^{-1}\|(\|A\| + 1)$$
$$\to 0 \quad \text{as} \quad n \to \infty \quad \blacksquare$$

The following theorem shows that to prove generalized convergence one need only show convergence of the resolvents at one point off the real axis.

Theorem VIII.19   Let $\{A_n\}_{n=1}^\infty$ and A be self-adjoint operators, and let λ_0 be a point in $\mathbb{C}$.

(a) If $\operatorname{Im} \lambda_0 \neq 0$ and $\|R_{\lambda_0}(A_n) - R_{\lambda_0}(A)\| \to 0$, then $A_n \to A$ in the norm resolvent sense.

(b) If $\operatorname{Im} \lambda_0 \neq 0$ and if $R_{\lambda_0}(A_n)\varphi - R_{\lambda_0}(A)\varphi \to 0$, for all $\varphi \in \mathcal{H}$, then $A_n \to A$ in the strong resolvent sense.

Proof (a) Both $R_\lambda(A)$ and $R_\lambda(A_n)$ are analytic in the half-plane containing λ_0 and have power series around λ_0,

$$R_\lambda(A) = \sum_{m=0}^{\infty} (\lambda_0 - \lambda)^m [R_{\lambda_0}(A)]^{m+1}$$

$$R_\lambda(A_n) = \sum_{m=0}^{\infty} (\lambda_0 - \lambda)^m [R_{\lambda_0}(A_n)]^{m+1}$$

which converge in norm in the circle $|\lambda - \lambda_0| < |\operatorname{Im} \lambda_0|^{-1}$. Since $R_{\lambda_0}(A_n) \to R_{\lambda_0}(A)$ in norm, $R_\lambda(A_n) \to R_\lambda(A)$ in norm for λ in this circle. Therefore, by repeating this process, we get convergence for all λ in the half-plane containing λ_0. Furthermore, since

$$\|R_{\bar\lambda_0}(A_n) - R_{\bar\lambda_0}(A)\| = \|(R_{\lambda_0}(A_n) - R_{\lambda_0}(A))^*\|$$
$$= \|R_{\lambda_0}(A_n) - R_{\lambda_0}(A)\|$$
$$\to 0 \quad \text{as} \quad n \to \infty$$

the same argument shows that the resolvents converge in norm in the half-plane containing $\bar\lambda_0$.

(b) The proof is the same as the proof of (a) except for two things. First, one considers the *vector-valued* functions $R_\lambda(A_n)\varphi$ and $R_\lambda(A)\varphi$. Secondly, since the map $T \to T^*$ is not continuous in the strong topology, one needs a separate argument to get from one half-plane to the other. Suppose that λ_0 is in the lower half-plane. Then, as in (a), we get convergence everywhere in the lower half-plane, in particular at $\lambda = -i$. The formula

$$(A_n - i)^{-1} - (A - i)^{-1}$$
$$= [(A_n + i)(A_n - i)^{-1}][(A_n + i)^{-1} - (A + i)^{-1}][(A + i)(A - i)^{-1}]$$

which follows from elementary manipulations, can then be used to prove that $(A_n - i)^{-1}$ converges strongly to $(A - i)^{-1}$. The above argument then shows that $R_\lambda(A_n)$ converges strongly to $R_\lambda(A)$ everywhere in the upper half-plane. ∎

For alternative ways of proving that strong convergence, $R_\lambda(A_n) \xrightarrow{s} R_\lambda(A)$, in one half-plane implies strong convergence in the other half-plane, see Theorem VIII.26 or Problem 20b.

We will investigate several aspects of generalized convergence. First, we ask how resolvent convergence is related to the convergence of other bounded functions of A_n and A. Secondly, we investigate the relationship between the spectra of A_n and the spectrum of A if $A_n \to A$ in a generalized sense. Finally, we give criteria on the operators A_n, A themselves which are sufficient to guarantee that $A_n \to A$ in a generalized sense.

Theorem VIII.20 Let A_n and A be self-adjoint operators.

(a) If $A_n \to A$ in the norm resolvent sense and f is a continuous function on $\mathbb{R}$ vanishing at ∞, then $\|f(A_n) - f(A)\| \to 0$.

(b) If $A_n \to A$ in the strong resolvent sense and f is a bounded continuous function on $\mathbb{R}$, then $f(A_n)\varphi \to f(A)\varphi$ for all $\varphi \in \mathcal{H}$.

Proof By the Stone–Weierstrass theorem, polynomials in $(x + i)^{-1}$ and $(x - i)^{-1}$ are dense in $C_\infty(\mathbb{R})$, the continuous functions vanishing at infinity. Thus, given $\varepsilon > 0$, we can find a polynomial $P(s, t)$ so that

$$\|f(x) - P((x + i)^{-1}, (x - i)^{-1})\|_\infty \leq \varepsilon/3$$

Therefore,

$$\|f(A_n) - P((A_n + i)^{-1}, (A_n - i)^{-1})\| \leq \varepsilon/3$$

and

$$\|f(A) - P((A + i)^{-1}, (A - i)^{-1})\| \leq \varepsilon/3$$

If $A_n \to A$ in the norm resolvent sense, then

$$P((A_n + i)^{-1}, (A_n - i)^{-1}) \to (P((A + i)^{-1}, (A - i)^{-1})$$

in norm as $n \to \infty$, and thus for n large enough, $\|f(A_n) - f(A)\| \leq \varepsilon$. This proves (a).

To prove (b) we first note that the same proof as above shows that if $A_n \to A$ in the strong resolvent sense and $h \in C_\infty(\mathbb{R})$, then $h(A_n)\varphi \to h(A)\varphi$. Let $\psi \in \mathcal{H}$ and $\varepsilon > 0$ be given and define $g_m(x) = \exp(-x^2/m)$. Since $g_m(x) \uparrow 1$ pointwise, $g_m(A)\psi \to \psi$ by Theorem VIII.5, so we can find an m with $\|g_m(A)\psi - \psi\| \leq \varepsilon(6\|f\|_\infty)^{-1}$. Furthermore since $g_m \in C_\infty(\mathbb{R})$, $g_m(A_n)\psi \to g_m(A)\psi$ by the remark above, so we can find an N_0, so that $n \geq N_0$ implies $\|g_m(A_n)\psi - g_m(A)\psi\| \leq \varepsilon(6\|f\|_\infty)^{-1}$. Therefore, if $n \geq N_0$,

$$\|g_m(A_n)\psi - \psi\| \leq \varepsilon(3\|f\|_\infty)^{-1}$$

Since fg_m is continuous and goes to zero at ∞, there is an N_1 so that $n \geq N_1$ implies

$$\|f(A_n)g_m(A_n)\psi - f(A)g_m(A)\psi\| \leq \varepsilon/3$$

Let $N = \max\{N_0, N_1\}$. Then for $n \geq N$,

$$\begin{aligned}\|f(A_n)\psi - f(A)\psi\| &\leq \|f(A_n)g_m(A_n)\psi - f(A)g_m(A)\psi\| \\ &\quad + \|f(A_n)\| \|g_m(A_n)\psi - \psi\| \\ &\quad + \|f(A)\| \|g_m(A)\psi - \psi\| \\ &\leq \varepsilon\end{aligned}$$

Since ψ and ε were arbitrary, this proves (b). ∎

As an example of an application of part (a) let $\{A_n\}$ and A be positive self-adjoint operators. Then, if $A_n \to A$ in the norm resolvent sense e^{-tA_n} converges in norm to e^{-tA} for each positive t. To see that part (a) does not extend to all of $C(\mathbb{R})$, notice that on $L^2(\mathbb{R})$ the operators $A_n = (1 - 1/n)x$ converge to the operator $A = x$ in the norm resolvent sense but $\|e^{iA_n} - e^{iA}\| = 1$ for all n.
A very important application of part (b) is the following theorem.

Theorem VIII.21 (Trotter) Let $\{A_n\}$ and A be self-adjoint operators. Then $A_n \to A$ in the strong resolvent sense if and only if e^{itA_n} converges strongly to e^{itA} for each t.

Proof Since e^{itx} is a bounded continuous function of x, Theorem VIII.20 implies that if $A_n \to A$ in the strong resolvent sense, then $e^{itA_n} \to e^{itA}$ strongly for each t.

To prove the theorem in the other direction, we first derive a formula for the resolvent of a self-adjoint operator A. Suppose that Im $\mu < 0$. Then, by the functional calculus

$$(\psi, R_\mu(A)\varphi) = \int_{-\infty}^{\infty} \left(\frac{1}{\mu - \lambda}\right) d(\psi, P_\lambda \varphi)$$

$$= \int_{-\infty}^{\infty} \left(i \int_0^{\infty} e^{-it\mu} e^{it\lambda} dt\right) d(\psi, P_\lambda \varphi)$$

$$= i \int_0^{\infty} e^{-it\mu} \left(\int_{-\infty}^{\infty} e^{it\lambda} d(\psi, P_\lambda \varphi)\right) dt$$

$$= i \int_0^{\infty} e^{-it\mu} (\psi, e^{itA} \varphi) dt$$

$$= \left(\psi, i \int_0^{\infty} e^{-it\mu} e^{itA} \varphi \, dt\right)$$

Therefore,

$$R_\mu(A)\varphi = i \int_0^{\infty} e^{-it\mu} e^{itA} \varphi \, dt \qquad \text{(VIII.9)}$$

where the integral is a Riemann integral. The third step in the computation uses Fubini's theorem. Applying (VIII.9) to the operators A_n and A

$$\|R_\mu(A_n)\varphi - R_\mu(A)\varphi\| \leq \int_0^{\infty} e^{(\text{Im } \mu)t} \|e^{itA_n}\varphi - e^{itA}\varphi\| \, dt$$

so if $e^{itA_n}\varphi \to e^{itA}\varphi$ for each t,

$$\|R_\mu(A_n)\varphi - R_\mu(A)\varphi\| \to 0$$

by the Lebesgue dominated convergence theorem. Using a formula similar to (VIII.9) one concludes in the same way that

$$\|R_\mu(A_n)\varphi - R_\mu(A)\varphi\| \to 0 \quad \text{for} \quad \text{Im}\,\mu > 0 \quad \blacksquare$$

We remark that it is possible to show (Problem 21) that if $A_n \to A$ in the strong resolvent sense, then $e^{itA_n}\varphi \to e^{itA}\varphi$ for each φ, *uniformly* for t in any finite interval. Next, we prove a related theorem.

Theorem VIII.22 (Trotter–Kato) Let A_n be a sequence of self-adjoint operators. Suppose that there exist points, λ_0 in the upper half-plane and μ_0 in the lower half-plane, so that $R_{\lambda_0}(A_n)\varphi$ and $R_{\mu_0}(A_n)\varphi$ converge for each $\varphi \in \mathscr{H}$. Suppose further that one of the limiting operators, T_{λ_0} or T_{μ_0}, has a dense range. Then there exists a self-adjoint operator A so that $A_n \to A$ in the strong resolvent sense.

Proof Since $\|R_{\lambda_0}(A_n)\| \leq |\text{Im}\,\lambda_0|^{-1}$, $\|T_{\lambda_0}\| \leq |\text{Im}\,\lambda_0|^{-1}$, and so

$$T_\lambda = \sum_{n=0}^{\infty} (\lambda_0 - \lambda)^n (T_{\lambda_0})^{n+1}$$

is well defined for $|\lambda - \lambda_0| \leq |\text{Im}\,\lambda_0|^{-1}$.

Furthermore, since $R_{\lambda_0}(A_n)\varphi \to T_{\lambda_0}\varphi$, $R_\lambda(A_n)\varphi \to T_\lambda\varphi$ in the same circle. Continuing in this way we can define an analytic operator valued function T_λ in the half-plane containing λ_0 which is the strong limit of $R_\lambda(A_n)$. Since the half-plane is simply connected, the determination of T_λ at a point λ is independent of the path taken from λ_0 to λ. The same argument for the half-plane containing μ_0 shows that we can extend the definition of T_λ to that half-plane so that

$$T_\lambda \varphi = \lim_{n \to \infty} R_\lambda(A_n)\varphi \quad \forall \lambda \text{ with } \text{Im}\,\lambda \neq 0$$

The T_λ commute, satisfy the first resolvent equation, and $T_\lambda^* = T_{\bar{\lambda}}$ since these statements are true for each $R_\lambda(A_n)$. It follows from the first resolvent formula and the commutativity that the ranges of all the T_λ are equal; we denote this common range by D. By hypothesis, D is dense and this implies that the kernel of T_λ is empty, since $\text{Ker}\,T_\lambda = (\text{Ran}\,T_\lambda^*)^\perp = (\text{Ran}\,T_{\bar{\lambda}})^\perp = D^\perp = \{0\}$. We can therefore define $A = \lambda I - T_\lambda^{-1}$ on D and a short calculation with the resolvent equation shows that this definition is independent of which λ, with $\text{Im}\,\lambda \neq 0$, is chosen. Since $\text{Ran}(A \pm i) = \text{Ran}(-T_{\mp i}^{-1}) = \mathscr{H}$, A is self-adjoint. It is clear that the resolvent of A is T_λ. $\blacksquare$

Notice that in the Trotter–Kato theorem we need convergence at *two* points, one in the upper half-plane and one in the lower half-plane. For, we cannot use Theorem VIII.19b until we know that the limiting operator is self-adjoint, and the self-adjointness proof depends on the convergence in both half-planes.

The Trotter–Kato theorem is important since its hypotheses do *not* assume the a priori existence of a limiting operator A. It can be used to assert the existence of a generalized limit of a sequence of self-adjoint operators. This can also be done with the one-parameter groups (see Problem 23). To see why it is necessary to use the resolvents or groups rather than the operators themselves to prove such an existence theorem consider the following example: Let A be a closed symmetric operator which is not self-adjoint but which has a self-adjoint extension $\tilde{A}$. Let P_n be the spectral projection of $\tilde{A}$ corresponding to the interval $[-n, n]$. Then $P_n \tilde{A} P_n$ are bounded self-adjoint operators (and therefore essentially self-adjoint on $D(A)$) such that for all $\varphi \in D(A)$

$$P_n \tilde{A} P_n \varphi \to \tilde{A}\varphi = A\varphi$$

Thus the $P_n \tilde{A} P_n$ are essentially self-adjoint on $D(A)$ and the strong limit exists but the limit is not essentially self-adjoint.

One of the most useful aspects of generalized convergence is that the spectra and projections of the A_n are related to the spectrum and projections of A. For applications of the following two theorems see Sections X.2 and XI.5.

Theorem VIII.23 Let $\{A_n\}_{n=1}^{\infty}$ and A be self-adjoint operators and suppose that $A_n \to A$ in the norm resolvent sense. Then

(a) If $\mu \notin \sigma(A)$, then $\mu \notin \sigma(A_n)$ for n sufficiently large and

$$\|R_\mu(A_n) - R_\mu(A)\| \to 0$$

(b) Let $a, b \in \mathbb{R}$, $a < b$, and suppose that $a \in \rho(A)$, $b \in \rho(A)$. Then

$$\|P_{(a,b)}(A_n) - P_{(a,b)}(A)\| \to 0$$

Proof (a) We need only consider the case where μ is real. Since $\mu \in \rho(A)$, there is a $\delta > 0$ so that $(\mu - \delta, \mu + \delta) \cap \sigma(A) = \varnothing$. Thus, by the functional calculus, $\|R_{\mu + i\delta/3}(A)\| < 1/\delta$. Now, we can find N so that

$$\|R_{\mu + i\delta/3}(A_n)\| \leq 2/\delta$$

for $n \geq N$, which implies that the power series for $R_\lambda(A_n)$ about $\mu + i\delta/3$ has radius of convergence at least $\delta/2$. We already know that where the series converges it is an inverse for A_n. So, $\mu \in \rho(A_n)$ for $n \geq N$ and

$$\|R_\mu(A_n) - R_\mu(A)\| \to 0$$

To prove (b), we note that since $a, b \in \rho(A)$, there exists $\varepsilon < \frac{1}{2}(b-a)$ and an N, so that

$$\sup_{n \geq N} \{ \|(A_n - a)^{-1}\|, \|(A_n - b)^{-1}\| \} \leq 1/\varepsilon$$

Therefore, by the functional calculus, $\sigma(A_n) \cap (a, b) \subset (a + \varepsilon, b - \varepsilon)$ for $n \geq N$. Let f be a continuous function which equals one on $(a + \varepsilon, b - \varepsilon)$ and is equal to zero outside (a, b). Then

$$P_{(a,b)}(A_n) = f(A_n) \qquad P_{(a,b)}(A) = f(A)$$

and so by Theorem VIII.20,

$$\|P_{(a,b)}(A_n) - P_{(a,b)}(A)\| \to 0 \quad \blacksquare$$

Theorem VIII.24 Let $\{A_n\}_{n=1}^\infty$ and A be self-adjoint operators and suppose that $A_n \to A$ in the strong resolvent sense. Then

(a) If $a, b \in \mathbb{R}$, $a < b$, and $(a, b) \cap \sigma(A_n) = \emptyset$ for all n, then $(a, b) \cap \sigma(A) = \emptyset$. That is, if $\lambda \in \sigma(A)$, then there exists $\lambda_n \in \sigma(A_n)$ so that $\lambda_n \to \lambda$.

(b) If $a, b \in \mathbb{R}$, $a < b$, and $a, b \notin \sigma_{pp}(A)$, then $P_{(a,b)}(A_n)\varphi \to P_{(a,b)}(A)\varphi$ for all $\varphi \in \mathcal{H}$.

Proof By the functional calculus, the statement that $(a, b) \cap \sigma(A) = \emptyset$ is equivalent to the statement that

$$\|(A - \lambda_0)^{-1}\| \leq \frac{\sqrt{2}}{b - a}$$

where

$$\lambda_0 = \frac{a+b}{2} + i\left(\frac{b-a}{2}\right)$$

But $(A_n - \lambda_0)^{-1}$ converges strongly to $(A - \lambda_0)^{-1}$ so we have

$$\|(A - \lambda_0)^{-1}\| \leq \varliminf_n \|(A_n - \lambda_0)^{-1}\| \leq \frac{\sqrt{2}}{b - a}$$

This proves (a).

To prove (b), we find uniformly bounded sequences of continuous functions f_n and g_n so that $0 \leq f_n \leq \chi_{(a,b)}$, $f_n(x) \nearrow \chi_{(a,b)}(x)$ pointwise and $\chi_{[a,b]} \leq g_n$, $g_n(x) \searrow \chi_{[a,b]}(x)$ pointwise. Then $f_n(A) \to P_{(a,b)}(A)$ and $g_n(A) \to P_{[a,b]}(A)$ strongly.

Since $a, b \notin \sigma_{pp}(A)$, $P_{[a,b]}(A) = P_{(a,b)}(A)$ which means that given ψ and $\varepsilon > 0$, we can find continuous functions f, g, with $f \leq \chi_{(a,b)} \leq \chi_{[a,b]} \leq g$, so that $\|f(A)\psi - g(A)\psi\| \leq \varepsilon/5$. By Theorem VIII.20b we can find N so that $n \geq N$ implies

$$\|f(A_n)\psi - f(A)\psi\| \leq \varepsilon/5 \quad \|g(A_n)\psi - g(A)\psi\| \leq \varepsilon/5$$

so by an $\varepsilon/3$ argument

$$\|f(A_n)\psi - g(A_n)\psi\| \leq 3\varepsilon/5$$

Since the functional calculus implies

$$\|f(A)\psi - P_{(a,b)}(A)\psi\| \leq \|f(A)\psi - g(A)\psi\|$$

another $\varepsilon/3$ argument implies

$$\|P_{(a,b)}(A_n)\psi - P_{(a,b)}(A)\psi\| \leq \varepsilon \quad \blacksquare$$

Part (a) of Theorem VIII.24 says that the spectrum of the limiting operator cannot suddenly expand. It can, however, contract rather spectacularly as the following example shows: Let $A_n = x/n$ on $L^2(\mathbb{R})$; then A_n converges to the zero operator in the strong resolvent sense. For each n, $\sigma(A_n) = \mathbb{R}$, but the spectrum of the limiting operator contains only the origin. An easy application of part (a) is the statement that if the A_n are positive and $A_n \to A$ in the strong resolvent sense, then A is positive.

If A_n converges to A in *norm* resolvent sense, Theorem VIII.23 tells us that the spectrum of the limiting operator cannot suddenly contract in the sense that if $\lambda \in \sigma(A_n)$ for all sufficiently large n, then $\lambda \in \sigma(A)$. Notice that in the example $A_n = x/n$ above, A_n does not converge to A in the norm resolvent sense.

The principle of noncontraction of the spectrum under norm resolvent convergence remains true even when A_n and A are not self-adjoint. But the principle of nonexpansion of the spectrum in the strong resolvent limit is not always valid for general not-necessarily-self-adjoint operators. In fact, there exists a *norm* convergent sequence of uniformly bounded operators $A_n \to A$ with $\sigma(A_n)$ the unit circle in $\mathbb{C}$ for each n and $\sigma(A)$ the entire unit disc. Thus the reader should be careful to apply Theorem VIII.24 only in the self-adjoint case.

In applications, one is usually given the operators $\{A_n\}$ and A on domains of self-adjointness or essential self-adjointness and it may be very difficult to compute the resolvents. Thus, in order to use Theorems VIII.23 and VIII.24 one must have criteria on the operators A_n, A themselves which guarantee norm or strong resolvent convergence.

Theorem VIII.25 (a) Let $\{A_n\}_{n=1}^{\infty}$ and A be self-adjoint operators and suppose that D is a common core for all A_n, A. If $A_n \varphi \to A\varphi$ for each $\varphi \in D$, then $A_n \to A$ in the strong resolvent sense.

(b) Let $\{A_n\}_{n=1}^{\infty}$ and A be self-adjoint operators with a common domain, D. Norm D with $\|\varphi\|_A = \|A\varphi\| + \|\varphi\|$. If

$$\sup_{\|\varphi\|_A = 1} \|(A_n - A)\varphi\| \to 0$$

then $A_n \to A$ in the norm resolvent sense.

(c) Let $\{A\}_{n=1}^{\infty}$ and A be positive self-adjoint operators with a common form domain $\mathcal{H}_{+1}$ which we norm with

$$\|\psi\|_{+1} = \sqrt{(\psi, A\psi) + (\psi, \psi)}$$

If $A_n \to A$ in norm in the sense of maps from $\mathcal{H}_{+1}$ to $\mathcal{H}_{-1}$, that is, if

$$\sup_{0 \neq \psi, \varphi \in D} \frac{|(\varphi, (A - A_n)\psi)|}{\|\varphi\|_{+1} \|\psi\|_{+1}} = \sup_{0 \neq \psi \in D} \frac{|(\psi, (A - A_n)\psi)|}{(\psi, (A+1)\psi)} \longrightarrow 0$$

then $A_n \to A$ in the norm resolvent sense.

Proof (a) Let $\varphi \in D$, $\psi = (A + i)\varphi$, then

$$[(A_n + i)^{-1} - (A + i)^{-1}]\psi = (A_n + i)^{-1}(A - A_n)\varphi$$

converges to zero as $n \to \infty$, since $(A - A_n)\varphi \to 0$ and the $(A_n + i)^{-1}$ are uniformly bounded. Since D is a core for A, the set of such ψ is dense so

$$(A_n + i)^{-1}\varphi \to (A + i)^{-1}\varphi \quad \text{for all} \quad \varphi \in \mathcal{H}$$

A similar proof works for $(A_n - i)^{-1}$.

We sketch the proofs of (b) and (c). For (b), first one proves that the hypothesis is equivalent to $(A_n - A)(A + i)^{-1} \to 0$ in the ordinary $\mathcal{H}$-operator norm. Thus $(I + (A_n - A)(A + i)^{-1})^{-1}$ exists and converges to I in norm as $n \to \infty$. As a result

$$(A_n + i)^{-1} = (A + i)^{-1}(I + (A_n - A)(A + i)^{-1})^{-1} \to (A + i)^{-1}$$

in norm. Similarily $(A_n - i)^{-1} \to (A - i)^{-1}$.

To prove (c), one first proves that the hypothesis is equivalent to

$$(A + I)^{-1/2}(A_n - A)(A + I)^{-1/2} \to 0$$

in the ordinary operator norm. Using

$$(A_n + I)^{-1} = (A + I)^{-1/2}(I + (A + I)^{-1/2}(A_n - A)(A + I)^{-1/2})^{-1}(A + I)^{-1/2}$$

one then follows the proof of (b). ∎

VIII.7 Convergence of unbounded operators

Finally, we introduce graph limits and compare them with generalized limits. Graph limits will be discussed again in Section X.8.

Definition Let A_n be a sequence of operators on a Hilbert space $\mathcal{H}$. We say that $\langle \psi, \varphi \rangle \in \mathcal{H} \times \mathcal{H}$ is in the **strong graph limit** of A_n if we can find $\psi_n \in D(A_n)$ so that $\psi_n \to \psi$, $A_n \psi_n \to \varphi$. We denote the set of pairs in the strong graph limit by Γ_∞^s. If Γ_∞^s is the graph of an operator A we say that A is the **strong graph limit** of A_n and write $A = \text{st. gr.-lim } A_n$.

First, we consider the case where all the A_n are self-adjoint *and A is also self-adjoint*.

Theorem VIII.26 Suppose that $\{A_n\}$ and A are self-adjoint operators. Then $A_n \to A$ in the strong resolvent sense if and only if $A = \text{st. gr.-lim } A_n$.

Proof Suppose first that $(A_n + i)^{-1} \to (A + i)^{-1}$ strongly. Suppose $\varphi \in D(A)$. Then $\varphi_n \equiv (A_n + i)^{-1}(A + i)\varphi \to \varphi$ and $A_n \varphi_n = (A + i)\varphi - i\varphi_n \to A\varphi$, so $\langle \varphi, A\varphi \rangle \in \Gamma_\infty^s$. Thus $\Gamma(A) \subset \Gamma_\infty^s$. On the other hand, suppose $\varphi_n \in D(A_n)$, $\varphi_n \to \varphi$ and $A_n \varphi_n \to \psi$. If we let $\eta_n = (A + i)^{-1}(A_n + i)\varphi_n \in D(A)$, then

$$\eta_n - \varphi_n = [(A + i)^{-1} - (A_n + i)^{-1}][(A_n + i)\varphi_n]$$
$$= [(A + i)^{-1} - (A_n + i)^{-1}][(A_n + i)\varphi_n - \psi - i\varphi]$$
$$+ [(A + i)^{-1} - (A_n + i)^{-1}][\psi + i\varphi]$$
$$\to 0 \quad \text{as} \quad n \to \infty$$

Thus, $\eta_n \to \varphi$ and $A\eta_n = (A_n + i)\varphi_n - i\eta_n \to \psi$, so since A is closed $\langle \varphi, \psi \rangle \in \Gamma(A)$. Thus, $\Gamma(A) = \Gamma_\infty^s$.

Conversely, suppose that $A = \text{st. gr.-lim } A_n$. Let $\varphi \in D(A)$. Then there exist $\varphi_n \in D(A_n)$ so that $\varphi_n \to \varphi$ and $A_n \varphi_n \to A\varphi$. Thus,

$$[(A_n + i)^{-1} - (A + i)^{-1}](A + i)\varphi$$
$$= (A_n + i)^{-1}[(A + i)\varphi - (A_n + i)\varphi_n] - (\varphi - \varphi_n)$$
$$\to 0 \quad \text{as} \quad n \to \infty$$

since $\|(A_n + i)^{-1}\| \leq 1$, $(A_n + i)\varphi_n \to (A + i)\varphi$, and $\varphi_n \to \varphi$. Since $\text{Ran}(A + i) = \mathcal{H}$, the strong convergence of $(A_n + i)^{-1}$ to $(A + i)^{-1}$ follows. ∎

Thus, we see that *if the limit is self-adjoint*, then strong graph and strong resolvent convergence are the same. It is in the case when we do not know a priori that the limit is self-adjoint that strong graph limits are particularly interesting. For example, in Section X.8 we will see that the existence of graph limits can sometimes be combined with other information to prove that the limit is self-adjoint.

Theorem VIII.27 Let $\{A_n\}$ be a sequence of symmetric operators.

(a) Let $D_\infty^s = \{\psi \,|\, \langle \psi, \varphi\rangle \in \Gamma_\infty^s \text{ for some } \varphi\}$. If D_∞^s is dense, then Γ_∞^s is the graph of an operator.

(b) Suppose that D_∞^s is dense and let $A = \text{st. gr.-lim } A_n$.

Then A is symmetric and closed.

Proof We will prove (a); the proof of (b) is left as an exercise (Problem 24). Suppose $\varphi_n, \varphi_n' \in D(A_n)$ and $\varphi_n \to \varphi$, $\varphi_n' \to \varphi$, $A_n \varphi_n \to \psi$, and $A_n \varphi_n' \to \psi'$. Let $\eta \in D_\infty^s$. Then there is an $\eta_n \in D(A_n)$, so that $\eta_n \to \eta$ and $A_n \eta_n \to \rho$. Thus,

$$(\psi - \psi', \eta) = \lim_{n \to \infty} (A_n(\varphi_n - \varphi_n'), \eta_n)$$
$$= \lim_{n \to \infty} (\varphi_n - \varphi_n', A_n \eta_n)$$
$$= 0$$

so $\psi = \psi'$, since D_∞^s is dense. ∎

One can also define weak graph limits. We give the definition and state one theorem.

Definition Let A_n be a sequence of operators on $\mathscr{H}$. We say that $\langle \psi, \varphi \rangle \in \mathscr{H} \times \mathscr{H}$ is in the **weak graph limit** Γ_∞^w if we can find $\psi_n \in D(A_n)$ so that $\psi_n \xrightarrow{\|\cdot\|} \psi$ and $A_n \psi_n \to \varphi$ weakly. If Γ_∞^w is the graph of an operator, A, we say that A is the **weak graph limit** of A_n; $A = \text{w. gr.-lim } A_n$.

Theorem VIII.28 Let A_n be a sequence of symmetric operators. If

$$D_\infty^w = \{\psi \,|\, \langle\psi, \varphi\rangle \in \Gamma_\infty^w \text{ for some } \varphi\}$$

is dense, then Γ_∞^w is the graph of a symmetric operator.

Finally we note that if A_n is a uniformly bounded sequence of operators then $A = \text{w. gr.-lim } A_n$ if and only if $A_n \to A$ in the weak operator topology (Problem 26). This fact combined with Problems 20 and 28 shows that the notions of weak graph limit and weak resolvent convergence are distinct. It is not true that weak graph limits are necessarily closed if each A_n is symmetric.

VIII.8 The Trotter product formula†

In this section we prove a useful approximation theorem for $\exp t(A + B)$ in terms of $\exp tA$ and $\exp tB$. It is illuminating to first consider finite-dimensional matrices for which one has the classical theorem of Lie:

Theorem VIII.29 (Lie product formula)   Let A and B be finite-dimensional matrices. Then
$$\exp(A + B) = \lim_{n \to \infty} [\exp(A/n) \exp(B/n)]^n$$

Proof   Let $S_n = \exp[(A + B)/n)]$, $T_n = \exp(A/n) \exp(B/n)$. Then
$$S_n^n - T_n^n = \sum_{m=0}^{n-1} S_n^m (S_n - T_n) T_n^{n-1-m}$$
so
$$\|S_n^n - T_n^n\| \leq n(\max\{\|S_n\|, \|T_n\|\})^{n-1} \|S_n - T_n\|$$
$$\leq n \|S_n - T_n\| \exp(\|A\| + \|B\|)$$
Since
$$\|S_n - T_n\| = \left\| \sum_{m=0}^{\infty} \frac{1}{m!} \left(\frac{A+B}{n}\right)^m - \left(\sum_{m=0}^{\infty} \frac{1}{m!} \left(\frac{A}{n}\right)^m\right)\left(\sum_{m=0}^{\infty} \frac{1}{m!} \left(\frac{B}{n}\right)^m\right) \right\|$$
$$\leq C/n^2 \quad (C \text{ depends on } \|A\| \text{ and } \|B\|)$$
we conclude that $\|S_n^n - T_n^n\| \to 0$. ∎

This theorem *and its proof* can be extended to the case where A and B are unbounded self-adjoint operators and $A + B$ is *self-adjoint* on $D(A) \cap D(B)$.

Theorem VIII.30   Let A and B be self-adjoint operators on $\mathcal{H}$ and suppose that $A + B$ is self-adjoint on $D = D(A) \cap D(B)$. Then
$$\text{s-}\lim_{n \to \infty} [e^{itA/n} e^{itB/n}]^n = e^{it(A+B)}$$

Proof   Let $\psi \in D$. Then
$$s^{-1}(e^{isA} e^{isB} - I)\psi = s^{-1}(e^{isA} - I)\psi + s^{-1} e^{isA}(e^{isB} - I)\psi \to iA\psi + iB\psi$$
and
$$s^{-1}(e^{is(A+B)} - I)\psi \to i(A + B)\psi$$

† A supplement to this section begins on p. 377.

as $s \to 0$. Letting $K(s) = s^{-1}(e^{isA}e^{isB} - e^{is(A+B)})$ we see that $K(s)\psi \to 0$ as $s \to 0$, for each $\psi \in D$. Since $A + B$ is self-adjoint on D, D is a Banach space under the norm

$$\|\psi\|_{A+B} = \|(A + B)\psi\| + \|\psi\|$$

Each of the maps $K(s)\colon D \to \mathscr{H}$ is bounded and $K(s)\psi \xrightarrow{\mathscr{H}} 0$ as $s \to 0$ or ∞ for each $\psi \in D$. Thus, we conclude from the uniform boundedness theorem that the $K(s)$ are uniformly bounded, that is, there is a constant C so that

$$\|K(s)\psi\| \leq C \|\psi\|_{A+B} \quad \text{for all} \quad s \in \mathbb{R} \text{ and } \psi \in D$$

Therefore, an $\varepsilon/3$ argument shows that on $\|\cdot\|_{A+B}$ compact subsets of D, $K(s)\psi \to 0$ uniformly.

Since $A + B$ is self-adjoint on D, $e^{is(A+B)}\psi \in D$ if $\psi \in D$. Moreover, $s \to e^{is(A+B)}\psi$ is a continuous map of $\mathbb{R}$ into D when D is given the $\|\cdot\|_{A+B}$ norm topology. Thus $\{e^{is(A+B)}\psi \mid s \in [-1, 1]\}$ is a $\|\cdot\|_{A+B}$ compact set in D for each fixed ψ.

We are now ready to mimic the proof of the Lie product formula. We know that

$$t^{-1}[e^{itA}e^{itB} - e^{it(A+B)}]e^{is(A+B)}\psi \to 0$$

uniformly for $s \in [-1, 1]$. Therefore, we write

$$[(e^{itA/n}e^{itB/n})^n - (e^{it(A+B)/n})^n]\psi$$
$$= \sum_{k=0}^{n-1}(e^{itA/n}e^{itB/n})^k[e^{itA/n}e^{itB/n} - e^{it(A+B)/n}][e^{it(A+B)/n}]^{n-1-k}\psi$$

The norm of the right hand side is less than or equal to

$$|t| \max_{|s|<t} \left\|\left(\frac{t}{n}\right)^{-1}(e^{it(A+B)/n} - e^{itA/n}e^{itB/n})e^{is(A+B)}\psi\right\|$$

and so we conclude that

$$(e^{itA/n}e^{itB/n})^n\psi \xrightarrow{\mathscr{H}} e^{it(A+B)}\psi, \quad \text{as } n \to \infty \quad \text{if } \psi \in D$$

Since D is dense and the operators are bounded by one, this statement holds on all of $\mathscr{H}$. ∎

The above proof shows that on a fixed vector the convergence is uniform for t in a compact subset of $\mathbb{R}$.

The same argument can be used to show that

$$\operatorname*{s-lim}_{n \to \infty} (e^{-tA/n}e^{-tB/n})^n = e^{-t(A+B)}$$

if A and B satisfy the same hypotheses and in addition are semibounded. The following result is considerably stronger than Theorem VIII.30 since it only requires *essential* self-adjointness of $A + B$ on $D(A) \cap D(B)$. The proof is quite different from the proof of Theorem VIII.30 (see the notes for references).

Theorem VIII.31 (the Trotter product formula) If A and B are self-adjoint operators and $A + B$ is essentially self-adjoint on $D(A) \cap D(B)$, then

$$\operatorname*{s-lim}_{n \to \infty} (e^{itA/n} e^{itB/n})^n = e^{i(A+B)t}$$

Moreover, if A and B are bounded from below, then

$$\operatorname*{s-lim}_{n \to \infty} (e^{-tA/n} e^{-tB/n})^n = e^{-t(A+B)}$$

For applications of the Trotter product formula the reader should see Section X.10 (Feynman path integrals), Section X.7 (hypercontractive semigroups), or Chapter XIX (the section on constructive quantum field theory).

VIII.9 The polar decomposition for closed operators

In Section VI.4, we saw that an arbitrary bounded operator T can be written $T = U|T|$ where $|T|$ is positive and self-adjoint and U is a partial isometry. Moreover, the conditions that $\operatorname{Ker}|T| = \operatorname{Ker} T$ and that the initial space of U equals $(\operatorname{Ker} T)^\perp$ uniquely determine $|T|$ and U. In this section we want to extend this result to *closed* unbounded operators. As in the bounded case, U is easy to construct once $|T|$ has been constructed and, as in the bounded case, we will let $|T| = \sqrt{T^*T}$. In the bounded case, the hard part was the construction of the square root. Now that we have the spectral theorem, it is easy to construct $\sqrt{T^*T}$ if we can prove that T^*T is a positive self-adjoint operator. It is this fact that is hard in the unbounded case. A priori, it is not clear that $\{\psi \mid \psi \in D(T) \text{ and } T\psi \in D(T^*)\}$ is different from $\{0\}$. In fact, this set is dense (Problem 45), but our approach using the theory of semi-bounded quadratic forms does not require us to prove this.

Theorem VIII.32 (the polar decomposition) Let T be an arbitrary closed operator on a Hilbert space $\mathcal{H}$. Then, there is a positive self-adjoint operator $|T|$, with $D(|T|) = D(T)$ and a partial isometry U with initial space,

(Ker $T)^\perp$, and final space, $\overline{\operatorname{Ran} T}$, so that $T = U|T|$. $|T|$ and U are uniquely determined by these properties together with the additional condition $\operatorname{Ker}(|T|) = \operatorname{Ker}(T)$.

Proof Define the quadratic form s on $D(T)$ by $s(\psi, \phi) = (T\psi, T\phi)$. s is clearly positive. Now suppose $\|\psi_n - \psi_m\|_{+1} \to 0$. Then $\|\psi_n - \psi_m\| \to 0$ and $\|T(\psi_n - \psi_m)\| \to 0$. Since T is closed there is a $\psi \in D(T)$ with $\|\psi_n - \psi\| + \|T(\psi_n - \psi)\| \to 0$, i.e. $\|\psi_n - \psi\|_{+1} \to 0$. Thus s is a closed form. Therefore, by Theorem VIII.15, there is a unique, positive self-adjoint operator S with $Q(S) = D(T)$ and $s(\psi, \phi) = (\psi, S\phi)$ in the sense of forms. Let $|T| = S^{1/2}$. Then $D(|T|) = Q(S) = D(T)$ and by construction $\||T|\psi\|^2 = s(\psi, \psi) = \|T\psi\|^2$ so $\ker |T| = \ker T$. Define $U: \operatorname{Ran} |T| \to \operatorname{Ran} T$ by $U|T|\psi = T\psi$. Since $\||T|\psi\| = \|T\psi\|$, U is well defined and norm preserving. Thus U extends to a partial isometry from $\overline{\operatorname{Ran} |T|}$ to $\overline{\operatorname{Ran} T}$. Finally, since $|T|$ is self-adjoint, $\overline{\operatorname{Ran} |T|} = (\operatorname{Ker} |T|)^\perp = (\operatorname{Ker} T)^\perp$. Uniqueness is left to the reader (Problem 44). ∎

VIII.10 Tensor products

In this section we describe some aspects of the theory of tensor products of operators on Hilbert spaces. Let A and B be densely defined operators on Hilbert spaces $\mathcal{H}_1$ and $\mathcal{H}_2$ respectively. We will denote by $D(A) \otimes D(B)$ the set of finite linear combinations of vectors of the form $\phi \otimes \psi$ where $\phi \in D(A)$ and $\psi \in D(B)$. $D(A) \otimes D(B)$ is dense in $\mathcal{H}_1 \otimes \mathcal{H}_2$. We define $A \otimes B$ on $D(A) \otimes D(B)$ by

$$(A \otimes B)(\phi \otimes \psi) = A\phi \otimes B\psi$$

and extend by linearity.

Proposition The operator $A \otimes B$ is well defined. Further, if A and B are closable, so is $A \otimes B$.

Proof Suppose that $\sum c_i \phi_i \otimes \psi_i$ and $\sum d_j \phi'_j \otimes \psi'_j$ are two representations of the same vector $f \in D(A) \otimes D(B)$. Using Gram–Schmidt orthogonalization we can obtain bases $\{\eta_k\}$ and $\{\theta_\ell\}$ for the spaces spanned by $\{\phi_i\} \cup \{\phi'_i\}$ and $\{\psi_j\} \cup \{\psi'_j\}$ respectively so that $\eta_k \in D(A)$ and $\theta_\ell \in D(B)$. $\phi_i \otimes \psi_i$ and $\phi'_j \otimes \psi'_j$ can be expressed

$$\phi_i \otimes \psi_i = \sum \alpha^i_{k\ell} \eta_k \otimes \theta_\ell$$
$$\phi'_j \otimes \psi'_j = \sum \beta^j_{k\ell} \eta_k \otimes \theta_\ell$$

Since the two expressions for f give the same vector, $\sum_i c_i \alpha_{k\ell}^i = \sum_j d_j \beta_{k\ell}^j$ for each pair $\langle k, \ell \rangle$.
Thus,

$$(A \otimes B) \sum c_i(\phi_i \otimes \psi_i) = \sum_{k\ell} (\sum_i c_i \alpha_{k\ell}^i)(A\eta_k \otimes B\theta_\ell)$$

$$= \sum_{k\ell} (\sum_j d_j \beta_{k\ell}^j)(A\eta_k \otimes B\theta_\ell)$$

$$= (A \otimes B) \sum d_j(\phi_j' \otimes \psi_j')$$

so $A \otimes B$ is well defined.

If g is any vector in $D(A^*) \otimes D(B^*)$, then $(A \otimes Bf, g) = (f, A^* \otimes B^* g)$ so

$$D(A^*) \otimes D(B^*) \subset D((A \otimes B)^*)$$

If A and B are closable, $D(A^*)$ and $D(B^*)$ are dense. Therefore, in that case $(A \otimes B)^*$ is densely defined which proves that $A \otimes B$ is closable. ∎

Similarly, if A and B are closable then $A \otimes I + I \otimes B$, defined on $D(A) \otimes D(B)$, is closable.

Definition Let A and B be closable operators on Hilbert spaces $\mathscr{H}_1$ and $\mathscr{H}_2$. The **tensor product** of A and B is the closure of the operator $A \otimes B$ defined on $D(A) \otimes D(B)$. We will denote the closure by $A \otimes B$ also. Usually $A + B$ will denote the closure of $A \otimes I + I \otimes B$ on $D(A) \otimes D(B)$.

Proposition Let A and B be bounded operators on Hilbert spaces $\mathscr{H}_1$ and $\mathscr{H}_2$. Then $\|A \otimes B\| = \|A\| \|B\|$.

Proof Let $\{\phi_k\}$ and $\{\psi_\ell\}$ be orthonormal bases for $\mathscr{H}_1$ and $\mathscr{H}_2$ and suppose $\sum c_{k\ell} \phi_k \otimes \psi_\ell$ is a finite sum. Then

$$\|(A \otimes I) \sum c_{k\ell}(\phi_k \otimes \psi_\ell)\|^2 = \sum_\ell \|\sum_k c_{k\ell} A\phi_k\|^2$$

$$\leq \sum_\ell \|A\|^2 \sum_k |c_{k\ell}|^2$$

$$= \|A\|^2 \|\sum c_{k\ell} \phi_k \otimes \psi_\ell\|^2$$

Since the set of such finite sums is dense in $\mathscr{H}_1 \otimes \mathscr{H}_2$ (Section II.4, Proposition 2), we conclude that $\|A \otimes I\| \leq \|A\|$. Thus

$$\|A \otimes B\| \leq \|A \otimes I\| \|I \otimes B\| \leq \|A\| \|B\|$$

Conversely, given $\varepsilon > 0$, there exist unit vectors $\phi \in \mathcal{H}_1, \psi \in \mathcal{H}_2$ so that $\|A\phi\| \geq \|A\| - \varepsilon$ and $\|B\psi\| \geq \|B\| - \varepsilon$. Then,

$$\|(A \otimes B)(\phi \otimes \psi)\| = \|A\phi\| \|B\psi\|$$
$$\geq \|A\| \|B\| - \varepsilon\|A\| - \varepsilon\|B\| + \varepsilon^2$$

Since $\varepsilon > 0$ is arbitrary $\|A \otimes B\| \geq \|A\| \|B\|$ which concludes the proof. ∎

We remark that both of the above propositions have natural generalizations to arbitrary finite tensor products of operators. This can be proven directly or by using the associativity of the tensor product of Hilbert spaces.

We turn now to questions of self-adjointness and spectrum. Let $\{A_k\}_{k=1}^N$ be a family of operators, A_k self-adjoint on $\mathcal{H}_k$. We will denote the closure of $I_1 \otimes \cdots \otimes A_k \otimes \cdots I$ on $D = \otimes D(A_k)$ by A_k also. Let $P(x_1, \ldots x_N)$ be a polynomial with real coefficients of degree n_k in x_k. Then, the operator $P(A_1, \ldots, A_N)$ makes sense on $\otimes_k D(A^{n_k})$ since $D(A^{n_k}) \subset D(A^\ell)$ for all $\ell \leq n_k$. In fact, P is essentially self-adjoint on that domain.

Theorem VIII.33 Let A_k be a self-adjoint operator on $\mathcal{H}_k$. Let $P(x_1, \ldots, x_N)$ be a polynomial with real coefficients of degree n_k in the kth variable and suppose that D_k^e is a domain of essential self-adjointness for $A_k^{n_k}$. Then,

(a) $P(A_1, \ldots, A_N)$ is essentially self-adjoint on

$$D^e = \bigotimes_{k=1}^N D_k^e$$

(b) The spectrum of $\overline{P(A_1, \ldots, A_N)}$ is the closure of the range of P on the product of the spectra of the A_k. That is

$$\sigma(\overline{P(A_1, \ldots, A_n)}) = \overline{P(\sigma(A_1), \ldots, \sigma(A_N))}$$

Proof We will first prove that $P(A_1, \ldots, A_N)$ is essentially self-adjoint on $D = \otimes_{k=1}^N D(A_k^{n_k})$. By the spectral theorem, there is a measure space $\langle M_k, \mu_k \rangle$ so that A_k is unitarily equivalent to multiplication by a real-valued measurable function f_k on $L^2(M_k, d\mu_k)$. By Proposition 3 in Section VIII.3 we may assume that μ_k is finite and that $f_k \in \bigcap_{1 \leq p < \infty} L^p(M_k, d\mu_k)$. Furthermore, by Theorem II.10(a), $\otimes_{k=1}^N L^2(M_k, d\mu_k)$ is naturally isomorphic to $L^2(\bigtimes_{k=1}^N M_k, \otimes_{k=1} d\mu_k)$. Under this isomorphism $P(A_1, \ldots, A_N)$ corresponds to multiplication by $P(f_1, \ldots, f_N)$ and D corresponds to the set of finite linear combinations of finite linear combinations of functions $\phi_1(m_1)\phi_2(m_2) \cdots \phi_N(m_N)$ such that $f_k^{n_k}\phi_k \in L^2(M_k, d\mu_k)$.

To prove essential self-adjointness we use Proposition 2 of Section VIII.3. First, since μ_k is finite and $f_k^{n_k} \in L^p(M_k, d\mu_k)$ we conclude that $f_k^l \in L^p(M_k, d\mu_k)$ for $1 \leq p < \infty$. From this it follows immediately that $P(f_1, \ldots, f_N)$ is in L^p for all such p; in particular $P(f_1, \ldots, f_N) \in L^4(\bigtimes_{k=1}^N M_k, \otimes_{k=1}^N d\mu_k)$. Since $f_k^{n_k}$ is self-adjoint on D_k, D_k contains the characteristic functions of measurable sets in M_k. Thus D contains all finite linear combinations of the characteristic functions of rectangles. The remarks on product measures at the end of Section I.4 show that the characteristic function of any measurable set in $\bigtimes_{k=1}^N M_k$ is equal to such a finite linear combination except on a set of arbitrarily small $\otimes_{k=1}^N d\mu_k$ measure. Thus the simple functions on $\bigtimes_{k=1}^N M_k$ can be approximated in the L^p sense ($1 \leq p < \infty$) by elements of D. In particular D is dense in $L^4(\bigtimes_{k=1}^N M_k, \otimes_{k=1}^N d\mu_k)$. Essential self-adjointness now follows from Proposition 2.

To show that P is essentially self-adjoint on D^e we need only show (by Problem 14) that $\overline{P \upharpoonright D^e}$ extends $P \upharpoonright D$. Suppose $\otimes_{k=1}^N \phi_k \in D$. Then $\phi_k \in D(A_k^{n_k})$, so since D_k^e is a domain of essential self-adjointness of $A_k^{n_k}$ there is a sequence $\{\phi_k^\ell\}_{\ell=1}^\infty$ so that $\phi_k^\ell \to \phi_k$, and $A_k^{n_k} \phi_k^\ell \to A_k^{n_k} \phi_k$. An easy estimate shows that this implies that $A_k^m \phi_k^\ell \to A_k^m \phi_k$ for all $1 \leq m \leq n_k$. Therefore $\otimes_{k=1}^N \phi_k^\ell \to \otimes_{k=1}^N \phi_k$ and $P(A_1, \ldots, A_N)(\otimes_{k=1}^N \phi_k^\ell) \to P(A_1, \ldots, A_N)(\otimes_{k=1}^N \phi_k)$. The same argument works for finite linear combinations of vectors of the form $\otimes_{k=1}^N \phi_k$ so $\overline{P \upharpoonright D^e}$ extends $P \upharpoonright D$. This completes the proof of (a).

To prove (b), suppose that $\lambda \in \overline{P(\sigma(A_1), \ldots, \sigma(A_N))}$. If I is any open interval about λ then $P^{-1}(I)$ contains a product $\bigtimes_{k=1}^N I_k$ of open intervals so that $I_k \cap \sigma(A_k) \neq \phi$. Since $\sigma(A_k) = \text{ess range } f_k^{n_k}$, $\mu_k[(f_k^{n_k})^{-1}(I_k)] \neq 0$ so

$$\mu[P(f_1, \ldots, f_N)^{-1}(I)] \neq 0.$$

That is, $\lambda \in \text{ess range } P(f_1, \ldots, f_N)$ which equals $\sigma(\overline{P(A_1, \ldots, A_N)})$ by the first proposition in Section VIII.3. Conversely if $\lambda \notin \overline{P(\sigma(A_1), \ldots, \sigma(A_N))}$ then $(\lambda - P(f_1, \ldots, f_N))^{-1}$ is bounded a.e. on $\bigtimes_{k=1}^N M_k$ so $\lambda \in \rho(\overline{P(A_1, \ldots, A_N)})$. ∎

If $A_1, \ldots, A_N$ are bounded, $P(\sigma(A_1), \ldots, \sigma(A_n))$ is closed, but in general it is not (Problem 43). The following corollary displays the two most important special case of Theorem VIII.33.

Corollary Let $A_1, \ldots, A_N$ be self-adjoint operators on $\mathcal{H}_1, \ldots, \mathcal{H}_N$ and suppose that, for each k, D_k is a domain of essential self-adjointness for A_k. Then,

(a) The operators $A_\pi = A_1 \otimes \cdots \otimes A_N$ and $A_\Sigma = A_1 + \cdots + A_N$ are essentially self-adjoint on $D = \otimes_{k=1}^N D_k$.

(b) $\sigma(\overline{A_\pi}) = \prod_{k=1}^N \sigma(\overline{A_k})$ and $\sigma(\overline{A_\Sigma}) = \overline{\sum_{k=1}^N \sigma(\overline{A_k})}$.

Example 1 Suppose that $V(x)$ is a potential so that $H_1 = -\Delta_x + V(x)$ is essentially self-adjoint on $\mathscr{S}(\mathbb{R}^3)$. Then $H_2 = -\Delta_x + V(x) + -\Delta_y + V(y)$ is essentially self-adjoint on the set of finite sums of products $\phi(x)\psi(y)$, with $\phi, \psi \in \mathscr{S}(\mathbb{R}^3)$. Further $\sigma(H_2) = \overline{\sigma(H_1) + \sigma(H_1)}$.

Example 2 (second quantization of the free Hamiltonian) Let $\mathscr{H}$ be a Hilbert space, $\mathscr{F}(\mathscr{H})$ the associated Fock space over $\mathscr{H}$ (see Section II.4). Suppose that A is a self-adjoint operator on $\mathscr{H}$ with a domain of essential self-adjointness D. Corresponding to each such A we can define an operator $d\Gamma(A)$ on $\mathscr{F}(\mathscr{H})$ as follows. Let $A^{(n)} = A \otimes I \cdots \otimes I + I \otimes A \otimes \cdots \otimes I \cdots + I \otimes \cdots \otimes A$ on $\otimes_{k=1}^n D$. Let $D_A \subset \mathscr{F}(\mathscr{H})$ be the set of $\psi = \{\psi_0, \psi_1, \ldots\}$ such that $\psi_n = 0$ for n large enough and $\psi_n \in \otimes_{k=1}^n D$ for each n. D_A is dense in $\mathscr{F}(\mathscr{H})$ since D is dense in $\mathscr{H}$. Define $A^{(0)} = 0$ and $d\Gamma(A) = \sum_{n=0}^\infty A^{(n)}$. $d\Gamma(A)$ makes sense on D_A and easily seen to be symmetric. By Theorem VIII.33, $A^{(n)}$ is essentially self-adjoint on $\otimes_{k=1}^n D$. Thus $A^{(n)} + \mu i$ has a dense range on $\otimes_{k=1}^n D$ whenever $\mu \in \mathbb{R}$ and $\mu \neq 0$. From this it follows quickly that $d\Gamma(A) \pm i$ has a dense range on D_A. Thus $d\Gamma(A)$ is essentially self-adjoint on D_A. If A is the quantum mechanical operator which corresponds to the free energy, $d\Gamma(A)$ is called the *second quantization* of the free energy. $d\Gamma(A)$ commutes with the projections onto the symmetric and antisymmetric Fock spaces and it follows that $d\Gamma(A) \restriction \mathscr{F}_s(\mathscr{H})$ and $d\Gamma(A) \restriction \mathscr{F}_a(\mathscr{H})$ are essentially self-adjoint on $D \cap \mathscr{F}_s(\mathscr{H})$ and $D \cap \mathscr{F}_a(\mathscr{H})$ respectively.

Part (b) of Theorem VIII.33 holds when $A_1, \ldots, A_N$ are allowed to be arbitrary bounded operators. Because of the techniques involved, we delay the proof until Chapter XIII, where we will also discuss some cases where $A_1, \ldots, A_N$ are unbounded but not self-adjoint.

VIII.11 Three mathematical problems in quantum mechanics

Our purpose in this short section is to describe briefly the mathematical model of quantum mechanics and to describe three general mathematical problems which arise. In the Notes we discuss how the model can be "derived" from various axiomatic schemes.

Quantum mechanical systems are described by operators and vectors in a separable Hilbert space $\mathscr{H}$. Corresponding to every vector of norm one in $\mathscr{H}$ there is a physical state. Two such vectors correspond to the same state

VIII.11 Three mathematical problems in quantum mechanics

if and only if they differ by a complex multiple of absolute value one. Corresponding to each observable, there is a self-adjoint operator A on $\mathscr{H}$. If the system is in the state φ and we measure the observable corresponding to A, the probability distribution of the value of the measurement is $d(\varphi, P_\lambda \varphi)$ where P_Ω is the projection-valued measure associated to A. That is, the probability that the value of the measurement will lie in the interval $[a, b]$, $a, b \in \mathbb{R}$, is $(\varphi, P_{[a,b]}\varphi)$. The dynamics of the system is given by a continuous one-parameter group of unitary operators $U(t)$. If the system is in the state φ at time $t = 0$, then it is in the state $U(t_0)\varphi$ at time $t = t_0$. For most systems, there is a particularly useful realization of $\mathscr{H}$ as $L^2(M, d\mu)$ and a simple correspondence between classical observables and their quantum-mechanical counterparts, self-adjoint operators on $L^2(M, \mu)$ (see the example).

The self-adjoint generator H of the group $U(t)$ is of special interest. It is called the Hamiltonian and is the operator corresponding to the classical energy observable. For vectors $\varphi \in D(H)$ we have

$$\frac{d}{dt}[U(t)\varphi] = iH[U(t)\varphi]$$

which is called the Schrödinger equation. The point spectrum of H is of interest because the corresponding eigenfunctions are stationary states of the system; the typical reaction of the system to outside stimuli is to move from one stationary state to another emitting light whose frequency is proportional to the difference between the corresponding point spectra.

There are three general mathematical problems which arise in any quantum mechanical model:

(1) *Self-adjointness*: In most cases, physical reasoning gives a formal expression for the Hamiltonian and other observables as operators on a realization of $\mathscr{H}$ as $L^2(M, d\mu)$. We use the word "formal" because domains are not specified. It is usually easy to find a domain on which a given formal expression is a well-defined symmetric operator. The first mathematical problem is to prove essential self-adjointness or, if the operator is not essentially self-adjoint, to investigate the various self-adjoint extensions and choose the "right one" to be the observable.

(2) *Spectral analysis*: The second problem is to investigate the spectra of the observables (in particular, the Hamiltonian) and to estimate the position and multiplicity of the point spectra.

(3) *Scattering theory*: The third problem is to describe in some way the behavior of the system for large t.

The development and application of techniques for resolving these problems occupies much of Volumes II–IV. Self-adjointness is studied in Chapter X, spectral analysis in Chapters XII and XIII, and scattering theory in Chapter XI.

We do not mean to imply that all interesting mathematical problems associated with quantum mechanics fall under one of these three headings, far from it. But, these three problems are central to a rigorous mathematical description of quantum mechanics.

Example (*n*-electron atom)   We will very briefly describe an approximate model for the *n*-electron atom. The classical energy of the n electrons is

$$h = \sum_{k=1}^{n} \frac{(p_x^k)^2 + (p_y^k)^2 + (p_z^k)^2}{2m} - \sum_{k=1}^{n} \frac{ne^2}{|r_k|} + \sum_{\substack{k,\ell=1 \\ k<\ell}}^{n} \frac{e^2}{|r_k - r_\ell|}$$

where p_x^k, p_y^k, p_z^k are the x, y, and z components of the momentum of the kth electron, $r_k = \langle x_k, y_k, z_k \rangle$ its position, m and e its mass and charge. The term $-ne^2/|r_k|$ is the potential energy of the kth electron due to the attraction of the protons in the nucleus; the term $e^2/|r_k - r_\ell|$ is the contribution to the potential energy due to the repulsion between the kth and ℓth electrons.

We take as our Hilbert space $\mathcal{H} = L^2(\mathbb{R}^{3n})$ and make the following correspondence between the classical observables and operators on $\mathcal{H}$ (we choose units in which the rationalized Planck's constant $\hbar$, is equal to one):

$$p_x^k \to \frac{1}{i}\frac{\partial}{\partial x_k} \qquad p_y^k \to \frac{1}{i}\frac{\partial}{\partial y_k} \qquad p_z^k \to \frac{1}{i}\frac{\partial}{\partial z_k}$$

x_k, y_k, z_k correspond to multiplication by x_k, y_k, z_k respectively,

$$h \to H = -\sum_{k=1}^{m} \frac{1}{2m}\Delta_k + V(r_1, \ldots, r_k)$$

where

$$\Delta_k = \frac{\partial^2}{\partial x_k^2} + \frac{\partial^2}{\partial y_k^2} + \frac{\partial^2}{\partial z_k^2}$$

and V denotes the operator which acts by multiplication by the function

$$-\sum_{k=1}^{n} \frac{ne^2}{|r_k|} + \sum_{k<\ell}^{n} \frac{e^2}{|r_k - r_\ell|}$$

All of the operators are essentially self-adjoint on $\mathscr{S}(\mathbb{R}^{3n})$, though the proof in the case of H is not at all immediate (see Section X.2). The dynamics is given by the unitary group $U(t) = e^{-itH}$. If $\varphi \in \mathcal{H}$, $\|\varphi\| = 1$, is the state of the system at $t = 0$, then

$$\int_{x_k=a}^{b} \int_{\mathbb{R}^{3n-1}} |\varphi(x_1, \ldots, z_n)|^2 \, dx_1 \cdots dz_n$$

is the probability at $t = 0$, that the x coordinate of the kth particle will be in the interval (a, b) and

$$\int_{x_k=a}^{b} \int_{\mathbb{R}^{3n-1}} |(U(t_0)\varphi)(x_1, \ldots, z_n)|^2 \, dx_1 \cdots dz_n$$

is the probability at $t = t_0$. It is clear that the spectral analysis of H and the large t behavior of e^{-itH} are difficult mathematical problems.

We remark that this model is a rather crude approximation to the n-electron atom for several reasons: We have ignored electron spin and the Pauli exclusion principle. We have also ignored the motion of the nucleus, treating it as fixed. And finally, the model is nonrelativistic.

NOTES

Section VIII.1. The theory of unbounded operators was stimulated by attempts in the late 1920s to put quantum mechanics on a rigorous mathematical foundation. The systematic development of the theory is due to von Neumann, "Allgemeine Eigenwerttheorie Hermitescher Functionaloperatoren," *Math. Ann.* **102**, 49–131 (1929–1930) and M. Stone, "Linear Transformations in Hilbert Spaces and their Applications to Analysis," *Amer. Math. Soc. Colloq. Publ.* **15**, New York, 1932. The technique of using the graph to analyze unbounded operators was introduced by von Neumann in "Über Adjungierte Funktionaloperatoren," *Ann. Math.* (2) **33** (1936), 294–310.

A function on $[a, b] \subset \mathbb{R}$ is said to be *absolutely continuous* if, given $\varepsilon > 0$, there is a $\delta > 0$ that

$$\sum_{k=1}^{n} |f(x_i') - f(x_i)| < \varepsilon$$

every finite collection of disjoint intervals $[x_i, x_i']$ satisfying

$$\sum_{i=1}^{n} |x_i' - x_i| < \delta.$$

then have the *Fundamental Theorem of Calculus*: If f is absolutely continuous on $[a, b]$, then f is differentiable a.e., $f'(x) \in L^1[a, b]$, and f is the indefinite integral of $f'(x)$. Conversely, $(x) \in L^1[a, b]$, then the indefinite integral, $G(x)$, of $g(x)$ is absolutely continuous and $() = g(x)$ a.e.

Section VIII.2. Theorem VIII.3 which is due to von Neumann (in the first paper cited above) is a special case of Theorem X.2 and its corollary. In his paper, von Neumann attributes the isolation of the notion of self-adjointness to E. Schmidt. We remark that von Neumann calls symmetric operators "Hermitian" and self-adjoint operators "hypermaximal Hermitian."

Section VIII.3. The fact that Hilbert's spectral resolution would not carry through for arbitrary symmetric operators was made clear by Carleman's book *Sur les équations intégrales singulières à noyau réel et symétrique*, Almquist and Wilesells, Uppsala, 1923. Von Neumann first discussed the spectral decomposition of unbounded operators in his investigation of the mathematical problems of quantum theory. The systematic proofs first appeared in the works of Von Neumann and Stone mentioned above and in F. Reisz, "Über die linearen Transformation des komplexen Hilbertschen Raumes," *Acta Sci. Math. (Szeged)* **5** (1930–1932), 23–54. Many of the ideas of spectral analysis already appear in matrix form in the work of A. Wintner.

One can develop a theory of integration with respect to vector-valued measures. This theory can then be applied to the projection-valued measure associated to any self-adjoint operator A. In particular, one can prove that for any $\varphi \in D(A)$,

$$A\varphi = \int \lambda \, d(P_\lambda \varphi)$$

where this integral converges strongly, that is, the Riemann–Stieltjes sums converge in norm to $A\varphi$. This is a stronger notion of convergence that we used in Section VIII.3 where the integral converged weakly.

Section VIII.4. Stone's theorem was announced in his paper "Linear Transformations in Hilbert Space, III," *Proc. Nat. Acad. Sci. U.S.A.* **15** (1929), 198–200, and was proved in "On One-Parameter Unitary Groups in Hilbert Space," *Ann. Math.* (2) **33** (1932), 643–648. Theorem VIII.9 appeared in von Neumann, "Über einen Satz von Herrn M. H. Stone," *Ann. Math.* (2) **33** (1932), 567–573. Our proof of Stone's theorem is due to Gårding and Wightmann (unpublished). The idea of using the group invariance to prove essential self-adjointness is due to Nelson, "Analytic Vectors," *Ann. Math.* **70** (1959), 572–614.

Let G be a locally compact Lie group and let $U(g)$ be a continuous unitary representation of G on $\mathscr{H}$, dg Haar measure on G. Then the set D of finite linear combinations of vectors of the form

$$\varphi_f = \int_G f(g) U(g) \varphi \, dg, \qquad \varphi \in \mathscr{H}, \quad f \in C_0^\infty(G)$$

is a dense set contained in the domains of the generators of all one-parameter subgroups of G and these generators take D into itself. This statement is due to L. Gårding, "Notes on Continuous Representations of Lie Groups," *Proc. Nat. Acad. Sci. U.S.A.* **33** (1947), 331–332 and D is often called the "Gårding domain." The existence of the Gårding domain is important because it allows one to get a representation of the Lie algebra of G on D.

Section VIII.5. The example due to Nelson is unpublished but a similar example appears in the "Analytic Vectors" paper cited above. Nelson also proves that if A and B are symmetric, D a dense set contained in $D(A) \cap D(B)$ and invariant under A and B, $AB\varphi - BA\varphi = 0$ for $\varphi \in D$, and $A^2 + B^2$ is essentially self-adjoint on D, then A and B are essentially self-adjoint on D and their closures commute. The original proof of Theorem VIII.14 may be found in J. von Neumann, "Die Eindeutigkeit der Schrödingerschen Operatoren," *Math. Ann.* **104** (1931), 570–578. A more modern proof can be found in Kastler, "The C^*-Algebra of a Free Boson Field, I," *Commun. Math. Phys.* **1** (1965), 14–48. The Weyl relations were introduced in H. Weyl, "Quantenmechanik und Gruppentheorie," *Z. Phys.* **46** (1927), 1–46. We prove Theorem VIII.14 in Chapter XIV (see also Problem 30 in Chapter X).

Concerning the question of a converse to the corollary to Theorem VIII.14 it is known that if P and Q are symmetric on D and (a) and (b) hold *and* $P^2 + Q^2$ is essentially self-adjoint on D, then P and Q are essentially self-adjoint on D and the groups satisfy the Weyl relations. For a proof and a discussion of the generalization to n degrees of freedom see J. Dixmier, "Sur la Relation $i(PQ - QP) = I$," *Compos. Math.* **13** (1956), 263–269. Also of interest is B. Fuglede, "On the Relation $PQ - QP = iI$," *Math. Scand.* **20** (1967), 79–88.

Section VIII.6. The spectral theory for bounded operators was originally expressed in terms of quadratic forms. It is an interesting historical fact that the simple properties of quadratic forms which are semibounded were not really appreciated until twenty-five years after the spectral theory had been extended to unbounded objects by using operators instead of forms. The idea of the relations of forms and operators is implicit in the work of Friedrichs discussed in the notes to Section X.3 (especially in Freudenthal's proof) but the Friedrichs extension theorem was always expressed in terms of operators before 1950. In the 1950s, Theorems VIII.15 and VIII.16 were independently discussed and discovered by a variety of authors: particularly by P. Lax and A. Milgram in "Parabolic Equations," *Ann. Math. Study* **33** (1954), 167–190; T. Kato in "Quadratic forms in Hilbert spaces and asymptotatic perturbation series," Tech. Rep. No. 9, Univ. of Calif. (1955), and by J. Lions in his work *Équations différentielles opérationnelles et problèmes aux limites*, Springer, New York, 1961. The discussion and proof of Theorem VIII.15 in terms of scales of spaces was emphasized by Nelson, see, e.g. E. Nelson, *Topics in Dynamics*, Vol. I, Princeton Univ. Press, Princeton, New Jersey, 1970 for additional discussion. There is also an exhaustive study of quadratic forms in Chapter 6 of T. Kato, *Perturbation for Linear Operators*, Springer-Verlag, Berlin and New York, 1966, and a nice discussion in Chapter 12 of M. Schechter, *Principles of Functional Analysis*, Academic Press, New York, and London 1971.

The term "accretive" originally appeared in K. Friedrichs "Symmetric Positive Linear Differential Equations," *Comm. Pure. Appl. Math.* **11** (1958), 333–418 as something of a joke, but it caught on. It referred to operators with $\text{Re}(u, Au) \geq 0$ for all $u \in D(A)$. The study of the such operators was essentially initiated by R. S. Phillips in "Perturbation Theory for Semigroups of Linear Operators," *Trans. Amer. Math. Soc.* **74** (1954), 199–221 (Phillips studied dissipative operators, that is, operators A with $\text{Re}(u, Au) \leq 0$ for all $u \in D(A)$). Sectorial was then used if $\{(u, Au)\} \subset \{z \mid |\arg(z - w)| < \theta\}$ for some w and some $\theta < \pi/2$. These definitions are often carried over to the forms, that is, what we have called strictly accretive is often called sectorial. Since rotated as well as translated sectors often arise in applications, we have introduced the term "strictly accretive" and extended the notion of sectorial."

m-accretive operators are maximal accretive operators in the same sense that self-adjoint operators are maximal symmetric operators. We return to this question in Section .6.

Section VIII.7. For much of this section, the reader can consult Kato's book cited above for additional discussion.

The notion of norm resolvent convergence is the restriction to self-adjoint operators of a natural topology on closed operators from one Banach space to another. Krein and coworkers introduced a natural metric on closed subspaces of a Banach space in the 1940s. Explicitly, given M and N in a Banach space X, one defines

$$d(M, N) = \sup_{u \in M, \|u\| = 1} (\inf_{v \in N, \|v\| = 1} \|u - v\|)$$

Then $M \subset N$ if and only if $d(M, N) = 0$. If $\hat{d}(M, N) = \max[d(M, N), d(N,M)]$, $\hat{d}$ is a metric on all closed subspaces. If $\Gamma(T)$ is the graph of T, one can introduce a metric on all closed operators from X to Y, by $\rho(T, S) = \hat{d}(\Gamma(T), \Gamma(S))$, where, say, $\|\langle x, y\rangle\|_{X \times Y} = \|x\|_X + \|y\|_Y$. This produces a topology on the closed operators first introduced by J. Newburgh in "A Topology for Closed Operators," *Ann. Math.* **53** (1951), 250–255. When restricted to the self-adjoint operators, this is precisely the topology of norm resolvent convergence.

Theorems VIII.21 and VIII.22 are stated most naturally in terms of the theory of semigroups of operators on an arbitrary Banach space. Theorem VIII.21 seems to have been first proven explicitly (in the general semigroup language) in H. Trotter, "Approximation of Semigroups of Operators" *Pacific J. Math.* **8** (1959), 887–919 although it was something of a folk theorem at the time. Theorem VIII.22 was also proven by Trotter in the above paper although one point in his proof was clarified by T. Kato in "Remarks on Pseudo-Resolvent and Infinitesimal Generators of Semigroups," *Proc. Jap. Acad.* **35** (1959), 467–468. Theorem VIII.21 is sometimes called the Trotter–Kato theorem also. For a discussion of norm and strong convergence of operators which are not necessarily self-adjoint, see Kato's book Section IV.2 (norm), Sections VIII.1 and VIII.3 (strong).

Theorems like Theorems VIII.23 and VIII.24 were first proven in F. Rellich, "Störungstheorie der Spektralzerlegung, II," *Math. Ann.* **113** (1936), 667–685. Extensions of Rellichs results appear in Sz.-Nagy "Perturbations des transformations autoadjointes dans l'espace de Hilbert," *Comm. Math. Helv.* **19** (1946–1947), 347–366; and E. Heinz, "Beiträge zur Störungstheorie der Spektralzerlegung," *Math. Ann.* **123** (1951), 415–438.

The systematic study of graph limits was begun in J. Glimm and A. Jaffe, "Singular Perturbations of Self-Adjoint Operators," *Comm. Pure Appl. Math.* **22** (1969), 401–414. We return to their ideas in Section X.10.

Section VIII.8. The extension of Lie's theorem to infinite dimensional cases was first made by H. Trotter, "On the Product of Semigroups of Operators," *Proc. Amer. Math. Soc.* **10** (1959), 545–551. He proved Theorem VIII.31 in the general context of semigroups on Banach space. His proof was later simplified in P. R. Chernoff, "Note on Product Formula for Operator Semigroups," *J. Func. Anal.* **2** (1968), 238–242.

The proof we give to Theorem VIII.30 is given in E. Nelson, "Feynman Integrals and the Schrödinger Equation," *J. Math. Phys.* **5** (1964), 332–343.

Extensions of Trotters formula to various special cases where $A + B$ is not essentially self-adjoint but has a definition as a sum of forms have been given by W. Faris, "The Product Formula for Semigroups Defined by Friedrichs Extensions," *Pacific J. Math.* **22** (1967) 47–70 and P. R. Chernoff, "Semigroup Product Formulas and Addition of Unbounded Operators," *Bull. Amer. Math. Soc.* **76** (1970), 395.

Section VIII.10. The first mathematical treatment of second quantization may be found in J. Cook, "The Mathematics of Second Quantization," *Trans. Amer. Math. Soc.* **74**. 222–245. For more information see I. Segal, "Tensor Algebras over Hilbert Spaces, I," *Trans. Amer. Math. Soc.* **81** (1956) 106–134.

The notation "$d\Gamma$" arises in the following way. $\mathscr{F}(\mathscr{H})$ is an algebra in a natural way with a product defined so that $(\psi_1 \otimes \cdots \otimes \psi_n) \cdot (\psi_{n+1} \otimes \cdots \otimes \psi_{n+k}) = \psi_1 \otimes \cdots \otimes \psi_{n+k}$. This product is denoted by $\otimes$. Thus $\psi \otimes \phi$ is defined for all $\psi, \phi \in \mathscr{F}(\mathscr{H})$. The natural "automorphisms" of $\mathscr{F}(\mathscr{H})$ are invertible linear, norm preserving maps, V, obeying $V(\psi \otimes \phi) = V\psi \otimes V\phi$. The natural automorphisms of $\mathscr{H}$ are just the unitaries. With each unitary, U, one can associate uniquely an automorphism, $\Gamma(U)$ on $\mathscr{F}(\mathscr{H})$ obeying $\Gamma(U) = U$ on $\mathscr{H}$ by requiring that on $\mathscr{H}^n = \otimes_{k=1}^n \mathscr{H}$, $\Gamma(U)$ be just $U \otimes \cdots \otimes U$ (n times). Thus Γ maps the group of unitaries on $\mathscr{H}$ into the group of automorphisms on $\mathscr{F}(\mathscr{H})$ in a strongly continuous manner. $d\Gamma$ is then defined by requiring $e^{itd\Gamma(A)} = \Gamma(e^{itA})$; that is for any self-adjoint A on $\mathscr{H}$, $d\Gamma(A)$ is the infinitesimal generator of the strongly continuous unitary group $\Gamma(e^{itA})$. In the language of Lie theory, $d\Gamma$ is differential of Γ as a map on the "Lie algebra" of the group of unitaries in $\mathscr{H}$ to the Lie algebra of the group of unitaries on $\mathscr{F}(\mathscr{H})$. That the $d\Gamma$ defined in the text has a closure equal to $d\Gamma$ as defined in the above paragraph is a simple exercise.

In the usual physicist's notation, if $\mathscr{H} = L^2(\mathbb{R}^3, dx)$ and A is defined by $(Af)(x) = \omega(x)f(x)$, then $d\Gamma(A)$ is what is written $\int \omega(x) a^*(x) a(x) \, dx$.

The proof of Theorem VIII.33 shows how the spectral theorem allows one to use L^p techniques on abstract Hilbert space problems. By using the spectral theorem one can often formulate the given problem in terms of $L^2(M, d\mu)$ for some convenient measure space $\langle M, \mu \rangle$. After this has been done the standard theorems and estimates of L^p theory can often be used.

Section VIII.11. Attempts at an a priori justification of the quantum mechanical picture we discuss go back to von Neumann's famous *Mathematische Grundlagen der Quantenmechanik*, Springer, New York, 1932 (reissued by Springer-Verlag, Berlin and New York, 1968; English transl. published by Princeton Univ. Press, Princeton, New Jersey, 1955). The approach of G. Mackey in *The Mathematical Foundations of Quantum Mechanics*, Benjamin, New York, 1963 and *Induced Representations of Groups and Quantum Mechanics*, Benjamin, New York, 1968 emphasizes the similarity with classical statistical mechanics and isolates the ad hoc elements of any axiomatic approach to quantum mechanics.

Mackey presents the following picture of classical statistical mechanics: The basic states of a classical mechanical system are points in a "phase space," M. Statistical states are just measures of total mass 1 on M. Observables are just measurable functions on M. Given a state μ and an observable f the measure $\nu_{\mu, f}$ on $\mathbb{R}$ given by $\nu_{\mu, f}(\Omega) = \mu(f^{-1}(\Omega))$ represents the probability a measurement of f, produces a value in Ω. The first remark of Mackey is that the points of M do not really enter in this picture; rather the family $\mathfrak{L}$ of Borel sets enter as the basic objects. The states μ are really functions on $\mathfrak{L}$ and to form $\nu_{\mu, f}$ one only needs f^{-1} which is a function from the Borel sets of $\mathbb{R}$ to $\mathfrak{L}$. The abstract structure on $\mathfrak{L}$ needed for $\nu_{\mu, f}$ to be a probability measure (total mass 1) on $\mathbb{R}$ is:
(i) $\mathfrak{L}$ has a partial order $\leq (A \leq B$ if A is a subset of B) with a largest element, 1 (we have $M = 1$) and a smallest element 0 (we have $\emptyset = 0$).
(ii) $\mathfrak{L}$ has a complementation ' (A' is just the complement $M \setminus A$), with the basic properties $(A')' = A$; $A \leq B$ if and only if $B' \leq A'$; $1' = 0$.
(iii) $\mathfrak{L}$ is a countable lattice; that is, given $A_1, \ldots, A_n, \ldots \in \mathfrak{L}$ there exists

$$A = \bigvee_{n=1}^{\infty} A_n$$

with $A \geq A_n$ for all n and $A \leq B$ whenever $B \geq A_n$ for all n.
(iv) The lattice and complementation structures are related by $A \vee A' = 1$.

VIII: UNBOUNDED OPERATORS

An abstract set $\mathfrak{L}$ with properties (i)–(iv) is called an **orthocomplemented lattice**. When one has such an object, $A, B \in \mathfrak{L}$ are called *disjoint* if $A \leq B'$. A *measure on* $\mathfrak{L}$ is a map $\mu: \mathfrak{L} \to \mathbb{R}^+ = \{a \in \mathbb{R} \mid a \geq 0\}$ with $\mu(1) = 1$, and

$$\mu\left(\bigvee_{n=1}^{\infty} A_n\right) = \sum_{n=1}^{\infty} \mu(A_n)$$

if A_i and A_j are disjoint for all i and j. An $\mathfrak{L}$-*valued measure* is a map $P: \mathscr{B} \to \mathfrak{L}$ where $\mathscr{B}$ is the family of Borel sets of $\mathbb{R}$ obeying $P(\mathbb{R}) = 1$,

$$P\left(\bigcup_{i=1}^{\infty} A_i\right) = \bigvee_{i=1}^{\infty} A_i \quad \text{and} \quad P(\mathbb{R}\backslash A) = P(A').$$

Thus Mackey's description of classical statistical mechanics leads to an abstract notion of *statistical system* as an orthocomplemented lattice. The observables are then $\mathfrak{L}$-valued measures and the states are measures on $\mathfrak{L}$. Given a state μ and an observable θ the Borel measure $\nu_{\mu,\theta}(\Omega) = \mu(\theta(\Omega))$ on $\mathbb{R}$ is interpreted as the probability that a measurement of the observable θ in the state μ will yield a value in Ω.

In the first quoted Mackey reference, there is a description of a set of reasonable axioms for the notion of measurement in a statistical system which lead to this orthocomplemented lattice. To obtain quantum mechanics, one might add the *ad hoc postulate* to the basic lattice picture: $\mathfrak{L}$ is just the family, $\mathfrak{L}_{\mathscr{H}}$, of closed subspaces of a separable Hilbert space $\mathscr{H}$ with the operations: $A \leq B$ if and only if $A \subseteq B$; $A' = A^\perp$, and $\bigvee_{n=1}^{\infty} A_n = \sum_{n=1}^{\infty} A_n$; $1 = \mathscr{H}$; $0 = \{0\}$. One is thus left with the problem of trying to justify this ad hoc postulate. Important partial progress towards this result was obtained in C. Piron, "Axiomatique Quantique," *Helv. Phys. Acta* **37** 439–468 (1964). See also the discussion in the Jauch monograph mentioned below.

Once one has the ad hoc postulate, one can determine the states and observables in a more explicit form. A. Gleason in "Measures on the Closed Subspaces of a Hilbert Space," *J. Math. Mech.* **6**, 885–894 (1957) proved that every measure on $\mathfrak{L}_{\mathscr{H}}$ has the following form: Each subspace $A \in \mathfrak{L}_{\mathscr{H}}$ is naturally associated with an orthogonal projection P with Ran $P = A$. Gleason proved that every measure μ on $\mathfrak{L}_{\mathscr{H}}$ has $\mu(A) = \text{tr}(\rho P)$ where ρ is some positive trace class operator with tr $\rho = 1$. By Theorems VI.17 and VI.21, we can find an orthonormal basis, $\{\Phi_n\}_{n=1}^{\infty}$ and $\alpha_1, \ldots, \alpha_n, \ldots \geq 0$ with $\sum_{n=1}^{\infty} \alpha_n = 1$, so that $\rho = \sum_{n=1}^{\infty} \alpha_n(\Phi_n, \cdot)\Phi_n$. Thus, arbitrary states are just sums of *vector states*, $\mu(P) = (\Phi_n, P\Phi_n)$ (in the language of Section XIV.1, these vector states are just the extreme points of the family of all states). As a result, one may analyze all states by considering only vector states.

$\mathfrak{L}_{\mathscr{H}}$-valued measures are precisely projection valued measures! Thus, by the spectral theorem (Theorem VIII.6), an observable is naturally associated with a self-adjoint operator A. The probability of obtaining a value in Ω, if A is measured in the vector state ψ is just $(\psi, P_\Omega \psi)$ where P_Ω is the p.v.m. for A.

Thus we see how the basic static elements of the picture in Section VIII.11 arise from the ad hoc postulate. The picture of the dynamics enters from the following analysis†:

(1) For each time t there should be a map α_t on the set of all states, so that α_t takes a state at time s into the state at time $s + t$. Since $\alpha_t \alpha_{-t} = \alpha_0 = I$, each α_t should be a bijection. Moreover $\alpha(\rho_1 + \rho_2)$ should be $\alpha(\rho_1) + \alpha(\rho_2)$.

† For further discussion of this picture, see B. Simon: "From Automorphism to Hamiltonian," pp. 305–326 in *Studies in Mathematical Physics* (E. H. Lieb, B. Simon, and A. S. Wightman, eds.) Princeton Univ. Press, Princeton, New Jersey, 1976.

(2) Because any state is a sum of vector states, we only need to know how α behaves on vector states. Vector states are uniquely determined by the property that they are extreme points, so α must take vector states into themselves since α is invertible. Thus, α is a map of unit rays (that is, families of vectors of the form $\{e^{i\theta}\psi \mid \theta \in [0, 2\pi]\}$) into themselves. It is not quite arbitrary since there are vector states $\rho_{\Phi_1}, \ldots, \rho_{\Phi_4}$ with

$$\tfrac{1}{2}\rho_{\Phi_1} + \tfrac{1}{2}\rho_{\Phi_2} = \tfrac{1}{2}\rho_{\Phi_3} + \tfrac{1}{2}\rho_{\Phi_4}$$

and it is necessary that

$$\tfrac{1}{2}\rho_{\alpha(\Phi_1)} + \tfrac{1}{2}\rho_{\alpha(\Phi_2)} = \tfrac{1}{2}\rho_{\alpha(\Phi_3)} + \tfrac{1}{2}\rho_{\alpha(\Phi_4)}$$

For example, if Φ_1 and Φ_2 are orthogonal and

$$\Phi_3 = 2^{-1/2}(\Phi_1 + \Phi_2) \qquad \Phi_4 = 2^{-1/2}(\Phi_1 - \Phi_2)$$

we can use the above to conclude $\alpha(\Phi_1)$ is orthogonal to $\alpha(\Phi_2)$. In general, one shows that a map on the rays yields an automorphism of all the states, if and only if $|(\alpha(\Phi_1), \alpha(\Phi_2))| = |(\Phi_1, \Phi_2)|$ for all unit rays Φ_1, Φ_2.

(3) An analysis of E. Wigner in *Group Theory and Its Applications to the Quantum Mechanics of Atomic Spectra*, Academic Press, New York, 1959 [see also V. Bargmann, "Note on Wigner's Theorem on Symmetry Operations," *J. Math. Phys.* **5**, 862–868 (1964)] uses the analysis of (2) to prove that every automorphism of the rays is of the form $\alpha(\Phi) = U\Phi$ with U unitary or antiunitary. Up to an overall change of phase, $U \to e^{i\theta}U$, U is uniquely determined by α.

(4) Since $\alpha_t = (\alpha_{t/2})^2$, the U_t of (3) must be unitary. It is natural to suppose that $t \to \alpha_t(\rho)$ is continuous. By a theorem of V. Bargmann "On the Unitary Ray Representations of Continuous Groups," *Ann. Math.* **59**, 1–46, (1954) and E. Wigner "Unitary Representations of the Inhomogeneous Lorentz Group," *Ann. Math.* **40**, 149–204 (1939), it is possible to choose the phases left arbitrary in (3) so that U_t is strongly continuous in t.

(5) Since $\alpha_t \alpha_s = \alpha_{t+s}$ we can conclude $U_t U_s = \lambda(t, s)U_{t+s}$ where $|\lambda(t, s)| = 1$. (There is still some arbitrariness left in phases.) A further analysis of Bargmann and Wigner (see the references cited above) implies that $\lambda(t, s) = \mu(t + s)\mu(t)^{-1}\mu(s)^{-1}$ for some measurable function μ with $|\mu(t)| = 1$. Letting $V(t) = \mu(t)U(t)$ we obtain a strongly continuous one-parameter group of unitaries which can then be analyzed by Stone's theorem as discussed in the text.

There is one further element of the picture which one can justify on the basis of more basic assumptions: this is the particular realization of $\mathcal{H}$ as $L^2(\mathbb{R}^n)$ with $p = i^{-1}\partial/\partial x$ etc. and the free Hamiltonian given by $H_0 = -(2m)^{-1}\Delta$. The realization of $\mathcal{H}$ as $L^2(\mathbb{R}^n)$ is connected, of course, with von Neumann's theorem on the uniqueness of solutions of the canonical commutation relations (see Section VIII.5). On a more fundamental level, it is connected with Euclidean invariance (symmetry under spatial translation and rotation) and position operators: this is discussed in the second Mackey reference and in A. Wightman, "On the Localizability of Quantum Mechanical Systems," *Rev. Mod. Phys.* **34**, 845–872. (1962). The fact that H_0 is $-(2m)^{-1}\Delta$ is connected with Gallilean invariance as discussed in the second Mackey reference, Bargmann's Ann. Math. article (above), E. Inönu and E. Wigner " Representations of the Galilei Group," *Nuovo Cimento* **9** (1952), 705–718 and C. Piron, "Sur le quantification du système de deux particules," *Helv. Phys. Acta.* **38** (1965), 104–108.

We have just discussed in detail one approach to quantum axiomatics. For additional discussion and other approaches, see G. Birkhoff and J. von Neumann, "The Logic of Quantum Mechanics," *Ann. Math.* **37** (1936), 823–843; G. Dahn, "Attempt of an Axiomatic

Foundation of Quantum Mechanics and More General Theories, IV," *Commun. Math. Phys.* **9** (1968), 192–211; E. B. Davies, "Quantum Stochastic Processes," *Commun. Math. Phys.* **15** (1969), 277–304; E. B. Davies and J. T. Lewis, "An Operational Approach to Quantum Probability," *Commun. Math. Phys.* **17** (1970), 239–260; C. M. Edwards, "The Operational Approach to Algebraic Quantum Theory, I," *Commun. Math. Phys.* **16** (1970), 207–230; J. Gunson, "On the Algebraic Structure of Quantum Mechanics," *Commun. Math. Phys.* **6** (1967), 262–285; K. E. Hellwig and K. Kraus, "Operations and Measurements, I, II," *Commun. Math. Phys.* **11** (1969), 214–220; **16** (1970), 142–147; J. Jauch, *Foundations of Quantum Mechanics*, Addison–Wesley, Reading, Massachusetts, 1968; P. Jordan, J. von Neumann, and E. Wigner "On the Algebraic Generalization of the Quantum Mechanical Formalism," *Ann. Math.* **35** (1934), 29–64; G. Ludwig, "Attempt at an Axiomatic Foundation of Quantum Mechanics and more General Theories, I–III," *Z. Phys.* **181** (1964), 223–260; *Commun. Math. Phys.* **4** (1967), 331–348; **9** (1968), 1–12; B. Mielnile, "Geometry of Quantum States," *Commun. Math. Phys.* **9** (1968), 55–80; R. J. Plymon, "A Modification of Piron's Axioms," *Helv. Phys. Acta* **41** (1968), 69–74; R. J. Plymon, "C^*-Algebras and Mackey's Axioms," *Commun. Math. Phys.* **8** (1968), 132–146; J. Pool, "Baer *-Semigroups and the Logic of Quantum Mechanics," *Commun. Math. Phys.* **9** (1968), 118–141; J. Pool, "Semimodularity and the Logic of Quantum Mechanics," *Commun. Math. Phys.* **9** (1968), 212–228; E. Prugovečki, *Quantum Mechanics in Hilbert Spaces*, Academic Press, New York, 1971; I. Segal, "Postulates for General Quantum Mechanics," *Ann. Math.* **48** (1947), 930–940; V. Varadarajan, *Geometry of Quantum Theory*, Van Nostrand–Reinhold, Princeton, New Jersey, 1968; H. Weyl, *The Theory of Groups and Quantum Mechanics*, Dover, New York, 1931; N. Zierler, "Axioms for Non-relativistic Quantum Mechanics," *Pacific J. Math.* **11** (1961), 1151–1169.

Despite the rather enormous literature on these "first level" foundations of quantum theory, there is no definitive theory of quantum axiomatics. Probably the most important result of the attempts to axiomatize quantum theory is the mathemetical "spin-off" that has resulted: the theory of unbounded self-adjoint operators, Jordan algebras, and the C^*-algebra approach to quantum theory (discussed in Chapters XIX and XX), all have their origin in attempts at quantum axiomatization.

PROBLEMS

1. Let $\{\varphi_n\}$ be an orthonormal basis for a Hilbert space $\mathscr{H}$ and let e_∞ be a vector in $\mathscr{H}$ which is not a finite linear combination of the φ_n. Let D be the set of finite linear combination of elements of $\{\varphi_n\}$ and e_∞, and on D define

$$T(be_\infty + \sum_{i=1}^{N} c_i \varphi_i) = be_\infty$$

Show that $\overline{\Gamma(T)}$ contains both $\langle e_\infty, e_\infty \rangle$ and $\langle e_\infty, 0 \rangle$ and thus is not the graph of a linear operator.

2. Let S be an operator from $D(S)$ to $\mathscr{H}$ which is injective. Consider the following additional statements about S:

   (1) S is a closed operator.

(2) Ran S is dense.
(3) Ran S is closed.
(4) For some constant C and all $\psi \in D(S)$, $\|S\psi\| \geq C\|\psi\|$.
  (a) Prove that (1)–(3) imply (4) (*Hint*: Apply the closed graph theorem to S^{-1}).
  (b) Prove that (2)–(4) imply (1)
  (c) Prove that (1) and (4) imply (3)
  Remark. Let T be a closed operator. Applying (a) to $\lambda - T$ we see that $\lambda \in \rho(T)$ if and only if $\lambda - T$ is bijective. (b) also has a "translation" of this sort.

†3. Prove that the operators in Example 5 of Section VIII.1 are closed.

4. (a) Suppose that C is a symmetric operator, $C \supset A$ and that $\operatorname{Ran}(A + i) = \operatorname{Ran}(C + i)$. Prove that $C = A$.
   (b) Suppose that A is a symmetric operator with $\operatorname{Ran}(A + i) = \mathscr{H}$ but $\operatorname{Ran}(A - i) \neq \mathscr{H}$. Prove that A has no self-adjoint extensions.

5. Let $\mathscr{H} = \ell_2$. Let $D(A) = \{a \in \mathscr{H} \mid \text{for some } N, \sum_{m=0}^{N} a_n = 0 \text{ and } a_n = 0 \text{ if } n > N\}$. For $a \in D(A)$, define $Aa \in \mathscr{H}$ by

$$(Aa)_n = i\left[\sum_{m=0}^{n-1} a_m + \sum_{m=0}^{n} a_m\right]$$

   (a) Prove that $D(A)$ is dense in $\mathscr{H}$.
   (b) Prove that A is symmetric. *Hint*: if $\sum_{m=0}^{N} a_n = 0$, then

$$(Aa)_n = i\left[\sum_{m=0}^{n-1} a_m - \sum_{m=n+1}^{N} a_m\right]$$

   (c) Prove that $\operatorname{Ran}(A + i)$ is dense in ℓ_2.
   (d) Prove that $(1, 0, 0, \ldots) \in D(A^*)$ and that $(A^* + i)(1, 0, 0, \ldots) = 0$
   (e) Prove that A has no self-adjoint extensions. (*Hint*: Apply Problem 4b to $\bar{A}$).

†6. Prove that the operator T in the example of Section VIII.2 is closed.

†7. Prove that the operators T_α in the example of Section VIII.2 are self-adjoint. (*Hint*: It follows from what was already proven that $\psi \in D(T_\alpha^*)$ implies that $\psi \in AC[0, 1]$ and $T_\alpha^* \psi = i\, d\psi/dx$.)

8. Consider $T = -d^2/dx^2$ as an operator on $L^2(\mathbb{R})$ with domain $C_0^\infty(\mathbb{R})$. What is the adjoint of T? Is T essentially self-adjoint?

9. Consider $T = i\, d/dx$ as an operator on $L^2(0, \infty)$ with domain $C_0^\infty[0, \infty]$, the infinitely differentiable functions with compact support away from the origin. Is T essentially self-adjoint?

10. Suppose that A is a densely defined symmetric operator which is positive, that is, $(\varphi, A\varphi) \geq 0$ if $\varphi \in D(A)$.
    (a) Prove that $\|(A + I)\varphi\|^2 \geq \|\varphi\|^2 + \|A\varphi\|^2$.
    (b) Show that $\operatorname{Ran}(A + I)$ is closed if A is a closed operator.
    (c) Prove that A is essentially self-adjoint if and only if the equation $A^*\psi = -\psi$ has no nonzero solutions.

†11. Prove part (c) of Theorem VIII.7.

†*12*. Prove Theorem VIII.11 directly without using Theorem VIII.10.

13. Find two dense, linear subspaces of $L^2(\mathbb{R})$, D_1 and D_2, with $D_1 \cap D_2 = \{0\}$, so that x is essentially self-adjoint on D_1 and x^2 is essentially self-adjoint on D_2.

†*14*. Let A be a symmetric operator with domain $D \subset \mathscr{H}$. Let $D_1 \subset D$ be a dense linear subset of $\mathscr{H}$ and suppose that $A \restriction D_1$ is essentially self-adjoint. Prove that A is essentially self-adjoint and $\bar{A} = \overline{A \restriction D_1}$.

†*15*. (a) Prove that an operator A is closed if and only if its domain, $D(A)$, is complete under the norm
$$\|\psi\|_A = \|A\psi\| + \|\psi\|.$$

(b) Prove that a semibounded quadratic form is closed if and only if whenever
$$\varphi_n \in Q(q), \quad \varphi_n \xrightarrow{\mathscr{H}} \varphi, \quad \text{and} \quad q(\varphi_n - \varphi_m, \varphi_n - \varphi_m) \xrightarrow[n,m \to \infty]{} 0,$$
then $\varphi \in Q(q)$ and $q(\varphi_n - \varphi, \varphi_n - \varphi) \to 0$.

†*16*. (a) Show that the quadratic form q associated with a semibounded self-adjoint operator A (Example 2 of Section VIII.6) is closed.

(b) Prove that any operator core for A is a form core for q.

†*17*. Prove the statements (a)–(d) in Example 3 of Section VIII.6.

†*18*. Fill in the details of Theorem VIII.16.

19. Let $A_n = (1 - 1/n)x$ on $L^2(\mathbb{R})$ and $A = x$. Show that one can choose domains of essential self-adjointness for A_n and A which have no nonzero vector in common but that $A_n \to A$ in the norm resolvent sense.

20. (a) Let $\{A_n\}$ and A be self-adjoint operators and suppose that for all $\varphi, \psi \in \mathscr{H}$ and all λ with Im $\lambda \neq 0$, $(R_\lambda(A_n)\varphi, \psi) \to (R_\lambda(A)\varphi, \psi)$. Prove that $A_n \to A$ in the strong resolvent sense. (*Hint*: You will need the first resolvent fomula.)

(b) Let $\{A_n\}$ and A be self-adjoint operators. Use part (a) to prove that if $R_\lambda(A_n)$ converges strongly to $R_\lambda(A)$ in the lower half-plane then $R_\lambda(A_n)$ also converges strongly to $R_\lambda(A)$ in the upper half-plane.

†*21*. Extend the proofs of Theorems VIII.20 and VIII.21 to show that if $A_n \to A$ in the strong resolvent sense, then $e^{itA_n}\varphi \to e^{itA}\varphi$ uniformly for t in any finite interval.

†*22*. Fill in the details of the proofs of parts (b) and (c) of Theorem VIII.25.

23. (a) Let A_n be a sequence of self-adjoint operators and suppose that for each $\varphi \in \mathscr{H}$ and each $t \in \mathbb{R}$, $e^{itA_n}\varphi$ converges in $\mathscr{H}$. Prove that there exists a self-adjoint operator A so that $A_n \to A$ in the strong resolvent sense. (*Hint*: You will need to use von Neumann's theorem from Section VIII.4.)

(b) Give an example to show that the conclusion of (a) may not hold if e^{itA_n} converges weakly instead of strongly.

†*24*. Prove Theorem VIII.27b.

†*25*. Prove Theorem VIII.28.

26. Let $\{A_n\}$ be a uniformly bounded sequence of self-adjoint operators. Let A be a bounded self-adjoint operator. Prove that $A = \text{w.gr.-lim } A_n$ if and only if $A_n \to A$ in the weak operator topology.

27. Let $\{A_n\}$ and A be positive self-adjoint operators. Prove that $A_n \to A$ in the strong resolvent sense if and only if $(A_n + I)^{-1} \to (A + I)^{-1}$ strongly.

†28. Prove that if $\{A_n\}$ and A are uniformly bounded self-adjoint operators, then $A_n \to A$ in the strong resolvent sense if and only if $A_n \to A$ strongly.

29. Let A be self-adjoint.
    (a) Prove that $tA \to t_0 A$ in the norm resolvent sense as $t \to t_0 \neq 0$.
    (b) Prove that $e^{itA} \to e^{it_0 A}$ in norm if and only if A is bounded.

30.†(a) Prove that if $\{A_n\}$ and A are uniformly bounded self-adjoint operators and $A_n \xrightarrow{w} A$ but $A_n \xrightarrow{s} \!\!\!\!\!/\; A$, then A_n does not converge to A in the weak resolvent sense.
    (b) Where does the weak analogue of Theorem VIII.18 breakdown?

31. Is the form analogue of part (a) of Theorem VIII.25 true?

32. Prove that $A_n = nI$ has a strong graph limit as $n \to \infty$ which is not the graph of an operator.

33. Let $\{A_n\}$ be a *constant* sequence of symmetric operators (that is, $A_n = B$, $\forall n$). Show that the strong graph limit of the A_n is the closure of B.

34. Let R be the right shift operator on ℓ_2. Prove that the w.gr.-lim R^n is the zero operator while the st.gr.-lim R^n is the graph $\{\langle 0, 0 \rangle\}$.

35. Prove directly (without using the Fourier transform) that $i^{-1}(d/dx)$ is essentially self-adjoint on $\mathscr{S}(\mathbb{R})$.

†36. Prove the corollary to Theorem VIII.14.

†37. Fill in all the details of the proof of Theorem VIII.22.

38. (a) Let $\{A_n\}$ and A be positive self-adjoint operators and suppose that $e^{-tA_n} \to e^{-tA}$ strongly for each $t > 0$. Prove that $A_n \to A$ in the strong resolvent sense.
    (b) Prove the analogue of (a) if strong convergence is replaced by norm convergence.

39. Let $U(t)$, $t \geq 0$, be a family of self-adjoint operators satisfying (i) $\|U(t)\| \leq e^{Et}$ for some $E \in \mathbb{R}$, (ii) $U(t)U(s) = U(t + s)$, (iii) the map $t \to U(t)$ is strongly continuous, (iv) $U(0) = I$. Then
    (a) By mimicking our proof of Stone's theorem prove that $U(t) = e^{-At}$ for a unique self-adjoint operator A.
    (b) Reach the same conclusion using the functional calculus.
    (c) Prove that $A \geq -E$.

40. Let $\langle M, \mu \rangle$ be a measure space and $\mathscr{H} = L^2(M, d\mu)$. A map T of $L^2(M, d\mu)$ into itself will be called **positivity preserving** if $(Tf)(x) \geq 0$ a.e. whenever $f(x) \geq 0$ a.e. Let A, B be self-adjoint operators on $\mathscr{H}$ and suppose that e^{-iAt} and e^{-iBt} are positivity preserving for all $t \in \mathbb{R}$ and that $A + B$ is essentially self-adjoint on $D(A) \cap D(B)$. Prove that $e^{i(At+Bt)}$ is positivity preserving for each $t \in \mathbb{R}$.

41. Let H_0 and V be closed, positive quadratic forms and let $\beta \in \mathbb{C}\setminus(-\infty, 0]$. Suppose that $Q(H_0) \cap Q(V)$ is dense in $\mathcal{H}$. Prove that $H_0 + \beta V$ defined as a quadratic form on $Q(H_0) \cap Q(V)$ is closed and sectorial.

42. Let T be a self-adjoint densely defined operator on a Hilbert space $\mathcal{H}$. Suppose that for some $\lambda_0 \in \rho(T)$, $R_{\lambda_0}(T)$ is a compact operator. Prove that $R_\lambda(T)$ is compact for all $\lambda \in \rho(T)$ and discuss the various kinds of spectra that T can have.

    Remark. We study operators with compact resolvent in Chapter XIII.

43. (a) Give an example to show that the *closure* of the range of P on $\mathsf{X}_{k=1}^n \sigma(A_k)$ may be needed to obtain the whole spectrum of $P(A_1, \ldots, A_N)$.
    (b) Prove that if $A_1, \ldots, A_N$ are bounded then the closure is *not* necessary.

44. Prove the uniqueness statement of Theorem VIII.32.

45. Let T be a closed operator. Prove that $M = \{\psi \mid \psi \in D(T), T\psi \in D(T^*)\}$ is dense and that T^*T defined on M is self-adjoint (*Hint*: Let S be the operator constructed in Theorem VIII.32. Show that $D(S) \subset M$ and that T^*T is a symmetric extension of S).

46. Let T be a closed operator on a Hilbert space. Define, $N(T)$, the numerical range of T by $N(T) = \{(\psi, T\psi) \mid \psi \in D(T) : \|\psi\| = 1\}$.
    (a) Prove $\sigma(T) \subset \overline{N(T)} \cup N(T^*)^*$ where $N(T^*)^*$ is the set $\overline{\{(\psi, T^*\psi) \mid \psi \in D(T)\}}$.
    (b) Find a T with $\sigma(T) \not\subset \overline{N(T)}$ and thus with $N(T) \neq N(T^*)^*$. (*Hint*: Take T symmetric!)

47. Let A be self-adjoint on $D(A)$ in $\mathcal{H}_1$ and B be self-adjoint on $D(B)$ in $\mathcal{H}_2$. Use Theorem VIII.10 to prove that $A \otimes I + I \otimes B$ is essentially self-adjoint on $D(A) \otimes D(B)$. (*Hint*: $e^{itA} \otimes e^{itB}$ leaves $D(A) \otimes D(B)$ invariant.)

48. Let A be a closed symmetric operator with $A \neq A^*$. Let a and b be the quadratic forms with

    $$Q(a) = D(A) \qquad a(\psi, \varphi) = (A\psi, A\varphi)$$
    $$Q(b) = D(A^*) \qquad b(\psi, \varphi) = (A^*\psi, A^*\varphi)$$

    Show that $a \subset b$ but $a \neq b$ despite the fact that a and b are positive symmetric forms.

49. A self-adjoint operator, A, is said to have **purely discrete spectrum** in (a, b) if $(a, b) \cap \sigma(A) = (a, b) \cap \sigma_{\text{disc}}(A)$.
    (a) Prove that A has purely discrete spectrum in (a, b) if and only if $P_{(a+\varepsilon, b-\varepsilon)}$ is compact for all small ε where $\{P_\Omega\}$ is the family of spectral projections for A.
    (b) Prove that A has purely discrete spectrum in (a, b) if and only if $f(A)$ is compact for every C^∞ function f with $\operatorname{supp} f \subset (a, b)$. (*Hint*: Use Problem 45 of Chapter VI.)
    (c) Let $A_n \to A$ in the norm resolvent sense. Suppose that each A_n has purely discrete spectra in (a, b). Prove that A has purely discrete spectra in (a, b).

50. Let A be a positive self-adjoint operator
    (a) Prove that $\|(A + w)^{-1}\| \leq w^{-1}$ for $w > 0$
    (b) Prove that
    $$\int_0^\infty w^{-\frac{1}{2}}(A + w)^{-1}\varphi \, dw$$
    exists for all $\varphi \in \operatorname{Ran} A$.

(c) Prove that
$$A^{+1/2}\psi = \left[\frac{1}{\pi}\int_0^\infty w^{-1/2}(A+w)^{-1}\,dw\right]A\psi \quad \text{for any} \quad \psi \in D(A).$$

(d) In the same way prove that
$$A^\alpha \psi = \frac{\sin \pi\alpha}{\pi}\int_0^\infty \left[w^{\alpha-1}(A+w)^{-1}\,dw\right]A\psi \quad \text{for any} \quad \psi \in D(A),$$
if $0 < \alpha < 1$.

(e) Prove that
$$\left(\lim_{\alpha \downarrow 0} \frac{A^\alpha - 1}{\alpha}\right)\psi = (\log A)\psi \quad \text{for any} \quad \psi \in D(A)$$

51. If A and B are self-adjoint operators with $A, B \geq 0$, we say $A \geq B$ if $Q(B) \supset Q(A)$ and $\langle \psi, B\psi \rangle \leq \langle \psi, A\psi \rangle$, for all $\psi \in D(A)$.
    (a) If $0 \leq A \leq B$, prove that $A(A+w)^{-1} \leq B(B+w)^{-1}$ if $w \geq 0$.
    (b) Prove that $A^\alpha \leq B^\alpha$ if $A \leq B$ and $0 < \alpha < 1$.
    (c) Prove that $\log A \leq \log B$ if $A \leq B$.

*52. Extend the proof of the spectral theorem in Problem 32 of Chapter VII to the unbounded case by using the polar decomposition for closed operators. Be sure to check that you can prove the polar decomposition without using the spectral theorem in the unbounded case. (*Hint*: You will need Problem 50).

Reference for Problem 52: Kato's book, pp. 281–282, 322–324, 334–335.

The Fourier Transform

We have therefore the equation of condition

$$F(x) = \int dq \, Q \cos qx$$

If we substituted for Q any function of q, and conducted the integration from $q = 0$ to $q = \infty$, we should find a function of x: it is required to solve the inverse problem, that is to say, to ascertain what function of q, after being substituted for Q, gives as a result the function $F(x)$, a remarkable problem whose solution demands attentive examination.   Joseph Fourier

IX.1 The Fourier transform on $\mathscr{S}(\mathbb{R}^n)$ and $\mathscr{S}'(\mathbb{R}^n)$, convolutions

The Fourier transform is an important tool of both classical and modern analysis. We begin by defining it, and the inverse transform, on $\mathscr{S}(\mathbb{R}^n)$, the Schwartz space of C^∞ functions of rapid decrease.

Definition   Suppose $f \in \mathscr{S}(\mathbb{R}^n)$. The **Fourier transform** of f is the function $\hat{f}$ given by

$$\hat{f}(\lambda) = \frac{1}{(2\pi)^{n/2}} \int_{\mathbb{R}^n} e^{-i\mathbf{x}\cdot\lambda} f(\mathbf{x}) \, d\mathbf{x}$$

where $\mathbf{x} \cdot \lambda = \sum_{i=1}^{n} x_i \lambda_i$. The **inverse Fourier transform** of f, denoted by $\check{f}$, is the function

$$\check{f}(\lambda) = \frac{1}{(2\pi)^{n/2}} \int_{\mathbb{R}^n} e^{i\mathbf{x}\cdot\lambda} f(\mathbf{x}) \, d\mathbf{x}$$

We will occasionally write $\hat{f} = \mathscr{F}f$.

Since every function in Schwartz space is in $L^1(\mathbb{R}^n)$, the above integrals make sense. Many authors begin by discussing the Fourier transform on $L^1(\mathbb{R}^n)$. We start with Schwartz space for two reasons: First, the Fourier transform is a one-to-one map of Schwartz space onto itself (Theorem IX.1). This makes it particularly easy to talk about the inverse Fourier transform, which of course turns out to be the inverse map. That is, on Schwartz space, it is possible to deal with the transform and the inverse transform on an equal footing. Though this is also true for the Fourier transform on $L^2(\mathbb{R}^n)$ (see Theorem IX.6), it is not possible to define the Fourier transform on $L^2(\mathbb{R}^n)$ directly by the integral formula since $L^2(\mathbb{R}^n)$ functions may not be in $L^1(\mathbb{R}^n)$; a limiting procedure must be used. Secondly, once we know that the Fourier transform is a one-to-one, bounded map of $\mathscr{S}(\mathbb{R}^n)$ onto $\mathscr{S}(\mathbb{R}^n)$, we can easily extend it to $\mathscr{S}'(\mathbb{R}^n)$. It is this extension that is fundamental to the applications in Sections 5, 6, and 8.

We will use the standard multi-index notation. A multi-index

$$\alpha = \langle \alpha_1, \ldots, \alpha_n \rangle$$

is an n-tuple of nonnegative integers. The collection of all multi-indices will be denoted by I_+^n. The symbols $|\alpha|$, x^α, D^α, and x^2 are defined as follows:

$$|\alpha| = \sum_{i=1}^n \alpha_i$$

$$x^\alpha = x_1^{\alpha_1} x_2^{\alpha_2} \cdots x_n^{\alpha_n}$$

$$D^\alpha = \frac{\partial^{|\alpha|}}{\partial x_1^{\alpha_1} \partial x_2^{\alpha_2} \cdots \partial x_n^{\alpha_n}}$$

$$x^2 = \sum_{i=1}^n x_i^2$$

In preparation for the proof that $\hat{}$ and $\check{}$ are inverses, we prove:

Lemma The maps $\hat{}$ and $\check{}$ are continuous linear transformations of $\mathscr{S}(\mathbb{R}^n)$ into $\mathscr{S}(\mathbb{R}^n)$. Furthermore, if α and β are multi-indices, then

$$((i\lambda)^\alpha D^\beta \hat{f})(\lambda) = \widehat{D^\alpha((-ix)^\beta f(x))} \tag{IX.1}$$

Proof The map $\hat{}$ is clearly linear. Since

$$(\lambda^\alpha D^\beta \hat{f})(\lambda) = \frac{1}{(2\pi)^{n/2}} \int_{\mathbb{R}^n} \lambda^\alpha (-ix)^\beta e^{-i\lambda \cdot x} f(x) \, dx$$

$$= \frac{1}{(2\pi)^{n/2}} \int_{\mathbb{R}^n} \frac{1}{(-i)^\alpha} (D_x^\alpha e^{-i\lambda \cdot x})(-ix)^\beta f(x) \, dx$$

$$= \frac{(-i)^\alpha}{(2\pi)^{n/2}} \int_{\mathbb{R}^n} e^{-i\lambda \cdot x} D_x^\alpha ((-ix)^\beta f(x)) \, dx$$

320 THE FOURIER TRANSFORM

We conclude that

$$\|\hat{f}\|_{\alpha,\beta} = \sup_\lambda |\lambda^\alpha (D^\beta f)(\lambda)| \le \frac{1}{(2\pi)^{n/2}} \int |D^\alpha_x(x^\beta f)|\, dx < \infty$$

so $\hat{\ }$ takes $\mathscr{S}(\mathbb{R}^n)$ into $\mathscr{S}(\mathbb{R}^n)$, and we have also proven (IX.1). Furthermore, if k is large enough, $\int (1+x^2)^{-k}\, dx < \infty$ so that

$$\|\hat{f}\|_{\alpha,\beta} \le \frac{1}{(2\pi)^{n/2}} \int_{\mathbb{R}^n} \frac{(1+x^2)^{-k}}{(1+x^2)^{-k}} |D^\alpha_x(-ix)^\beta f(x)|\, dx$$

$$\le \frac{1}{(2\pi)^{n/2}} \left(\int (1+x^2)^{-k}\, dx\right) \sup_x \{(1+x^2)^{+k} |D^\alpha_x(-ix)^\beta f(x)|\}$$

Using Leibnitz's rule we easily conclude that there exist multi-indices α_j, β_j and constants c_j so that

$$\|\hat{f}\|_{\alpha,\beta} \le \sum_{i=1}^M c_j \|f\|_{\alpha_j, \beta_j}$$

Thus, $\hat{\ }$ is bounded and by Theorem V.4 is therefore continuous. The proof for $\check{\ }$ is the same. ∎

We are now ready to prove the Fourier inversion theorem. The proof we give uses the original idea of Fourier.

Theorem IX.1 (Fourier inversion theorem) The Fourier transform is a linear bicontinuous bijection from $\mathscr{S}(\mathbb{R}^n)$ onto $\mathscr{S}(\mathbb{R}^n)$. Its inverse map is the inverse Fourier transform, i.e., $\overset{\times}{\hat{f}} = f = \overset{\circ}{\hat{f}}$.

Proof We will prove that $\overset{\times}{\hat{f}} = f$. The proof that $\overset{\circ}{\hat{f}} = f$ is similar. $\overset{\circ}{\hat{f}} = f$ implies that $\hat{\ }$ is surjective and $\overset{\times}{\hat{f}} = f$ implies that $\hat{\ }$ is injective. Since $\hat{\ }$ and $\check{\ }$ are continuous maps of $\mathscr{S}(\mathbb{R}^n)$ into $\mathscr{S}(\mathbb{R}^n)$, it is sufficient to prove that $\overset{\times}{\hat{f}} = f$ for f contained in the dense set $C_0^\infty(\mathbb{R}^n)$. Let C_ε be the cube of volume $(2/\varepsilon)^n$ centered at the origin in $\mathbb{R}^n$. Choose ε small enough so that the support of f is contained in C_ε. Let

$$K_\varepsilon = \{\mathbf{k} \in \mathbb{R} \mid \text{each } k_i/\pi\varepsilon \text{ is an integer}\}$$

Then

$$f(x) = \sum_{k \in K_\varepsilon} ((\tfrac{1}{2}\varepsilon)^{n/2} e^{i k \cdot x}, f)(\tfrac{1}{2}\varepsilon)^{n/2} e^{i k \cdot x}$$

is just the Fourier series of f which converges uniformly in C_ε to f since f is continuously differentiable (Theorem II.8). Thus

$$f(x) = \sum_{k \in K_\varepsilon} \frac{\hat{f}(k) e^{i k \cdot x}}{(2\pi)^{n/2}} (\pi\varepsilon)^n \qquad (IX.2)$$

Since $\mathbb{R}^n$ is the disjoint union of the cubes of volume $(\pi\varepsilon)^n$ centered about the points in K_ε, the right-hand side of (IX.2) is just a Riemann sum for the integral of the function $\hat{f}(k)e^{ik\cdot x}/(2\pi)^{n/2}$. By the lemma, $\hat{f}(k)e^{ik\cdot x} \in \mathscr{S}(\mathbb{R}^n)$, so the Riemann sums converge to the integral. Thus $\check{\hat{f}} = f$. ∎

Corollary Suppose $f \in \mathscr{S}(\mathbb{R}^n)$. Then

$$\int_{\mathbb{R}^n} |f(x)|^2 \, dx = \int_{\mathbb{R}^n} |\hat{f}(k)|^2 \, dk$$

Proof This is really a corollary of the proof rather than the statement of Theorem IX.1. If f has compact support, then for ε small enough,

$$f(x) = \sum_{k \in K_\varepsilon} ((\tfrac{1}{2}\varepsilon)^{n/2} e^{ik\cdot x}, f(x))(\tfrac{1}{2}\varepsilon)^{n/2} e^{ik\cdot x}$$

Since $\{(\tfrac{1}{2}\varepsilon)^{n/2} e^{ik\cdot x}\}_{k \in K_\varepsilon}$ is an orthonormal basis for $L^2(C_\varepsilon)$,

$$\int_{\mathbb{R}^n} |f(x)|^2 \, dx = \int_{C_\varepsilon} |f(x)|^2 \, dx$$
$$= \sum_{k \in K_\varepsilon} |((\tfrac{1}{2}\varepsilon)^{n/2} e^{ik\cdot x}, f(x))|^2$$
$$= \sum_{k \in K_\varepsilon} |\hat{f}(k)|^2 (\pi\varepsilon)^n$$
$$\xrightarrow[\varepsilon \to 0]{} \int_{\mathbb{R}^n} |\hat{f}(k)|^2 \, dk$$

This proves the corollary for $f \in C_0^\infty$. Since $\hat{}$ and $\|\cdot\|_2$ are continuous on $\mathscr{S}$ and C_0^∞ is dense, the result holds for all of $\mathscr{S}$. ∎

Example 1 We compute the Fourier transform of $f(x) = e^{-\alpha x^2/2} \in \mathscr{S}(\mathbb{R})$ where $\alpha > 0$.

$$\hat{f}(\lambda) = \frac{1}{\sqrt{2\pi}} \int_{\mathbb{R}} e^{-\alpha x^2/2} e^{-i\lambda \cdot x} \, dx$$
$$= \frac{1}{\sqrt{2\pi}} \int_{\mathbb{R}} \sqrt{\frac{2}{\alpha}} \exp\left(-t^2 - it\lambda \sqrt{\frac{2}{\alpha}}\right) dt$$
$$= \frac{e^{-\lambda^2/2\alpha}}{\sqrt{\alpha\pi}} \int_{\mathbb{R}} \exp -\left(t + i\frac{\lambda}{\sqrt{2\alpha}}\right)^2 dt$$
$$= \frac{e^{-\lambda^2/2\alpha}}{\sqrt{\alpha\pi}} \int_{\mathbb{R}} e^{-t^2} \, dt = \frac{e^{-\lambda^2/2\alpha}}{\sqrt{\alpha}}$$

The next to last step follows from the Cauchy integral formula and the exponential decrease of e^{-z^2} along lines parallel to the x axis.

We now define the Fourier transform on $\mathscr{S}'(\mathbb{R}^n)$.

Definition Let $T \in \mathscr{S}'(\mathbb{R}^n)$. Then the Fourier transform of T, denoted by $\hat{T}$, is the tempered distribution defined by $\hat{T}(\varphi) = T(\hat{\varphi})$.

Suppose that $h, \varphi \in \mathscr{S}(\mathbb{R}^n)$, then by the polarization identity and the corollary to Theorem IX.1 we have $(h, \varphi) = (\hat{h}, \hat{\varphi})$. Substituting $\bar{\hat{g}} = \check{\bar{g}}$ for h, we obtain

$$T_{\hat{g}}(\varphi) = \int \hat{g}(x)\varphi(x)\, dx = \int g(x)\hat{\varphi}(x)\, dx = T_g(\hat{\varphi}) = \hat{T}_g(\varphi)$$

where $T_{\hat{g}}$ and T_g are the distributions corresponding to the functions $\hat{g}$ and g respectively. This shows that the Fourier transform on $\mathscr{S}'(\mathbb{R}^n)$ extends the transform we previously defined on $\mathscr{S}(\mathbb{R}^n)$.

Theorem IX.2 The Fourier transform is a one-to-one linear bijection from $\mathscr{S}'(\mathbb{R}^n)$ to $\mathscr{S}'(\mathbb{R}^n)$ which is the unique weakly continuous extension of the Fourier transform on $\mathscr{S}(\mathbb{R}^n)$.

Proof If $\varphi_n \xrightarrow{\mathscr{S}} \varphi$, then by Theorem IX.1, $\hat{\varphi}_n \xrightarrow{\mathscr{S}} \hat{\varphi}$, so $T(\hat{\varphi}_n) \to T(\hat{\varphi})$ for each T in $\mathscr{S}'(\mathbb{R}^n)$. Thus $\hat{T}(\varphi_n) \to \hat{T}(\varphi)$, which shows that $\hat{T}$ is a continuous linear functional on $\mathscr{S}(\mathbb{R}^n)$. Furthermore, if $T_n \xrightarrow{w} T$, then $\hat{T}_n \xrightarrow{w} \hat{T}$ because $T_n(\hat{\varphi}) \to T(\hat{\varphi})$ implies $\hat{T}_n(\varphi) \to \hat{T}(\varphi)$. Thus $T \mapsto \hat{T}$ is weakly continuous.

The remaining properties of $\hat{}$ follow immediately from the corresponding statements on $\mathscr{S}(\mathbb{R}^n)$ (see Problem 19 in Chapter V). ∎

Example 2 We compute the Fourier transform of the derivative of the delta function at $b \in \mathbb{R}$:

$$\hat{\delta}'_b(\varphi) = \delta'_b(\hat{\varphi})$$

$$= \delta_b\left(-\frac{d}{d\lambda}\hat{\varphi}(\lambda)\right)$$

$$= \delta_b\left(\frac{-1}{(2\pi)^{1/2}}\int e^{-i\lambda x}(-ix)\varphi(x)\, dx\right)$$

$$= \int \left(\frac{ixe^{-ibx}}{\sqrt{2\pi}}\right)\varphi(x)\, dx$$

So, the Fourier transform of δ'_b is the function $ixe^{-ibx}/\sqrt{2\pi}$.

* * *

IX.1 The Fourier transform on $\mathscr{S}(\mathbb{R}^n)$ and $\mathscr{S}'(\mathbb{R}^n)$, convolutions

We now introduce a new operation on functions.

Definitions Suppose that $f, g \in \mathscr{S}(\mathbb{R}^n)$. Then the **convolution** of f and g, denoted by $f * g$, is the function

$$(f * g)(y) = \int_{\mathbb{R}^n} f(y - x)g(x)\,dx$$

The convolution arises in many circumstances (we have already used it in discussing closed operators in Section VIII.1). In Section 4 we use interpolation theorems to prove L^p estimates on the convolution $f * g$ in terms of f and g. In this section we concentrate on the properties of the convolution as a map from $\mathscr{S}(\mathbb{R}^n) \times \mathscr{S}(\mathbb{R}^n)$ to $\mathscr{S}(\mathbb{R}^n)$. Using these properties we show that the convolution can be extended to a map from $\mathscr{S}'(\mathbb{R}^n) \times \mathscr{S}(\mathbb{R}^n)$ to O_M^n, the polynomially bounded C^∞ functions. Convolutions frequently occur when one uses the Fourier transform because the Fourier transform takes products into convolutions (Theorem IX.3b and Theorem IX.4c).

Theorem IX.3

(a) For each $f \in \mathscr{S}(\mathbb{R}^n)$, $g \to f * g$ is a continuous map of $\mathscr{S}(\mathbb{R}^n)$ into $\mathscr{S}(\mathbb{R}^n)$.
(b) $\widehat{fg} = (2\pi)^{-n/2}\hat{f} * \hat{g}$ and $\widehat{f * g} = (2\pi)^{n/2}\hat{f}\hat{g}$.
(c) For f, g, h, in $\mathscr{S}(\mathbb{R}^n)$, $f * g = g * f$ and $f * (g * h) = (f * g) * h$.

Proof From the polarization identity and the corollary to Theorem IX.1 we find that $(\varphi, \psi) = (\hat\varphi, \hat\psi)$ for $\varphi, \psi \in \mathscr{S}(\mathbb{R}^n)$. Letting $y \in \mathbb{R}^n$ be fixed, we apply this identity to $e^{iy \cdot x}\overline{f}(x)$ and g obtaining $(e^{iy \cdot x}\overline{f}, g) = (\widehat{e^{iy \cdot x}\overline{f}}, \hat{g})$. But

$$(e^{iy \cdot x}\overline{f}, g) = \int_{\mathbb{R}^n} e^{-iy \cdot x} f(x) g(x)\,dx$$

and

$$(\widehat{e^{iy \cdot x}\overline{f}}, \hat{g}) = \int_{\mathbb{R}^n} \left((2\pi)^{-n/2} \int_{\mathbb{R}^n} e^{-i\lambda \cdot x + iy \cdot x}\overline{f(x)}\,dx\right) \hat{g}(\lambda)\,d\lambda$$

$$= \int_{\mathbb{R}^n} \hat{f}(y - \lambda)\hat{g}(\lambda)\,d\lambda$$

which proves that $\widehat{fg} = (2\pi)^{-n/2}\hat{f} * \hat{g}$. Using the inverse Fourier transform this formula may be stated as

$$(2\pi)^{n/2}\widehat{\check{f}\check{g}} = f * g$$

This shows that convolution is the composition of the inverse Fourier transform, multiplication by $(2\pi)^{n/2}\hat{f}$, and the Fourier transform. It follows that convolution is continuous.

The statements in (c) follow trivially from (b). ∎

In order to extend the map $C_f : g \to f * g$ to $\mathscr{S}'$, we look for a continuous map $\tilde{C}_f : \mathscr{S}' \to \mathscr{S}'$ so that $\tilde{C}'_f \upharpoonright \mathscr{S} = C_f$. We then define $\tilde{C}'_f$ to be convolution on $\mathscr{S}'$.

Definition Suppose that $f \in \mathscr{S}(\mathbb{R}^n)$, $T \in \mathscr{S}'(\mathbb{R}^n)$ and let $\tilde{f}(x)$ denote the function, $f(-x)$. Then, the **convolution** of T and f, denoted $T * f$, is the distribution in $\mathscr{S}'(\mathbb{R}^n)$ given by

$$(T * f)(\varphi) = T(\tilde{f} * \varphi)$$

for all $\varphi \in \mathscr{S}(\mathbb{R}^n)$.

The fact that $g \to \tilde{f} * g$ is a continuous transformation guarantees that $T * f \in \mathscr{S}'(\mathbb{R}^n)$. The following theorem summarizes the properties of this extended convolution.

Let f_y denote the function $f_y(x) = f(x - y)$ and $\tilde{f}_y$ the function $f(y - x)$. When f is given by a large expression $(\cdots)$, we will sometimes write $(\cdots)^\sim$ rather than $(\widetilde{\cdots})$.

Theorem IX.4 For each $f \in \mathscr{S}(\mathbb{R}^n)$ the map $T \to T * f$ is a weakly continuous map of $\mathscr{S}'(\mathbb{R}^n)$ into $\mathscr{S}'(\mathbb{R}^n)$ which extends the convolution on $\mathscr{S}(\mathbb{R}^n)$. Furthermore,

(a) $T * f$ is a polynomially bounded C^∞ function, i.e. $T * f \in O_M^n$. In fact, $(T * f)(y) = T(\tilde{f}_y)$ and

$$D^\beta(T * f) = (D^\beta T) * f = T * D^\beta f \qquad \text{(IX.3)}$$

(b) $(T * f) * g = T * (f * g)$

(c) $\widehat{T * f} = (2\pi)^{n/2} \hat{f} \hat{T}$

Proof Since $T \to T * f$ is defined as the adjoint of a bounded map from $\mathscr{S}$ to $\mathscr{S}$, it is automatically weakly continuous. The fact that it extends the convolution on $\mathscr{S}$ is just a change of variables. The statements (IX.3), (b), and (c), all follow immediately from the corresponding statements for $T \in \mathscr{S}$ and the facts that $\mathscr{S}$ is weakly dense in $\mathscr{S}'$ and that $\mathscr{F}$, D^β, multiplication by $\hat{f}$, and convolution are all weakly continuous on $\mathscr{S}'$.

IX.1 The Fourier transform on $\mathscr{S}(\mathbb{R}^n)$ and $\mathscr{S}'(\mathbb{R}^n)$, convolutions

It remains to prove the first part of (a). Since $T \in \mathscr{S}'(\mathbb{R}^n)$, it follows from the regularity theorem (Theorem V.10) that there is a bounded continuous function h, a positive integer r, and a multi-index β so that

$$T(\tilde{f}_y) = \int_{\mathbb{R}^n} h(x)(1 + x^2)^r (D^\beta f)(y - x)\, dx$$

Since $D^\beta f \in \mathscr{S}$, $T(\tilde{f}_y)$ is an infinitely differentiable function of y. The change of variables $\tau = y - x$ shows that

$$|T(\tilde{f}_y)| \leq \|h\|_\infty \int_{\mathbb{R}^n} (1 + x^2)^r |(D^\beta f)(y - x)|\, dx$$

$$= \|h\|_\infty \int_{\mathbb{R}^n} (1 + (y - \tau)^2)^r |D^\beta f(\tau)|\, d\tau$$

from which it follows easily that $y \mapsto T(\tilde{f}_y)$ is polynomially bounded. A similar proof works for the derivatives of $y \mapsto T(\tilde{f}_y)$. Thus $T(\tilde{f}_y) \in O_M^n$.

Suppose that a distribution $S \in \mathscr{S}'(\mathbb{R}^n)$ is given by a polynomially bounded continuous function s. Then, using Fubini's theorem we find that for $\varphi \in \mathscr{S}(\mathbb{R}^n)$

$$(S * f)(\varphi) \equiv S(\tilde{f} * \varphi)$$

$$= \int s(x)\left(\int \tilde{f}(x - y)\varphi(y)\, dy\right) dx$$

$$= \int \left(\int s(x)\tilde{f}_y(x)\, dx\right)\varphi(y)\, dy$$

$$= (S(\tilde{f}_y))(\varphi)$$

so $S * f = S(\tilde{f}_y)$. By the regularity theorem $T = D^\alpha S$ for some such S. Thus by (IX.3)

$$T * f = (D^\alpha S) * f = S * D^\alpha f$$
$$= S((D^\alpha f)\tilde{\,}_y)$$
$$= (-1)^{|\alpha|} S(D^\alpha(\tilde{f}_y))$$
$$= D^\alpha S(\tilde{f}_y)$$
$$= T(\tilde{f}_y)$$

This completes the proof. ∎

Theorem IX.5 Let $T \in \mathscr{S}'(\mathbb{R}^n)$ and $f \in \mathscr{S}(\mathbb{R}^n)$. Then $\widehat{f T} \in O_M^n$ and $\widehat{fT}(k) = (2\pi)^{-n/2} T(fe^{-ik \cdot x})$. In particular, if T has compact support and $\psi \in \mathscr{S}(\mathbb{R}^n)$ is identically one on a neighborhood of the support of T, then

$$\hat{T}(k) = (2\pi)^{-n/2} T(\psi e^{-ik \cdot x})$$

Proof By Theorem IX.4c and the Fourier inversion formula we have $\widehat{fT} = (2\pi)^{-n/2}\hat{f} * \hat{T}$. Thus $\widehat{fT} \in O_M^n$ and

$$\widehat{fT}(k) = (2\pi)^{-n/2} \hat{T}(\tilde{\hat{f}}_k)$$
$$= (2\pi)^{-n/2} T(e^{-ik \cdot x}f) \blacksquare$$

We remark that one can also define the convolution of a distribution $T \in \mathscr{D}'(\mathbb{R}^n)$ with an $f \in \mathscr{D}(\mathbb{R}^n)$ by $(T * f)(y) = T(\tilde{f}_y)$. A proof similar to the proof of Theorem IX.4 shows that $T * f$ is a (not necessarily polynomially bounded) C^∞ function and that (IX.3) holds.

We have already introduced the term "approximate identity" in Section VIII.1; we now define it formally.

Definition Let $j(x)$ be a positive C^∞ function whose support lies in the sphere of radius one about the origin in $\mathbb{R}^n$ and which satisfies $\int j(x) \, dx = 1$. The sequence of functions $j_\varepsilon(x) = \varepsilon^{-n} j(x/\varepsilon)$ is called an **approximate identity**.

Proposition Suppose $T \in \mathscr{S}'(\mathbb{R}^n)$ and let $j_\varepsilon(x)$ be an approximate identity. Then $T * j_\varepsilon \to T$ weakly as $\varepsilon \to 0$.

Proof If $\varphi \in \mathscr{S}(\mathbb{R}^n)$, then $(T * j_\varepsilon)(\varphi) = T(\tilde{j}_\varepsilon * \varphi)$, so it is sufficient to show that $\tilde{j}_\varepsilon * \varphi \xrightarrow{\mathscr{S}(\mathbb{R}^n)} \varphi$. To do this it is sufficient to show that $(2\pi)^{n/2}\hat{j}_\varepsilon \hat{\varphi} \xrightarrow{\mathscr{S}} \hat{\varphi}$. Since $\hat{j}_\varepsilon(\lambda) = \hat{j}(\varepsilon\lambda)$ and $\hat{j}(0) = (2\pi)^{-n/2}$, it follows that $(2\pi)^{n/2}\hat{j}_\varepsilon(x)$ converges to 1 uniformly on compact sets and is uniformly bounded. Similarly, $D^\alpha \hat{j}_\varepsilon$ converges uniformly to zero. We conclude that $(2\pi)^{n/2}\hat{j}_\varepsilon \hat{\varphi} \xrightarrow{\mathscr{S}} \hat{\varphi}$. $\blacksquare$

IX.2 The range of the Fourier transform: Classical spaces

We have defined the Fourier transform on $\mathscr{S}(\mathbb{R}^n)$ and $\mathscr{S}'(\mathbb{R}^n)$. In this section, Section IX.3, and Section IX.9, we investigate the range of the Fourier transform when it is restricted to various subsets of $\mathscr{S}'(\mathbb{R}^n)$. These

IX.2 The range of the Fourier transform: Classical spaces

questions are natural and have historical interest, but more important, characterizing the range of the Fourier transform is very useful. One is often able to obtain information about the Fourier transform of a function and one would like to know what this says about the function itself. We begin with two theorems which follow easily from the work that we have already done in Section IX.1.

Theorem IX.6 (the Plancherel theorem)   The Fourier transform extends uniquely to a unitary map of $L^2(\mathbb{R}^n)$ onto $L^2(\mathbb{R}^n)$. The inverse transform extends uniquely to its adjoint.

Proof   The corollary to Theorem IX.1 states that if $f \in \mathscr{S}(\mathbb{R}^n)$, then $\|f\|_2 = \|\hat{f}\|_2$. Since $\mathscr{F}[\mathscr{S}] = \mathscr{S}$, $\mathscr{F}$ is a surjective isometry on $L^2(\mathbb{R}^n)$. ∎

Theorem IX.7 (the Riemann–Lebesgue lemma)   The Fourier transform extends uniquely to a bounded map from $L^1(\mathbb{R}^n)$ into $C_\infty(\mathbb{R}^n)$, the continuous functions vanishing at ∞.

Proof   For $f \in \mathscr{S}(\mathbb{R}^n)$, we know that $\hat{f} \in \mathscr{S}(\mathbb{R}^n)$ and thus $\hat{f} \in C_\infty(\mathbb{R}^n)$. The estimate

$$\|\hat{f}\|_\infty \leq (2\pi)^{-n/2} \|f\|_1$$

is trivial. The Fourier transform is thus a bounded linear map from a dense set of $L^1(\mathbb{R}^n)$ into $C_\infty(\mathbb{R}^n)$. By the B.L.T. theorem, extends uniquely to a bounded linear transformation of $L^1(\mathbb{R}^n)$ into $C_\infty(\mathbb{R}^n)$. ∎

We remark that the Fourier transform takes $L^1(\mathbb{R}^n)$ into, but not onto $C_\infty(\mathbb{R}^n)$ (Problem 16).

A simple argument with test functions shows that the extended transform on $L^1(\mathbb{R}^n)$ and $L^2(\mathbb{R}^n)$ is the restriction of the transform on $\mathscr{S}'(\mathbb{R}^n)$, but it is useful to have an explicit integral representation. For $f \in L^1(\mathbb{R}^n)$, this is easy since we can find $f_m \in \mathscr{S}(\mathbb{R}^n)$ so that $\|f - f_m\|_1 \to 0$. Then, for each λ,

$$\hat{f}(\lambda) = \lim_{m \to \infty} (\hat{f}_m(\lambda))$$

$$= \lim_{m \to \infty} \left\{ \frac{1}{(2\pi)^{n/2}} \int_{\mathbb{R}^n} e^{-i\lambda \cdot x} f_m(x) \, dx \right\}$$

$$= \frac{1}{(2\pi)^{n/2}} \int_{\mathbb{R}^n} e^{-i\lambda \cdot x} f(x) \, dx$$

So, the Fourier transform of a function in $L^1(\mathbb{R}^n)$ is given by the usual formula.

Next, suppose $f \in L^2(\mathbb{R}^n)$ and let

$$\chi_R(x) = \begin{cases} 1 & |x| \leq R \\ 0 & |x| > R \end{cases}$$

Then $\chi_R f \in L^1(\mathbb{R}^n)$ and $\chi_R f \xrightarrow[R \to \infty]{L^2} f$, so by the Plancherel theorem $\widehat{\chi_R f} \xrightarrow[R \to \infty]{L^2} \hat{f}$. For $\chi_R f$ we have the usual formula; thus

$$\hat{f}(\lambda) = \text{l.i.m.}_{R \to \infty} (2\pi)^{-n/2} \int_{|x| \leq R} e^{-i\lambda \cdot x} f(x) \, dx$$

where by "l.i.m." we mean the limit in the L^2-norm. Sometimes we will dispense with $|x| \leq R$ and just write

$$\hat{f}(\lambda) = \text{l.i.m.} (2\pi)^{-n/2} \int e^{-i\lambda \cdot x} f(x) \, dx$$

for functions $f \in L^2(\mathbb{R}^n)$.

We have proven above that $L^2(\mathbb{R}^n) \xrightarrow{\hat{}} L^2(\mathbb{R}^n)$ and $L^1(\mathbb{R}^n) \xrightarrow{\hat{}} L^\infty(\mathbb{R}^n)$ and in both cases $\hat{}$ is a bounded operator. It is exactly in situations like this that one can use the interpolation theorems which we will prove in the Appendix to Section 4.

Theorem IX.8 (Hausdorff–Young inequality) Suppose $1 \leq q \leq 2$, and $p^{-1} + q^{-1} = 1$. Then the Fourier transform is a bounded map of $L^q(\mathbb{R}^n)$ to $L^p(\mathbb{R}^n)$ and its norm is less than or equal to $(2\pi)^{n(1/2 - 1/q)}$.

Proof We use the Riesz–Thorin theorem (Theorem IX.17) with $q_0 = 2 = p_0$, $p_1 = \infty$, and $q_1 = 1$. Since $\|\hat{f}\|_2 = \|f\|_2$ and $\|\hat{f}\|_\infty \leq (2\pi)^{-n/2} \|f\|_1$, we conclude that $\|\hat{f}\|_{p_t} \leq C_t \|f\|_{q_t}$ where $p_t^{-1} = (1-t)/2$, $q_t^{-1} = (1-t)/2 + t = 1 - p_t^{-1}$, and $\log C_t = t \log(2\pi)^{-n/2}$. ∎

* * *

We now come to another natural question. What are the Fourier transforms of the finite positive measures on $\mathbb{R}^n$? Suppose that we define

$$\hat{\mu}(\lambda) = (2\pi)^{-n/2} \int_{\mathbb{R}^n} e^{-i\lambda \cdot x} \, d\mu(x)$$

IX.2 The range of the Fourier transform: Classical spaces

Then, if $\varphi \in \mathscr{S}(\mathbb{R}^n)$,

$$\int_{\mathbb{R}^n} \hat{\mu}(\lambda)\varphi(\lambda)\, d\lambda = (2\pi)^{-n/2} \int_{\mathbb{R}^n} \left(\int_{\mathbb{R}^n} e^{-i\lambda \cdot x}\, d\mu(x) \right) \varphi(\lambda)\, d\lambda$$

$$= (2\pi)^{-n/2} \int_{\mathbb{R}^n} \left(\int_{\mathbb{R}^n} e^{-i\lambda \cdot x} \varphi(\lambda)\, d\lambda \right) d\mu(x)$$

$$= \int_{\mathbb{R}^n} \hat{\varphi}(x)\, d\mu(x)$$

so this definition coincides with the restriction of the Fourier transform on $\mathscr{S}'(\mathbb{R}^n)$ to the positive measures. Suppose $\lambda_1, \ldots, \lambda_N \in \mathbb{R}^n$ and $\xi = \langle \xi_1, \xi_2, \ldots, \xi_N \rangle \in \mathbb{C}^N$. Then

$$\sum_{i,j=1}^{N} \hat{\mu}(\lambda_i - \lambda_j)\bar{\xi}_j \xi_i = \int \left| \sum_{i=1}^{N} \xi_i e^{-i\lambda_i \cdot \mathbf{x}} \right|^2 d\mu(\mathbf{x}) \geq 0$$

This shows that the function $\hat{\mu}(\lambda)$ has the property that for any $\lambda_1, \ldots, \lambda_N \in \mathbb{R}^n$, $\{\hat{\mu}(\lambda_i - \lambda_j)\}$ is the matrix of a positive operator on $\mathbb{C}^N$. Furthermore, by the dominated convergence theorem, $\hat{\mu}$ is continuous, and since

$$|\hat{\mu}(\lambda)| \leq (2\pi)^{-n/2} \int_{\mathbb{R}^n} |e^{-i\lambda \cdot \mathbf{x}}|\, d\mu(\mathbf{x})$$

$$= (2\pi)^{-n/2} \mu(\mathbb{R}^n)$$

$\hat{\mu}(\cdot)$ is also bounded.

Definition A complex-valued, bounded, continuous function f on $\mathbb{R}^n$ that has the property that $\{f(\lambda_i - \lambda_j)\}_{i,j}$ is a positive matrix on $\mathbb{C}^N$ for each N and all $\lambda_1, \ldots, \lambda_N \in \mathbb{R}^n$ is called a **function of positive type**.

There are three properties of functions of positive type which follow easily from the definition. Letting $N = 1$, $x \in \mathbb{R}^N$,

(1) $\quad f(0) \geq 0$

since $f(0)$ is a positive operator on $\mathbb{C}^1$. Letting $N = 2$, and choosing $\lambda_1 = x$, $\lambda_2 = 0$, we see that the matrix

$$\begin{pmatrix} f(0) & f(x) \\ f(-x) & f(0) \end{pmatrix}$$

must be positive and therefore self-adjoint with positive determinant. This implies that

(2) $\quad f(x) = \overline{f(-x)}$
(3) $\quad |f(x)| \leq f(0)$

Notice that in proving these three properties we did not use the fact that $f(x)$ is bounded, so we could have left out the word *bounded* in the definition and recovered boundedness from (3) above. It is clear that any convex combinations or scalar multiples of functions of positive type again give functions of positive type, so these functions form a cone.

Theorem IX.9 (Bochner's theorem) The set of Fourier transforms of the finite, positive measures on $\mathbb{R}^n$ is exactly the cone of functions of positive type.

Proof We do not give Bochner's original proof but rather an easy, interesting argument based on Stone's theorem. We have already shown that the Fourier transforms of finite positive measures are functions of positive type. We need to prove the converse. Suppose f is of positive type. Let $\mathscr{K}$ denote the set of complex-valued functions on $\mathbb{R}^n$ which vanish except at a finite number of points. Then

$$(\psi, \varphi)_f = \sum_{\mathbf{x}, \mathbf{y} \in \mathbb{R}^n} f(\mathbf{x} - \mathbf{y})\overline{\psi(\mathbf{x})}\varphi(\mathbf{y})$$

has all the properties of a well-defined inner product except that we may have $(\varphi, \varphi)_f = 0$ for some $\varphi \neq 0$. If we let $\mathscr{N}$ be the set of such φ, then $\mathscr{K}/\mathscr{N}$ is a well-defined pre-Hilbert space under $(\cdot, \cdot)_f$. Suppose that $\mathbf{t} \in \mathbb{R}^n$ and define $U_\mathbf{t}$ on $\mathscr{K}$ by $(U_\mathbf{t}\varphi)(\mathbf{x}) = \varphi(\mathbf{x} - \mathbf{t})$. Since $U_\mathbf{t}$ preserves the form $(\cdot, \cdot)_f$, it takes equivalence classes into equivalence classes and thus restricts to an isometry on $\mathscr{K}/\mathscr{N}$. Since the same is true of $U_{-\mathbf{t}}$, this isometry has dense range and thus extends to a unitary operator $\tilde{U}_\mathbf{t}$ on $\mathscr{H} = \overline{\mathscr{K}/\mathscr{N}}$. Furthermore, $\tilde{U}_{\mathbf{t}+\mathbf{s}} = \tilde{U}_\mathbf{t}\tilde{U}_\mathbf{s}$, $\tilde{U}_0 = I$, and because of the continuity of f, $\tilde{U}_\mathbf{t}$ is strongly continuous. Thus the map $\mathbf{t} \to \tilde{U}_\mathbf{t}$ satisfies the hypotheses of Theorem VIII.12 (the generalization of Stone's theorem). Therefore, there is a projection-valued measure P_λ, on $\mathbb{R}^n$ so that

$$(\varphi, \tilde{U}_\mathbf{t}\psi)_f = \int_{\mathbb{R}^n} e^{i\mathbf{t} \cdot \lambda} \, d(\varphi, P_\lambda \psi)_f$$

Let $\tilde{\varphi}_0$ denote the equivalence class containing the function

$$\varphi_0(\mathbf{x}) = \begin{cases} 1, & \mathbf{x} = 0 \\ 0, & \mathbf{x} \neq 0 \end{cases}$$

Then

$$f(\mathbf{t}) = (\tilde{U}_\mathbf{t}\tilde{\varphi}_0, \tilde{\varphi}_0)_f = (\tilde{\varphi}_0, \tilde{U}_{-\mathbf{t}}\tilde{\varphi}_0)_f = \int e^{-i\mathbf{t} \cdot \lambda} \, d(\tilde{\varphi}_0, P_\lambda \tilde{\varphi}_0)_f$$

so we have displayed f as the Fourier transform of a finite positive measure. ∎

IX.2 The range of the Fourier transform: Classical spaces 331

The notion of positive type may be generalized to distributions. If $f(x)$ is a bounded continuous function, then $f(x)$ will be of positive type if and only if

$$\int \int f(x - y)\overline{\varphi(y)}\varphi(x)\, dx\, dy \geq 0 \qquad (IX.4)$$

for all $\varphi \in C_0^\infty(\mathbb{R}^n)$. To see this one need only approximate the integral in (IX.4) by a Riemann sum. This condition can be rewritten as

$$\int\int f(\tau)\overline{\varphi(x - \tau)}\varphi(x)\, d\tau\, dx = \int f(\tau)(\tilde{\bar{\varphi}} * \varphi)(\tau)\, d\tau \geq 0 \qquad (IX.5)$$

where $\tilde{\varphi}$ is the function $\tilde{\varphi}(x) = \varphi(-x)$. This suggests the following definition.

Definition A distribution $T \in \mathscr{D}'(\mathbb{R}^n)$ is said to be of **positive type** if $T(\tilde{\bar{\varphi}} * \varphi) \geq 0$ for all $\varphi \in \mathscr{D}(\mathbb{R}^n)$.

The following generalization of Bochner's theorem is due to Schwartz. This theorem is particularly interesting since it implies that certain ordinary distributions must be tempered. The proof is sketched in Problem 20 (or see the Notes for a reference).

Theorem IX.10 (the Bochner–Schwartz theorem) A distribution $T \in \mathscr{D}'(\mathbb{R}^n)$ is a distribution of positive type if and only if $T \in \mathscr{S}'(\mathbb{R}^n)$ and T is the Fourier transform of a positive measure of at most polynomial growth.

If $f(x)$ is a function of positive type, then this theorem implies that the weak derivatives $(-\Delta)^m f$ are all distributions of positive type. For $\hat{f} = \mu$, a finite measure by Theorem IX.9, and $\widehat{(-\Delta)^m f} = |x|^{2m}\mu$, a positive measure of polynomial growth.

Finally, we determine which bounded measurable functions are distributions of positive type. A bounded measurable function f on $\mathbb{R}^n$ is said to be of **weak positive type** if (IX.4) holds. Since (IX.5) follows from (IX.4), the distribution

$$T_f(\varphi) = \int f(x)\varphi(x)\, dx$$

is of positive type and therefore $\hat{T}_f = \mu$, a polynomially bounded positive measure. If $j_\varepsilon(x)$ is an approximate identity that is symmetric about the origin, then

$$\|f\|_\infty \geq T_f(j_\varepsilon * j_\varepsilon) = \widehat{T_f((j_\varepsilon * j_\varepsilon))}$$
$$= (2\pi)^{n/2}\mu(|\check{j}_\varepsilon(x)|^2)$$
$$= (2\pi)^{n/2}\int |\check{j}_\varepsilon(x)|^2\,d\mu(x)$$

On each compact subset of $\mathbb{R}^n$, $\check{j}_\varepsilon(x)$ converges uniformly to $(2\pi)^{-n/2}$ as $\varepsilon \to 0$, so the μ-measure of any compact set is less than $(2\pi)^{n/2}\|f\|_\infty$, so μ is finite.

We now come to the interesting point. Since μ is finite, its Fourier transform is a continuous function of positive type. Since μ and f must coincide a.e., we have proven:

Proposition A bounded function of weak positive type is equal almost everywhere to a continuous function of positive type.

IX.3 The range of the Fourier transform: Analyticity

In this section we investigate the connection between the decay properties of a function or distribution at infinity and the analyticity properties of its Fourier transform. The most extreme form of decay at infinity is to have compact support. We will prove the Paley–Wiener and Schwartz theorems which characterize explicitly the Fourier transforms of C^∞ functions and distributions with compact support. We then state two theorems relating exponential decay to analyticity properties of the Fourier transform. We close the section by characterizing the Fourier transforms of tempered distributions whose supports lie in symmetric cones. There are many other theorems of this genre; some of them are discussed in the Notes.

Suppose that $f \in C_0^\infty(\mathbb{R}^n)$. Then for all $\zeta = \langle \zeta_1, \ldots, \zeta_n \rangle \in \mathbb{C}^n$, the integral

$$\hat{f}(\zeta) = (2\pi)^{-n/2}\int e^{-i\zeta \cdot x}f(x)\,dx$$

is well defined. Furthermore, $\hat{f}(\zeta)$ is an entire analytic function of the n complex variables $\zeta_1, \zeta_2, \ldots, \zeta_n$ since we can differentiate under the

IX.3 The range of the Fourier transform: Analyticity

integral sign. In addition, if the support of f is contained in the sphere of radius R, then an integration by parts yields

$$\prod_{i=1}^{n}(i\zeta_i)^\alpha \hat{f}(\zeta) = (2\pi)^{-n/2} \int_{|x|\leq R} e^{-i\zeta \cdot x} D^\alpha f(x)\,dx$$

Taking the absolute value of both sides and using the fact that $\hat{f}(\zeta)$ is bounded on the set $\{\zeta \,|\, |\text{Im}\,\zeta| < \varepsilon\}$, we easily conclude that for each N,

$$|\hat{f}(\zeta)| \leq \frac{C_N e^{R|\text{Im}\,\zeta|}}{(1+|\zeta|)^N} \quad \text{for all } \zeta \in \mathbb{C}^n$$

where C_N is a constant that depends on N and f. The interesting fact is that these estimates are not only necessary but also sufficient for f to be in $C_0^\infty(\mathbb{R}^n)$.

Theorem IX.11 (the Paley–Wiener theorem) An entire analytic function of n complex variables $g(\zeta)$ is the Fourier transform of a $C_0^\infty(\mathbb{R}^n)$ function with support in the ball $\{x \,|\, |x| \leq R\}$ if and only if for each N there is a C_N so that

$$|g(\zeta)| \leq \frac{C_N e^{R|\text{Im}\,\zeta|}}{(1+|\zeta|)^N} \tag{IX.6}$$

for all $\zeta \in \mathbb{C}^n$.

Proof We have already proven the "only if" part. Suppose that g is entire and satisfies estimates of the form (IX.6). Let $\zeta = \lambda + i\eta$, where $\lambda, \eta \in \mathbb{R}^n$. Then for each η, $g(\lambda + i\eta)$ is in $\mathcal{S}(\mathbb{R}^n)$ as a function of λ, since the derivatives fall off polynomially by (IX.6) and the Cauchy formula. Let

$$f(x) = (2\pi)^{-n/2} \int_{\mathbb{R}^n} e^{ix\cdot\lambda} g(\lambda)\,d\lambda \tag{IX.7}$$

Then by Theorem IX.1, $f \in \mathcal{S}(\mathbb{R}^n)$ and $g(\lambda) = \hat{f}(\lambda)$. We want to show that $f(x)$ has support in the ball of radius R. Because of the estimates (IX.6) and Cauchy's theorem, we can shift the region of integration in (IX.7) so that

$$f(x) = (2\pi)^{-n/2} \int_{\mathbb{R}^n} e^{i(\lambda+i\eta)\cdot x} g(\lambda + i\eta)\,d\lambda \tag{IX.8}$$

Thus, by (IX.6)

$$|f(x)| \leq e^{R|\eta|-x\cdot\eta}(2\pi)^{-n/2}\int \frac{C_N}{(1+|\lambda+i\eta|)^N}\,d\lambda$$

$$\leq e^{R|\eta|-x\cdot\eta}(2\pi)^{-n/2}\int \frac{C_N}{(1+|\lambda|)^N}\,d\lambda$$

where we have chosen N large enough so that the integral on the right is finite. Now, $f(x)$ does not depend on η, so if we let $\eta \to \infty$ in an appropriate direction, we conclude that $|f(x)| = 0$ if $|x| > R$. ∎

This theorem has a natural generalization to the distributions with compact support. Recall that a distribution $T \in \mathscr{S}'(\mathbb{R}^n)$ has support in a closed set K if and only if $T(\varphi) = 0$ for every test function φ with support in $\mathbb{R}^n \backslash K$. If K is compact, then T is said to have **compact support**. The set of distributions with compact support is the dual space of $\mathscr{E}(\mathbb{R}^n)$ (see Problems 39 and 40 of Chapter V).

Theorem IX.12 A distribution $T \in \mathscr{S}'(\mathbb{R}^n)$ has compact support if and only if $\hat{T}$ has an analytic continuation to an entire analytic function of n variables $\hat{T}(\zeta)$ that satisfies

$$|\hat{T}(\zeta)| \leq C(1 + |\zeta|)^N e^{R|\operatorname{Im} \zeta|} \tag{IX.9}$$

for all $\zeta \in \mathbb{C}^n$ and some constants C, N, R. Moreover, if (IX.9) holds, the support of T is contained in the ball of radius R.

Proof Suppose that $T \in \mathscr{S}'(\mathbb{R}^n)$ has compact support and let φ be a $C_0^\infty(\mathbb{R}^n)$ function which is equal to one on the support of T. Define $F(\zeta) = T[(2\pi)^{-n/2} e^{-i\zeta \cdot x} \varphi(x)]$. By Theorem IX.5, $F(\lambda + i0)$ is the Fourier transform of T. Furthermore, since

$$\left(\frac{\exp(-i(x_j(\zeta_j + h_j) + \sum_{k \neq j} \zeta_k x_k))\varphi(x) - e^{-i\zeta \cdot x}\varphi(x)}{h_j} \right)$$

$$\xrightarrow{\mathscr{S}(\mathbb{R}^n)} -ix_j e^{-i\zeta \cdot x} \varphi(x)$$

and $T \in \mathscr{S}'(\mathbb{R}^n)$, $F(\zeta)$ is differentiable in the complex sense in each variable and is thus entire.

Since $T \in \mathscr{S}'(\mathbb{R}^n)$,

$$|T(f)| \leq C_1 \sum_{\substack{|\alpha| \leq N \\ |\beta| \leq N}} \|x^\alpha D^\beta f\|_\infty$$

for some N and C_1 and all $f \in \mathscr{S}(\mathbb{R}^n)$. Thus, if φ has support in the sphere of radius R, then

$$|F(\zeta)| \leq C_2 (1 + R^n)(1 + |\zeta|^N) e^{|\operatorname{Im} \zeta| R}$$

Conversely, suppose that $F(\zeta)$ is an entire function satisfying the estimate (IX.9). Then $F(\lambda + i0) \in \mathscr{S}'(\mathbb{R}^n)$, so it is the Fourier transform of some $T \in \mathscr{S}'(\mathbb{R}^n)$. Let $j_\varepsilon(x)$ be an approximate identity. Then by Theorem IX.4,

IX.3 The range of the Fourier transform: Analyticity

$\widehat{T * j_\varepsilon} = (2\pi)^{-n/2} \hat{j}_\varepsilon(\lambda) F(\lambda)$. Since j_ε has compact support in $\{x \mid |x| \leq \varepsilon\}$, we know by the Paley–Wiener theorem, that for each M we can find a constant C_M so that

$$|\hat{j}_\varepsilon(\zeta)| \leq \frac{C_M}{(1 + |\zeta|)^{N+M}} e^{\varepsilon |\mathrm{Im}\, \zeta|}$$

Therefore

$$|(2\pi)^{-n/2} \hat{j}_\varepsilon(\zeta) F(\zeta)| \leq \frac{C_M \, C e^{(R+\varepsilon)|\mathrm{Im}\, \zeta|}}{(1 + |\zeta|)^N}$$

which implies (again by the Paley–Wiener theorem) that the support of $T * j_\varepsilon$ is contained in the sphere of radius $R + \varepsilon$. Since ε is arbitrary and $(T * j_\varepsilon) \to T$ weakly, we conclude that the support of T is contained in the sphere of radius R about the origin. ∎

One natural way to extend the above theorems is to replace "compact support" with some weaker notion of decay at infinity. The following pair of theorems (whose proofs are outlined in Problem 76) will be used in Chapter XIII to prove the exponential decay of bound states of atomic Hamiltonians.

Theorem IX.13 Let f be in $L^2(\mathbb{R}^n)$. Then $e^{b|x|} f \in L^2(\mathbb{R}^n)$ for all $b < a$ if and only if $\hat{f}$ has an analytic continuation to the set $\{\zeta \mid |\mathrm{Im}\, \zeta| < a\}$ with the property that for each $\eta \in \mathbb{R}^n$ with $|\eta| < a$, $\hat{f}(\cdot + i\eta) \in L^2(\mathbb{R}^n)$ and for any $b < a$

$$\sup_{|\eta| \leq b} \|\hat{f}(\cdot + i\eta)\|_2 < \infty$$

Theorem IX.14 Let T be in $\mathscr{S}'(\mathbb{R}^n)$. Suppose that $\hat{T}$ is a function with an analytic continuation to the set $\{\zeta \mid |\mathrm{Im}\, \zeta| < a\}$ for some $a > 0$. Suppose further that for each $\eta \in \mathbb{R}^n$ with $|\eta| < a$, $\hat{T}(\cdot + i\eta) \in L^1(\mathbb{R}^n)$ and for any $b < a$, $\sup_{|\eta| < b} \|\hat{T}(\cdot + i\eta)\|_1 < \infty$. Then T is a bounded continuous function and for any $b < a$, there is a constant C_b so that

$$|T(x)| \leq C_b e^{-b|x|}$$

Paley–Wiener theorems are useful for understanding certain kinds of analytic completion theorems in the theory of several complex variables. The basic phenomenon is illustrated in the following simple example: Let D_r be the polydisk $\{\langle z, w\rangle \,|\, |z| < r, |w| < r\}$ in $\mathbb{C}^2$. Suppose that f is an analytic function of two variables in the "polyannulus" $D_1 \backslash \bar{D}_{\frac{1}{2}}$. Then for each z with $\frac{1}{2} < |z| < 1$, $g_z(w) \equiv f(z, w)$ is analytic in the unit disk and for each z with $|z| \leq \frac{1}{2}$, $g_z(w)$ is analytic in the annulus $\frac{1}{2} < w < 1$. Thus for each z there is a Laurant expansion

$$g_z(w) = \sum_{n=-\infty}^{\infty} a_n(z) w^n$$

with $a_n(z) = 0$ for $n < 0$ and $1 > |z| > \frac{1}{2}$. But the $a_n(z)$ are easily seen to be analytic functions of z so that $a_n(z) = 0$ for $n < 0$ and $|z| < 1$. It follows that f has a continuation from $D_1 \backslash \bar{D}_{\frac{1}{2}}$ to all of D_1! This is in striking contrast to the case of one variable where given any open connected set $\Omega \subset \mathbb{C}$, one can find an f that is analytic on Ω and on no larger set.

Definition An open connected set $\Omega \subset \mathbb{C}^n$ is called a **holomorphy domain** if and only if for any $w \notin \Omega$, there exists a function f analytic in Ω with no continuation to w. Given open connected sets $\Omega \subset \hat{\Omega} \subset \mathbb{C}^n$, we say that $\hat{\Omega}$ is the **analytic completion** or **holomorphy envelope** of Ω if and only if:

(i) $\hat{\Omega}$ is a holomorphy domain;
(ii) every function analytic in Ω has a continuation to all of $\hat{\Omega}$.

With this definition, not every open connected set will have an analytic completion, although if one extends the definition to allow $\hat{\Omega}$ to have a multisheeted structure, then every Ω does have a unique analytic completion. In the example above D_1 is the analytic completion of $D_1 \backslash \bar{D}_{\frac{1}{2}}$. We have already seen that D_1 obeys property (ii). To see that (i) holds, we note:

Proposition If $\Omega \subset \mathbb{C}^n$ is an open, connected, convex set, then Ω is a holomorphy domain.

Proof Given $w \notin \Omega$, we can find (by the Hahn–Banach theorem) a real linear functional ℓ from $\mathbb{C}^n \to \mathbb{R}$ so that $\ell(w) > \ell(z)$ for all $z \in \Omega$. Let

$$L(z) = \ell(z) - i\ell(iz)$$

and $f(z) = (L(z) - L(w))^{-1}$. Then f is analytic on Ω since Re $L(z) \neq$ Re $L(w)$ for $z \in \Omega$, but f is singular at w. ∎

Paley–Wiener ideas are useful in discussing the analytic completions of certain special domains.

IX.3 The range of the Fourier transform: Analyticity

Definition Let $S \subset \mathbb{R}^n$ be open. The **tube over** S, denoted by $\mathcal{T}(S)$, is defined by:

$$\mathcal{T}(S) = \{x + iy \,|\, x \in \mathbb{R}^n, y \in S\}$$

Theorem IX.14.1 (Bochner's tube theorem) Let $S \subset \mathbb{R}^n$ be open and let $\hat{S}$ be its (open) convex hull. Then $\mathcal{T}(\hat{S})$ is the analytic completion of $\mathcal{T}(S)$.

Sketch of Proof It follows from the proposition that each $\mathcal{T}(\hat{S})$ is a holomorphy domain, so it suffices to show property (i). We shall sketch a proof of the slightly weaker fact that any polynomially bounded function $f(z)$ on $\mathcal{T}(S)$ extends to $\mathcal{T}(\hat{S})$. Since we could replace f by, for eaxmple, $e^{-z^2}f(z)$, this is not really much of a restriction, and one can recover the full result from this special case. By translating, we can suppose, without loss, that $0 \in S$. Let $g_y(x) = f(x + iy)$. For each $y \in S$, g_y is a tempered distribution, so let $T_y = \check{g}_y$. By the Paley–Wiener method, $T_y(x) = e^{x \cdot y} T_0(x)$. Thus T_0 is a tempered distribution with $e^{x \cdot y} T_0(x)$ tempered for all $y \in S$. By the lemma below, this is true for all $y \in \hat{S}$ and thus, turning the Paley–Wiener argument around, $\hat{T}_0$ is analytic in $\mathcal{T}(\hat{S})$. ∎

Lemma Let $T \in \mathscr{S}'(\mathbb{R}^n)$. Then the set of y so that $e^{x \cdot y} T(x) \in \mathscr{S}'(\mathbb{R}^n)$ is a convex set.

Proof Since convexity is defined in terms of line segments, it is not hard to see that it suffices to show that if T and $e^\lambda T(\lambda)$ are in $\mathscr{S}'(\mathbb{R})$, then $e^{\theta\lambda} T(\lambda)$ is in $\mathscr{S}'(\mathbb{R})$ for all $\theta \in (0, 1)$. Let φ_n be the nth Hermite function of the Appendix to Section V.3 (where the sequence space representation of $\mathscr{S}$ and $\mathscr{S}'$ is discussed). Define

$$g_n(z) = \int T(\lambda) e^{iz\lambda} \varphi_n(\lambda) \, d\lambda$$

which is entire since $e^{iz\lambda}\varphi_n(\lambda) \in \mathscr{S}$. Since $T, e^\lambda T(\lambda) \in \mathscr{S}'$, we have that

$$|g_n(x)| \leq C(1 + n^2)^m (2 + x^2)^m$$
$$|g_n(x - i)| \leq C(1 + n^2)^m (2 + x^2)^m$$

for suitable C and m. Applying the maximum principle to $(2 + z^2)^{-m-1} g_n(z)$, we see that $|g_n(-i\theta)| \leq D(1 + n^2)^m$ for all θ in $(0, 1)$ so $e^{\theta\lambda} T(\lambda) \in \mathscr{S}'$. ∎

This method is ideal for dealing with certain "degenerate cases." In the result below the reader should keep in mind the case $S = \{\langle y_1, y_2 \rangle \,|\, y_1 = 0$ or $y_2 = 0; |y_1| + |y_2| < 1\}$, which is *not* an open set.

Theorem IX.14.2 (degenerate tube theorem) Let $T_1(x_1, z_2)$ and $T_2(z_1, x_2)$ be tempered distributions in the x variables ($x \in \mathbb{R}^n$) and polynomially bounded analytic functions in the z variables in $\{z \in \mathbb{C}^n \mid |\operatorname{Im} z| < 1\}$, i.e., for each z_2, $T_1(x_1, z_2)$ is a distribution in x_1 and $\int g(x_1) T(x_1, z_2)\, dx_1$ is analytic. Suppose that $T_1(x_1, x_2 + i0) = T_2(x_1 + i0, x_2)$ as distributions in the two variables. Then there exists a function $f(z_1, z_2)$ analytic in $\mathcal{T}(\tilde{S})$ with $\tilde{S} = \{\langle y_1, y_2 \rangle \mid |y_1| + |y_2| < 1\}$ so that

$$T_1(x_1, z_2) = f(x_1 + i0, z_2) \quad \text{and} \quad T_2(z_1, x_2) = f(z_1, x_2 + i0)$$

Proof Let $g(\lambda) = \check{T}_1(\cdot, \cdot + i0)$. Then, by hypothesis, $e^{\theta \lambda_1} g(\lambda)$ and $e^{\theta \lambda_2} g(\lambda)$ are in $\mathscr{S}'$ for $-1 < \theta < 1$ so $e^{a \cdot \lambda} g(\lambda) \in \mathscr{S}'$ for $|a_1| + |a_2| < 1$, proving the result. ∎

NOTES

Section IX.1 J. Fourier's original argument for the inversion formula appears in his classic *La Théorie Analytique de Chaleur*, Didot, Paris, 1822. Although his argument would not be considered a "rigorous proof" by modern standards, it contained the main ideas of the proof we have presented. The approach of first defining the Fourier transform on $\mathscr{S}(\mathbb{R}^n)$ and $\mathscr{S}'(\mathbb{R}^n)$ and then restricting to the classical L^p spaces is due to L. Schwartz and is described in *Théorie des Distributions*, Vol. II, Hermann, Paris, 1954. Schwartz's lucid book is the basic reference for the study of the Fourier transform on spaces of distributions and the theory of convolutions of distributions.

The Hermite expansion discussed in the Appendix to Section V.3 can be used to provide short proofs of the Fourier inversion and Plancherel theorems since $\hat{\varphi}_n(k) = (-i)^n \varphi_n(k)$.

Section IX.2 The Riemann–Lebesgue lemma was first proven for a restricted class of functions in B. Riemann, "Ueber der Darstellbarkeit einer Function durch einen trigonometrische Reihe" in *Math. Werke*, Teubner, 1876, pp. 213–253, and for all of L^1 in H. Lebesgue, "Sur les Séries Trigonométriques," *Ann. Sci. Ecole Norm. Sup.* **20** (1903), 453–485. The Plancherel theorem appears in M. Plancherel, "Contribution à l'étude de la représentation d'un fonction arbitraire par des intégrales définies," *Rend. Circ. Mat. Palermo* **30** (1910), 289–335. The Hausdorff–Young theorem was first proven in W. Young: "Sur la généralisation du théorème de Parseval," *C. R. Acad. Sci. Paris Sér. A-B* **155** (1912), 30–33; and extended in F. Hausdorff, "Eine Ausdehnung des Parselvalschen Satzes über Fourierreihen," *Math. Z.* **16** (1923), 163–169.

The original proof of Bochner's theorem appears in S. Bochner, *Vorlesungen über Fouriersche Integrale*, Akademie-Verlag, Berlin, 1932. For a proof of the generalization to distributions, see Schwartz's book. The proof of Bochner's theorem which we give depends on Stone's theorem. Conversely, it is possible to derive Stone's theorem from Bochner's theorem, see E. Hopf: *Ergodentheorie*, Springer-Verlag, Berlin, 1937, or F. Riesz and B. Sz.-Nagy: *Functional Analysis*, Ungar, New York, 1955.

In some sense, the "natural" setting for the L^p theory of the Fourier transform is on an arbitrary locally compact abelian group; see Chapters XIV and XV.

Section IX.3 The close relationship between the support properties of a function and the analyticity properties of its Fourier transform was first developed by R. Paley and N. Wiener in *Fourier Transforms in the Complex Domain*, Amer. Math. Soc. Colloqium Publication, Providence, Rhode Island, 1934. Their work concerned L^2 functions and L^2 boundary values (see below). Nevertheless, a whole class of theorems relating support properties to analyticity properties are usually called Paley–Wiener theorems. The connection between analyticity and the Fourier Transform was further studied by E. C. Titchmarsh, *Introduction to the Theory of Fourier Integrals*, Oxford Univ. Press (Clarendon), London and New York, 1937. The generalization to distributions with compact support was first proven in L. Schwartz, "Transformation de Laplace des distributions," *Comm. Sém. Math. Lund*, tome suppl. dédié à M. Riesz (1952). A more detailed relation between support and analyticity for functions supported on compact, convex, balanced sets is given in Problem 22.

For additional discussion of the phenomena of analytic completion, see: S. Bochner and W. T. Martin, *Several Complex Variables*, Princeton Univ. Press, Princeton, New Jersey, 1948; R. Gunning and H. Rossi, *Analytic Functions of Several Complex Variables*, Prentice-Hall, Englewood Cliffs, New Jersey, 1965; L. Nachbin, *Holomorphic Functions, Domains of Holomorphy and Local Properties*, North Holland, Amsterdam, 1970; L. Hörmander, *An Introduction to Complex Analysis in Several Variables*, Van Nostrand, Princeton, New Jersey, 1966. For applications to quantum field theory, see the Epstein article and Wightman's *J. Indian Math. Soc.* article quoted in the Notes to Section IX.8.

Bochner's tube theorem is due to S. Bochner, "A Theorem on Analytic Continuation of Functions in Several Variables," *Ann. Math.* **39** (1938), 14–19. The connection between tube theorems and the Paley–Wiener theorem has been noted by E. Stein and collaborators; see E. M. Stein and G. Weiss, *Introduction to Fourier Analysis on Euclidean Spaces*, Princeton Univ. Press, Princeton, New Jersey, 1971. In particular, the degenerate tube theorem was proven using these ideas by R. A. Kunze and E. M. Stein in "Uniformly Bounded Representations, II," *Amer. J. Math.* **83** (1961), 723–786 (see Lemma 21). Independently, Malgrange and Zerner proved this result using more classical methods; see H. Epstein's article.

PROBLEMS

1. Find the Fourier transform of $3x^2 + 1$.

†*2.* Give the details of the convergence of the Riemann sum to the integral at the end of the proof of Theorem IX.1.

3. (a) Let R be a rotation and R^t its transpose. Let $f \in \mathcal{S}$. Prove that $\widehat{f \circ R} = \hat{f} \circ R^t$.

(b) Let D_λ be the map $D_\lambda x = \lambda x$ on $\mathbb{R}^n$. Let $f \in \mathcal{S}(\mathbb{R}^n)$. Prove that

$$\widehat{f \circ D_\lambda} = \lambda^{-n} \hat{f} \circ D_{\lambda^{-1}}$$

(c) Let $T \in \mathcal{S}'(\mathbb{R}^n)$. Prove that

$$\widehat{(T \circ R)} = \hat{T} \circ R^t, \qquad \widehat{T \circ D_\lambda} = \lambda^{-n} \hat{T} \circ D_{\lambda^{-1}}$$

4. Compute the Fourier transform of $\mathcal{P}(1/x)$, the Cauchy principle part, by using Equation (V.4).

5. Compute the Fourier transform of $f(x) = e^{-\alpha x^2/2}$ as follows:
(a) Prove that $-\lambda \hat{f}(\lambda) = \alpha\, d\hat{f}(\lambda)/d\lambda$ and conclude that $\hat{f}(\lambda) = c e^{-\lambda^2/2\alpha}$

(b) Use the Plancherel theorem to prove $c = 1/\sqrt{\alpha}$.
(c) Check the Fourier inversion formula explicitly in this example.

6. Let $\mathscr{H} = L^2(\mathbb{R}, e^{-x^2}\,dx)$ and define $\psi_n = x^n \in \mathscr{H}$ for $n = 0, 1, \ldots$.
   (a) Prove that $\sum_{m=0}^{M} ((ik)^m/m!)\psi_m \xrightarrow[M\to\infty]{} e^{ikx}$ in the norm topology on $\mathscr{H}$.
   (b) Suppose $\eta \in \mathscr{H}$ and $(x^m, \eta) = 0$ for all m. Prove $\varphi = 0$. (*Hint*: show that $\eta e^{-x^2} = 0$.)
   (b') Reach the conclusion of (b) without recourse to the Fourier transform. (*Hint*: Use the fact that the functions $(x \pm i)^{-n}$ are total in $C_\infty(\mathbb{R})$ and the formula $(x + i)^{-1} = i\int_0^\infty e^{-s}e^{isx}\,ds$).
   (c) Let $\{H_n\}$ be the orthonormal set obtained from $\{\psi_n\}$ by Gram–Schmidt orthogonalization. Prove that $\{H_n\}$ is a basis for $\mathscr{H}$.
   (d) Prove that $\{H_n(x)e^{-x^2/2}\}_{n=0}^\infty$ is an orthonormal basis for $L^2(\mathbb{R}, dx)$.
   (e) Prove that $H_n(x)e^{-x^2/2}$ is just the nth Hermite function (defined in the Appendix to Section V.3).

7. Let $\{A_n(\lambda)\}$ be the polynomials determined by the formula

$$\sum_{n=0}^{\infty} A_n(\lambda) \frac{\alpha^n}{n!} = e^{-\alpha^2 + 2\alpha\lambda}$$

Define $\phi_n(\lambda) = (2^n n!)^{-1/2} A_n(\lambda) e^{-\lambda^2/2}$.
(a) Prove that

$$\phi_n(\lambda) = \frac{(-1)^n}{\sqrt{2^n n!}} e^{\lambda^2/2} \left(\frac{d}{d\lambda}\right)^n e^{-\lambda^2}$$

so that the $\phi_n(\lambda)$ are just the Hermite functions of the Appendix to Section V.3.
(b) If $f \in L^2(\mathbb{R}, dx)$ and $(f, \phi_n) = 0$ for all $n = 0, 1, \ldots$, prove that for all a,

$$\int f(x) e^{-(x-a)^2/2}\,dx = 0$$

(c) Use the Fourier transform to show that if $\int_\mathbb{R} f(x) e^{-(x-a)^2/2}\,dx = 0$ for all a, then $f = 0$.
(d) Conclude that $\{\phi_n\}$ is a basis for $L^2(\mathbb{R}, dx)$.

8. The purpose of this problem is to prove the Plancherel theorem and the inversion formula by using the Hermite functions. Let

$$A = \frac{1}{\sqrt{2}}\left(x + \frac{d}{dx}\right) \quad \text{and} \quad A^\dagger = \frac{1}{\sqrt{2}}\left(x - \frac{d}{dx}\right)$$

(a) If $f \in \mathscr{S}(\mathbb{R}^n)$ prove $\widehat{A^\dagger f}(\lambda) = -i(A^\dagger \hat{f})(\lambda)$.
(b) Prove that $\hat{\phi}_n = (-i)^n \phi_n$.
(c) Supposing the fact that the Hermite functions are a basis for $L^2(\mathbb{R})$, prove the Plancherel theorem and the inversion formula.

9. Suppose that C is a continuous map of $\mathscr{S}(\mathbb{R}^n)$ into $C^\infty(\mathbb{R}^n)$ which commutes with translations. Prove that there is a $T \in \mathscr{S}'(\mathbb{R}^n)$ so that $C(\varphi) = T * \varphi$, for all $\varphi \in \mathscr{S}(\mathbb{R}^n)$. (*Hint*: If $T \in \mathscr{S}'(\mathbb{R}^n)$ and $\varphi \in \mathscr{S}(\mathbb{R}^n)$, then $T(\varphi) = (T * \tilde{\varphi})(0)$.)

10. Prove directly (without using the Fourier transform) that for fixed $f \in \mathscr{S}(\mathbb{R}^n)$, the map $g \mapsto f * g$ is a continuous linear transformation of $\mathscr{S}(\mathbb{R}^n)$ into $\mathscr{S}(\mathbb{R}^n)$.

11. Let μ_1 and μ_2 be finite Borel measures on $\mathbb{R}^n$ and define

$$(\mu_1 * \mu_2)(E) = \int_{\mathbb{R}^n} \mu_1(E - y)\, d\mu_2(y)$$

(a) Prove that $\mu_1 * \mu_2$ is a finite Borel measure on $\mathbb{R}^n$, that $\mu_1 * \mu_2 = \mu_2 * \mu_1$, and that for any $f \in C_\infty(\mathbb{R}^n)$

$$(\mu_1 * \mu_2)(f) = \int_{\mathbb{R}^n} \int_{\mathbb{R}^n} f(x + y)\, d\mu_1(x)\, d\mu_2(y)$$

(b) Prove that $\mu_1 * \mu_2$ is absolutely continuous with respect to Lebesgue measure if either μ_1 or μ_2 is absolutely continuous with respect to Lebesgue measure. Give an example where $\mu_1 * \mu_2$ is not absolutely continuous.

12. The Fourier transforms of Borel measures of mass one on $\mathbb{R}^n$ are sometimes called "characteristic functions." A characteristic function $E(\lambda)$ is said to be **infinitely divisible** if for all positive integers n, there exists a characteristic function $E_n(\lambda)$ so that $E(\lambda) = (E_n(\lambda))^n$.

(a) Let μ be a Borel measure of mass one on $\mathbb{R}$ and let E be the corresponding characteristic function. Prove that E is infinitely divisible if and only if for all n there is a Borel measure of mass one, μ_n, so that

$$\mu = \underbrace{\mu_n * \mu_n * \cdots * \mu_n}_{n \text{ times}}$$

(b) Show that

$$E(\lambda) = \exp(i\alpha\lambda - \tfrac{1}{2}\beta\lambda^2) + \int_{-\infty}^{\infty} (e^{i\lambda x} - 1)\, d\rho$$

is an infinitely divisible characteristic function if $\alpha \in \mathbb{R}$, $\beta \geq 0$, and ρ is a Borel measure of finite mass on $\mathbb{R}$. Give a (convolution) formula for the corresponding measure in terms of α, β, and ρ. What is the corresponding measure if $\rho = 0$? What is the corresponding measure if $\alpha = 0 = \beta$ and $\rho = \delta(x - x_0)$?

Remark: There is a characterization of the Fourier transforms of all infinitely divisible distributions known as the Lévy-Khinchin formula. See, for example, L. Breiman, *Probability*, Addison Wesley, Reading, Massachusetts, 1968, 193–195.

13. Let Ω be an open set in $\mathbb{R}^n$, K a compact subset of Ω. Prove that there is a function in $C_0^\infty(\Omega)$ which is equal to one on K. (*Hint*: See Problem 61 of Chapter V.)

14. The purpose of this exercise is to prove the Fourier inversion formula by an alternative method. Suppose that $f \in \mathscr{S}(\mathbb{R}^n)$.
(a) Prove that $\lim_{\varepsilon \downarrow 0} \int_\varepsilon^{1/\varepsilon} ((\sin x)/x)\, dx$ exists. Call it d. Now, show that

$$\lim_{\varepsilon \downarrow 0} \int_\varepsilon^{1/\varepsilon} \frac{\sin Rx}{x}\, dx = d \qquad \text{for any } R > 0$$

(*Hint*: Use facts about telescoping series.)

(b) Prove that

$$\int_0^\infty \left[\frac{f(y-u)+f(y+u)}{2} - f(y)\right] \frac{\sin Ru}{u} \, du \xrightarrow[R\to\infty]{} 0$$

(Use the Riemann-Lebesgue lemma.)

(c) Using (b) conclude that

$$4df(y) = \lim_{R\to\infty} \int_{-\infty}^\infty \left(\int_{-R}^R e^{i(y-x)\cdot k} f(k) \, dk\right) dx$$

(d) Prove that $f(y) = (\sqrt{2\pi}/4d) \int_{-\infty}^\infty e^{iky} \hat{f}(y) \, dk$.

(e) By letting $f(x) = e^{-x^2/2}$, conclude that $d = \pi/2$.

15. The purpose of this exercise is to provide an alternative proof of the Plancherel theorem.
    (a) Prove directly that if $f, g \in \mathscr{S}(\mathbb{R}^n)$,

$$\widehat{f * g} = (2\pi)^{n/2} \hat{f} \hat{g}$$

(b) Letting $\tilde{f}(x) = f(-x)$, prove that

$$(f * \tilde{f})(y) = \int |\hat{f}(k)|^2 e^{iky} \, dk$$

(c) Set $y = 0$ and conclude that

$$\int |f(x)|^2 \, dx = \int |\hat{f}(k)|^2 \, dk$$

*16. Prove that the map $L^1(\mathbb{R}^n) \xrightarrow{\hat{}} C_\infty(\mathbb{R}^n)$ is not onto by exhibiting a function in $C_\infty(\mathbb{R}^n)$ which is not in its range.

17. The purpose of this problem is to develop the Fourier transform on $L^1(\mathbb{R}^n)$ without reference to $\mathscr{S}(\mathbb{R}^n)$.
    (a) If $f \in L^1(\mathbb{R}^n)$, prove directly that

$$\hat{f}(\lambda) = (2\pi)^{-n/2} \int_{\mathbb{R}^n} e^{-i\lambda x} f(x) \, dx$$

is a bounded continuous function. (*Hint*: Use the dominated convergence theorem.)

(b) If $f \in L^1(\mathbb{R}^n)$, prove that $\hat{f}(\lambda) \to 0$ as $|\lambda| \to \infty$. (*Hint*: Prove that $2\hat{f}(\lambda) = (2\pi)^{-n/2} \int e^{-ix\lambda}(f(x) - f(x - \pi\lambda/|\lambda|^2)) \, dx$.

(c) Prove directly that $(2\pi)^{n/2} \hat{f} \hat{g} = \widehat{f * g}$.

18. Find a function $f(x)$ that satisfies all the conditions in the definition of "functions of positive type" except continuity. To which function of positive type is $f(x)$ equal a.e.?

19. Display a distribution of positive type that is not a function. What is its Fourier transform?

*20. Prove the Bochner-Schwartz theorem (Theorem IX.10). (*Hint*: Mimic our proof of Bochner's theorem using the inner product $(\varphi, \psi) = T(\tilde{\overline{\varphi}} * \psi)$ and the formula $T(\tilde{\overline{\varphi}} * \varphi_x) = (T * \overline{\varphi} * \tilde{\varphi})(x)$.)

21. What does the generalization of the Paley–Wiener theorem to distributions with compact support say about the Fourier transform of distributions with support at the origin? Compute the same result directly by using Theorem V.11.

22. Let C be a convex, compact, balanced set in $\mathbb{R}^n$. Let
$$C^\circ = \{k \mid k \cdot x \geq -1 \text{ for all } x \in C\}$$
be its polar. Let ρ be the Minkowski functional of C°, i.e.,
$$\rho(\eta) = \sup_{x \in C}(\eta \cdot x) = \inf_{\lambda > 0}\{\lambda \mid \lambda \in \mathbb{R}, \lambda\eta \in C^\circ\}$$
Prove the following version of the Paley–Wiener theorem:

A function $f \in \mathscr{S}(\mathbb{R}^n)$ has support in C if and only if $\hat{f}$ is the restriction to $\mathbb{R}^n$ of an entire function $\hat{f}(z)$ which obeys the condition that for any n there is a constant D_n so that
$$|\hat{f}(z)| \leq D_n(1 + |z|^2)^{-n} e^{-\rho(\operatorname{Im} z)}$$

Supplementary Material

Supplement to II.2   Applications of the Riesz lemma

There are a number of applications of the Riesz lemma (Theorem II.4) illustrating the power of abstract methods.

Application 1 (Von Neumann's proof of the Radon–Nikodym theorem) The following result implies both the Radon–Nikodym theorem (Theorem I.19) and the Lebesgue decomposition theorem (Theorem I.20) when all of the underlying measures are finite. To go from the finite to the σ-finite case is not hard (Problem 1).

Theorem S.1   Let μ and ν be finite measures on a measure space $\langle M, \mathcal{R} \rangle$. Then, there exists a set $A \in \mathcal{R}$ with $\nu(M \setminus A) = 0$ and $f \in L^1(M, d\nu)$ so that $f \geq 0$ and
$$\mu(B) = \int_B f(x)\, d\nu(x)$$
for all $B \subset A$.

Applications of the Riesz lemma 345

Proof Let α be the measure $\mu + \nu$. Define the linear functional L on $L^2(M, d\alpha)$ by

$$L(g) = \int g \, d\mu$$

By the Schwarz inequality,

$$|L(g)| \leq \left(\int |g|^2 \, d\mu\right)^{\frac{1}{2}} \mu(M)^{\frac{1}{2}}$$

$$\leq \left(\int |g|^2 \, d\alpha\right)^{\frac{1}{2}} \mu(M)^{\frac{1}{2}}$$

so L is a bounded linear functional on $L^2(M, d\alpha)$. Thus, by the Riesz lemma, there is a function F with

$$\mu(C) = L(\chi_C) = \int_C F(x) \, d\alpha(x) \tag{S.1}$$

Fix $\gamma > 1$ and let $C_\gamma = \{x \mid F(x) \geq \gamma\}$. Then, by (S.1),

$$\mu(C_\gamma) \geq \gamma(\mu(C_\gamma) + \nu(C_\gamma))$$

from which it follows that

$$\mu(C_\gamma) = \nu(C_\gamma) = 0$$

so that $F \leq 1$ a.e. (with respect to $\alpha, \mu,$ and ν). Since (S.1) holds with ν replacing μ and $1 - F$ replacing F, we see that $F \geq 0$ a.e. also.

Let

$$A_n = \{x \mid 1 - n^{-1} \leq F(x) < 1 - (n+1)^{-1}\}$$

and

$$A = \bigcup_{n=1}^{\infty} A_n$$

Then, since $F(x) = 1$ on $M \setminus A$,

$$\mu(M \setminus A) = \int_{M \setminus A} F \, d\alpha = \mu(M \setminus A) + \nu(M \setminus A)$$

so $\nu(M \setminus A) = 0$.

Define $g_n(x) = F(x)(1 - F(x))^{-1}$ for $x \in A_n$, and $g_n(x) = 0$ for $x \notin A_n$. Then for any B and $h = \chi_B$, the characteristic function of B,

$$\int g_n h \, d\nu = \int h F (1-F)^{-1} \chi_{A_n} (1-F) \, d\alpha$$

$$= \int h F \chi_{A_n} \, d\alpha = \int h \chi_{A_n} \, d\mu$$

$$= \mu(B \cap A_n)$$

Taking $B = M$, we see that $\sum_{n=0}^{\infty} \int g_n \, d\nu = \mu(A) < \infty$, so $f = \sum_{n=0}^{\infty} g_n$ is in L^1 by the monotone convergence theorem. Moreover, if $B \subset A$,

$$\int_B f \, d\nu = \sum_n \int g_n h \, d\nu = \sum_n \mu(B \cap A_n) = \mu(B) \quad \blacksquare$$

Application 2 (the Bergmann reproducing kernel) Let Ω be a bounded open subset of $\mathbb{C}$. We want to introduce a natural object associated with the analytic structure of Ω. Write $z = x + iy$ and let $d^2 z = dx \, dy$. Let $\mathscr{A}(\Omega)$ be the set of functions $f \in L^2(\Omega, d^2 z)$ that are analytic in Ω.

Lemma For all $f \in \mathscr{A}(\Omega)$ and $z \in \Omega$,

$$|f(z)| \leq \pi^{-\frac{1}{2}} [\text{dist}(z, \partial \Omega)]^{-1} \|f\|_2 \tag{S.2}$$

and $\mathscr{A}(\Omega)$ is a closed subset of $L^2(\Omega)$.

Proof First, suppose $\Omega = \{z \,|\, |z| < 1\}$. Then, by the Cauchy integral formula,

$$f(0) = \frac{1}{2\pi} \int_0^{2\pi} f(re^{i\theta}) \, d\theta$$

for any $r < 1$. It follows that

$$|f(0)| = \frac{1}{\pi} \left| \int_0^1 \int_0^{2\pi} f(re^{i\theta}) r \, dr \, d\theta \right|$$

$$\leq \|f\|_2 \pi^{-1} \pi^{\frac{1}{2}} = \pi^{-\frac{1}{2}} \|f\|_2$$

By scaling and translation, if $\{z \,|\, |z - z_0| < r\} \subset \Omega$, then

$$|f(z_0)| \leq \pi^{-\frac{1}{2}} r^{-1} \|f\|_2$$

so that (S.2) follows.

(S.2) implies that whenever $f_n \to f$ in L^2 and $f_n \in \mathscr{A}$, then f_n converges to f uniformly on compact subsets of Ω. Therefore, by a theorem of Weierstrass (which follows from the Cauchy integral formula), f is in $\mathscr{A}$. ∎

Theorem S.2 Let Ω be a bounded open subset of $\mathbb{C}$. Then, there exists a function $K(z, w)$ on $\Omega \times \Omega$ so that:

(a) K is analytic in z and $\bar{w}$
(b) $|K(z, w)| \leq \pi^{-1}[\mathrm{dist}(z, \partial\Omega)]^{-1}[\mathrm{dist}(w, \partial\Omega)]^{-1}$
(c) $\overline{K(z, w)} = K(w, z)$
(d) For any fixed z, $\overline{K(z, w)} \in \mathscr{A}(\Omega)$
(e) For any $f \in \mathscr{A}(\Omega)$

$$f(z) = \int_\Omega K(z, w) f(w) \, d^2w \tag{S.3}$$

Proof By the lemma, $\mathscr{A}(\Omega)$ is a Hilbert space. By (S.2), $f \to f(z)$ is a bounded linear functional on $\mathscr{A}$ for each fixed $z \in \Omega$. So, by the Riesz lemma, there exists $h_z \in \mathscr{A}(\Omega)$ such that

$$f(z) = (h_z, f) \tag{S.4}$$

for any $f \in \mathscr{A}$ and

$$\|h_z\| \leq \pi^{-\frac{1}{2}}[\mathrm{dist}(z, \partial\Omega)]^{-1} \tag{S.5}$$

Let P be the orthogonal projection from $L^2(\Omega)$ to $\mathscr{A}(\Omega)$ (projections are defined in Section VI.2, but their existence depends only on Theorem II.3). Since $h_z \in \mathscr{A}$, for any $g \in L^2$,

$$(h_z, g) = (Ph_z, g) = (h_z, Pg) = (Pg)(z)$$

It follows (using Theorem VI.4) that $z \to h_z$ is antianalytic in z. Now define

$$K(z, w) = (h_z, h_w)$$

Using (S.5) and the above, (a)–(c) are obvious. Moreover, by (S.4),

$$K(z, w) = \overline{(h_w, h_z)} = \overline{h_z(w)}$$

so (d) is evident and (S.4) implies (S.3). ∎

Because of (S.3), K is called the **Bergmann reproducing kernel** for Ω.

Example Let $\{g_n(z)\}_{n=0}^{\infty}$ be an orthonormal basis for $\mathscr{A}(\Omega)$ (orthonormal bases are defined in Section II.3). Then

$$K(z, w) = \sum_{n=0}^{\infty} (h_z, g_n)(g_n, h_w)$$

$$= \sum_{n=0}^{\infty} g_n(z)\overline{g_n(w)} \quad \text{(S.6)}$$

By Theorem II.6, the expression (S.6) converges to $K(z, w)$ for each fixed z and w. But, since the finite sums are uniformly bounded on compact subsets and analytic in z and $\bar{w}$, the Vitali convergence theorem implies that (S.6) is uniformly convergent on compact subsets of Ω.

Consider the case $\Omega = \{z \mid |z| < 1\}$. Then $\{z^n(2n+2)^{\frac{1}{2}}(2\pi)^{-\frac{1}{2}}\}_{n=0}^{\infty}$ is an orthonormal set, and it is not hard to see that it is a basis (Problem 2). Thus, by (S.6),

$$K(z, w) = \sum_{n=0}^{\infty} \pi^{-1}(n+1)(z\bar{w})^n = \pi^{-1}(1 - z\bar{w})^{-2}$$

in the case where Ω is the unit disk.

Supplement to III.1 Basic properties of L^p spaces

In this supplement we prove some of the basic facts about L^p spaces which are presented without proof in Sections III.1 and III.2.

Theorem S.3 (Hölder's inequality) Let $1 < p < \infty$, $p^{-1} + q^{-1} = 1$. Let $f \in L^p$, $g \in L^q$. Then $fg \in L^1$ and

$$\|fg\|_1 \leq \|f\|_p \|g\|_q \quad \text{(S.7)}$$

Moreover, if neither f nor g are a.e. zero, equality holds in (S.7) if and only if $|g|$ is a.e. a constant multiple of $|f(x)|^{p-1}$.

Proof We begin by proving an inequality on positive numbers:

$$ab \leq a^p p^{-1} + b^q q^{-1} \quad \text{(S.8)}$$

To prove this, consider the graph of the function

$$a = F(b) \equiv b^{q-1}$$

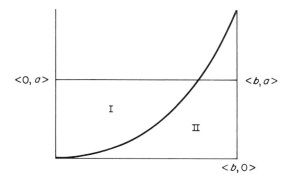

FIGURE S.1 The graph of $a = b^{q-1}$, $b = a^{p-1}$.

or equivalently

$$b = F^{-1}(a) = a^{p-1}$$

since $(p-1)(q-1) = 1$. (S.8) is geometrically obvious from Figure S.1: for the area of region I is clearly equal to $\int_0^a a^{p-1}\, da = a^p/p$ and the area of region II is less than $\int_0^b b^{q-1}\, db = b^q/q$.

Now we can prove (S.7). It is easy to prove if f or g is a.e. zero. By replacing f by $f\|f\|_p^{-1}$ and g by $g\|g\|_q^{-1}$, we need only prove (S.7) when $\|f\|_p = \|g\|_q = 1$. Then, by (S.8),

$$\int |f(x)g(x)|\, dx \leq p^{-1}\int |f(x)|^p\, dx + q^{-1}\int |g(x)|^q\, dx$$

$$= p^{-1} + q^{-1} = 1$$

This proves (S.7). That equality holds when $|g(x)| = |f(x)|^{p-1}$ is easy. That it only holds in this case is left to the reader (Problem 4). ∎

Corollary 1 Let $f \in L^p$, $1 \leq p \leq \infty$. Then

$$\|f\|_p = \sup\{\|fg\|_1 \,|\, \|g\|_q = 1\} \tag{S.9}$$

Proof. That $\|f\|_p$ is greater than or equal to the sup follows from (S.7). To get equality, take g to be a suitable multiple of $|f(x)|^{p-1}$. ∎

Corollary 2 (Minkowski's inequality) Let $1 \leq p \leq \infty$. Then

$$\|f+g\|_p \leq \|f\|_p + \|g\|_p$$

Proof By (S.9) and the triangle inequality for $\|\cdot\|_1$,

$$\begin{aligned}\|f+g\|_p &= \sup\{\|(f+g)h\|_1 \,|\, \|h\|_q = 1\}\\ &\leq \sup\{\|fh\|_1 + \|gh\|_1 \,|\, \|h\|_q = 1\}\\ &\leq \sup\{\|fh\|_1 \,|\, \|h\|_q = 1\} + \sup\{\|gh\|_1 \,|\, \|h\|_q = 1\}\\ &= \|f\|_p + \|g\|_p \quad \blacksquare\end{aligned}$$

Theorem S.4 Let $\langle X, \mu\rangle$ be a σ-finite measure space, $1 \leq p < \infty$, and L be an element of $L^p(X, d\mu)^*$. Then there exists $g \in L^q(X, d\mu)$ with $\|g\|_q = \|L\|_{(L^p)^*}$ so that

$$L(f) = \int g(x) f(x)\, d\mu(x) \tag{S.10}$$

Thus $(L^p)^*$ is isometrically isomorphic to L^q.

Proof By Theorem S.3, the L in (S.10) is a bounded linear functional on L^p if $g \in L^q$ with norm $\|L\|_{(L^p)^*} \leq \|g\|_q$. In fact, $\|L\|_{(L^p)^*} = \|g\|_q$ by Corollary 1. Thus, we need only prove the existence of g, given L.

Suppose that $\mu(X) < \infty$ and that we are given an $L \in (L^p)^*$ which is **positive** in the sense that $L(f) \geq 0$ for $f \geq 0$. Define a function ν on measurable sets by

$$\nu(A) = L(\chi_A)$$

where χ_A is the characteristic function of A. Since $p < \infty$, if $A = \bigcup_{n=1}^\infty A_n$ with A_i disjoint, then $\sum_{n=0}^N \chi_{A_n} \to \chi_A$ in L^p by the monotone convergence theorem. Thus

$$\nu(A) = \lim_{N \to \infty} L\left(\sum_{n=1}^\infty \chi_{A_n}\right) = \sum_{n=1}^\infty \nu(A_n)$$

so ν defines a measure. If $\mu(A) = 0$, then $\chi_A = 0$ in L^p so $\nu(A) = 0$. Therefore, by the Radon–Nikodym theorem, there exists $g \in L^1$ with $g \geq 0$ so that

$$L(\chi_A) = \int g(x) \chi_A(x)\, d\mu(x)$$

for any A. Now let h be bounded and positive. One can find h_n so that $h_n(x) \uparrow h(x)$, $\|h_n - h\|_p \to 0$, and each h_n is a finite linear combination of χ_A's. Thus, by the monotone convergence theorem,

$$L(h) = \int h(x) g(x)\, d\mu(x)$$

Fix K and let $h(x) = |g(x)|^{q-1}$ (resp. 0) if $|g(x)| \leq K$ (resp. $|g(x)| > K$). Then

$$\int_{|g(x)| \leq K} |g(x)|^q \, d\mu(x) = L(h) \leq \|L\| \left(\int_{|g(x)| \leq K} |g(x)|^{(q-1)p} \, d\mu(x) \right)^{1/p}$$

$$= \|L\| \left(\int_{|g(x)| \leq K} |g(x)|^q \, d\mu(x) \right)^{1 - 1/q}$$

Therefore,

$$\int_{|g(x)| \leq K} |g(x)|^q \, d\mu(x) \leq \|L\|^q$$

Since $\|L\| < \infty$, and g is a.e. finite, we conclude that $g \in L^q$.
This proves the result if L is positive and $\mu(X) < \infty$. The argument in the proof of Theorem IV.16 shows that any $L \in (L^p)^*$ is a finite linear combination of positive L's (Problem 6), and it is also easy to extend the result to the σ-finite case (Problem 7). ∎

Supplement to IV.3   Proof of Tychonoff's theorem

In this supplement we provide a proof of Tychonoff's theorem, Theorem IV.5, fleshing out the proof sketched in the notes to Section IV.3.

Definition   A net $\{x_\alpha\}_{\alpha \in I}$ of points in some set X is called a **universal net** if and only if for *every* subset $A \subset X$ either eventually $x_\alpha \in A$ or else eventually $x_\alpha \in X \setminus A$.

Notice that the notion of universal net makes no mention of topology. Also notice that this is a very strange notion. For example, the sequence $\{x_n = 1/n\}$ is not universal, for suppose that $A = \{1/2n\}$. In fact, no *sequence* which is not eventually constant is a universal net (Problem 8). The point of the definition is the following obvious result.

Lemma 1   Any cluster point of a universal net is the limit of the net.

Universal nets are so strange that it is not clear that any exist. Indeed, the existence of a universal subnet for the sequence $\{x_n = n\}_{n \in \mathbb{Z}}$ is equivalent to the axiom of choice. Using this axiom, we can construct lots of universal nets.

Clearly, for every $A \subset X$, we must make a choice of either A or $X \backslash A$. These choices are not quite independent, since if x_α is eventually in both A and B, then it is eventually in $A \cap B$ so $A \cap B \neq \emptyset$. This motivates:

Definition A **filter** is a family $\mathscr{F}$ of subsets of X obeying:

(i) $\emptyset \in \mathscr{F}$;
(ii) if $A, B \in \mathscr{F}$, then $A \cap B \in \mathscr{F}$;
(iii) if $A \subset B$ and $A \in \mathscr{F}$, then $B \in \mathscr{F}$.

An **ultrafilter** is a maximal filter, i.e., a filter that is not strictly contained in any other filter.

Example 1 Fix x_0. Then $\{A \,|\, x_0 \in A\}$ is an ultrafilter.

Example 2 Let $\{x_\alpha\}_{\alpha \in I}$ be any net. Then $\mathscr{F} = \{A \,|\, x_\alpha \text{ is eventually in } A\}$ is a filter.

Lemma 2 Any filter is contained in an ultrafilter.

Proof A simple application of Zorn's lemma. ∎

The following lemma shows the connection between ultrafilters and universal nets.

Lemma 3 If $\mathscr{U}$ is an ultrafilter, then for any $A \subset X$, either $A \in \mathscr{U}$ or $X \backslash A \in \mathscr{U}$.

Proof Suppose that $\mathscr{F}$ is any filter and $B \subset X$ obeys $B \cap C \neq \emptyset$ for all $C \in \mathscr{F}$. Then $\{C \cap D \,|\, C \in \mathscr{F}, B \subset D\}$ is a filter containing $\mathscr{F}$ and B. Therefore, if $\mathscr{U}$ is an ultrafilter, and $B \notin \mathscr{U}$, then there is a $C \in \mathscr{U}$ with $B \cap C = \emptyset$.

Thus if $A \notin \mathscr{U}$ and $X \backslash A \notin \mathscr{U}$, we can find $C, D \in \mathscr{U}$ with $A \cap C = \emptyset$ and $(X \backslash A) \cap D = \emptyset$. Therefore,

$$\begin{aligned} C \cap D &= C \cap D \cap X \\ &= [(C \cap D) \cap A] \cup [(C \cap D) \cap (X \backslash A)] \\ &= \emptyset \end{aligned}$$

which is not allowed since C and D are in $\mathscr{U}$. Thus, to avoid contradiction, either $A \in \mathscr{U}$ or $X \backslash A \in \mathscr{U}$. ∎

The following lemma contains all the hard work in the proof of Tychonoff's theorem.

Lemma 4 Any net has a universal subnet.

Proof Let $\{x_\alpha\}_{\alpha \in I}$ be the net and let $\mathscr{U}$ be an ultrafilter containing the filter, $\mathscr{F}$ of Example 2. Let $J = \{\langle \alpha, A \rangle \mid \alpha \in I, A \in \mathscr{U}, x_\alpha \in A\}$. Order J by $\langle \alpha, A \rangle > \langle \beta, B \rangle$ if $\alpha > \beta$ and $A \subset B$. The following argument shows that J is directed: If $A \in \mathscr{U}$, then x_α is frequently in A for if it were not, then $X \setminus A \in \mathscr{F} \subset \mathscr{U}$. Thus, given any A, B, α, β, there is $\gamma > \alpha$, β so that $x_\gamma \in A \cap B$, i.e., J is directed. Let $F: J \to I$ by $F(\alpha, A) = \alpha$. Then $y_{\langle \alpha, A \rangle} = x_\alpha$ is a subnet of $\{x_\alpha\}_{\alpha \in I}$. Moreover, given any $A \in \mathscr{U}$, find α with $x_\alpha \in A$. If $\langle \beta, B \rangle > \langle \alpha, A \rangle$, then $y_{\langle \beta, B \rangle} \in A$ so y is eventually in A. By Lemma 3, we conclude that y is a universal net. ∎

Lemma 5 A Hausdorff topological space X is compact if and only if all universal nets converge.

Proof If X is compact, then by the Bolzano–Weierstrass theorem, any universal net has a cluster point and so, by Lemma 1, a limit. Conversely, by Lemma 4, if all universal nets converge, then every net has a convergent subnet so X is compact by the Bolzano–Weierstrass theorem. ∎

Proof of Tychonoff's theorem (Theorem IV.5) Let $\{x^{(\beta)}\}_{\beta \in I}$ be a universal net in $X = \bigtimes X_\alpha$ where each X_α is compact. Then $\{x_\alpha^{(\beta)}\}_{\beta \in I}$ is universal in X_α, so by Lemma 5, $x_\alpha^{(\beta)} \to x_\alpha^{(\infty)}$ for some $x_\alpha^{(\infty)} \in X_\alpha$. Thus, there is $x^{(\infty)}$ in X with $x^{(\beta)} \to x^{(\infty)}$. By Lemma 5 again, X is compact. ∎

Supplement to IV.4 The Riesz–Markov theorem for $X = [0, 1]$

Functions of bounded variation and Riemann–Stieltjes integrals are defined in Problem 11 of Chapter 1. In this supplement we prove:

Theorem S.5 Let $L \in (C_{\mathbb{R}}[0, 1])^*$. Then, there exists a function α of bounded variation so that

$$L(f) = \int_0^1 f \, d\alpha$$

for any $f \in C_{\mathbb{R}}[0, 1]$.

The questions of the uniqueness of α and the positivity of $d\alpha$ if L is positive are left to the reader (Problems 9 and 10). Our proof is very special in that it does not extend naturally to more general spaces. We give it for two reasons: Most applications that we give, in particular the spectral theorem, require only the case $X = [0, 1]$ or the case $X = (-\infty, \infty)$ which is a simple extension. Secondly, the proof is a nice illustration of the power of the Hahn–Banach theorem.

Proof of Theorem S.5 Let $PC_\mathbb{R}[0, 1]$ be the space defined in Section I.2. By the Hahn–Banach theorem, L extends to a bounded linear functional $\tilde{L}$ on $PC_\mathbb{R}[0, 1]$. Let $\chi_{[x, y)}$ denote the characteristic function of $[x, y)$. Define

$$\alpha(x) = \tilde{L}(\chi_{[0, x)}) \quad \text{if} \quad x < 1$$
$$\alpha(1) = \tilde{L}(\chi_{[0, 1]})$$

We first claim that α is of bounded variation. Fix $0 = x_0 < x_1 < \cdots < x_n = 1$ and let $\chi_i = \chi_{[x_{i-1}, x_i)}$ except for $i = n$, for which we take $\chi_n = \chi_{[x_{n-1}, x_n]}$. Then, for any $a_1, \ldots, a_n \in \mathbb{R}$,

$$|\sum a_i[\alpha(x_i) - \alpha(x_{i-1})]| = |\tilde{L}(\sum a_i \chi_i)|$$
$$\leq \|\tilde{L}\| \sup_i |a_i|$$

From this it follows that

$$\sum |\alpha(x_i) - \alpha(x_{i-1})| \leq \|\tilde{L}\|$$

by taking each $a_i = \pm 1$. Thus α has bounded variation.

Now, given $f \in C[0, 1]$, let $\chi_i^{(n)}$ be the above functions for $x_i = i/n$ and let

$$f_n = \sum_{i=1}^n f\left(\frac{i}{n}\right) \chi_i^{(n)}$$

Then, $f_n \xrightarrow{PC[0,1]} f$. Moreover, $\tilde{L}(f_n) = \int f_n \, d\alpha$ by construction. Thus

$$L(f) = \lim_n \tilde{L}(f_n) = \lim_n \int f_n \, d\alpha = \int f \, d\alpha \quad \blacksquare$$

Supplement to IV.5 Minimization of functionals

In this supplement we want to show how the Banach–Alaoglu theorem can be used to prove that certain functionals on Banach spaces take on their minimum value. Such minimization problems often occur, for example, in

Minimization of functionals 355

the problem of finding geodesics, i.e., curves of minimum length. They also enter in the context of solving equations of the form

$$T(u) = 0 \quad (S.11)$$

where T is a map from X to X^*. For suppose that T is the gradient of a map F from X to $\mathbb{R}$ in the sense that

$$\left.\frac{d}{d\lambda} F(u + \lambda v)\right|_{\lambda=0} = (T(u))(v)$$

Then, if u minimizes F, it obeys (S.11).

An attractive method for showing that a function takes its minimum value is to show that it is a continuous function on a compact set. For this, the norm topology is not usually suitable, since only rather small subsets of Banach spaces are compact in the norm topology. On the other hand, the Banach–Alaoglu theorem suggests that, when X is reflexive, the weak topology is an attractive possibility. The only problem with this is that lots of natural functions are *not* weakly continuous. For example, the norm is not weakly continuous for if it were, then $\{x \mid \|x\| = 1\}$ would be weakly closed and it is not (Problem 40 of Chapter IV). However, if $x_n \to x$ weakly, then

$$\|x\| = \sup\{|\ell(x)| \mid \ell \in X^*; \|\ell\| = 1\}$$
$$= \sup\{\lim |\ell(x_n)| \mid \ell \in X^*, \|\ell\| = 1\}$$
$$\leq \sup\{\underline{\lim}\|\ell\| \, \|x_n\|\} = \underline{\lim}\|x_n\|$$

This shows that $\|\cdot\|$ is weakly lower semicontinuous where:

Definition A function F from a topological space X to $\mathbb{R}$ is called **lower semicontinuous** (l.s.c.) if and only if $x_\alpha \to x$ implies that $F(x) \leq \underline{\lim} \, F(x_\alpha)$.

One can show (Problem 11) that F is l.s.c. if and only if $\{x \mid F(x) \leq \alpha\}$ is closed for all α. The point of this definition is the following:

Proposition A lower semicontinuous function F on a compact space is bounded from below and takes its minimum value.

Proof Let $A = \inf_x F(x)$. If $A > -\infty$, pick x_n with

$$F(x_n) \leq A + n^{-1}$$

and if $A = -\infty$, pick x_n with $F(x_n) \leq -n$. Let x be a cluster point of the x_n and let $x_{n(\alpha)}$ be a subnet converging to x. Then, by lower semicontinuity,
$$F(x) \leq \varliminf F(x_{n(\alpha)}) = A$$
Thus $A \neq -\infty$ and $F(x) = A$. ∎

The Banach–Alaoglu theorem therefore implies the following basic result:

Theorem S.6 Let M be a weakly closed subset of a reflexive Banach space X. Suppose that F is a function from M to $\mathbb{R}$ obeying:

(i) for some a, $\{x \in M \mid F(x) \leq a\}$ is a *bounded* nonempty subset of X;
(ii) F is weakly lower semicontinuous.

Then, F is bounded from below and there exists $x_0 \in M$ with
$$F(x_0) = \min_{x \in M} F(x)$$

Notice that there are no changes in the above if we allow F to take the value $+\infty$. Henceforth we allow this.

It is important to have a large supply of weakly l.s.c. functions. The following is often useful:

Definition A function F on a vector space V is called **convex** if
$$F(\lambda x + (1-\lambda)y) \leq \lambda F(x) + (1-\lambda)F(y)$$
for all $x, y \in V$ and $\lambda \in [0, 1]$.

Theorem S.7 Let F be a function from a Banach space X to $(-\infty, \infty]$ (with $+\infty$ allowed). Suppose that $\{x \in X \mid F(x) < \infty\} = V$ is a subspace, that F is convex on V, and that F is norm l.s.c. Then F is weakly l.s.c.

Proof Since F is convex, $\{x \mid F(x) \leq a\}$ is convex. It is norm closed by Problem 11 since F is norm l.s.c. By general principles (Problem 12) such a set is weakly closed. Thus, by Problem 11 again, F is weakly l.s.c. ∎

Example 1 Let $G: \mathbb{C} \to [0, \infty)$ be convex and let $\langle M, \mu \rangle$ be a measure space. Then the functional
$$F(u) = \int G(u(x)) \, d\mu(x)$$

Minimization of functionals   357

(F may be $+\infty$) is weakly l.s.c. on all the L^p spaces, $p < \infty$. For F is clearly convex and therefore it suffices to prove that F is norm l.s.c. Let $u_n \to u$ in L^p norm. Pick a subsequence with $\lim_i F(u_{n(i)}) = \underline{\lim}\, F(u_n)$ and then a further subsequence so that $u_{n(i)}(x) \to u(x)$ pointwise a.e. Then $G(u_{n(i)}(x)) \to G(u(x))$ pointwise since a convex function is automatically continuous. So, by Fatou's lemma (Theorem I.17),

$$F(u) \leq \underline{\lim}\, F(u_{n(i)})$$

This establishes that F is weakly l.s.c. Notice that a special case of this example is the fact that the L^r norm is weakly l.s.c. on L^p for any p and $r < \infty$.

We want to sketch some applications of these ideas. Occasionally, we shall use terminology not defined until Chapter V or Volume II.

Example 2 (a nonlinear elliptic equation)   Given a real-valued function $f \in C_0^\infty(\mathbb{R}^\nu)$, we shall show the existence of a real-valued function $u \in L^2(\mathbb{R}^\nu)$ such that

$$-\Delta u + u|u| + u = f \tag{S.12}$$

where $-\Delta u$ is interpreted in the sense of distributions. Once one knows that this equation has a solution in L^2, the regularity methods of Section IX.6 show that u is, in fact, C^3. To solve (S.12) we define $F: L^2 \to (-\infty, \infty]$ by

$$F(u) = \tfrac{1}{2}\int |\nabla u|^2\, dx + \tfrac{1}{3}\int |u|^3\, dx + \tfrac{1}{2}\int |u|^2\, dx - \int f(x) u(x)\, dx$$

where we set $F(u) = \infty$ if $u \notin L^3$ or if $\nabla u \notin L^2$. If $g \in C_0^\infty(\mathbb{R}^\nu)$ and $F(u) < \infty$, then $F(u + \lambda g)$ is differentiable in λ and

$$\left.\frac{d}{d\lambda} F(u + \lambda g)\right|_{\lambda=0} = \int g(x)[-\Delta u(x) + |u(x)| u(x) + u(x) - f(x)]\, dx$$

If u minimizes F, then the left-hand side is zero. Since this is true for all $g \in C_0^\infty$, u satisfies (S.12). Thus, we wish to show that F takes on its minimum value.

By the Schwarz inequality,

$$F(u) \geq \tfrac{1}{2}\|u\|_2^2 - \|u\|_2 \|f\|_2 \geq \tfrac{1}{4}\|u\|_2^2 - \|f\|_2^2$$

Thus, $\{u \in L^2 \mid F(u) \leq 1\}$ is bounded, and it is nonempty since $F(0) = 0$. As a result, by Theorem S.6, it suffices to prove that F is weakly l.s.c. The last term is obviously weakly continuous, and the second and third terms are weakly l.s.c. by Theorem S.7. As for the first term, let $u_n \to u$ weakly; then

$$\begin{aligned}
\|\nabla u\|_2 &= \sup\{|(f, \nabla u)|\,|\,f \in C_0^\infty, \|f\|_2 = 1\} \\
&= \sup\{|(\nabla \cdot f, u)|\,|\,f \in C_0^\infty, \|f\|_2 = 1\} \\
&\leq \sup\left\{\lim_n |(\nabla \cdot f, u_n)|\,\Big|\,f \in C_0^\infty, \|f\|_2 = 1\right\} \\
&= \sup\left\{\lim_n |(f, \nabla u_n)|\,\Big|\,f \in C_0^\infty, \|f\|_2 = 1\right\} \\
&\leq \varliminf_n \|\nabla u_n\|_2
\end{aligned}$$

This proves the existence of solutions of (S.12).

The proof that we used above depends very strongly on the form of the nonlinear term $g(u) = (|u| + 1)u$. In particular, it depends on the fact that g is the derivative of a convex function. This is certainly necessary for the method to work. However, our proof above would not work if $g(u) = |u|u$, even though $|z|^3$ is convex, since we would not know that $\{u\,|\,F(u) \leq 1\}$ is bounded in L^2. Also it would not work if $g(u) = u^3 + u$, for in that case we could not argue that $u \in L^2$ implies that $g(u)$ is a distribution (i.e., an L^1_{loc} function). However, both of these difficulties are easy to overcome (Problem 15) if one realizes that L^2 is not sacred. In fact, for $|g(u)| \leq |u|^{p-1}$, the natural space to take is $L^p \cap H_1$ where $H_1 = \{u \in L^2\,|\,\nabla u \in L^2\}$ with the norm $(\|\nabla u\|_2^2 + \|u\|_2^2)^{\frac{1}{2}}$. That this space is the dual space of $L^q + H_1^*$ follows from general principles (Problem 14).

Example 3 (the Thomas–Fermi equations) Fix $z_1, \ldots, z_k > 0$ and $\mathbf{R}_1, \ldots, \mathbf{R}_k \in \mathbb{R}^3$ and let

$$V(\mathbf{r}) = \sum_{i=1}^{k} z_i |\mathbf{r} - \mathbf{R}_i|^{-1}$$

The **Thomas–Fermi equations** are

$$\rho = \max(\varphi, 0)^{\frac{3}{2}}$$

$$\varphi(\mathbf{x}) = V(\mathbf{x}) - \int \rho(\mathbf{y})|\mathbf{x} - \mathbf{y}|^{-1}\,d\mathbf{y} \tag{S.13}$$

Occasionally, this is written as the differential equation

$$-\Delta\varphi + 4\pi[\max(\varphi, 0)]^{\frac{3}{2}} = 4\pi \sum_{i=1}^{k} z_i \delta(r - R_i)$$

To solve this, we let

$$\mathscr{E}(\rho) \equiv \frac{3}{5} \int \rho^{\frac{5}{3}}(x)\, dx - \int V(x)\rho(x)\, dx + \frac{1}{2} \int \frac{\rho(x)\rho(y)}{|x-y|}\, dx\, dy \quad (S.14)$$

defined on $\{\rho \mid \rho \in L^{\frac{5}{3}}, \rho \in K, \rho \geq 0\}$, where K is the completion of C_0^∞ in the norm

$$\|\rho\|_K \equiv \left(\int \frac{\rho(x)\rho(y)}{|x-y|}\, dx\, dy \right)^{\frac{1}{2}}$$

That the integral in question is positive follows from the fact that $\|\rho\|_K^2 = \lim_{a \downarrow 0} 4\pi(\rho, (-\Delta + a)^{-1}\rho)$. By general principles (Problem 14), $L^{\frac{5}{3}} \cap K$ is the dual space of the sum of functionals in $L^{\frac{5}{2}}$ and K^*. The first and third terms of (S.14) are weakly l.s.c. by general arguments (Theorem S.7). We leave it to the reader (Problem 16) to prove that the middle term is a linear functional in $L^{\frac{5}{2}} + K^*$. It follows that $\mathscr{E}(\rho)$ takes its minimum value and, as in Example 2, the minimizing ρ solves (S.13). The reason that $\max\{\varphi, 0\}$ enters is that we know that

$$\left. \frac{d}{d\lambda} \mathscr{E}(\rho + \lambda g) \right|_{\lambda=0} = 0$$

only for g with $|g| \leq c\rho$. For general $g \geq 0$, we only know that

$$\left. \frac{d}{d\lambda} \mathscr{E}(\rho + \lambda g) \right|_{\lambda=0} \geq 0$$

It can be proven using special features of this problem that $\varphi > 0$ if φ is the minimizing function (see the reference in the notes).

Example 4 (Hartree equations for helium) We seek solutions of the equations

$$\begin{aligned} -\Delta u - 2|x|^{-1}u + W_1(x)u &= \varepsilon_1 u \\ -\Delta v - 2|x|^{-1}v + W_2(x)v &= \varepsilon_2 v \end{aligned} \quad (S.15)$$

$x \in \mathbb{R}^3$, with the subsidiary conditions

$$\|u\|_2 = \|v\|_2 = 1 \quad (S.16)$$

and

$$W_1(x) = \int |x-y|^{-1} v(y)\, dy$$

$$W_2(x) = \int |x-y|^{-1} u(y)\, dy$$

The physical significance of the equations is that formally (S.15) comes from minimizing

$$\mathcal{E}(u, v) = \int (|\nabla u|^2 + |\nabla v|^2)\, dx - 2 \int |x|^{-1}(|u|^2 + |v|^2)$$

$$+ \int |u(x)|^2 |v(y)|^2 |x - y|^{-1}\, dx\, dy \qquad \text{(S.17)}$$

with the subsidiary condition (S.16); the ε_i are just Lagrange multipliers. For u and v obeying (S.16), one sees that

$$\mathcal{E}(u, v) = (\psi, H\psi)$$
$$\psi(x, y) = u(x)v(y)$$

where

$$H = -\Delta_x - \Delta_y - \frac{2}{|x|} - \frac{2}{|y|} + \frac{1}{|x - y|}$$

so that solutions of (S.15), (S.16) can be expected to give approximate solutions of $H\psi = E\psi$ in some vague sense.

The new feature of this example is the subsidiary condition (S.16), for the set of functions obeying (S.16) is *not* weakly closed. Thus, once one shows that $\mathcal{E}$ is weakly lower semicontinuous on $H_1 \oplus H_1$ with $H_1 = \{u\,|\,\nabla u \in L^2, u \in L^2\}$ (see the reference in the Notes for this), one can use Theorem S.6 to show only that $\mathcal{E}$ takes its minimum subject to the subsidiary conditions

$$\|u\|_2 \leq 1, \qquad \|v\|_2 \leq 1 \qquad \text{(S.18)}$$

for this set is the weak closure of the set obeying (S.16). However, one can actually show that the minimizing function subject to (S.18) obeys (S.16). For, let $u^{(0)}$ and $v^{(0)}$ minimize $\mathcal{E}$ subject to (S.18). Then

$$\mathcal{E}(u, v^{(0)}) = c + (u, hu)$$

where

$$h = -\Delta + V(x)$$

$$V(x) = -\frac{2}{|x|} + \int |v^{(0)}(y)|^2 |x - y|^{-1}\, dy$$

Let $\tilde{u}$ be the function that is the lowest eigenvector of $\tilde{h} = -\Delta - |x|^{-1}$. Then

$$(\tilde{u}, h\tilde{u}) = (\tilde{u}, \tilde{h}\tilde{u}) + (v^{(0)}, \tilde{W}v^{(0)})$$

where

$$\tilde{W}(x) = -|x|^{-1} + \int |\tilde{u}(y)|^2 |x-y|^{-1} dy$$

$$= -|x|^{-1} + \int |\tilde{u}(y)|^2 [\max\{|y|, |x|\}]^{-1} dy \le 0$$

since $\tilde{u}$ is spherically symmetric. Thus

$$(\tilde{u}, h\tilde{u}) < 0$$

It follows that $(u^{(0)}, hu^{(0)}) < 0$, for if not, then we would have $\mathscr{E}(\tilde{u}, v^{(0)}) < \mathscr{E}(u^{(0)}, v^{(0)})$. If $\|u^{(0)}\|_2 = a$, then

$$\mathscr{E}(a^{-1}u^{(0)}, v^{(0)}) = \mathscr{E}(u^{(0)}, v^{(0)}) + (a^{-2} - 1)(u^{(0)}, hu^{(0)})$$

Since $\langle u^{(0)}, v^{(0)} \rangle$ minimizes $\mathscr{E}(u^{(0)}, v^{(0)})$ over all $\langle u^{(0)}, v^{(0)} \rangle$ with $\|u^{(0)}\| \le 1$, $\|v^{(0)}\| \le 1$, we have that $a = 1$, so $\|u^{(0)}\|_2 = 1$.

The above is typical of applications with subsidiary conditions in that extra structure must be taken into account. Sometimes the extra structure is not present. For example, if one tries to minimize $\mathscr{E}(\rho)$, given by (S.14), with the condition $\int \rho(x)\,dx = \lambda$, then a minimum exists if and only if $\lambda \le \sum_{i=1}^{k} z_i$ (see the reference in the Notes).

Example 5 (existence of geodesics) Let M be a closed subset of $\mathbb{R}^\nu$, for example, a smoothly embedded surface. A continuous curve $\gamma: [0, 1] \to M \subset \mathbb{R}^\nu$ is called **rectifiable** if and only if its distributional derivative (thinking of γ as a function on $[0, 1]$ with values in $\mathbb{R}^\nu$), $\nabla\gamma$, lies in $L^1([0, 1], \mathbb{R}^\nu)$. For any rectifiable curve, one defines its length $L(\gamma)$ by

$$L(\gamma) = \int_0^1 |\nabla\gamma(s)|\,ds$$

We want to prove:

Theorem S.8 Let M be a closed subset of $\mathbb{R}^\nu$ and let x and y be in M. Suppose that there is a rectifiable curve lying in M with $\gamma(0) = x$, $\gamma(1) = y$. Then there is a curve $\tilde{\gamma}$ with these properties so that

$$L(\tilde{\gamma}) = d_{x,y} \equiv \inf\{L(\gamma) \mid \gamma \text{ lies in } M, \gamma(0) = x, \gamma(1) = y\}.$$

Such a minimizing curve is called **geodesic**.

Proof One cannot try to use Theorem S.6 directly on $L(\gamma)$ because L^1 is not reflexive. In fact, there is good reason why one cannot do this. For consider a reparametrization. Let f be a positive L^1 function on $[0, 1]$ with $\int_0^1 f(s)\,ds = 1$. Define

$$\tau(s) = \int_0^s f(t)\,dt$$

$$\tilde{\gamma}(s) = \gamma(\tau(s))$$

Then $\nabla\tilde{\gamma} = [(\nabla\gamma) \circ \tau]f$. Thus

$$L(\tilde{\gamma}) = \int_0^1 |\nabla\tilde{\gamma}|\,ds = \int_0^1 |(\nabla\gamma) \circ \tau|\,d\tau = L(\gamma)$$

Of course, this invariance of length is what makes length such a natural object. But it means the following: Let γ_n be a sequence of curves with $L(\gamma_n) \to d_{x,y}$. Let $\tilde{\gamma}_n$ be a curve which is a reparametrization of γ_n with the property that $\int_0^{1/n} |\nabla\tilde{\gamma}_n|\,ds = \frac{1}{2}$ (this can be done with a piecewise linear reparametrization). Then $L(\tilde{\gamma}_n)$ also converges to $d_{x,y}$, but $\tilde{\gamma}_n$ cannot have a reasonable limit. This says that one cannot hope to deal directly with $L(\gamma)$ by just being clever in the choice of space.

The way around this difficulty is what mathematicians often call "Dirichlet's principle" and what a physicist would describe by saying that "a free particle on a curved surface moves along geodesics at constant velocity." We define

$$E(\gamma) = \int_0^1 |\nabla\gamma(s)|^2\,ds$$

We shall use Theorem S.6 to find a curve γ_0 that minimizes E and then verify that $L(\gamma_0) \leq L(\gamma)$ for any γ (i.e., γ_0 is a geodesic). Along the way we shall show that $|\nabla\gamma_0(s)|$ is a constant (constant speed).

We begin by setting

$$Y \equiv \left\{ f \in L^2([0, 1]; \mathbb{R}^\nu, dx) \,\middle|\, x + \int_0^s f(t)\,dt \equiv \gamma^{(f)}(s) \in M \right.$$

$$\left. \text{for all } s \in [0, 1], \text{ and } \gamma^{(f)}(1) = y \right\}$$

and

$$\tilde{E}(f) \equiv \int_0^1 |f(s)|^2\,ds = E(\gamma^{(f)})$$

Clearly $\tilde{E}(f)$ is weakly lower semicontinuous and for any a, $\{f \mid \tilde{E}(f) \leq a\}$ is bounded. We defer the proof that it is nonempty for some a. Notice that

$$f \mapsto \int_0^s f(t)\,dt$$

is a continuous linear functional on $L^2([0, 1]; \mathbb{R}^v, dx)$ so, since M is closed, we have that Y is weakly closed. Hence there exists a γ_0 minimizing E.

Given any rectifiable curve γ define a function T by

$$T(s) \equiv L(\gamma)^{-1} \int_0^s |\nabla \gamma(t)|\,dt$$

We claim there is a continuous curve γ^* so that

$$\gamma^*(T(s)) = \gamma(s)$$

To see this, note that $T(s_0) = T(s_1)$ if and only if γ is constant on $[s_0, s_1]$. γ^* is called the **geodesic reparametrization** of γ, although in the special case that $|\nabla \gamma| = 0$ on an interval, it is not a reparametrization! However,

$$L(\gamma^*) = L(\gamma)$$

since γ is always a reparametrization of γ^*. Next notice that

$$|\nabla \gamma^*(s)| = L(\gamma)$$

since $\int_0^{T(s)} |\nabla \gamma^*(t)|\,dt = T(s)L(\gamma)$ for any $T(s)$. In particular, $E(\gamma^*) < \infty$, so that if there are rectifiable curves, there are also curves of finite energy. This shows that the set Y above is nonempty.

Next notice that by the Schwarz inequality, for any γ,

$$E(\gamma) \geq L(\gamma)^2 = L(\gamma^*)^2 = E(\gamma^*)$$

with equality only if $\gamma = \gamma^*$! Thus, the curve of minimum energy has $|\nabla \gamma_0| =$ constant. Moreover, for any other γ,

$$L(\gamma) = E(\gamma^*)^{\frac{1}{2}} \geq E(\gamma_0)^{\frac{1}{2}} = L(\gamma_0)$$

so γ_0 is indeed a geodesic. ∎

Supplement to V.5 Proofs of some theorems in nonlinear functional analysis

In this supplement we shall sketch the proof of the Leray–Schauder–Tychonoff theorem in the case where C has a separable topology, and we shall state and prove the implicit function theorem. The Leray–Schauder–Tychonoff theorem is already highly nontrivial in the finite-dimensional case. We use some machinery which we take from algebraic topology. We state the following Theorem without proof:

Theorem S.9 For $n = 0, 1, 2, \ldots$, there is a map H_n from the set of topological spaces $\mathcal{T}$ to the set of abelian groups $\mathcal{A}$ so that for each $X \in \mathcal{T}$, $Y \in \mathcal{T}$ and continuous function $f: X \to Y$, there is a group homomorphism f_* from $H_n(X)$ to $H_n(Y)$ satisfying:

(i) If $f: X \to Y$ and $g: Y \to Z$, then $(g \circ f)_* = g_* \circ f_*$.
(ii) If $id: X \to X$ is the identity map on X, then $(id)_*$ is the identity map on $H_n(X)$.
(iii) If D^m is the m-dimensional disk, $\{x \in \mathbb{R}^m \,|\, |x| \leq 1\}$, and S^{m-1} is its boundary, then for $m \geq 2$, $H_{m-1}(D^m) = \{0\}$, the group consisting of a single element, and $H_{m-1}(S^{m-1}) = \mathbb{Z}$, the integers. For $n = 1$, $H_0(D^1) = \mathbb{Z}$ and $H_0(S^0) = \mathbb{Z} \oplus \mathbb{Z}$.

$H_n(X)$ is called the nth **homology group of** X. It is a measure of the number and type of $(n + 1)$-dimensional "holes" in X.

Theorem S.10 (Brouwer fixed point theorem) Let C be a compact convex subset of $\mathbb{R}^n$ and suppose that $f: C \to C$ is continuous. Then f has a fixed point.

Proof Suppose first that $C = D^n$, the n-disk, and suppose that f has no fixed point. Given $x \in C$, let $g(x)$ be the point at which the line from $f(x)$ to x intersects S^{n-1}; i.e., $g(x)$ is determined by the conditions $g(x) = f(x) + \alpha(x - f(x))$ for some $\alpha > 0$, and $|g(x)| = 1$. Since $f(x)$ is assumed to be always distinct from x, it is easy to see that $x \to g(x)$ is a continuous map of D^n into S^{n-1}, and it is clear that this map leaves points in S^{n-1} fixed. Let h be the natural map of S^{n-1} into D^n obtained by inclusion. Then $g \circ h = id$. Thus, by parts (i) and (ii) of Theorem S.9, $g_* \circ h_* = id_*$ on homology. But for $n > 1$, $h_*: \mathbb{Z} \to \{0\}$ and $g_*: \{0\} \to \mathbb{Z}$ so that $g_* \circ h_*$ could not possibly be the identity on $\mathbb{Z}$. For $n = 1$, $h_*: \mathbb{Z} \oplus \mathbb{Z} \to \mathbb{Z}$ and $g_*: \mathbb{Z} \to \mathbb{Z} \oplus \mathbb{Z}$ so again $g_* \circ h_*$ cannot be the identity. Thus, we have a contradiction and conclude that f must have a fixed point if $C = D^n$.

Given a compact, convex $C \subset \mathbb{R}^n$, we can find a plane P of minimal dimension m and a vector v_0 so that $C \subset v_0 + P$. It is not hard to prove (Problem 20) that if we think of C as a topological subspace of $v_0 + P$, it has an interior point x_0. Thus, for any $v \in P$, $|v| = 1$, the set $\{t \geq 0 \,|\, x_0 + tv \in C\}$ is a nonempty closed interval $[0, h(v)]$ since C is convex and closed. Further, $h(v) > 0$ since x_0 is an interior point of C in $v_0 + P$. h is a continuous map of the unit sphere S^{m-1} to $\mathbb{R}_+$ and the map

$$F: y \to [h((y - x_0)/|y - x_0|)]^{-1}(y - x_0)$$

is a homeomorphism of C onto D^m. Thus, the theorem for disks implies the general finite-dimensional case since all compact convex sets are homeomorphic to disks. ∎

Proof of Theorem V.19 (in the case where C has a countable base)   Let $\{x_n\}$ be the countable dense set that we are assuming exists. For each n and m, we can find by the Hahn–Banach theorem an $\ell_{nm} \in X^*$ so that $\ell_{nm}(x_n) \neq \ell_{nm}(x_m)$. Since each ℓ_{nm} is bounded on C we can normalize the ℓ_{nm} so that $\sup_{y \in C} |\ell_{nm}(y)| \leq 2^{-n-m}$. Introduce a metric d on C by

$$d(x, y) = \left[\sum_{n, m} |\ell_{nm}(x - y)|^2\right]^{\frac{1}{2}}$$

Since d separates the x_n from one another, it is easy to see that it separates points. Moreover, d is continuous. Thus, the identity map from C with its original topology to C with the metric topology is a continuous bijection from a compact space to a Hausdorff space, and hence a homeomorphism.

By the definition of d, it is easy to see that it obeys the parallelogram law since this law is obeyed by each $|\ell_{nm}(x - y)|$ term separately. By mimicking the proof of the lemma to Theorem II.3, one sees that for each compact convex subset D of C and each $x \in C$ there is a unique point $\pi_D(x)$ in D closest to C and that π_D is continuous.

Now, given $\varepsilon > 0$, we can find, by compactness, a finite subset $y_1, \ldots, y_N$ of C so that any point in C lies within ε of some y_i. Let D be the set of points

$$\left\{\theta_1 y_1 + \cdots + \theta_n y_n \,\Big|\, 0 \leq \theta_i \leq 1;\, \sum_{i=1}^n \theta_i = 1\right\}$$

D is a compact convex subset of C homeomorphic to a convex subset of some $\mathbb{R}^m$ ($m \leq n$). Moreover,

$$d(\pi_D(x), x) \leq \varepsilon \qquad (S.19)$$

for all $x \in C$. Let $g: D \to D$ by $g(y) = \pi_D(f(y))$. By Theorem S.10, g has a fixed point x_ε. By (S.19), $d(f(x_\varepsilon), x_\varepsilon) \leq \varepsilon$. If x_0 is a limit point of the sequence $x_{1/n}$, then $d(f(x_0), x_0) = 0$, i.e., x_0 is a fixed point of f. ∎

* * *

To illustrate the uses of the contraction mapping theorem, we shall prove the implicit function theorem in Banach spaces. First, we define:

Definition Let X and Y be Banach spaces. We say that a function f from a neighborhood N_0 of x_0 in X to Y is differentiable at x_0 if and only if there exists a linear map $T \in \mathscr{L}(X, Y)$ so that

$$\lim_{x \to 0} \|x\|^{-1}[f(x_0 + x) - f(x_0) - Tx] = 0$$

We write $T \equiv (Df)(x_0)$. If N is open, if f is differentiable at all points in N, and if $x \to (Df)(x)$ is a norm continuous map of N into $\mathscr{L}(X, Y)$, we say that f is C^1 on N.

Theorem S.11 (implicit function theorem) Let X and Y be Banach spaces and let M_0 and N_0 be neighborhoods of $x_0 \in X$ and $y_0 \in Y$, respectively. Let $F: M_0 \times N_0 \to Y$ be a C^1 function. From the definitions it follows that for each $\langle x_1, y_1 \rangle \in M_0 \times N_0$, there are bounded operators $D_1 \in \mathscr{L}(X, Y)$ and $D_2 \in \mathscr{L}(Y, Y)$ so that $DF(x_1, y_1)$ is the linear operator given by $\langle x, y \rangle \to D_1 x + D_2 y$. At the point $\langle x_1, y_1 \rangle = \langle x_0, y_0 \rangle$, we set $T = D_1$, $S = D_2$.

Suppose that $F(x_0, y_0) = 0$ and that S has a bounded inverse. Then, there are neighborhoods $M_1 \subset M_0$ and $N_1 \subset N_0$ of x_0 and y_0 and a C^1 map $g: M_1 \to N_1$ so that $F(x, g(x)) = 0$ for all $x \in M_1$ and for each $x \in M_1$, $g(x)$ is the unique point y in N_1 satisfying $F(x, y) = 0$.

Proof We shall prove the existence and uniqueness of g. The verification that g is C^1 is left to the reader (Problem 21). First, choose ε_1 and δ so that

$$M_2 = \{x \,|\, \|x - x_0\| < \varepsilon_1\} \subset M_0, \quad N_1 = \{y \,|\, \|y - y_0\| < \delta\} \subset N_0$$

and, for $\langle x, y \rangle \in M_2 \times N_1$,

$$\|I - S^{-1}D_2\| < \tfrac{1}{2}$$

Such a choice is possible since $S^{-1}D_2$ is norm continuous as $\langle x, y \rangle$ varies and equals I at $\langle x_0, y_0 \rangle$. Now, set

$$\alpha = \sup_{\langle x, y \rangle \in M_2 \times N_1} \|S^{-1}D_1\|$$

and let $M_1 = \{x \,|\, \|x - x_0\| < \varepsilon_2\}$ where ε_2 is chosen so that $\varepsilon_2 < \varepsilon_1$ and $\alpha \varepsilon_2 < \tfrac{1}{4}\delta$. We claim that for any fixed $x_1 \in M_1$, there exists exactly one $y \in N_1$ with $F(x_1, y) = 0$. By the fundamental theorem of calculus, for any C^1 function G,

$$\|G(a, b) - G(a', b')\| \le c_1 \|a - a'\| + c_2 \|b - b'\| \tag{S.20}$$

where
$$c_1 = \sup_{0 \leq \theta \leq 1} \|D_1 G(\theta a + (1-\theta)a', b')\|$$

and
$$c_2 = \sup_{0 \leq \theta \leq 1} \|D_2 G(a, \theta b + (1-\theta)b')\|$$

Given $x_1 \in M_1$ as above, let
$$H(y) \equiv y - S^{-1} F(x_1, y)$$

Using (S.20), we see that
$$\|H(y_0) - y_0\| \leq \alpha \|x_1 - x_0\| < \tfrac{1}{4}\delta$$

Moreover, by (S.20) again,
$$\|H(y) - H(y')\| \leq \tfrac{1}{2}\|y - y'\|$$

if y and y' are in N_1. Thus, H maps N_1 into itself and is a contraction. Notice that if N'_1 is the open ball about y_0 of radius $\delta' < \delta$ where $|\delta' - \delta|$ is small, then this same argument shows that H takes N'_1 into itself. Thus H has a unique fixed point y_1 in $\overline{N}'_1 \subset N_1$. Since $H(y_1) = y_1$, we have $F(x_1, y_1) = 0$. ∎

Corollary 1 (inverse function theorem)   Let Y be a Banach space and suppose that $f: N \to Y$ is a C^1 mapping where N is a neighborhood of $y_0 \in Y$. If $(Df)(y_0)$ is invertible, then there are neighborhoods N_1 of $f(y_0)$ and N_2 of y_0 and a C^1 map $g: N_1 \to N_2$ so that $f \circ g = I$ and $g(y)$ is the only point in N_2 with $f(g(y)) = y$.

Proof   Let $F: N \times N \to Y$ by $F(x, y) = x - f(y)$. Then $D_2 F = -Df$ is invertible at $\langle f(y_0), y_0 \rangle$. Now apply the implicit function theorem. ∎

Corollary 2 (implicit function theorem—extended form)   Let X, Y, and Z be Banach spaces and let N_0 and M_0 be neighborhoods of $x_0 \in X$ and $y_0 \in Y$, respectively. Let $F: M_0 \times N_0 \to Z$ be a C^1 function. Thus if $T \equiv (Df)(x_0, y_0)$, there are $R \in \mathscr{L}(X, Z)$ and $S \in \mathscr{L}(Y, Z)$ so that $T\langle x, y\rangle = Rx + Sy$. Suppose that $F(x_0, y_0) = 0$.

Suppose that S is onto Z and that there is a closed subspace $Y_1 \subset Y$ so that $Y_1 + \operatorname{Ker} S = Y$, $Y_1 \cap \operatorname{Ker} S = \{0\}$. Then, there is a neighborhood M_1 of x_0 and a C^1 function $g: M_1 \to Y$ with $g(x_0) = y_0$ so that $F(x, g(x)) = 0$ for all $x \in M_1$.

Proof Pick ε so that $\{y \mid \|y - y_0\| \leq \varepsilon\} \subset N_0$ and let G be the mapping of $M_0 \times \{y \in Y_1 \mid \|y\| < \varepsilon\}$ into Z given by $G(x, y) = F(x, y_0 + y)$. Then $D_2 G|_{\langle x_0, 0 \rangle} = S \upharpoonright Y_1$. By construction, $S \upharpoonright Y_1$ is one to one, onto, and continuous. Thus Y_1 is isomorphic to Z and $S \upharpoonright Y_1$ is invertible. By the basic implicit function theorem applied to G, g exists. ∎

Notice that the $g(x)$ in this last corollary need not be unique; it is unique if one adds the condition $g(x) - y_0 \in Y_1$. Two cases where a suitable Y_1 exists whenever S is onto are the case $\dim Y < \infty$ and the case where Y is a Hilbert space.

Supplement to VI.5 Applications of compact operators

In Section VI.4 we outlined the application of compact operators to the solution of the Dirichlet problem for bounded regions in $\mathbb{R}^2$. In this supplement we give two simpler applications. Consider the following initial-boundary value problem in partial differential equations (q is continuous):

$$u_t(x, t) - u_{xx}(x, t) + q(x)u(x, t) = 0, \quad 0 \leq x \leq 1$$
$$u(x, 0) = f(x) \tag{S.21}$$
$$u(0, t) = 0 = u(1, t)$$

There are various physical situations that are modeled by this equation. For example, $u(x, t)$ could be the density at x at time t of a gas in a pipe of length one. If $q(x)$ is zero, then we have the usual diffusion equation with diffusion constant one. If $q(x)$ is positive, then $q(x)u(x, t)$ units of gas per unit length per unit time are removed at $\langle x, t \rangle$ by a chemical reaction with other material in the pipe. It is reasonable that the amount removed at $\langle x, t \rangle$ be dependent on the local density $u(x, t)$ and on the position in the pipe through $q(x)$. The condition $u(x, 0) = f(x)$ gives the gas density at $t = 0$, and the boundary conditions reflect a situation where the gas can freely escape at the ends so that the density at the ends is effectively zero.

Following the method of eigenfunction expansions, one first tries to find solutions of the differential equation and boundary conditions of the form $\psi(t)\varphi(x)$. Plugging into the differential equation, we find that

$$\frac{\psi'(t)}{\psi(t)} = \frac{\varphi''(x)}{\varphi(x)} - q(x)$$

Applications of compact operators   369

for all x and t, so both sides equal a constant, call it $-\lambda$. Thus, we must solve the differential equation

$$\psi'(t) + \lambda\psi(t) = 0$$

and the eigenvalue problem

$$-\varphi''(x) + q(x)\varphi(x) - \lambda\varphi(x) = 0 \tag{S.22}$$

$$\varphi(0) = 0 = \varphi(1)$$

Recall what happens in the case $q(x) \equiv 0$. In that case we can solve the differential equation explicitly, and we find that the eigenvalue problem has a solution if and only if $\lambda = \lambda_n \equiv (n\pi)^2$, $n = 1, 2, \ldots$, in which case

$$\varphi(x) = \varphi_n(x) \equiv \sqrt{2}\sin n\pi x \quad \text{and} \quad \psi(t) = \psi_n(t) \equiv e^{-(n\pi)^2 t}$$

Thus, for each n, $\psi_n(t)\varphi_n(x)$ solves the differential equation and boundary conditions in (S.21). Since the differential equation is linear and the boundary conditions are homogeneous, any linear combination of these solutions is again a solution of the differential equation and the boundary conditions. So, one tries to write the solution of (S.21) in the form

$$u(x, t) = \sum_{n=1}^{\infty} a_n \varphi_n(x)\psi_n(t)$$

and tries to choose the coefficients $\{a_n\}$ so that

$$f(x) = u(x, 0) = \sum_{n=1}^{\infty} a_n \varphi_n(x) \tag{S.23}$$

The fact that one can do this is just the classical fact that every (sufficiently nice) function $f(x)$ can be represented by a convergent Fourier (sine) series.

Now, consider the case where $q(x)$ is not identically zero. At first sight the problem seems almost impossible. It is not immediately evident that the eigenvalue problem will have solutions for a discrete set of λ_n. Even if it did, we would not expect to be able to compute the λ_n or the corresponding eigenfunctions φ_n explicitly. And, even if we could compute the φ_n explicitly, how could we expect them to obey the many identities for sine and cosine which are used in proving the classical Fourier series theorem? Notice, however, that if the eigenvalue problem (S.22) does have solutions λ_n, φ_n, then

$$\int_0^1 \varphi_n(x)\varphi_m(x)\, dx = 0$$

if $n \neq m$ and $\lambda_n \neq \lambda_m$. This can easily be checked (Problem 22) by substituting using the differential equation, and integrating by parts. Thus, the eigenfunctions, if they exist, are orthogonal in the sense of (real) $L^2[0, 1]$. This suggests that we reformulate our problem as a Hilbert space problem. What we are really asking is whether (S.22) has a set of solutions $\{\varphi_n(x)\}$, with corresponding eigenvalues λ_n, which are an orthonormal basis for $L^2[0, 1]$. If this is true, then every $f \in L^2[0, 1]$ can be written in the form (S.22). Further, the $\varphi_n(x)$ are automatically twice continuously differentiable. Also, one can show by integration by parts (Problem 22) that any λ_n that do exist must satisfy $\lambda_n \geq \inf q(x)$. In the proof below it turns out that the λ_n can have no finite limit point, so only finitely many of them can be nonpositive. Therefore the series

$$u(x, t) = \sum_{n=1}^{\infty} a_n e^{-\lambda_n t} \varphi_n(x)$$

will converge very rapidly for $t > 0$, so that one can check that $u(x, t)$ satisfies the differential equation (S.21) for $t > 0$. $u(x, t)$ satisfies the boundary conditions since the $\varphi_n(x)$ do, and u satisfies the initial condition in the sense that

$$\|u(x, t) - f(x)\|_2 \to 0$$

as $t \downarrow 0$. Thus, we can solve our original problem in partial differential equations if we show that (S.22) has a family of solutions $\{\varphi_n\}$ which are a basis for $L^2[0, 1]$.

For simplicity, we shall assume that $q(x)$ is continuous and that $q(x) \geq 0$. The positivity hypotheses is easily removed (see Problem 22). Let $u_0(x)$ be a nonzero solution of $-u''(x) + q(x)u(x) = 0$ that satisfies $u_0(0) = 0$ and let $u_1(x)$ be a nonzero solution that satisfies $u_1(1) = 0$. The Wronskian of the two solutions $W = u_0'(x)u_1(x) - u_0(x)u_1'(x)$ is constant on $[0, 1]$ and equals zero if and only if $u_0(x) = \alpha u_1(x)$ for some α. But if this is so, $u_0(0) = 0 = u_0(1)$ so

$$\int_0^1 (u_0')^2 \, dx = -\int_0^1 u_0'' u_0 \, dx = -\int_0^1 q(x) u_0^2 \, dx$$

which implies that $u_0'(x) \equiv 0$ so $u_0(x)$ is constant. But since $u_0(0) = 0$, we would have $u_0(x) \equiv 0$. This contradiction shows that $W \neq 0$. Thus, for $y \in [0, 1]$, we can define

$$k(x, y) = \begin{cases} u_0(x)u_1(y)W^{-1}, & x \leq y \\ u_0(y)u_1(x)W^{-1}, & x > y \end{cases}$$

and set

$$(T\psi)(x) = \int_0^1 k(x, y)\psi(y) \, dy$$

Applications of compact operators 371

Then, by Theorem VI.23, T is a Hilbert–Schmidt (and therefore compact) operator on $L^2[0, 1]$. We shall show that T is a right inverse to $L = -d^2/dx^2 + q(x)$. Suppose that $\psi(x) \in L^2[0, 1]$ and write

$$(T\psi)(x) = W^{-1}u_1(x)\int_0^x u_0(y)\psi(y)\,dy + W^{-1}u_0(x)\int_x^1 u_1(y)\psi(y)\,dy$$

Since $\psi \in L^2$, the indefinite integrals are absolutely continuous and

$$\frac{d}{dx}(T\psi)(x) = W^{-1}u_1'(x)\int_0^x u_0(y)\psi(y)\,dy + W^{-1}u_0'(x)\int_x^1 u_1(y)\psi(y)\,dy$$

Again we see that $(d/dx)(T\psi)(x)$ is absolutely continuous and

$$\frac{d^2}{dx^2}(T\psi)(x) = W^{-1}u_1''(x)\int_0^x u_0(y)\psi(y)\,dy + W^{-1}u_0''(x)\int_x^1 u_1(y)\psi(y)\,dy$$

$$+ W^{-1}u_1'(x)u_0(x)\psi(x) - W^{-1}u_0'(x)u_1(x)\psi(x)$$

$$= q(x)\int_0^1 k(x, y)\psi(y)\,dy - \psi(x)$$

Thus, $(T\psi)(x)$ is twice differentiable, with second derivative in $L^2[0, 1]$ and

$$-\frac{d^2}{dx^2}(T\psi)(x) + q(x)(T\psi)(x) = \psi(x)$$

that is, $LT\psi = \psi$. Notice also that, because $u_0(0) = 0 = u_1(1)$, $T\psi$ satisfies $(T\psi)(0) = 0 = (T\psi)(1)$. In the terminology of Chapter VIII, we would say that T takes $L^2[0, 1]$ into the domain of $-(d^2/dx^2) + q(x)$.

As pointed out above, T is compact and since $k(x, y) = k(y, x)$, T is self-adjoint. Thus by the Hilbert–Schmidt theorem (Theorem VI.16), there is an orthonormal basis $\{\varphi_n\}$ and a sequence of numbers $\{\mu_n\}$ so that

$$T\varphi_n = \mu_n \varphi_n$$

$LT = I$, so we have

$$\varphi_n = LT\varphi_n = \mu_n L\varphi_n$$

which shows that $\mu_n \neq 0$ and

$$-\varphi_n''(x) + q(x)\varphi_n(x) = \frac{1}{\mu_n}\varphi_n$$

and, since φ_n is in the range of T, we automatically have

$$\varphi_n(0) = 0 = \varphi_n(1)$$

Notice that $\mu_n \to 0$ since T is compact. Thus the λ_n cannot have any finite limit point. Since they are bounded below (Problem 22), we have $\lambda_n \to \infty$. We summarize:

Theorem S.12 Suppose that $q(x)$ is a continuous function on $[0, 1]$. Then there exists an orthonormal basis $\{\varphi_n\}$ for $L^2[0, 1]$ consisting of twice continuously differentiable functions and a sequence of numbers $\lambda_n \to \infty$ such that

$$-\varphi_n''(x) + q(x)\varphi_n(x) = \lambda_n \varphi_n(x)$$

In Problem 23 some of these ideas are extended to space dimension three. The unbounded operator L discussed above has a compact resolvent at $\lambda = 0$ and therefore, by the first resolvent formula, L has a compact resolvent for all $\lambda \in \rho(L)$. Operators with compact resolvent are discussed in Section XIII.14.

As a second application of compact operators we prove the following classical theorem of Weyl:

Theorem S.13 Suppose that A and B are self-adjoint operators on a Hilbert space such that $A - B$ is compact. Then $\sigma_{\text{ess}}(A) = \sigma_{\text{ess}}(B)$.

Proof $\sigma_{\text{ess}}(\cdot)$ is defined in Section VII.3 and according to Weyl's criterion (Theorem VII.12), $\lambda \in \sigma_{\text{ess}}(A)$ if and only if there is an orthonormal sequence $\{\psi_n\}$ such that $\|(A - \lambda)\psi_n\| \to 0$. Suppose that this is the case. Since the ψ_n are orthonormal, they converge weakly to zero. By Theorem VI.11, this implies that $\|(A - B)\psi_n\| \to 0$ since $A - B$ is compact. Thus,

$$\|(B - \lambda)\psi_n\| \leq \|(A - \lambda)\psi_n\| + \|(A - B)\psi_n\| \to 0$$

Therefore, by Weyl's criterion, $\lambda \in \sigma_{\text{ess}}(B)$. This shows that $\sigma_{\text{ess}}(A) \subset \sigma_{\text{ess}}(B)$. By symmetry, $\sigma_{\text{ess}}(B) \subset \sigma_{\text{ess}}(A)$. ∎

Various generalizations of this theorem as well as applications are discussed in Section XIII.4.

Supplement to VIII.7 Monotone convergence for forms

In this supplement we prove some very useful approximation theorems for positive self-adjoint operators.

Definition Let t and s be two positive quadratic forms on a Hilbert space $\mathscr{H}$. We say that $t \leq s$ if and only if $Q(t) \supset Q(s)$ and $t(\varphi, \varphi) \leq s(\varphi, \varphi)$ for all $\varphi \in Q(s)$.

Note that this definition says that if t is an extension of s to a larger domain, then $t \leq s$. In particular, $\bar{t} \leq t$ for any closable form t.

Theorem S.14 Let $t_1, t_2, \ldots, t_n, \ldots$ be a sequence of closed, positive quadratic forms satisfying $0 \leq t_1 \leq \cdots \leq t_n \leq \cdots$. Suppose that

$$Q(t_\infty) = \left\{ \varphi \in \mathscr{H} \,\Big|\, \sup_n t_n(\varphi, \varphi) < \infty \right\}$$

is dense. Then the quadratic form

$$t_\infty(\varphi, \varphi) = \lim_{n \to \infty} t_n(\varphi, \varphi) = \sup_n t_n(\varphi, \varphi)$$

with domain $Q(t_\infty)$ is closed. Moreover, if T_n, T_∞ are the operators corresponding to these forms, then $T_n \to T_\infty$ in strong resolvent sense.

We give two proofs of this theorem below. Note that by the Schwarz inequality applied to $t_n(\varphi, \varphi)$, $Q(t_\infty)$ is a vector space and that, by polarization, the values of $t_\infty(\varphi, \varphi)$ determine those of $t_\infty(\varphi, \psi)$.

To state the second convergence theorem, we need some preliminaries. We say that a form t is **closable** if and only if it has a closed extension. Then there always exists a smallest closed extension which we call $\bar{t}$, that is, a closed extension whose domain is contained in the domains of all other closed extensions.

Theorem S.15 Let t be any positive quadratic form. Then, there exists a largest closable quadratic form t_r that is smaller than t.

We can now state:

Theorem S.16 Let $t_1, \ldots, t_n, \ldots$ be a sequence of closed, positive, quadratic forms with $t_1 \geq t_2 \geq \cdots \geq 0$. Let $Q(t_\infty) = \bigcup_n Q(t_n)$ and

$$t_\infty(\varphi, \varphi) = \lim_n t_n(\varphi, \varphi) = \inf_n t_n(\varphi, \varphi)$$

Let T_n be the operator corresponding to t_n and let S correspond to the closure of $(t_\infty)_r$. Then $T_n \to S$ in strong resolvent sense.

Notice that part of the first convergence theorem has no analog in the second.

Example Let $Q = \{f \in \mathscr{H} = L^2(-\infty, \infty) \mid f' \in L^2(-\infty, \infty)\}$. It can be shown (see Example 3 in Section X.2) that any $f \in Q$ is continuous and that the t_n defined on Q by

$$t_n(f, g) = n^{-1} \int_{-\infty}^{\infty} \overline{f'(x)} g'(x)\, dx + \overline{f(0)} g(0)$$

are closed, positive, quadratic forms. Clearly $t_1 \geq t_2 \geq \cdots$ and $t_\infty(f, g) = \overline{f(0)} g(0)$. But t_∞ is *not* closed or even closable. In fact $(t_\infty)_r = 0$ so $S = 0$ in this case.

The following result, which is of independent interest, is relevant to the monotone convergence theorems.

Theorem S.17 Let S and T be two positive self-adjoint operators and let s, t be the corresponding quadratic forms. Then $s \leq t$ if and only if $(T + 1)^{-1} \leq (S + 1)^{-1}$.

Proof Suppose that $s \leq t$. Then for any φ, $(T + 1)^{-1}\varphi \in D(T) \subset Q(t) \subset Q(s)$, so

$$\begin{aligned}
(\varphi, (T + 1)^{-1}\varphi) &= ((S + 1)^{-\frac{1}{2}}\varphi, (S + 1)^{\frac{1}{2}}(T + 1)^{-1}\varphi) \\
&\leq (\varphi, (S + 1)^{-1}\varphi)^{\frac{1}{2}}((T + 1)^{-1}\varphi, (S + 1)(T + 1)^{-1}\varphi)^{\frac{1}{2}} \\
&\leq (\varphi, (S + 1)^{-1}\varphi)^{\frac{1}{2}}((T + 1)^{-1}\varphi, (T + 1)(T + 1)^{-1}\varphi)^{\frac{1}{2}} \\
&\leq (\varphi, (S + 1)^{-1}\varphi)^{\frac{1}{2}}(\varphi, (T + 1)^{-1}\varphi)^{\frac{1}{2}}
\end{aligned}$$

proving that $(T + 1)^{-1} \leq (S + 1)^{-1}$. The first inequality is just the Schwarz inequality and the second uses $s \leq t$.

Conversely, suppose that $(T + 1)^{-1} \leq (S + 1)^{-1}$. Let $\eta \in Q(s)$. Then

$$\begin{aligned}
\|(T + 1)^{-\frac{1}{2}}(S + 1)^{\frac{1}{2}}\eta\|^2 &= ((S + 1)^{\frac{1}{2}}\eta, (T + 1)^{-1}(S + 1)^{\frac{1}{2}}\eta) \\
&\leq ((S + 1)^{\frac{1}{2}}\eta, (S + 1)^{-1}(S + 1)^{\frac{1}{2}}\eta) \\
&= (\eta, \eta)
\end{aligned}$$

Thus $(T + 1)^{-\frac{1}{2}}(S + 1)^{\frac{1}{2}}$ is a bounded map of norm at most one from $Q(s)$ to $\mathscr{H}$. It follows that its adjoint is everywhere defined and has norm at most one, i.e., $\mathrm{Ran}((T + 1)^{-\frac{1}{2}}) = Q(t) \subset Q(s)$ and for any $\eta = (T + 1)^{-\frac{1}{2}}\varphi \in Q(t)$,

$$(s + 1)(\eta, \eta) = \|(S + 1)^{\frac{1}{2}}(T + 1)^{-\frac{1}{2}}\varphi\|^2 \leq \|\varphi\|^2 = (t + 1)(\eta, \eta) \quad \blacksquare$$

Our first proof of Theorem S.14 depends on the notion of lower semicontinuous function defined in the supplement to Section IV.5.

Theorem S.18 Let t be a positive quadratic form. Define $\tilde{t}(\varphi)$ on $\mathcal{H}$ by

$$\tilde{t}(\varphi) = \begin{cases} t(\varphi, \varphi) & \text{if } \varphi \in Q(t) \\ \infty & \text{if } \varphi \notin Q(t) \end{cases}$$

Then t is closed if and only if $\tilde{t}$ is lower semicontinuous.

Proof Given any positive quadratic form t, we can define $\mathcal{H}_{+1,t}$ to be the *completion* of $Q(t)$ in the norm $[t(\varphi, \varphi) + (\varphi, \varphi)]^{\frac{1}{2}}$. The bounded linear map $\varphi \mapsto \varphi$ of $Q(t)$ into $\mathcal{H}$ extends to a map i of $\mathcal{H}_{+1,t}$ into $\mathcal{H}$. It is not hard to see (Problem 24 and Section X.3) that t is closable if and only if i is one to one, and in that case $\tilde{t}$ has domain Ran i with $\tilde{t}(\varphi, \psi) = (i^{-1}(\varphi), i^{-1}(\psi))_{+1} - (\varphi, \psi)$.
If t is not closable, we can find $\varphi \in \mathcal{H}_{+1}$ with $i(\varphi) = 0$, i.e., φ_n in $Q(t)$ so that $\varphi_n \to 0$ and $t(\varphi_n - \varphi_m, \varphi_n - \varphi_m) \to 0$ but with $\lim_n t(\varphi_n, \varphi_n)$ nonzero, say 2. We can pass to a subsequence with $t(\varphi_n - \varphi_m, \varphi_n - \varphi_m) \leq \frac{1}{2}$ and $t(\varphi_n, \varphi_n) \geq 1$. Let $\eta_n = \varphi_1 - \varphi_n$. Then $\eta_n \to \varphi_1$ but $\tilde{t}(\varphi_1) \geq 1 > \frac{1}{2} \geq \varliminf \tilde{t}(\eta_n)$ so $\tilde{t}$ is not lower semicontinuous.
Suppose next that t is closable but not closed. Then, we can find $\varphi \notin Q(t)$, $\varphi_n \in Q(t)$ with $\varphi_n \to \varphi$ and $t(\varphi_n - \varphi_m, \varphi_n - \varphi_m) \to 0$. Then $\tilde{t}(\varphi_n)$ is Cauchy, and so it has a finite limit α; thus $\tilde{t}(\varphi) = \infty > \alpha = \varliminf \tilde{t}(\varphi_n)$. Again, $\tilde{t}$ is not lower semicontinuous.
Thus, we see that if $\tilde{t}$ is lower semicontinuous, then t is closed. Conversely, suppose that t is closed. Let T denote the corresponding self-adjoint operator. Let $T_n = TP_n$ where P_n is the spectral projection of T for the interval $[0, n)$. Let $\varphi_m \to \varphi$ and fix n. Noticing that $T_n \leq T$ and $T_n^{\frac{1}{2}}\varphi_m \to T_n^{\frac{1}{2}}\varphi$ since T_n is bounded, we have that

$$(\varphi, T_n\varphi) = \lim_m(\varphi_m, T_n\varphi_m) \leq \varliminf(\varphi_m, T\varphi_m)$$

Since $\tilde{t}(\varphi) = \lim_n(\varphi, T_n\varphi)$ by the monotone convergence theorem for the spectral measure $d\mu_\varphi$, we have that

$$\tilde{t}(\varphi) \leq \varliminf \tilde{t}(\varphi_m)$$

i.e., $\tilde{t}$ is lower semicontinuous. ∎

First Proof of Theorem S.14 By definition $\tilde{t}_\infty(\varphi) = \sup_n \tilde{t}_n(\varphi)$ for any φ where $\tilde{s}$ is defined in Theorem S.18. By that theorem, each $\tilde{t}_n$ is lower semicontinuous. Thus, if $\varphi_m \to \varphi$, then

$$\tilde{t}_n(\varphi) \leq \varliminf \tilde{t}_n(\varphi_m) \leq \varliminf \tilde{t}_\infty(\varphi_m)$$

since $\tilde{t}_n \leq \tilde{t}_\infty$. Taking the sup over n, we see that $\tilde{t}_\infty$ is lower semicontinuous and thus t_∞ is closed.

Now let T_∞ be the operator corresponding to t_∞. By Theorem S.15, $(\varphi, (T_n + 1)^{-1}\varphi) \geq (\varphi, (T_\infty + 1)^{-1}\varphi)$ and $(\varphi, (T_n + 1)^{-1}\varphi)$ is monotone decreasing. It follows that

$$\lim_n (\varphi, (T_n + 1)^{-1}\varphi) = \inf_n (\varphi, (T_n + 1)^{-1}\varphi)$$

has a nonzero value, so we can find a self-adjoint A with zero kernel so that $(T_n + 1)^{-1} \to A$ weakly. Let s be the quadratic form of $S \equiv A^{-1} - 1$. Since $(T_n + 1)^{-1} \geq A \geq (T_\infty + 1)^{-1}$, we have $t_n \leq s \leq t_\infty$. Since $\tilde{t}_\infty = \sup_n \tilde{t}_n$, we have that $\tilde{s} = \tilde{t}_\infty$, i.e., $A = (T_\infty + 1)^{-1}$. Thus $(T_n + 1)^{-1}$ converges weakly to $(T_\infty + 1)^{-1}$. By a similar argument this holds if 1 is replaced by any $c \geq 0$; and then by analyticity we have weak convergence of the resolvents on $\mathbb{C}\setminus[0, \infty)$. As noted in Section VIII.7, this implies strong resolvent convergence. ∎

Proof of Theorem S.15 Let $\mathcal{H}_{+1}$ and i be as defined in the proof of Theorem S.18. Let P be the projection (orthogonal in the natural inner product on $\mathcal{H}_{+1}$) onto Ker i and let $Q = 1 - P$. For $\varphi \in Q(t)$, let $j(\varphi)$ be its natural image in $\mathcal{H}_{+1}$, so that $i \circ j = 1$ and $(j(\varphi), j(\varphi))_{+1} = t(\varphi, \varphi) + (\varphi, \varphi)$. Define

$$t_r(\varphi, \psi) = (Qj(\varphi), j(\psi))_{+1} - (\varphi, \psi)$$
$$t_s(\varphi, \psi) = (Pj(\varphi), j(\psi))$$

We claim that t_r is closable; indeed, the $\mathcal{H}_{+1}$ associated to t_r is just Ran Q and the corresponding $\tilde{i}$ is just the above i restricted to Ran Q. Since Ran $Q \cap$ Ran $P = \{0\}$, this $\tilde{i}$ is one to one.

Next, we note that $t_r \geq 0$. For $iQ = i(1 - P) = i$ since Ran $P \subset$ Ker i. Thus, for any $\varphi \in Q(t)$,

$$(\varphi, \varphi) = (i \circ j(\varphi), i \circ j(\varphi))$$
$$= (iQj(\varphi), iQj(\varphi))$$
$$\leq (Qj(\varphi), Qj(\varphi))_{+1}$$

since i is norm decreasing from $\mathcal{H}_{+1}$ to $\mathcal{H}$.

Now suppose that h is a positive closable form and $h \leq t$. By the corollary to Theorem II.4, there exists a unique positive operator A on $\mathcal{H}_{+1}$ with $(h + 1)(\varphi, \psi) = (j(\varphi), Aj(\psi))_{+1}$. Let $\varphi \in \text{Ran } P \subset \mathcal{H}_{+1}$. Pick $\eta_n \in Q(t)$ with $j(\eta_n) \equiv \varphi_n \to \varphi$. Then $i(\varphi_n) = \eta_n \to i(\varphi) = 0$. Moreover, since the φ_n are Cauchy in $\mathcal{H}_{+1}$, they will be Cauchy in the h norm. Since h is closable, $h(\varphi_n, \varphi_n) \to 0$, i.e., $(\varphi, A\varphi) = 0$. It follows that $h(\varphi, \varphi) = h(Q\varphi, Q\varphi) \leq (Q\varphi, Q\varphi)_{+1}$, so that $h \leq t_r$. Thus t_r is the largest closable form less than t. ∎

Second Proof of Theorem S.14 We give a second proof only of the fact that t_∞ is closed. The rest of the proof is identical to the first proof.

Clearly $t_n \leq t_\infty$; and since t_n is closable, we have that $t_n \leq (t_\infty)_r$. But then $\sup_n t_n(\varphi, \varphi) \leq (t_\infty)_r(\varphi, \varphi)$, i.e., $t_\infty = (t_\infty)_r$, so t_∞ is closable and therefore $\bar{t}_\infty \leq t_\infty$. Moreover, given $\psi_m \in Q(t_\infty)$, such that $\psi_m \to \psi$ in the t_∞ norm, we have that $\psi_m \to \psi$ in t_n norm, so $t_n(\psi, \psi) \leq \bar{t}_\infty(\psi, \psi)$, i.e., $t_n \leq \bar{t}_\infty$. Thus, as above $t_\infty = \bar{t}_\infty$. ∎

Proof of Theorem S.16 As in the proof of Theorem S.14, T_n converges in strong resolvent sense to an operator S given by

$$(\varphi, (S + 1)^{-1}\varphi) = \sup_n (\varphi, (T_n + 1)^{-1}\varphi)$$

Let s be the quadratic form of S. By Theorem S.17, $s \leq t_n$ so $s \leq t_\infty$. Since $s = s_r, s \leq (t_\infty)_r$ and thus, as in the last proof, $s \leq \overline{(t_\infty)_r}$. On the other hand, if T_∞ corresponds to $\overline{(t_\infty)_r}$, then, since $\overline{(t_\infty)_r} \leq t_\infty \leq t_n$, we have that $(T_\infty + 1)^{-1} \geq (T_n + 1)^{-1}$. Taking limits in n, $(T_\infty + 1)^{-1} \geq (S + 1)^{-1}$ so $s \geq \overline{(t_\infty)_r}$, i.e. $s = \overline{(t_\infty)_r}$. ∎

Supplement to VIII.8 More on the Trotter product formula

In this section we want to consider positive self-adjoint operators A and B and prove that

$$\text{s-lim}_{n \to \infty}(e^{-tA/n}e^{-tB/n})^n = e^{-tC} \qquad (S.24)$$

for suitable C under various hypotheses. As we explain in the Notes, some of the ideas below can be used to prove the Trotter product formula for general contraction semigroups on Banach spaces including Theorem VIII.30.

Theorem S.19 (Chernoff's theorem; self-adjoint case) Let A be a positive self-adjoint operator and let $f(t)$ be a family of self-adjoint operators with $0 \leq f(t) \leq 1$. Define $S(t) = t^{-1}(1 - f(t))$. If $S(t)$ converges in strong resolvent sense to A as $t \downarrow 0$, then for any fixed t,

$$\text{s-lim}_{n \to \infty} f\left(\frac{t}{n}\right)^n = e^{-tA} \qquad (S.25)$$

Proof By hypothesis and Theorem VIII.20,

$$\text{s-lim}_{n\to\infty} \exp\left(-tS\left(\frac{t}{n}\right)\right) = e^{-tA}$$

so we need only show that

$$\text{s-lim}_{n\to\infty}\left[\exp\left(-tS\left(\frac{t}{n}\right)\right) - f\left(\frac{t}{n}\right)^n\right] \equiv \text{s-lim}_{n\to\infty} Q(n,t) = 0 \qquad (S.26)$$

But, since $tS(t/n) = n(1 - f(t/n))$ obeys $0 \leq tS(t/n) \leq n$, we have that

$$Q(n,t) = G_n\left(tS\left(\frac{t}{n}\right)\right)$$

where

$$G_n(x) = \begin{cases} e^{-x} - \left(1 - \dfrac{x}{n}\right)^n, & 0 \leq x \leq n \\ e^{-x}, & x \geq n \end{cases}$$

Thus (S.26) follows if we show that $\lim_n \|G_n\|_\infty = 0$. This is easily established by using the fact that $G_n(x) \to 0$ pointwise, and therefore, since

$$|G_n'(x)| = \left|e^{-x} - \left(1 - \frac{x}{n}\right)^{n-1}\right| \leq 1$$

it goes to zero uniformly on bounded intervals. Moreover,

$$\sup_{n \geq x}\left(1 - \frac{x}{n}\right)^n \leq \sup_{a \geq 1}\left[\left(1 - \frac{1}{a}\right)^a\right]^x \leq e^{-x}$$

since $u(x) = (1 - x^{-1})^x$ obeys

$$(\ln u)'(x) = \sum_{k=1}^\infty \frac{1}{x^k}\left(1 - \frac{1}{k}\right) > 0$$

on $(1, \infty)$ and $\lim_{x\to\infty} u(x) = e^{-1}$. ∎

It is now easy to prove the last half of Theorem VIII.31.

Theorem S.20 (Trotter product formula, self-adjoint case) Let A and B be positive self-adjoint operators such that $A + B$ is essentially self-adjoint on $D(A) \cap D(B)$. Then (S.24) holds where C is the closure of $A + B$.

Proof Let $f(t) = e^{-tA/2}e^{-tB}e^{-tA/2}$ and suppose that $\eta \in D(A) \cap D(B)$. Then,

$$t^{-1}(1 - f(t))\eta = e^{-tA/2}e^{-tB}[t^{-1}(1 - e^{-tA/2})\eta]$$
$$+ e^{-tA/2}[t^{-1}(1 - e^{-tB})\eta] + t^{-1}(1 - e^{-tA/2})\eta$$

Using the fact that the product of uniformly bounded, strongly convergent sequences of operators is strongly convergent,

$$t^{-1}(1 - f(t))\eta \to (A + B)\eta = C\eta$$

for $\eta \in D(A) \cap D(B)$. By Theorem VIII.25a, $t^{-1}(1 - f(t))$ converges to C in strong resolvent sense so, by Theorem S.19, s-lim $f(t/n)^n = e^{-tC}$.

Since s-lim$_{t \to 0} f(t) = I$, we have that $f(t/n)^n - f(t/n)^{n-1} = f(t/n)^{n-1} \times (f(t/n) - 1)$ converges strongly to zero. But

$$(e^{-tA/n}e^{-tB/n})^n = e^{-tA/2n}f(t/n)^{n-1}e^{-tA/2n}e^{-tB/n}$$

so (S.24) holds since products of uniformly bounded convergent sequences are convergent. ∎

Theorem S.20 can be strengthened:

Theorem S.21 (Kato's strong Trotter product formula)   Let A and B be positive self-adjoint operators with $Q(A) \cap Q(B)$ dense. Then the quadratic form $\langle \varphi, \psi \rangle \to (\varphi, A\psi) + (\varphi, B\psi)$ with domain $Q(A) \cap Q(B)$ is closed. Let C be the corresponding self-adjoint operator. Then (S.24) holds.

Proof That the given form is closed is easy and left to the problems (Problem 25). By the same argument as at the end of the last proof, it suffices to show that s-lim $f(t/n)^n = e^{-tC}$ where

$$f(t) = e^{-tA/2}e^{-tB}e^{-tA/2}$$

By Chernoff's theorem, we need only show that

$$S(t) \equiv t^{-1}(1 - f(t))$$

converges to C in strong resolvent sense. Define

$$A(t) = t^{-1}(1 - e^{-At}), \qquad B(t) = t^{-1}(1 - e^{-Bt})$$

and $C(t) = A(\tfrac{1}{2}t) + B(t)$. Notice that

$$\|tA(t)\| \leq 1; \qquad \|tB(t)\| \leq 1 \tag{S.27}$$

$$\text{s-lim}_{t \downarrow 0} tA(t) = \text{s-lim}_{t \downarrow 0} tB(t) = 0 \tag{S.28}$$

Moreover, since $t^{-1}(1 - e^{-xt}) = x \int_0^1 e^{-xts} \, ds$ is positive and monotone decreasing in t for x fixed, we see that $A(t)$ and $B(t)$ are monotone increasing to A and B, respectively, as $t \downarrow 0$. It follows from the first monotone convergence theorem for forms (Theorem S.14) that for $b > 0$,

$$(b + C(t))^{-1} - (b + C)^{-1} \xrightarrow{s} 0 \tag{S.29}$$

Rewriting $S(t)$ in terms of $A(t)$ and $B(t)$, we see that

$$S(t) = C(t) + E(t)$$

where

$$E(t) = \frac{t}{4} A\left(\frac{t}{2}\right)(1 - tB(t))A\left(\frac{t}{2}\right) - \frac{t}{2} A\left(\frac{t}{2}\right)B(t) - \frac{t}{2} B(t)A\left(\frac{t}{2}\right)$$

Fix $b > 0$ and let

$$Q(t) = (C(t) + b)^{-\frac{1}{2}} E(t)(C(t) + b)^{-\frac{1}{2}}$$

We claim that

$$(\varphi, Q(t)\varphi) \geq -\frac{\sqrt{2}}{2} \|\varphi\|^2 \tag{S.30}$$

and

$$\text{s-lim}_{t \downarrow 0} Q(t) = 0 \tag{S.31}$$

Accepting (S.30) and (S.31) for the moment, we can complete the proof. Since all operators involved are bounded, we have that

$$(S(t) + b)^{-1} = (C(t) + b)^{-\frac{1}{2}}(1 + Q(t))^{-1}(C(t) + b)^{-\frac{1}{2}}$$

Thus,

$$(S(t) + b)^{-1} - (C(t) + b)^{-1} = -(C(t) + b)^{-\frac{1}{2}}(1 + Q(t))^{-1} Q(t)(C(t) + b)^{-\frac{1}{2}}$$

By (S.30), $\|(1 + Q(t))^{-1}\| \leq 2/(2 - \sqrt{2})$ so since $\|(C(t) + b)^{-\frac{1}{2}}\| \leq b^{-\frac{1}{2}}$, we need only that s-lim$_{t \downarrow 0} Q(t)(C(t) + b)^{-\frac{1}{2}} = 0$. This follows from (S.29) and (S.31). Thus

$$\text{s-lim}_{t \downarrow 0}[(S(t) + b)^{-1} - (C(t) + b)^{-1}] = 0$$

so, by (S.29), $S(t)$ converges to C in strong resolvent sense.

Therefore, we need only prove (S.30) and (S.31). We begin by noting that the first term of $E(t)$ is positive, so

$$(\varphi, Q(t)\varphi) \geq -\operatorname{Re}\left((C(t) + b)^{-\frac{1}{2}}\varphi, tA\left(\frac{t}{2}\right)B(t)(C(t) + b)^{-\frac{1}{2}}\varphi\right)$$

$$\geq -\|t^{\frac{1}{2}}B(t)^{\frac{1}{2}}\| \left\|t^{\frac{1}{2}}A\left(\frac{t}{2}\right)^{\frac{1}{2}}\right\| \left\|A\left(\frac{t}{2}\right)^{\frac{1}{2}}(C(t) + b)^{-\frac{1}{2}}\varphi\right\|$$
$$\times \|B(t)^{\frac{1}{2}}(C(t) + b)^{-\frac{1}{2}}\varphi\|$$

$$\geq -\frac{\sqrt{2}}{2}\left((C(t) + b)^{-\frac{1}{2}}\varphi, \left[A\left(\frac{t}{2}\right) + B(t)\right](C(t) + b)^{-\frac{1}{2}}\varphi\right)$$

$$\geq -\frac{\sqrt{2}}{2}(\varphi, \varphi)$$

where in the last step we used $\|C(t)(C(t) + b)^{-1}\| \leq 1$, and in the step before that the inequality $xy \leq \frac{1}{2}(x^2 + y^2)$. This proves (S.30).

To prove (S.31), first note that

$$\left\|A\left(\frac{t}{2}\right)^{\frac{1}{2}}(C(t) + b)^{-\frac{1}{2}}\varphi\right\|^2 + \|B(t)^{\frac{1}{2}}(C(t) + b)^{-\frac{1}{2}}\varphi\|^2 + b\|(C(t) + b)^{-\frac{1}{2}}\varphi\|^2$$
$$= \|\varphi\|^2 \qquad (S.32)$$

(S.32) implies in the first place that

$$\left\|A\left(\frac{t}{2}\right)^{\frac{1}{2}}(C(t) + b)^{-\frac{1}{2}}\right\| \leq 1 \qquad \|B(t)(C(t) + b)^{-\frac{1}{2}}\| \leq 1 \qquad (S.33)$$

Moreover, the analog of (S.32) for A, B, C replacing $A(\frac{1}{2}t)$, $B(t)$, $C(t)$, shows that weak convergence of the three operators in (S.32) as $t \downarrow 0$ implies strong convergence (Problem 26). Because of (S.33), one only needs

$$(\varphi, B(t)^{\frac{1}{2}}(C(t) + b)^{-\frac{1}{2}}\psi) \to (\varphi, B^{\frac{1}{2}}(C + b)^{-\frac{1}{2}}\psi)$$

for φ in a dense set. This convergence for φ in $D(B)$ and any ψ follows from (S.29). Thus

$$\operatorname*{s-lim}_{t \downarrow 0} B(t)^{\frac{1}{2}}(C(t) + b)^{-\frac{1}{2}} = B^{\frac{1}{2}}(C + b)^{-\frac{1}{2}}$$

and similarly for $A(\frac{1}{2}t)$ and A. Now, to prove (S.31), take a typical term in the sum like

$$(C(t) + b)^{-\frac{1}{2}} \frac{t}{4} A\left(\frac{t}{2}\right)(1 - tB(t))A\left(\frac{t}{2}\right)(C(t) + b)^{-\frac{1}{2}}$$

and write it as $\alpha\beta\gamma\delta\varepsilon$ with

$$\alpha = (C(t) + b)^{-\frac{1}{2}} A\left(\frac{t}{2}\right)^{\frac{1}{2}}$$

$$\beta = \frac{1}{4}\left[tA\left(\frac{t}{2}\right)\right]^{\frac{1}{2}}$$

$$\gamma = (1 - tB(t))$$

$$\delta = \left[tA\left(\frac{t}{2}\right)\right]^{\frac{1}{2}}$$

and

$$\varepsilon = A(t)^{\frac{1}{2}}(C(t) + b)^{-\frac{1}{2}}$$

α, β, γ, δ, and ε are uniformly bounded and since ε is strongly convergent, and $\delta \to 0$ strongly as $t \to 0$, the product converges strongly to zero. ∎

Additional supplement **Uses of the maximum principle**

It is not evident from looking at the basic definitions in functional analysis that the theory of analytic functions should play any role in the subject at all. That complex variable techniques *are* applicable is due mainly to the analyticity of the resolvent (Theorem VI.5) and to the analyticity properties of the Fourier transforms of functions with restricted support (see Section IX.3). Since analytic functions have such strong properties, it is not surprising that theorems from complex analysis are useful, indeed central, when many of the basic objects under study have analyticity properties. One of the most useful of these theorems is the maximum principle which states: Let f be an analytic function on an open bounded subset D of $\mathbb{C}$ and suppose that f *is* continuous on $\bar{D}$. Then $|f|$ achieves its maximum on the boundary. Often, it is not the maximum principle itself which is applied, but a corollary of it or an extended maximum principle for an unbounded domain. See, for example, the three line lemma (in the Appendix to Section IX.4), Theorem XI.89, Theorem XII.18, the lemma to Corollary 4 in Section XIII.4 (this is equivalent to Example 2 below via a conformal mapping), Carlson's theorem (the lemma to Theorem XIII.61), and the Borel–Carathéodory theorem (Lemma 5 in Section XIII.17).

Uses of the maximum principle 383

In this section we want to acquaint the reader with a nice trick which is the main device for proving many of the extended maximum principles and corollaries of the maximum principle. Here is the idea: Let f be analytic in D and continuous on $\overline{D}$. If h is another analytic function in D which is continuous on $\overline{D}$ and which satisfies $h(z_0) = 1$ for some $z_0 \in D$, then

$$|f(z_0)| = |h(z_0)f(z_0)| \leq \sup_{w \in \partial D} |h(w)| \, |f(w)|$$

By choosing h cleverly one can arrange to give more weight to some parts of the boundary where we have more information. We can let h depend on z_0 and often it depends on an auxiliary parameter. We illustrate this idea by two examples.

Example 1 Let $f(z)$ be a function which is continuous in the region $R = \{z \mid 0 \leq \operatorname{Re} z \leq 1, \operatorname{Im} z \geq 0\}$, analytic in the interior. Suppose that $\sup_{z \in R} |f(z)| < \infty$ and that $\lim_{y \to \infty} |f(iy)| = 0$. Then, for each $\varepsilon > 0$,

$$\lim_{y \to \infty} \sup_{0 \leq x \leq 1 - \varepsilon} |f(x + iy)| = 0 \tag{S.34}$$

To prove this, let $C \equiv \sup_{z \in R} |f(z)|$ and let $z_0 = x_0 + iy_0$ be a point in R. Let the region $D(y_0)$ be given by

$$D(y_0) = \{z \in R \mid 0 < \operatorname{Re} z < 1, \tfrac{1}{2}y_0 < \operatorname{Im} z < \tfrac{3}{2}y_0\}$$

and take

$$h(z) = e^{(z - z_0)^2 - B(z - z_0)}$$

where B is a positive constant to be chosen in a moment. Since h is analytic and $h(z_0) = 1$, the maximum principle for D implies that

$$|f(z_0)| \leq \max\{a_1, a_2, a_3, a_4\}$$

where

$$a_1 = \max_{\frac{1}{2}y_0 \leq y \leq \frac{3}{2}y_0} |f(1 + iy)h(1 + iy)| \leq Ce^{-B(1 - x_0) + 1}$$

$$a_2 = \max_{0 \leq x \leq 1} |f(x + \tfrac{1}{2}iy_0)h(x + \tfrac{1}{2}iy_0)| \leq Ce^{B - (y_0^2/4) + 1}$$

$$a_3 = \max_{0 \leq x \leq 1} |f(x + \tfrac{3}{2}iy_0)h(x + \tfrac{3}{2}iy_0)| \leq Ce^{B - (y_0^2/4) + 1}$$

$$a_4 = \max_{\frac{1}{2}y_0 \leq y \leq \frac{3}{2}y_0} |f(iy)h(iy)| \leq Ce^{B + 1} \max_{y \geq \frac{1}{2}y_0} |f(iy)|$$

We now choose B (depending on y_0) so that

$$e^{-2B} = e^{-y_0^2/4} + \max_{y \geq \frac{1}{2}y_0} |f(iy)|$$

This implies that $B < y_0^2/8$, so $a_2(y_0) \to 0$ and $a_3(y_0) \to 0$ as $y_0 \to \infty$. Also, $B(y_0) \to \infty$ as $y_0 \to \infty$, so $a_4(y_0) \to 0$ since $e^B \max_{y \geq \frac{1}{2}y_0} |f(iy)| = e^{-B} - e^{B-(y_0^2/4)}$. Finally, $a_1(y_0) \to 0$ as $y_0 \to \infty$ uniformly for all $x_0 \leq 1 - \varepsilon$, so (S.34) holds.

Example 2 Let $\{f_n(z)\}_{n=1}^{\infty}$ be a sequence of functions which are analytic in the unit disc $R = \{z \mid |z| < 1\}$ and which satisfy $\sup_{n, |z|<1} |f_n(z)| \equiv C < \infty$. Fix some $\theta_0 > 0$ and suppose that for each n, f_n can be extended to be continuous on

$$R \cup \{z \mid |z| = 1, |\arg z| \leq \theta_0\}, \text{ and that } \lim_{n \to \infty} \sup_{|\theta| \leq \theta_0} |f_n(e^{i\theta})| = 0$$

We shall show that for each $\delta > 0$,

$$\lim_{n \to \infty} \sup_{|z| \leq 1-\delta} |f_n(z)| = 0 \qquad (S.35)$$

Let $D_{\theta_0} = \{z \in R \mid |\arg z| < \theta_0\}$ and define $g_n(z) = f_n(z^{2\theta_0/\pi})$ for $z \in D_{\pi/2} \cup \{z \mid |z| = 1, |\arg z| \leq \pi/2\}$. Then $\sup_{n, z \in D_{\pi/2}} |g_n(z)| \leq C$ and

$$\lim_{n \to \infty} \sup_{|\theta| \leq \pi/2} |g_n(e^{i\theta})| = 0$$

On the straight piece of the boundary of $D_{\pi/2}$ we have no information except boundedness of the boundary values of g_n. Take $z \in D_{\pi/2}$ and set $h(z) = e^{B(z-z_0)}$ for $B > 0$ fixed. Then, by the maximum principle for $D_{\pi/2}$,

$$|g_n(z_0)| \leq \sup_{z \in \partial D_{\pi/2}} |e^{B(z-z_0)} g_n(z)|$$

$$\leq C_1(B) \sup_{|\theta| \leq (\pi/2)} |g_n(e^{i\theta})| + Ce^{-Bx_0}$$

Thus, $\lim_{n \to \infty} |g_n(z_0)| \leq Ce^{-Bx_0}$ and since B is arbitrary, we conclude that $\lim_{n \to \infty} |g_n(z_0)| = 0$. Thus, $\lim_{n \to \infty} |f_n(z_0)| = 0$ for $z_0 \in D_{\theta_0}$. Standard methods in complex function theory now permit one to show easily that $f_n(z) \to 0$ uniformly on compact subsets of D_{θ_0} and analytic continuation then allows one to conclude (S.35). The reader is asked to provide these details in Problem 28.

We remark that in functional analysis the maximum principle and its extensions are typically applied to Banach space-valued analytic functions. This generalization causes no problems as the reader is asked to show in Problem 29. In Problem 30, the reader can use the technique of this section to prove a maximum principle for an unbounded domain.

NOTES

Supplement II.2 Theorem S.1 and its proof by these means is due to J. von Neumann, "On Rings of Operators, III," *Ann. Math.* **41** (1940), 127. The Bergmann kernel is named after the work of S. Bergman, "Ueber Hermitische unendliche Formen, die zu einem Bereich gehören, nebst Anwendungen auf Fragen der Abbildung durch Funktionen zweier komplexen Veränderlichen," *Math. Z.* **29** (1929), 640–677. For further discussion and applications to partial differential equations and several complex variables, see S. Bergman, *The Kernel Function and Conformal Mapping*, Amer. Math. Soc. Survey, No. 5 (2nd ed., 1970), Amer. Math. Soc., Providence, Rhode Island.

Supplement IV.5 Theorem S.6 is a basic technique in nonlinear functional analysis sometimes called the *direct method of the calculus of variations*. It is a descendent of ideas of Hilbert. The existence properties of the Thomas–Fermi equations are discussed in E. Lieb and B. Simon, "The Thomas–Fermi Theory of Atoms, Molecules and Solids," *Advances in Mathematics* **23** (1977), 22–116, and of the Hartree equation in E. Lieb and B. Simon, "The Hartree–Fock Theory for Coulomb Systems," *Commun. Math. Phys.* **53** (1977), 185–194.

For further discussion and examples of this method, see Chapter 6 of M. Berger, *Nonlinearity and Functional Analysis*, Academic Press, New York, 1977.

There is a readable discussion of the existence of geodesics which does not use Theorem S.6 but which uses instead ideas from differential geometry in M. Spivak, *A Comprehensive Introduction to Differential Geometry*, Vol. I, Publish or Perish, Cambridge, Massachusetts, 1970. Theorem S.8 is closely related to what is usually called the Hopf–Rinow theorem.

Our proof of the existence of geodesics actually shows that there is a minimizing path in each homotopy class (see Problem 18). For a beautiful application of this result to prove the existence of periodic orbits in certain classical mechanical systems, see p. 248 of V. I. Arnold, *Mathematical Methods of Classical Mechanics*, Springer-Verlag, New York, 1978.

Supplement V.5 For a discussion of homology and the Brouwer fixed point theorem, see any book on algebraic topology, e.g., the book of Hilton and Wylie quoted in the notes to Section V.5 or J. Vick, *Homology Theory*, Academic Press, New York, 1973. There is an "elementary" proof of the Brouwer theorem in J. Milnor, "Analytic proofs of the 'hairy ball theorem' and the Brouwer fixed point theorem," *Amer. Math. Mon.* **85** (1978), 521–524, with more details in J. Franklin, *Methods of Mathematical Economics*, Springer, New York, 1980.

Supplement VIII.7 Versions of the monotone convergence theorems that suppose that t_∞ is closed go back at least as far as Kato's book quoted in the Notes to Section VIII.6. That t_∞ is automatically closed in the increasing case is due to D. Robinson, *The Thermodynamic Pressure in Quantum Statistical Mechanics*, Springer Lecture Notes in Physics **9** (1971), Springer-Verlag, New York, where an essentially equivalent result appears. Unaware of this work, E. B. Davies found Theorem S.18 and the resulting proof of Theorem S.14 which we give as the first proof in "A Model for Absorbtion or Decay," *Helv. Phys. Acta* **48** (1975), 365–382, as did T. Kato (quoted in the next mentioned paper). See also B. Simon, "Lower Semicontinuity of Positive Quadratic Forms," *Proc. Roy. Soc. Edinburgh* **29** (1977), 267–273.

The decomposition Theorem S.15 and the identification of the limit in Theorem S.16 are due to B. Simon, "A Canonical Decomposition for Quadratic Forms with Applications to Monotone Convergence Theorems," *J. Functional Anal.* **28** (1978), 377–385.

There is a version of the first monotone convergence theorem even in the case where $Q(t_\infty)$ is not dense; see Simon's two papers quoted on page 385.

Supplement VIII.8 Chernoff's theorem was proven in his *J. Functional Anal.* paper quoted in the Notes to Section VIII.8. He proves it when A is the generator of any contraction semigroup on any Banach space and $f(t)$ any family of contractions (operators with $\|f(t)\| \leq 1$). One advantage of the form of his theorem is that it implies that for many real-valued functions F and G,

$$\left[F\left(\frac{tA}{n}\right)G\left(\frac{tB}{n}\right)\right] \to e^{-t(A+B)}$$

e.g., under the hypotheses of Theorem S.20,

$$\left[\left(1 + \frac{tA}{n}\right)^{-1}\left(1 + \frac{tB}{n}\right)^{-1}\right]^n \to e^{-t(A+B)}$$

The proofs in this section are such that it is easy to verify the uniformity of convergence in the t variable as t runs through compact subsets of $[0, \infty)$.

Kato's strong version of the product formula is proven in T. Kato, "Trotter's Product Formula for an Arbitrary Pair of Self-Adjoint Contraction Semigroups," in *Topics in Functional Analysis* (G. C. Rota, ed.), Academic Press, New York, 1978, pp. 185-195. Extension to more than two factors and some nonlinear operators can be found in T. Kato and K. Masuda, "Trotter's Product Formula for Nonlinear Semigroups Generated by the Subdifferentials of Convex Functionals," *J. Math. Soc. Japan* **30** (1978), 169-178. Extensions to include generators of holomorphic semigroups can be found in Kato's paper and a *kind* of Trotter product formula for unitary groups under the hypotheses of Theorem S.21 is proven in T. Ichinose, "A Product Formula and Its Application to the Schrödinger Equation," to appear.

Kato actually proves a stronger result than we prove in Theorem S.21. Namely, he proves that whenever A and B are positive self-adjoint operators on a Hilbert space, then s-lim$(e^{-A/n} e^{-B/n})^n = e^{-C} P$ where P is the projection onto $\overline{Q(A) \cap Q(B)}$ and C is the obvious operator on Ran P.

There has been a considerable amount of work on nonlinear product formulas. See, for example: H. Brezis and A. Pazy, "Semigroups of Nonlinear Contractions on Convex Sets," *J. Functional Anal.* **5** (1970), 237-281; H. Brezis and A. Pazy, "Convergence and Approximation of Semigroups of Nonlinear Operators in Banach Spaces," *J. Functional Anal.* **9** (1972), 63-74; Paul R. Chernoff, "Product Formulas, Nonlinear Semigroups, and Addition of Unbounded Operators," *Memoirs of the American Mathematical Society*, Number 140; A. J. Chorin, T. J. R. Hughes, M. F. McCracken, and J. E. Marsden, "Product Formulas and Numerical Algorithms," *Comm. Pure Appl. Math.* **XXXI** (1978), 205-256; M. G. Crandall and T. M. Liggett, "Generation of Semi-Groups of Nonlinear Transformations on General Banach Spaces," *Amer. J. Math.* **93** (1971), 265-298; J. Marsden, "On Product Formulas for Nonlinear Semigroups," *J. Functional Anal.* **13** (1973), 51-72; Eric Schechter, "Well-Behaved Evolutions and Trotter Products," Thesis, University of Chicago, 1978; G. F. Webb, "Exponential Representation of Solutions to an Abstract Semi-Linear Differential Equation," *Pacific J. Math.* **70** (1977), 269-279; and Fred B. Weissler, "Construction of Non-Linear Semi-Groups Using Product Formulas," *Israel J. Math.* **29** (1978), 265-275. These product formulas play an important role in the existence theorems for certain classes of nonlinear partial differential equations.

Additional Supplement The results of this section and Problem 25 are often called Phragmén–Lindelöf theorems. Some of the original papers are: J. Hadamard, "Sur les fonctions entières," *Bull. Soc. Math. France* **24** (1896), 186–187; E. Phragmén, "Sur une extension d'un théorème classique de la théorie des fonctions," *Acta Math.* **28** (1904), 351–368; E. Lindelöf and E. Phragmén, "Sur une extension d'un principe classique de l'analyse et sur quelque propriétés des fonctions monogènes dans le voisinage d'un point singulier," *Acta Math.* **31** (1908), 381–406; E. Lindelöf, "Sur un principe général d'analyse et ses applications à la theorie de la représentation conforme," *Acta Soc. Sci. Finn.* **46**(4) (1915), 1–35. For simple proofs see L. Ahlfors, "On Phragmén–Lindelöf's principle," *Trans. Amer. Math. Soc.* **41** (1937), 1–8.

PROBLEMS

1. Use Theorem S.1 to prove Theorems I.19 and I.20 in the general σ-finite case.

2. (a) Let $\Omega = \{z \mid |z| < 1\}$ and define $T_r: L^2(\Omega) \to L^2(\Omega)$ by
$$(T_r f)(z) = f(zr)$$
for $0 < r < 1$. Prove that $T_r f \to f$ as $r \to 1$ for any $f \in L^2$. (*Hint:* Prove it for $f \in C(\bar{\Omega})$ and use an $\varepsilon/3$ argument.)
   (b) If f is analytic in $\{z \mid |z| < 1 + \varepsilon\}$ for some $\varepsilon > 0$, prove that the Taylor series for f about $z = 0$ converges to f in $L^2(\Omega)$.
   (c) Conclude that $\{z^n\}_{n=0}^\infty$ is a basis for $\mathscr{A}(\Omega)$.

3. Let K be the Bergmann reproducing kernel for some set Ω.
   (a) Prove that $|K(z, w)| \leq K(z, z)^{\frac{1}{2}} K(w, w)^{\frac{1}{2}}$.
   (b) Prove that $K(z, z) + K(w, w) - K(z, w) - K(w, z) \geq 0$.
   (c) Fix $z \in \Omega$. Let
$$a_{ij}(z) = \frac{\partial}{\partial x_i} \frac{\partial}{\partial y_j} K(x_1 + ix_2, y_1 + iy_2)\bigg|_{x_1 + ix_2 = z = y_1 + iy_2}$$
   Prove that $a = \{a_{ij}\}$ is a positive definite matrix.

   Remark $\{a_{ij}(z)\}$ defines a metric in the sense of Riemann geometry and introduces a natural geometric structure into Ω. For example, if Ω is the unit disk, the resulting geometry is one of the standard non-Euclidean geometries.

4. (a) Prove that equality holds in (S.8) only if $a = b^{q-1}$.
   (b) Suppose that $\|f\|_p = \|g\|_q = 1$. Prove that equality in (S.7) holds if and only if $|g| = |f(x)|^{p-1}$.
   (c) Prove that equality holds in general in (S.7) only if g is zero (a.e.) or if
$$g(x) = \lambda \overline{f(x)} |f(x)|^{p-2}$$
   for some fixed $\lambda \in \mathbb{C}$.
   (d) Let $h \in L^1$. Prove that $\int h \, d\mu = \|h\|_1$ if and only if h is a.e. nonnegative.
   (e) Prove that $|\int f(x)g(x) \, d\mu(x)| \leq \|f\|_p \|g\|_q$ with equality only if $g = |f|^{p-1} \operatorname{sgn} f$ where $\operatorname{sgn} f = \bar{f}/|f|$ at points where $f \neq 0$ and $\operatorname{sgn} f = 0$ otherwise.

5. Use the method of the proof of (S.8) to prove Young's inequality: $xy \leq e^x + y \log y - y$ for all $x > 0, y > 0$.

†6. By mimicking the proof of Theorem IV.16 and its lemma, prove that any $L \in (L^p)^*$ is a finite linear combination of positive functionals.

†7. (a) Let X be σ-finite and let $Y \subset X$. Given $L \in L^p(X, d\mu)^*$, define $\tilde{L} \in L^p(Y, d\mu)^*$ by $\tilde{L}(f) = L(\tilde{f})$ where $\tilde{f}$ is the function on X obtained by extending f to be zero on $X \setminus Y$. Prove that $\|\tilde{L}\| \leq \|L\|$.
   (b) Given Theorem S.4 for finite measure spaces, prove the same result in the σ-finite case.

8. Prove that no sequence that is not eventually constant can be a universal net.

9. (a) Let α be a function of bounded variation on $[0, 1]$. Prove that for any $t \in [0, 1]$, $\alpha(t - 0) \equiv \lim_{s \uparrow t} \alpha(s)$ and $\alpha(t + 0) \equiv \lim_{s \downarrow t} \alpha(s)$ exist.
   (b) Show that $\alpha(t - 0) = \alpha(t + 0)$ for all but at most countably many t's.
   (c) Show that among all α's leading to the same Riemann–Stieltjes integral, there is precisely one with $\alpha(0) = 0$ and $\alpha(t) = \alpha(t - 0)$ for all t.
   (d) Show that if $\tilde{\alpha}$ obeys $\tilde{\alpha}(0) = 0$, $d\tilde{\alpha} = d\alpha$ where α is the function of part (c), and Variation (α) = Variation($\tilde{\alpha}$), then for any t, either
   $$\alpha(t - 0) \leq \tilde{\alpha}(t) \leq \alpha(t + 0) \quad \text{or} \quad \alpha(t + 0) \leq \tilde{\alpha}(t) \leq \alpha(t - 0)$$

10. Let α be of bounded variation. Show that
    (a) $(b - a)^{-1} \int_a^b (x - a) \, d\alpha = \alpha(b - 0) - [\int_a^b \alpha(x) \, dx](b - a)^{-1}$ and use Problem 9a to show that
    $$\lim_{b \downarrow a} (b - a)^{-1} \int_a^b (x - a) \, d\alpha = 0$$
    (b) Fix $a < c$ and for δ small let f_δ be given by
    $$f_\delta(x) = \begin{cases} 0, & x \leq a \text{ or } x \geq c \\ \delta^{-1}(x - a), & a \leq x \leq a + \delta \\ 1, & a + \delta \leq x \leq c - \delta \\ \delta^{-1}(c - x), & c - \delta \leq x < c \end{cases}$$
    Show that
    $$\lim_{\delta \downarrow 0} \int f_\delta(x) \, d\alpha(x) = \alpha(c - 0) - \alpha(a + 0)$$
    (c) Prove that if $\int \cdot \, d\alpha$ is a positive Riemann–Stieltjes integral on $C[0, 1]$, then the α with $\alpha(t) = \alpha(t - 0)$ is monotone nondecreasing so that corresponding Lebesgue–Stieltjes measure is positive.

†11. Let $F: X \to (-\infty, \infty]$. Show that F is lower semicontinuous if and only if for all a, $\{x \mid F(x) \leq a\}$ is closed.

†12. Let C be a convex subset of a locally convex space X. Use the separating hyperplane theorem (Theorem V.4) to show that C is closed in the topology of X if and only if C is closed in the weak ($\sigma(X, X^*)$) topology.

13. A function F on a vector space V is called *strictly convex* if and only if $F(\lambda x + (1 - \lambda)y) < \lambda F(x) + (1 - \lambda)F(y)$ for $x \neq y$ and $0 < \lambda < 1$. Show that a strictly convex function has at most one point where its minimum value is taken and that any stationary point of F is its minimizing point. Use these ideas to show that (S.12) has a unique L^2 solution.

14. Let $\|\cdot\|^{(0)}$ and $\|\cdot\|^{(1)}$ be consistent norms on a vector space V in the sense of Section IX.4. Let

$$\|x\|^+ = \inf\{\|y\|^{(0)} + \|z\|^{(1)} \mid x = y + z\}$$

Let X_0, X_1, X_+ be the completions of V in the norms $\|\cdot\|^{(0)}, \|\cdot\|^{(1)}, \|\cdot\|^+$. Let $\|\cdot\|_{(0)}, \|\cdot\|_{(1)}$ be the norm of X_0^*, X_1^*. Prove that $X_+^* = X_0^* \cap X_1^*$ with the norm

$$\|\ell\| = \max\{\|\ell\|_{(0)}, \|\ell\|_{(1)}\}$$

15. Let H be a C^1 convex function on $(-\infty, \infty)$ with $|H(u)| \leq C|u|^p$ for some $p \geq 1$. Let $f \in C_0^\infty$. Show that

$$-\Delta u + H'(u) = f$$

has a solution in $L^p \cap L^2$.

16. Let $h(x) = [\max\{1, |x|\}]^{-1}$ and $g(x) = |x|^{-1} - h(x)$. In the language of Example 3 of the supplement to Section IV.5, prove that $h \in K^*$ and that $g \in L^{\frac{3}{2}}$. Conclude that

$$f \to \int f(x) V(x) \, dx$$

is an element of $L^{\frac{3}{2}} + K^*$.

†17. Fill in the missing details of Examples 2 and 4 of the supplement to Section IV.5.

18. Let M be a closed subset of $\mathbb{R}^\nu$ and suppose that $\{\gamma \text{ in } M \mid \gamma(0) = x, \gamma(1) = y\}$ lies in several homotopy classes. Fix one such class c and let

$$Y_c = \{f \in Y \mid \gamma^{(f)} \in c\}$$

where Y and $\gamma^{(f)}$ are given in the proof of Theorem S.8. Show that Y_c is weakly closed and conclude that there is a curve in c minimizing the length among all curves in c.

19. Prove Theorem S.8 "directly" by showing that if γ_n is such that $L(\gamma_n) \to \inf\{L(\gamma)\}$ and that if γ_n is geodesically parametrized, then $\{\gamma_n\}$ is equicontinuous and so by the Ascoli–Arzela theorem, there exists a subsequence $\gamma_n \to \gamma_\infty$.

20. Let C be a compact convex subset of $\mathbb{R}^n$ so that $0 \in C$ and the vectors in C span $\mathbb{R}^n$. Prove that C has nonempty interior.

21. Verify that the map g in Theorem S.11 is C^1.

22. (a) Show that if $\lambda_n, \varphi_n(x)$ and $\lambda_m, \varphi_m(x)$ are solutions of (S.22) and $\lambda_n \neq \lambda_m$, then

$$\int_0^1 \varphi_n(x) \varphi_m(x) \, dx = 0$$

(b) Prove that any λ for which there is a $\varphi(x)$ satisfying (S.22) must satisfy

$$\lambda \geq \min_{0 \leq x \leq 1} q(x)$$

(c) Using part (b), extend the proof of Theorem S.12 to the case where $q(x)$ can be negative.

23. The purpose of this problem is to extend Theorem S.12 to three dimensions. Let $\Omega \subset \mathbb{R}^3$ be a bounded open region with smooth boundary $\partial\Omega$. Let $G_0(x, y) \equiv (4\pi|x - y|)^{-1}$.
  (a) Let $y \in \Omega$. We suppose that the Dirichlet problem is solvable (see Section VI.4) and let $H(x, y)$ be the solution of $-\Delta_x H(x, y) = 0$, $H(x, y) = G_0(x, y)$ for x on $\partial\Omega$. Let $G_D(x, y) = G_0(x, y) - H(x, y)$. Using the maximum principle, show that
$$0 \leq G_D(x, y) \leq G_0(x, y)$$
  (b) For $f \in C_0^\infty(\Omega)$ define $(T_0 f)(x) = \int_\Omega G_D(x, y) f(y)\, dy$. Prove that $(-\Delta)T_0 f = f = T_0(-\Delta)f$ and $(T_0 f)(x) = 0$ for x on $\partial\Omega$. Conclude that G_D is symmetric.
  (c) Prove that T_0 is Hilbert–Schmidt on $L^2(\Omega)$.
  (d) Show that any eigenvalue of T_0 is positive and, since T_0 is compact, conclude that T_0 is a positive operator.
  (e) Suppose that $q(z) > 0$ for all $z \in \Omega$ and that q is C^∞. Using the fact that T_0 is positive, prove that
$$f(x) = -\int_\Omega G_D(x, z) q(z) f(z)\, dz$$
  has no solutions. Conclude that there is a unique solution $k(x, y)$ of
$$k(x, y) = G_D(x, y) - \int_\Omega G_D(x, z) q(z) k(z, y)\, dz$$
  for all $y \in \Omega$.
  (f) Use the maximum principle to prove that $0 \leq k(x, y) \leq G_D(x, y)$ and conclude that $k(x, y)$ is the kernel of a Hilbert–Schmidt operator T.
  (g) Prove that T is a two-sided inverse for $-\Delta + q$ with the boundary condition $f \upharpoonright \partial\Omega = 0$. Show that the eigenvalue problem
$$-\Delta u(x) + q(x) u(x) - \lambda u(x) = 0 \quad \text{for} \quad x \in \Omega$$
$$u(x) = 0 \quad \text{for} \quad x \in \partial\Omega$$
  has an L^2 basis of eigenfunctions.

24. Fill in the details of the $\mathscr{H}_{+1,t}$ construction of Theorem S.18.

†25. Let A and B be closed semibounded quadratic forms with $Q \equiv Q(A) \cap Q(B)$ dense. Prove that the form $\langle \varphi, \psi \rangle \to (\varphi, A\psi) + (\varphi, B\psi)$ on $Q \times Q$ defines a *closed* semibounded quadratic form.

†26. (a) Let $\varphi_n \to \varphi$ weakly and $\|\varphi_n\| \to \|\varphi\|$. Prove that $\varphi_n \to \varphi$ in norm (*Hint*: Compute $\|\varphi - \varphi_n\|^2$.)
  (b) Let $\varphi_n^{(1)} \to \varphi^{(1)}, \ldots, \varphi_n^{(j)} \to \varphi^{(j)}$ weakly and
$$\sum_{i=1}^{j} \|\varphi_n^{(i)}\|^2 \to \sum_{i=1}^{j} \|\varphi^{(i)}\|^2$$
  Prove that $\varphi_n^{(i)} \to \varphi^{(i)}$ in norm.

27. Let A and B be semibounded self-adjoint operators with $A + B$ essentially self-adjoint on $D(A) \cap D(B)$. Let C be the operator associated to the form of Problem 25. Prove that $C = \overline{A + B}$.

† 28. Complete the complex variables argument outlined at the end of Example 2 in the additional supplement.

29. Let X be a Banach space and let D be a bounded open subset of $\mathbb{C}$. Let $f: \overline{D} \to X$ in such a way that for each $\ell \in X^*$, $\ell \circ f$ is analytic in D and continuous on $\overline{D}$. Prove that
$$\sup_{z \in D} \|f(z)\| = \sup_{z \in \partial D} \|f(z)\|$$

30. Suppose that f is analytic in the wedge $N = \{z \mid |z| > 0, \alpha < \arg z < \beta\}$, where $\beta - \alpha < \pi$, and continuous on $\overline{N}$. Suppose further that $|f(z)| \le C_1 e^{C_2 |z|}$ for all $z \in N$ and some constants C_1, C_2. Prove that f obeys an extended maximum principle
$$|f(z)| \le \max_r \{|f(re^{i\alpha})|, |f(re^{i\beta})|\}$$

(*Hint*: Consider the case $-\pi/2 < \alpha < \beta < \pi/2$ and let $h(z) = e^{-\varepsilon z^\mu}$ for an appropriate μ.)

List of Symbols

$A^{\#}$		142	l.s.c.	355
$AC[0,1]$		254	ℓ_2	40
$\mathbb{C}$	the complex numbers		ℓ_p	69
c_0		69	ℓ_∞	69
$C(X)$		102	L^1	17
$C_\mathbb{R}(X)$		101, 102	L^2	39
$C_\infty(X)$		111	L^∞	67
$C_0^\infty(\mathbb{R}^n)$		145	$L^p(X, d\mu)$	68
$d(\cdot, \cdot)$		4	$L^2(X, d\mu; \mathcal{H}')$	40
$D(\cdot)$ (domain)		249	$L^p(M, d\mu; E)$	89
$\mathscr{D}_\Omega, \mathscr{D}_{\mathbb{R}^n}, \mathscr{D}$		147	$\mathscr{L}^1$	16
$\mathscr{D}'_\Omega, \mathscr{D}'_{\mathbb{R}^n}, \mathscr{D}'$		148	$\mathscr{L}(\mathcal{H})$	182
$\mathscr{D}_{L^\infty(\mathbb{R}^n)}$		178	$\mathscr{L}(X, Y)$	69
$\mathscr{E}, \mathscr{E}'$		178	$\mathscr{L}_\mathcal{H}$	310
$\check{f}, \hat{f}$		318	$\mathscr{M}(X), \mathscr{M}_+(X), \mathscr{M}_{+,1}(X)$	109
$f(A)$ (continuous functional calculus)		222	$\mathscr{O}_D$	127
			$\mathscr{O}_M^n$	137
$f * g$		323	$P_\Omega(A), P_\Omega^A$	234
F_σ		31	$\mathscr{P}\left(\dfrac{1}{x}\right)$	136
$\mathscr{F}(\mathcal{H}), \mathscr{F}_a(\mathcal{H}), \mathscr{F}_s(\mathcal{H})$		53, 54		
$\mathscr{F}, \mathscr{F}^{-1}$		318	$Q(q)$	276
G_δ		31, 105	$\mathbb{R}$	the real numbers
$H_n(X)$ (homology group)		364	$R_\lambda(T)$ (resolvent)	188
$\mathcal{H}$		39	Ran	185
$\mathcal{H}_{pp}, \mathcal{H}_{ac}, \mathcal{H}_{sing}$		230	s	69
$\mathscr{I}_1$		207	$\mathscr{S}(\mathbb{R}^n)$	133
$\mathscr{I}_2$		208	$\mathscr{S}'(\mathbb{R}^n)$	134
Ker		185	$\mathrm{tr}(\cdot)$	207, 212

LIST OF SYMBOLS

$\beta(F, E)$	165
$\Gamma(T)$ (graph)	250
$\Gamma(\cdot)$	309
$d\Gamma(\cdot)$	302
$\eta(E^{**}, E)$	180
$\kappa(X)$	111
μ_ψ	225
$\rho(T)$	188
$\sigma(T)$	188
$\sigma(X, Y)$	113
$\sigma_{pp}, \sigma_{cont}, \sigma_{ac}, \sigma_{sing}$	231
$\sigma_{disc}, \sigma_{ess}$	236
$\tau(X, Y)$	163
ϕ_n (Hermite functions)	143
χ_A	2

$\|\cdot\|$	8
$\|\cdot\|_1$	9
$\|\cdot\|_\infty$	9
$\|\cdot\|_{\alpha, \beta}$	133
$\|\cdot\|_{\alpha, \beta, 2}$	141
$\|\cdot\|_{\alpha, \beta, \infty}$	141

$\oplus$		40, 78
$\otimes$	(measures)	26
$\otimes$	(Hilbert spaces)	49
$\otimes$	(functions)	141
$\otimes$	(operators)	299
$\leq$	(quadratic forms)	373
⁻	(closure)	92
$\circ$	(interior)	92
$\circ$	(polar)	167
$*$	(adjoint)	187
$*$	(dual space)	72
$*$	(convolution)	323
$'$	(adjoint)	185
Δ	(symmetric difference)	66
$\xrightarrow{\|\cdot\|}, \xrightarrow{w}, \xrightarrow{s}$		182, 183
$\|\cdot\|$	(absolute value of an operator)	196
$\perp$	(orthogonal complement)	41
$\setminus$	(set difference)	1
$/$	(quotient)	78, 79
$\langle \cdot, \cdot \rangle$	(ordered pair)	1
$(\cdot, \cdot)$	(inner product)	36

Index

A

Absolute value of an operator, 196
Absolutely continuous subspace, 230
Absorbing, 127
Adjoint, Banach space, 185
Adjoint, Hilbert space, 186
Adjoint of unbounded operator, 252
Affine linear map, 151
Almost everywhere (a.e.,), 17
Analytic completion, 336
Analytic Fredholm theorem, 201
Analytic function, vector-valued, 189–190
Approximate identity, 251, 326
Ascoli's theorem, 30
Atomic model, 304

B

Baire category theorem, 80
Baire functions, 105
Baire measure, 105, 110
Baire sets, 105, 110
Balanced, 127
Banach–Alaoglu theorem, 115
　applications, 354–363
Banach space, 67
Banach–Steinhaus principle, *see* Principle of uniform boundedness
Base for a topological space, 91
Bergmann kernel, 347
Bessel's inequality, 38
Bicontinuous, 92
Bijective, 2

Bipolar theorem, 168
B.L.T. theorem, 9
Bochner integral, 119
Bochner–Schwartz theorem, 331
Bochner's theorem, 330
Bochner's tube theorem, 337
Bolzano–Weierstrass theorem, 98
Borel function, 15
Borel sets, 14, 105
Boson Fock space, 53
Boundary, 92
Bounded linear transformation, 8
Bounded operator, 8
Bounded set, 165
Bounded variation, 33
Brouwer fixed point theorem, 364

C

Canonical form for compact operators, 203
Cantor function, 21
Cantor set, 20
Cartesian product, 1
Cauchy net, 125
Cauchy principal value, 136
Cauchy sequence, 5
Chernoff's theorem, 377
Circled, 127
Closable form, 373
Closable operator, 250, 252–253
Closed graph theorem, 83
Closed operator, 250
Closed quadratic form, 277
Closure, 92
　of an operator, 250

INDEX

Cluster point, 96
Commuting (unbounded) operators, 271–272
Compact operator, 199
  applications, 204–206, 368–372
Compact space, 98
Compact support, functions of, 111
Completely continuous operator, *see* Compact operator
Completion, 7, 9
Cone, 109
Connected, 95
Continuity of the functional calculus, 286–287
Continuous function, 6, 92
Continuous functional calculus, 222
Contraction mapping theorem, 151
Convex cone, 109
Convex function, 356
  strictly, 389
Convex set, 109
Convolution, 323, 324
Core, 256
Countable
  first, 94
  second, 94
Cyclic vector, 226

D

Degenerate tube theorem, 338
Dense, 6
Direct sum
  of Banach spaces, 78
  of Hilbert spaces, 40
Directed family of seminorms, 126
Directed system, 95
Dirichlet problem, 204–206
Dirichlet's principle, 362
Distribution, *see* Generalized function; Tempered distributions
Domain, 2
  of an unbounded operator, 249
Dominated convergence theorem, 17, 24
Dual space, 43, 72
Dunford functional calculus, 245
Dunford–Taylor formula, 316

E

Eigenvalue, 188
Eigenvector, 188
$\varepsilon/3$ argument, 26–27
Equicontinuous, 29, 28–30
Equivalence relation, 2
Equivalent family of seminorms, 126
Ergodic, 58
Ergodic theorem
  Birkhoff, 60
  von Neumann, 57
Essential range, 229
Essentially self-adjoint, 256
"Eventually," 96
Exaggeration, 60, 1–400
Extension of an operator, 250

F

Fatou's lemma, 24
Fermion Fock space, 54
Filter, 352
f.i.p., 98
First resolvent formula, 191
Fock space, 53
Form core, 277
Form domain, 276
  of operator, 277
Fourier coefficients, 46
Fourier inversion theorem, 320
Fourier transform, 318
Fréchet space, 132
Fredholm alternative, 203
"Frequently," 96
Fubini's theorem, 25–26
Functional calculus, 222, 225, 245, 263, 286–287
Functions of rapid decrease, 133

G

Gauge, *see* Minkowski functional
Generalized convergence, *see* Norm resolvent sense; Strong resolvent sense
Generalized function, 148
  of compact support, 334

Geodesic, 361
Graph, 83, 250
Graph limit, 293–294

H

Haar measure, 155
Hahn–Banach theorem, 75–77, 130
Hartree equations, 359
Hausdorff space, 94
Hausdorff–Young inequality, 328
Hellinger–Toeplitz theorem, 84
Hermitian, *see* Symmetric operator
Hilbert–Schmidt operators, 210
Hilbert–Schmidt theorem, 203
Hilbert space, 39
Hölder's inequality, 68, 84, 348
Holomorphy domain, 336
Holomorphy envelope, 336
Homeomorphism, 92
Homology group, 364

I

Implicit function theorem, 366
Infinitely divisible, 341
Injective, 2
Inner product, 36
Interior, 92
Inverse Fourier transform, 318
Inverse function theorem, 367
Inverse mapping theorem, 83
Isometric isomorphism, 71
Isometry, 7

K

Kakutani–Krein theorem, 104
Kato's strong Trotter product formula, 379
Kernel, 185, 198

L

Lebesgue decomposition theorem, 22–23, 25

Lebesgue measure, 15, 13–18
Lebesgue–Stieltjes integral, 19–21
Leray–Schauder–Tychonoff theorem, 151, 365
lim inf ($\underline{\lim}$), 11, 12
lim sup ($\overline{\lim}$), 11, 12
Linear transformation, 2
Linearly ordered, 3
Locally compact, 110
Locally convex spaces, 125
Lower semicontinuous, 355
L^p spaces, 68, 348–351
Lusts of the flesh, 249

M

Mackey–Arens theorem, 164, 167–169
Mackey topology, 163
Mapping, 1
Markov–Kakutani theorem, 152
Maximum principle, 382–384
Measurable functions, 15–16
Measure, 23, 104–111
  absolutely continuous, 22, 24
  continuous, 22
  pure point, 22
  singular, 22, 24
Measure class, 232
Metric space, 5
Metric transitivity, 59
Minimization of functionals, 354–363
Minkowski functional, 128
Minkowski's inequality, 68, 349
Mixing, 239
Monotone convergence theorems
  for forms, 372–377
  for functions, 17, 24
  for nets, 106
Montel space, 173
Multiplicity free operators, 231
Multiplicity theory, 231–234

N

Neighborhood, 91
Neighborhood base, 91
Nets, 96, 351

Neumann series, 191
Norm, 8
  equivalent, 71
Norm resolvent sense, convergence in, 284, 284–291
Normal operator, 246
Normal space, 94
Normed linear space, 8
Nuclear theorem, 141, 144

O

One-parameter unitary group, 265
Open function, 92
Open mapping theorem, 82, 132
Open set, 91
Operator, 2
  of uniform multiplicity, 233
Orthocomplemented lattice, 309–310
Orthogonal, 37
Orthogonal complement, 41
Orthonormal, 37
Orthonormal basis, 44, 44–46

P

Paley–Wiener theorem, 333
Parallelogram law, 38, 63
Parseval's relation, 45, 46
Partial isometry, 197
Partial ordering, 3
Pettis' theorem, 119
Phragmén–Lindelof theorems, 382–384, 391
Plancherel theorem, 327
Polar, 167
Polar decomposition, 197, 297–298
Polarization identity, 63
Positive linear functional, 106, 350
Positive operator, 195
Positive quadratic form, 276
Positive type
  distribution of, 331
  function of, 329
  weak, 331
Principle of uniform boundedness, 81, 132
Product topology, 94

Projection, 187
  orthogonal, 187
Projection theorem, 42
Projection-valued measure (p.v.m.), 234–235, 262–263
Pythagorean theorem, 37

Q

Quadratic form, 276
Quantum mechanics, 302–305
Quotient space, 78–79

R

Radon–Nikodym theorem, 25, 344
Range, 2
Rectifiable, 361
Reflexive space, 74, 167, 174
Regular space, 94
Regularity theorem for tempered distributions, 139, 144
Relative topology, 95
Relatively open, 95
Reproducing kernel, 347
Resolvent, 188, 253
Resolvent set, 188, 253
Riemann–Lebesgue lemma, 327
Riemann–Stieltjes integral, 33
Riesz lemma, 43, 41–44
  applications, 344–348
Riesz–Markov theorem, 107, 111, 353–354
Riesz–Schauder theorem, 203

S

Schrödinger representation, 274
Schwarz inequality, 38
Second dual, 74
Second quantization, 302, 309
Self-adjoint operator
  bounded, 187
  unbounded, 255
Self-adjointness, basic criterion for, 256–257
Semibounded quadratic form, 276
Seminorm, 125
Separable, 47, 95

Separating hyperplane theorem, 130–131
Sesquilinear form, 44
σ-field, 23
σ-finite, 23
σ-ring, 23
Singular subspace, 230
Singular value of a compact operator, 203–204
Spectral mapping theorem, 222
Spectral measures, 228
  associated with a vector, 225
Spectral projections, 234
Spectral radius, 192
  formula, 192
Spectral representation, 227
Spectral theorem
  functional calculus form, 225, 263
  multiplication operator form, 225, 263
  p.v.m. form, 235, 263–264
Spectrum, 188
  absolutely continuous, 231
  continuous, 231
  continuous singular, 231
  discrete, 236
  essential, 236
  point, 188, 231
  residual, 188
Square root lemma, 196
Stieltjes integral, see Lebesgue–Stieltjes integral
Stone's formula, 237
Stone's theorem, 266, 265–267
Stone–Weierstrass theorem, 102, 104
  complex, 102
Strict inductive limit, 146–147
Strict solution of a partial differential equation, 149
Strictly m-accretive form, 281
Strictly m-accretive operator, 281
Strictly m-sectorial form, 282
Strong graph limit, 293
Strong measurability, see Strongly measurable
Strong operator topology, 182
Strong resolvent sense, convergence in, 284, 284–291
Strong (dual) topology, 165
Strongly analytic, 189
Strongly measurable, 116
Subnet, 97

Support of a distribution, 139
Surjective, 2
Symmetric operator, 255
Symmetric quadratic form, 276

T

T_1, T_2, T_3, T_4, 94
Tempered distributions, 134
Tensor product
  of Hilbert spaces, 50, 49–54
  of operators, 299–302
Thomas–Fermi equations, 358
Tietze extension theorem, 102–103, 121
Topological dual, 129
Topological space, 90–91
Topology, 91
  weaker, 91
Trace, 207, 211–212
Trace class, 207, 207–210
Trotter–Kato theorem, 288
Trotter product formula, 295–297, 377–382
Trotter's theorem, 287
Tube, 337
Tube theorems, 337, 338
Tychonoff's theorem, 100, 118, 351–353

U

Ultrafilter, 352
Uniform boundedness principle, see Principle of uniform boundedness
Uniform operator topology, 182
Uniformly convex spaces, 87–88, 174
Unitary operator, 39
Universal net, 351
Urysohn's lemma, 101

V

Vague topology, 114
Von Neumann's theorem, 268
Von Neumann's uniqueness theorem, 275

W

Weak derivative, 138
Weak graph limit, 294
Weak measurability, *see* Weakly measurable function
Weak mixing, 239
Weak operator topology, 183
Weak solution of partial differential equation, 149
Weak topology, 93, 111
Weak-∗ topology, 113
Weakly analytic (vector-valued) function, 189
Weakly measurable (vector-valued) function, 114
Weyl's criterion, 237
Weyl's theorem, 372

Y

Young's inequality, 338

Z

Zorn's lemma, 3